ACTIVE GALACTIC NUCLEI

PRINCETON SERIES IN ASTROPHYSICS

Edited by Jeremiah P. Ostriker and David N. Spergel

Theory of Rotating Stars, *by Jean-Louis Tassoul*

Theory of Stellar Pulsation, *by John P. Cox*

Galactic Dynamics, *by James Binney and Scott Tremaine*

Dynamical Evolution of Globular Clusters, *by Lyman Spitzer, Jr.*

Supernovae and Nucleosynthesis: An Investigation of the History of Matter from the Big Bang to the Present, *by David Arnett*

Unsolved Problems in Astrophysics, *edited by John N. Bahcall and Jeremiah P. Ostriker*

Galactic Astronomy, *by James Binney and Michael R. Merrifield*

Active Galactic Nuclei: From the Central Black Hole to the Galactic Environ, *by Julian H. Krolik*

Julian H. Krolik

ACTIVE GALACTIC NUCLEI: FROM THE CENTRAL BLACK HOLE TO THE GALACTIC ENVIRONMENT

PRINCETON UNIVERSITY PRESS

Princeton, New Jersey

Published by Princeton University Press, 41 William Street,
Princeton, New Jersey 08540
In the United Kingdom: Princeton University Press,
Chichester, West Sussex

Library of Congress Cataloging-in-Publication Data

Krolik, Julian H., 1950 –
    Active galactic nuclei: from the central black hole to the galactic
    environment / Julian H. Krolik
        p.   cm.—(Princeton series in astrophysics)
    Includes bibliographical references and index.
    ISBN 0-691-01152-4 (cl: alk. paper)
    ISBN 0-691-01151-6 (pb: alk. paper)
    1. Active galactic nuclei. I. Title. II. Series
    QB858.3.K76   1999
    512.8—DC21   98-21857
                                    CIP

This book has been composed in New Times Roman by TEX

The publisher would like to acknowledge Julian Krolik for providing
the PostScript files from which this book was printed.

The paper used in this publication meets the minimum requirements
of ANSI/NISO Z39.48-1992 (R1197) (*Permanence of Paper*)

http://pup.princeton.edu

Printed in the United States of America

10 9 8 7 6 5 4 3 2 1
10 9 8 7 6 5 4 3 2 1
(Pbk.)

Contents

Preface   xv

Guide for Readers   xix

1. What Are Active Galactic Nuclei? And Why
Does Anyone Care?   3

  1.1 What Makes Them Interesting?   3

  1.2 What Exactly Are We Talking About? The Most Salient
  Properties of AGNs   4

      1.2.1 Very small angular size   5
      1.2.2 High luminosity   7
      1.2.3 Broad-band continuum emission   8
      1.2.4 Emission lines   11
      1.2.5 Variability   13
      1.2.6 Polarization   15
      1.2.7 Radio emission   17

  1.3 AGN Nomenclature   18

2. How to Find AGNs   22

  2.1 Optical Color   23

  2.2 Optical Emission Lines   26

  2.3 Radio Flux   29

      2.3.1 The 3C catalog and the confusion limit   29
      2.3.2 Optical identifications and the discovery of quasars   31
      2.3.3 Subsequent surveys   33

  2.4 X-rays   35

  2.5 Infrared   36

  2.6 γ-rays   37

2.7 Merged Catalogs and Their Flux Limits 38

3. Evolution 42

3.1 Source Counts 42

3.1.1 Formalism 42
3.1.2 AGN source counts in the real world 45
3.1.3 Radio source counts 47
3.1.4 Optical source counts 49
3.1.5 X-ray source counts 50
3.1.6 In which band do the counts go the "deepest"? 51

3.2 The V/V_{max} Test 52

3.3 Luminosity Function as a Function of Redshift 54

3.4 Parameterizing Evolution 63

3.5 Can One Solve the Population Equation? 66

4. Global Energetics and Black Holes 68

4.1 The Mean Energy Density of AGN Light 68

4.2 Energy Production Efficiency and Remnant Masses 71

4.3 Rees's Flow Chart 74

4.3.1 Causality bounds 74
4.3.2 Variability timescales 75
4.3.3 The flow chart 76

4.4 Evidence for Massive Black Holes in Galaxies 80

4.4.1 Expected response of stars and interstellar gas to a
large nuclear mass 81
4.4.2 Observational tests 86

5. Black Hole Physics 90

5.1 Particle Mechanics in the Vicinity of Black Holes 90

5.1.1 General properties of relativistic particle mechanics 90
5.1.2 Particle motion in a spherically symmetric potential 94
5.1.3 Motion in the Schwarzschild metric 96
5.1.4 Motion in the Kerr metric 103

Contents

5.2 Electromagnetic Fields   111

   5.2.1 Introductory comments   111
   5.2.2 Surface resistance   112
   5.2.3 The rotating black hole battery   113

5.3 Photon Propagation   115

6. Spherical Accretion   120

6.1 Bondi Accretion   121

6.2 The Eddington Limit   125

6.3 Nonadiabatic Accretion   128

   6.3.1 Critical point structure   129
   6.3.2 Optical depth and photon trapping   130
   6.3.3 Thermal instability   131

7. Accretion Disks and the Optical/Ultraviolet Continuum   135

7.1 Fundamental Equations of Thin Disks   135

7.2 What Is the Torque?   138

   7.2.1 Viscosity?   138
   7.2.2 Magnetized winds?   141
   7.2.3 Turbulent Maxwell stresses: the Velikhov-Chandra-sekhar-Balbus-Hawley instability   143

7.3 Time-Steady Disks   147

   7.3.1 Radial structure: angular momentum and energy conservation   147
   7.3.2 Relativistic effects   150
   7.3.3 LTE spectrum   153
   7.3.4 Vertical structure   155
   7.3.5 Advection-dominated disks   162
   7.3.6 Inside the marginally stable orbit   163

7.4 Time-Dependent Disks   164

   7.4.1 Characteristic timescales   164
   7.4.2 Stability   165

7.5 Emitted Spectrum   169

   7.5.1 Opacity mechanisms   169
   7.5.2 Thermalization   172
   7.5.3 Emergent spectrum from a vertically isothermal disk   173
   7.5.4 Emergent spectrum from a disk with vertical temperature gradients   178
   7.5.5 Angle dependence   183
   7.5.6 Polarization   185

7.6 Observational Tests   188

   7.6.1 The big bump and disk model fits   189
   7.6.2 Lyman edge features   191
   7.6.3 Polarization   192
   7.6.4 Variability   193

8. X-ray and γ-ray Emission   196

8.1 Observed Spectra   196

   8.1.1 Radio-quiet AGNs   196
   8.1.2 Radio-loud lobe-dominated AGNs   198
   8.1.3 Radio-loud core-dominated AGNs   200
   8.1.4 Observed variability   201

8.2 Radiation Physics: Bremsstrahlung and Inverse Compton Scattering   202

   8.2.1 Bremsstrahlung   203
   8.2.2 Inverse Compton scattering   206

8.3 Output Spectra from Inverse Compton Scattering   209

   8.3.1 Optically thin   210
   8.3.2 Optically thick   212

8.4 Electron-Positron Pair Physics   216

   8.4.1 Pair production processes and their rates   216
   8.4.2 Pair annihilation   220
   8.4.3 Peculiarities of pair dynamics   222

8.5 Self-Consistent Equilibria   223

   8.5.1 Thermal electrons   224
   8.5.2 Nonthermal electrons   232

Contents

8.6 Sources of Energetic Electrons   237

    8.6.1 Hot thermal electrons   237
    8.6.2 Relativistic electrons   238

8.7 X-ray and γ-ray Propagation   242

    8.7.1 The nature of soft X-ray opacity   242
    8.7.2 Fe Kα emission   245
    8.7.3 Consequences of Comptonization and pair production for γ-rays   247
    8.7.4 X-ray reflection   248

8.8 Comparison with Observations   250

    8.8.1 X-rays   250
    8.8.2 γ-rays   253

9. Radio Emission and Jets   256

9.1 Phenomenology   256

    9.1.1 Relative strength of radio emission   256
    9.1.2 Morphology   258
    9.1.3 Spectrum   263
    9.1.4 Polarization   266
    9.1.5 Variability   266

9.2 Synchrotron Theory   269

    9.2.1 Optically thin emission   270
    9.2.2 Polarization   272
    9.2.3 Estimating the magnetic field strength: the minimum energy argument   273
    9.2.4 Optically thick emission   276
    9.2.5 Spontaneous and stimulated Compton scattering in optically thick synchrotron sources   282

9.3 Jets   286

    9.3.1 Content   288
    9.3.2 Acceleration and collimation   289
    9.3.3 Propagation and morphology   293
    9.3.4 Jet radiation   301

10. Emission Lines — 309

  10.1 Observables — 309

    10.1.1 Line lists and relative strengths — 309
    10.1.2 Line profiles — 315
    10.1.3 Classification on the basis of emission line properties — 317

  10.2 Physical Overview — 319

  10.3 Line Production by Photo-ionization — 321

    10.3.1 Geometry and internal dynamics — 323
    10.3.2 Ionization equilibrium — 323
    10.3.3 Thermal equilibrium — 329
    10.3.4 Radiation transfer — 335
    10.3.5 Excited state population equilibrium, recombination cascade, and the approach to thermodynamic equilibrium — 343

  10.4 Conditions for Achieving Equilibrium — 352

  10.5 Results of Photo-ionization Calculations — 354

    10.5.1 The T–Ξ diagram — 354
    10.5.2 Generic models — 357
    10.5.3 Model construction — 365

  10.6 Shocks — 383

    10.6.1 Internal shock structure — 383
    10.6.2 Self-generated photo-ionization — 387
    10.6.3 Emergent spectrum — 388

  10.7 Line-Emitting Gas Dynamics and Life Cycle — 390

11. Intrinsic Absorption and Outflows — 399

  11.1 Phenomenology — 399

  11.2 Physical Conditions in the Absorbing Gas — 407

  11.3 The Magnitude of the Problem — 409

  11.4 Dynamics — 412

    11.4.1 Thermally driven winds — 412
    11.4.2 Magnetically driven winds — 414
    11.4.3 Radiation pressure driven winds — 414

Contents

12. Anisotropic Appearance and Unification of Disparate AGN Varieties

12. Anisotropic Appearance and Unification of Disparate AGN Varieties   424

12.1 General Principles   424

12.2 Mechanisms Producing Intrinsically Anisotropic Emission   427

     12.2.1 Nonspherical relativistic motion   427
     12.2.2 Angle-dependent disk emission   428

12.3 Angle-Dependent Obscuration Mechanisms   429

12.4 Mirrors and Screens: or, How to See AGNs from Another Side   432

     12.4.1. Polarization   433
     12.4.2 Reflection without dilution   437
     12.4.3 Emission lines   437
     12.4.4 Infrared continuum   439
     12.4.5 Spectra versus images   440

12.5 Unification of Flat-Spectrum Compact Radio Sources with Steep-Spectrum Extended Sources   441

     12.5.1 Luminosity function of compact radio sources   442
     12.5.2 Jet length and speed   446
     12.5.3 Depolarization   448

12.6 Unification of Radio-Quiet Broad Line and Narrow Line AGNs via Toroidal Obscuration   449

     12.6.1 Mirrors   450
     12.6.2 Screens   454
     12.6.3 Energy-dependent obscuration   454
     12.6.4 Relative numbers of type 1 and type 2 Seyfert galaxies   455
     12.6.5 Extrapolation to higher luminosity?   459

12.7 Unification of Radio-Loud Quasars and Radio Galaxies via Toroidal Obscuration   460

     12.7.1 Mirrors   460
     12.7.2 Screens: infrared reradiation   462
     12.7.3 Relative numbers of radio-loud quasars and radio galaxies   463

13. Properties of AGN Host Galaxies   464

13.1 Methodology   464

13.1.1 Imaging the host   464
13.1.2 Sample definition   467

13.2 Host Morphology   469

13.2.1 Sample selection   469
13.2.2 Host morphology and existence of an AGN   470
13.2.3 Host morphology and radio loudness   471
13.2.4 Fine points of host morphology: bars and other disturbances   473

13.3 Host Luminosity   476

13.4 Host Interstellar Medium   479

13.4.1 Signals   480
13.4.2 Surveys   485

13.5 Orientation   486

13.6 Evolutionary State   487

13.7 Environment   489

13.8 Impact of the Nucleus on the Host   492

14. Onset and Fueling   494

14.1 Introduction   494

14.2 The Timescale Problem   495

14.2.1 Creation and growth of the black hole   495
14.2.2 Heavy element abundances   499
14.2.3 Black hole centralization and multiple black holes   499

14.3 Basic Fueling Mechanisms   501

14.3.1 Gravitational forces   502
14.3.2 Hydrodynamic dissipation   505
14.3.3 Magnetic forces   506
14.3.4 Poynting-Robertson drag   507
14.3.5 Stellar-interstellar mass exchange   508
14.3.6 Gas heating and cooling   510

14.4 Triggers and Scenarios   511

Contents

14.4.1 Cooling flows   512

14.4.2 Axisymmetric collapse enhanced by magnetic braking   513

14.4.3 Bar-driven inflow   516

14.4.4 Encounters   524

14.5 Some Open Questions   527

15. Where We Stand   530

15.1 Accomplishments   530

15.2 The Future   531

Appendix A: Basic General Relativity   533

Appendix B: Basic Magnetohydrodynamics   537

Appendix C: Shocks and Other Discontinuities   542

Appendix D: Luminosity Functions and Other Population Statistics   552

D.1 More Sophisticated Estimates of the Luminosity Function   552

D.2 Correlations and Multivariate Probability Distributions   555

Appendix E: The Tensor Virial Theorem   559

Appendix F: Kinematics in an Expanding Universe   566

References   573

Index   589

Preface

We are both enabled and limited by the tools that we use. Because the techniques of observational astronomy change from band to band, too often we construct mental views of astronomical objects sadly limited by the unconscious donning of wavelength-sensitive intellectual blinders. This effect is particularly pernicious in the case of active galactic nuclei, whose emission is spread over a tremendous range of photon varieties. Similar criticisms can be (and often are) voiced about theoretical work—for example, because we understand atomic spectroscopy far better than plasma physics, we prefer to think about line formation regions rather than accretion dynamics. In writing this book, I have made a special effort, therefore, to be as fair to the subject as possible. That is, I have striven to treat all the different sorts of radiation on an equal footing, and I have attempted to point out where there are important questions we have been unsuccessful in answering, or have completely ignored, as well as to set out well-established results. It hardly needs be said that making the attempt is not the same thing as reaching the goal, but I hope that I have at least gone part way.

Writing any book about a lively subject of research is also an exercise in hitting a moving target. One wants to avoid ideas that may no longer be interesting a few years in the future, while at the same time one would like as much of the material as possible to be useful to readers five and ten years hence. One must, therefore, face Janus-like both past and future and make one's best guess about which old ideas will last and which new ones will thrive. My private articulation of this effort is to put in only those ideas that have at least a 50% probability of not being laughable in ten years. I await reader comments in 2008 judging to what extent I have been successful in this regard.

In line with the philosophy of the preceding paragraphs, the book is organized along what I hope is a robust train of logic. First, one cannot expect anyone to persevere without motivation. That is the job of Chapter 1. There is no subject, of course, unless one can find examples, and quantification of their properties follows immediately (Chapters 2 and 3). The most fundamental property of AGNs is their remarkable energy output; recognition of this fact leads immediately to the belief that a massive black hole lies at the center of every AGN (Chapter 4), so the student must acquire at least a modicum of knowledge about how they work (Chapter 5).

Accretion is how black holes generate power. Although it is an unrealistically simple picture, consideration of accretion in spherical symmetry leads to some powerful insights (Chapter 6). A minimally realistic view must include the effects of accretion with angular momentum; this topic leads into a discussion of how the single most important continuum component (the ultraviolet) is produced (Chapter 7). From there we move progressively outward. Chapter 8 presents what we know about X-ray and γ-ray radiation; Chapter 9 deals with radio emission and jets; and Chapter 10 tells the reader how we think the strong emission lines of AGNs are made. Somewhere outside the broad emission line region we find absorbing material that travels outward at speeds up to a few tenths of c; what little we know of this phenomenon is presented in Chapter 11. Up until this point we have pretended that all observers see the same picture, no matter what their point of view; Chapter 12 corrects this misapprehension and develops the dramatic consequences of anisotropic appearance. Chapter 13 takes up the largest scale, the relationship between an active nucleus and its host galaxy. The final substantive chapter, Chapter 14, deals with a problem that lies at the heart of the subject, but one we barely even know how to approach: what ignites AGNs, sustains them, and then extinguishes them. Finally, in Chapter 15 I give a very brief summary of where the field stands as of the date of writing.

The general level of presentation is intended to match the background possessed by a typical second-year graduate student specializing in astrophysics at an American university. That is, I assume the student has had courses in the basics of radiation transfer and elementary radiative processes, classical electromagnetism, and atomic physics. I also assume the student has a nodding acquaintance with fluid mechanics and has heard something about general relativity. In these latter areas, qualitative understanding is more important than the ability to do technical calculations. Very brief refreshers on general relativity, magneto-hydrodynamics, and fluid discontinuities (e.g., shocks) are provided in the appendices. There is also an appendix on the technicalities of estimating luminosity functions and some other, related, statistical issues, and another on the tensor virial theorem. The last appendix quickly summarizes kinematics in an expanding universe. Because the training of graduate students begins to split off into the different subfields of astronomy at about this level, I hope that the treatment will likewise be useful to professionals in astrophysics whose primary expertise is in other subject areas.

However, in recognition of the fact that there are several different communities of readers for this book, I have defined several specialized "tracks" for particular sets of readers. These are described in detail in the Guide for Readers.

Finally, no book can be complete without acknowledgments, for it is well-nigh impossible to write a book without receiving significant aid from others. Many colleagues—Marek Abramowicz, Eric Agol, Ski Antonucci, Mitch Begelman, Chris Done, Richard Green, Tim Heckman, Tim Kallman, Pawan Kumar, Chris O'Dea, Brad Peterson, Greg Shields, Mark Sincell, Mark Voit, Andrew Wilson, and Andrzej Zdziarski—did me the great service of closely reading drafts of selected chapters, helpfully pointing out—and correcting—numerous typos, oversights, and errors. Two people, Pawan Kumar and Greg Shields, used a draft of the manuscript when teaching classes and gave me valuable feedback on how it worked. Brian Boyle, Alan Bridle, Alessandro Capetti, Chris Carilli, Chris Done, Jim Dunlop, Paul Francis, Reinhard Genzel, Bob Goodrich, Bob Hartman, Anuradha Koratkar, Ari Laor, Joe Maslowski, Charlie Nelson, Paolo Padovani, Elena Pian, Paola Pietrini, Rita Sambruna, Mark Sincell, Zlatan Tsvetanov, Marie-Helene Ulrich, Meg Urry, Sylvain Veilleux, Kim Weaver, Andrzej Zdziarski, and Piotr Życki graciously produced "made-to-order" versions of figures or supplied me with proprietary data tables so that I could produce my own version of a published figure. Ron Allen insightfully (and patiently) answered my questions about the mechanics of radio observations. Jerry Kriss and the JHU FOS group gave me access to computer resources that substantially reduced the labor of figure preparation. Mark Voit provided much useful counsel about the tricky thickets of broad absorption line phenomenology and physics. John Mackenty, Brad Whitmore, and especially Tim Heckman were invaluable sounding boards and sources of carefully nuanced judgment on numerous observational questions. But most importantly, I owe a debt of gratitude to the many excellent collaborators with whom I've worked over the years, and from whom I've learned much of what I know. In rough chronological order, they have been Chris McKee, John Kwan, Tim Kallman, Mitch Begelman, Andrzej Zdziarski, Ed Pier, Mark Sincell, Chris Done, Paola Pietrini, and Tim Heckman.

Guide for Readers

While it may be an author's dream that every reader sits down and studies his book from cover to cover, it is not very likely that many readers will do so with this one. From the point of view of the graduate student, there is much more material here than can be covered in a one semester course. Practitioners already in the field, or physicists who would like to learn a little about what their astrophysicist colleagues are up to, are, for different reasons, unlikely to have the time or inclination to read it through in its entirety. To help readers such as these, I set out below a few sample tracks through the book, each catering to a specific audience.

- Track 1: A Feasible One-Semester Course

Instructors should, of course, tune their choices to the particular group of students they are teaching and the time allotted. With "rms" second-year graduate students at an American university, in 40-odd hours of lecture one might try covering:

Chap. 1 What Are Active Galactic Nuclei? And Why Does Anyone Care?

Chap. 2 How to Find AGNs

Chap. 3 Evolution

Chap. 4 Global Energetics and Black Holes, omitting §4.4.1

Chap. 5 Black Hole Physics, §5.1 only

Chap. 6 Spherical Accretion, §6.2 only

Chap. 7 Accretion Disks and the Optical/Ultraviolet Continuum

Chap. 8 X-ray and γ-ray Emission, omitting §8.6

Chap. 9 Radio Emission and Jets, omitting §§9.2.5, 9.3.1, and 9.3.2

Chap. 10 Emission Lines, §§10.1, 10.2, 10.5, and 10.7 only

Chap. 11 Intrinsic Absorption and Outflows, omitting §11.4

Chap. 12 Anisotropic Appearance and Unification of Disparate AGN Varieties

Chap. 15 Where We Stand

- Track 2: The Phenomenologist

For those people mostly interested in observations of AGN, and who wish for only a qualitative assessment of how well we understand them, the best path through the book is:

Chap. 1 What Are Active Galactic Nuclei? And Why Does Anyone Care?

Chap. 2 How to Find AGNs

Chap. 3 Evolution

Chap. 4 Global Energetics and Black Holes, omitting §4.4

Chap. 6 Spherical Accretion, §6.2 only

Chap. 7 Accretion Disks and the Optical/Ultraviolet Continuum, §§7.1 and 7.6 only

Chap. 8 X-ray and γ-ray Emission, §§8.1, 8.2.2, 8.3, 8.7, and 8.8 only

Chap. 9 Radio Emission and Jets, omitting §§9.2.5, 9.3.1, and 9.3.2

Chap. 10 Emission Lines, §§10.1, 10.2, and 10.5 only

Chap. 11 Intrinsic Absorption and Outflows, omitting §11.4

Chap. 12 Anisotropic Appearance and Unification of Disparate AGN Varieties

Chap. 13 Properties of AGN Host Galaxies

Chap. 15 Where We Stand

- Track 3: The Physicist

This track is intended for the physicist who might like to get a feel for what astrophysical problems look like. It concentrates on defining the context, and on laying bare the underlying physics issues. Applications of general relativity are found almost exclusively in Chapter 5, with a few additional points discussed in §§7.3.2 and 7.5.5.

Chap. 1 What Are Active Galactic Nuclei? And Why Does Anyone Care?

Chap. 4 Global Energetics and Black Holes

Chap. 5 Black Hole Physics

Chap. 6 Spherical Accretion, §6.2 only

Chap. 7 Accretion Disks and the Optical/Ultraviolet Continuum, §§7.1, 7.2, and 7.3 (through §7.3.3)

Guide for Readers

Chap. 8 X-ray and γ-ray Emission, §§8.4, 8.5, and 8.6 only

Chap. 9 Radio Emission and Jets, omitting §§9.1 and 9.3.4

Chap. 14 Onset and Fueling

Chap. 15 Where We Stand

ACTIVE GALACTIC NUCLEI

1 What Are Active Galactic Nuclei? And Why Does Anyone Care?

1.1 What Makes Them Interesting?

Active galactic nuclei (hereafter abbreviated "AGNs") are among the most spectacular objects in the sky. They produce prodigious luminosities (in some cases apparently as much as 10^4 times the luminosity of a typical galaxy) in tiny volumes (probably $\ll 1$ pc^3). This radiation can emerge over an extraordinarily broad range of frequencies: in at least one case the luminosity per logarithmic frequency interval, that is, the luminosity per "band," is roughly constant (to within factors of several) across thirteen orders of magnitude in frequency! Their line spectra are almost as remarkable as their continua: in the optical and UV, they often display emission (and occasionally absorption) lines whose total flux is several percent to tens of percent of the continuum flux, and whose widths suggest velocities ranging up to $\sim 10^4$ km s^{-1}.

Although in most cases we cannot as yet make images in which AGNs are resolved, in certain objects it is possible to do so at radio frequencies. In those cases, one often sees variable structure with apparent speeds in the sky plane of $\sim 10c$!

Active galaxies are also noteworthy in displaying very strong cosmological evolution. The most luminous active galaxies were a thousand times more numerous at redshift 2.5 than they are today. Because their high luminosities and distinctive spectra make them relatively easy to pick out, they are disproportionately represented in our tally of known high redshift sources. The fact that the luminosity of AGNs is such a strong function of redshift suggests strongly that there is something special about youthful galaxies that promotes the creation of active nuclei.

Another interesting connection between AGNs and the evolution of the Universe comes from their use as light sources for studying intervening gas clouds, galaxies, and clusters of galaxies. Our knowledge of the intergalactic medium comes almost exclusively from the study of absorption lines in the spectra of distant quasars. On the other hand, systems that are transparent, but possess moderately deep gravitational wells, like galaxies

3

and clusters of galaxies, can form gravitational lenses. The distorted—and sometimes multiple—images of distant AGNs formed by these lenses can be used to infer the nature of the gravitational potential—and therefore the mass—of the lenses.

To the degree that we do understand AGNs physically (a rather controversial point), their basic nature involves events of considerable physical drama. Most workers in the field now believe that the power for AGNs comes from accretion onto massive black holes. If this is true, their most basic properties depend on some of the most exotic physics we know: strong-field relativistic gravity.

This whole stew—exotic physics, photons detectable with virtually every sort of astronomical instrument, and a deep connection to cosmology—has made AGNs the focus of a significant fraction of the world astronomical community's attention for the past 30 years.

1.2 What Exactly Are We Talking About? The Most Salient Properties of AGNs

In order to start any discussion of AGNs, we must first define what we mean by the term *active galactic nuclei*. Unfortunately, although the black hole model has achieved widespread acceptance, it is not yet completely confirmed; moreover, and this is part of why its confirmation remains incomplete, direct signatures of accreting massive black holes are much harder to see than a variety of more indirect signals. Consequently, the only clear way to define AGN is operationally; all we can do is to list the observable phenomena we use to find them. This procedure is, admittedly, circular, but such a "bootstrap" approach is all that is possible until we achieve a more fundamental understanding.

It turns out that AGNs can be found in many ways, but not all AGNs have every single property. Thus, there is no single defining list of qualities to look for. A better way to think about the situation is to imagine a "menu" of phenomena from which they choose, with some items more popular than others. Note that in evaluating "popularity," one's terms of definition must be carefully chosen; in samples selected on the basis of a certain property, that property will always be very popular, whereas in samples selected otherwise it may be only a specialized taste.

Table 1.1 illustrates this point by listing some of the salient observational signatures of AGN, along with brief comments on how often these properties are found, and a few concise caveats. Each of these features is hardly ever seen in normal galaxies. In the following subsections we will discuss these signatures at somewhat greater length.

1.2.1 Very small angular size

The first property is certainly the most visually striking. When the AGN is near enough that a host galaxy of reasonable surface brightness can be seen, in optical images the nucleus often appears to be a bright point whose flux can often rival, or even exceed, the flux from all the rest of the host galaxy (e.g., NGC 1566, a nearby AGN, is shown in Plate 1).

However, this simple picture is a bit misleading. Our ability to see both a bright point and its surrounding host is a function of the luminosity contrast between the nucleus and its host galaxy, a quantity that varies both from case to case and as a function of wavelength. If the luminosity of the nucleus is too small relative to the host, it will not stand out, of course. Conversely, when the luminosity of the nucleus is much greater than that of its host, light from the nucleus can overwhelm any light from the host.

This latter effect is exacerbated by the kinematics of our expanding Universe. For a start, at $z \simeq 1$ the angular size of an average galaxy is only $\sim 1''$, comparable to the seeing disk even at a good ground-based observatory. The problems don't end there. The typical luminosity of the AGNs we can find at $z \simeq 2$ is ~ 100 times greater than at the present epoch, so the luminosity contrast between the nucleus and its host is very large. In addition, although the nucleus remains effectively a point, the bolometric surface brightness of the host falls as $(1 + z)^{-4}$ (see Appendix F). Finally, if the AGN has a redshift greater than about unity, the light we observe in the visible band was ultraviolet in the rest frame, and most galaxies are comparatively dim in the UV. All these effects combined make hosts of high-redshift AGNs very difficult to see (§13.1.1).

The picture we see also depends strongly on wavelength. As we will discuss at slightly greater length in the next section, many AGNs have a much greater ratio of X-ray luminosity to optical than does any normal galaxy. For this reason, their X-ray images are essentially pure points. On the other hand, radio emission, more often than not, extends over a sizable region, frequently much larger than a galaxy (§§ 1.2.7, 9.1.2).

In the long run we can hope to obtain images of even the smallest AGN structures, but the requirements are daunting. As we shall see later, there are AGN pieces spread over a very wide dynamic range of radial scale.

Table 1.1: The Menu

Property	Popularity	Comments and Exceptions
Very small angular size	Many	Wavelength-dependent
Galactic (or greater) luminosity	Many	Lower luminosity is hard to find; obscuration and beaming may mislead
Broad-band continuum	Most	Often $dL/d\log\nu \simeq const.$ from IR to X-rays; sometimes to γ-rays
Strong emission lines	Most	Sometimes very broad, sometimes not
Variable	Most	Modest amplitude; short wavelengths stronger, faster than long
Weakly polarized	Most	~ 1% linear; a minority much stronger
Radio emission	Minority	Sometimes, but not always, extended on enormous scales
Strongly variable and polarized	Small minority	Correlated with bright radio and high-energy γ-rays; in some cases emission lines absent

Equivalent elements in different AGNs tend to have roughly the same effective temperature (i.e., luminosity emitted per unit area $F = \sigma T_{\text{eff}}^4$, where σ is the Stefan-Boltzmann constant) because the effective temperature roughly characterizes the nature of the structural element. The effective temperature is also a rough guide to the typical wavelength emitted by that part of the AGN ($\lambda \sim ch/kT_{\text{eff}}$) if the radiation mechanism is roughly thermal, but nonthermal mechanisms can easily make the typical wavelength far shorter.

Even at the heart of the beast, T_{eff} is rarely much more than $\sim 10^5$ K (§§7.1, 7.3.3); discounting radio emission regions, at the outer edge of an AGN proper, it may fall to a few hundred K (§12.4.4). We can then roughly predict that in an AGN whose observed flux is F_{obs}, the angular size of the region whose effective temperature is T_{eff} is

$$\theta = 87(1+z)^2 \left(\frac{F_{\text{obs}}}{10^{-11} \text{ erg cm}^{-2} \text{ s}^{-1}} \right)^{1/2} \left(\frac{T_{\text{eff}}}{10^3 \text{ K}} \right)^{-2} \text{ microarcsecond,}$$

(1.1)

where the fiducial value of F_{obs} has been chosen at a level corresponding to roughly the tenth brightest AGN in the sky. Thus, the angular size of any particular AGN structure scales $\propto F_{\text{obs}}^{1/2}$, so that the apparently brightest AGNs are also the largest (in angular terms)—and even they are very small indeed.

1.2.2   High luminosity

We know of AGNs with luminosities all the way from $\sim 10^{42}$ to $\sim 10^{48}$ erg s^{-1}. To put this in perspective, the characteristic luminosity of the field galaxy distribution (L_*) is $\sim 10^{44}$ erg s^{-1}. In other words, we see AGNs whose power output ranges from as little as 1% of a typical galaxy to $\sim 10^4$ times as great. However, one must be careful in interpreting these luminosities. On the one hand, we cannot easily detect active nuclei much weaker than the host galaxy, and so there may be a large population of "mini-AGNs" that are as yet unknown. On the other hand, there is also reason to think (see Chap. 12) that in many AGNs the active nucleus is obscured by extremely thick dust extinction, so that we can be grossly misled as to its true luminosity if our only measure is the power output in optical or ultraviolet light. Relativistic beaming (§§9.3.4, 12.2.1) can also substantially distort the angular distribution of light from AGNs, and, of course, there is a strong selection effect in favor of observing those whose radiating material is moving toward us. Thus, in objects where beaming is likely to be significant, it is important to distinguish between the luminosity inferred assuming isotropic radiation and the true luminosity. Obscuration, or beaming directed away from us, can, of course, so weaken the light we see from an AGN that we may not even recognize it.

1.2.3 Broad-band continuum emission

It is best to begin this discussion with a digression about an issue of notation. In most fields of astronomy, the property of light that is measured is the specific flux (F_ν or F_λ), the rate at which energy arrives per unit area per unit frequency ν or per unit wavelength λ. The standard unit for F_ν is the Jansky, 10^{-23} erg cm^{-2} s^{-1} Hz^{-1}. In high-energy (i.e., X-ray and γ-ray) astronomy, where photon-counting devices prevail, the customary measured quantity is N_ϵ ($= F_\nu/(h\epsilon)$), the rate at which photons arrive per unit area per unit energy ϵ. However, when we speak of a "band" of the electromagnetic spectrum, whether it is radio, infrared, visible, or X-ray, we generally mean a span in the logarithm of the wavelength. The term *infrared*, for example, generally refers to a range of $\sim 10^2$ in wavelength, from ~ 1 to ~ 100 μ. Therefore, for describing which band is most important in terms of energetics, the most convenient quantity is $\nu F_\nu = dF/d\log\nu = dF/d\log\lambda = \lambda F_\lambda$, the energy flux per logarithmic bandwidth. In most of this book (starting with fig. 1.1), this will be the favored form for the presentation of spectra.

To understand what is meant by "broad-band" continuum radiation, one should first contrast the spectra of ordinary galaxies (see fig. 1.1). To a first approximation, galaxies are piles of stars. A good zeroth-order approximation to a stellar spectrum is that it is a blackbody, so the great majority of a star's luminosity comes out within a factor of three in frequency. The total span of stellar surface temperatures is only about a factor of ten, and in any particular galaxy, the stellar mix is usually such that a limited temperature range dominates the total power output. Thus, a typical galaxy emits nearly all its power within no more than one decade of frequency, and usually rather less. The only possible modification to this picture is due to interstellar dust. In many spiral galaxies the dust extinction is great enough that a significant fraction of the optical and ultraviolet light is absorbed by cool dust grains and reradiated in the far-infrared. Because there is a wide range in extinctions, the relative size of this secondary peak in the infrared varies substantially from one spiral galaxy to the next. In ellipticals, on the other hand (as shown in fig. 1.1), there is generally little dust, and hence at most weak infrared emission.

Most (but not all) AGN continuum spectra look spectacularly different from normal galaxy spectra. A particularly well observed example,

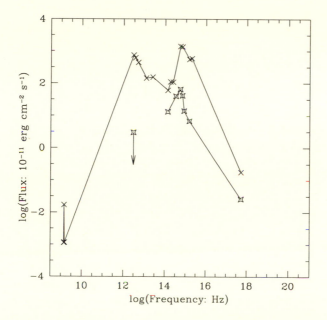

Fig. 1.1   The broad-band spectra of two typical galaxies, M 101 (Hubble type Sc: plot symbol ×) and NGC 4168 (Hubble type E2: plot symbol a four-pointed star). The star with an arrow is an upper limit. Connecting lines are meant only to guide the eye. In both cases, the intrinsic radiation is confined to a narrow range of optical frequencies, with only minor amounts of power emerging in the radio or X-ray bands. In the case of M 101, however, almost half the intrinsic power is absorbed by dust and reemitted at $\sim 100\ \mu$. The spike at 1.4 GHz in the spectrum of M 101 is the 21 cm HI line. At all frequencies but the near-infrared, the fluxes are integrated over the entire galaxy from maps; in that band the apertures used cover most, but not quite all, of the galaxy and therefore give fluxes that are slightly too low. Data sources are: White and Becker 1992, Rice et al. 1988, Fabbiano et al. 1992, and NED.

NGC 4151, is shown in figure 1.2. In terms of νF_ν, NGC 4151, like most AGNs, has a spectrum that is flat (to within factors of several) all the way from the mid-infrared to the hardest X-rays observed (the highest observed energy is anywhere from a few to a few hundred keV, depending on the object). Compared to normal galaxies, the fraction of the bolometric luminosity radiated in the radio band is generally an order of magnitude greater, but in some cases it is several orders of magnitude larger still; the fraction of the power that emerges in X-rays is three to four orders of magnitude larger in AGNs than in normal galaxies. NGC 4151 is so well observed because it is very nearby ($z = 0.003$), but its luminosity is relatively low,

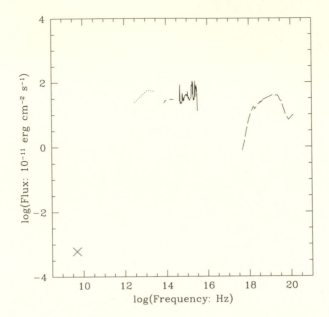

Fig. 1.2 The broad-band continuum spectrum of one of the nearest AGNs, NGC 4151, compiled from nonsimultaneous (!) data. Its flux per logarithmic bandwidth has comparable peaks in the infrared, the optical/ultraviolet, and the hard X-ray bands. Far less luminosity is radiated in radio frequencies. The dip in the soft X-ray region is due to intervening absorption, possibly in the AGN itself; the bumps in the optical/ultraviolet are smoothed versions of the strong emission lines seen in this object. The data are taken from: Ulvestad, Wilson, and Sramek (1981), the ×; Edelson and Malkan (1986), the dotted line; Kriss et al. (1995), the broken line in the optical band; the *HST* archive, the solid line; and Zdziarski, private communication, the dashed line in the X-ray band. In all cases but the four infrared points at 12 μ, 25 μ, 60 μ, and 100 μ, the apertures used were small enough that the AGN dominates the flux.

merely comparable to its host galaxy. Composite spectra illustrating what AGNs look like at high redshift and high luminosity are shown in figure 1.3.

All told, the range over which νF_ν is roughly flat is more than a factor of 10^5 in frequency, producing a spectrum far broader than any normal galaxy's. Although there are weak local maxima, it is clearly inappropriate to speak of any one frequency band dominating the output. These local maxima may, however, be signaling to us that the primary emission mechanisms change as functions of frequency (Chaps. 7, 8, 9, 12). In the case of one particular subclass of AGN, the flux in photons as hard as 1 GeV is at least as great as that in lower frequency bands, and there are a few

10

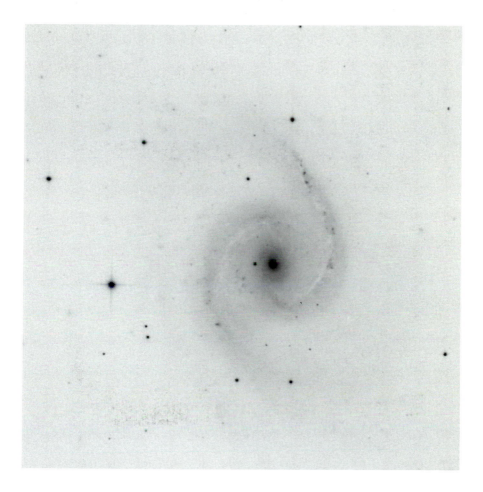

Plate 1 NGC 1566, a nearby active galaxy. It appears to be a fairly normal spiral galaxy, but for the extremely bright nucleus (image courtesy of Z. Tsvetanov).

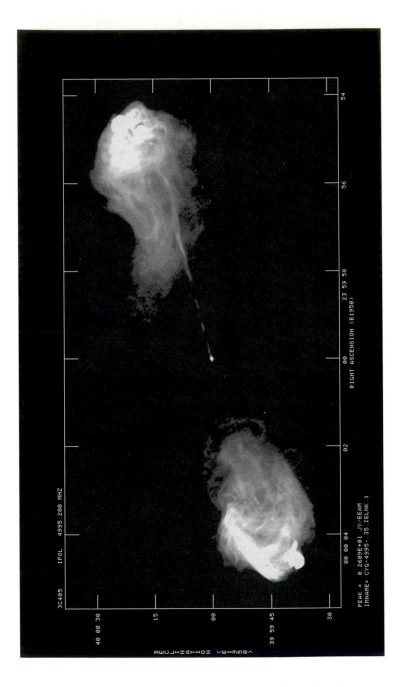

Plate 2   The original extragalactic radio source, Cygnus A, at 6 cm (image courtesy of
R. Perley). The total linear extent of this source is $\simeq$ 120 kpc.

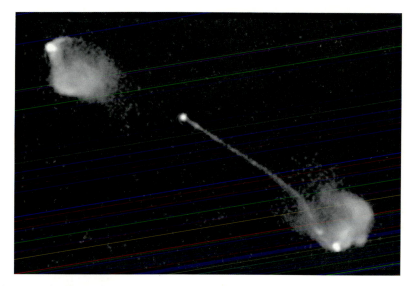

Plate 3   The FR2 radio galaxy 3C 175 at 4.9 GHz, with North to the right and East up (courtesy A. Bridle; originally published in Bridle et al. 1994). The surface brightness peaks in the outer rims of the lobes, especially in the bright hot spots. The jet, while visible, is relatively faint and can be seen on only one side of the nucleus. Even at this comparatively high frequency, the core is much dimmer than the lobes.

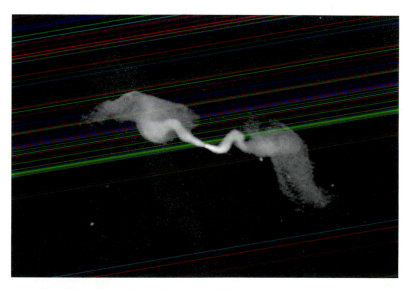

Plate 4   The FR1 radio galaxy 3C 31 at 1.4 GHz, oriented with North to the right and East to the top (courtesy A. Bridle). From one end of the radio emission to the other is $\simeq 400(h/0.75)^{-1}$ kpc. Note how the jet is clearly stronger on one side of the nucleus, and how the surface brightness declines with increasing distance from the nucleus on both sides. In addition, like many, but not all, FR1s, the jets make sharp bends after traveling out some distance from the nucleus.

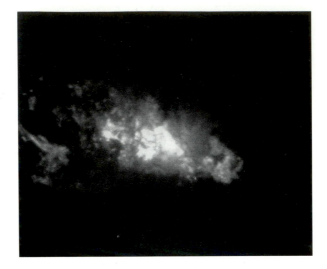

Plate 5   A $4'' \times 6''$ gray-scale *HST* image of [O III] 5007 emission in NGC 1068 (Macchetto et al. 1994). The nucleus is near the base of the plume, which is $\simeq 200(h/0.75)^{-1}$ pc long. Note that the vertical axis is rotated 50° from North; the [O III] plume is actually aligned fairly well with the total intensity plume shown in Plate 6.

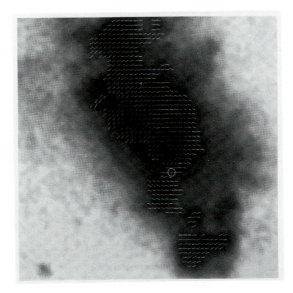

Plate 6   The innermost $3.3'' \times 2.9''$ of the polarizing mirror in NGC 1068 (Capetti et al. 1995). Unlike Plate 5, here North is up. Total intensity is shown in gray scale; the white lines show the local direction of the polarization $\vec{E}$-vector. Because the local polarization must always be exactly perpendicular to the direction to the source, the intersection of the $\vec{E}$-vector normals gives a very good estimate of the position of the nucleus; this is the white circle.

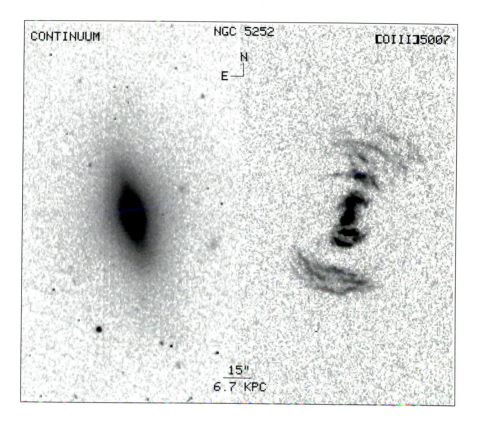

Plate 7 The "ionization cone" in the type 2 Seyfert galaxy NGC 5252 (courtesy Z. Tsvetanov). The left-hand panel shows a continuum image of the galaxy; the right-hand, an image in a filter centered on red-shifted [OIII] 5007. The line emission is entirely within a sharp-edged double cone canted at an oblique angle with respect to the galactic axis.

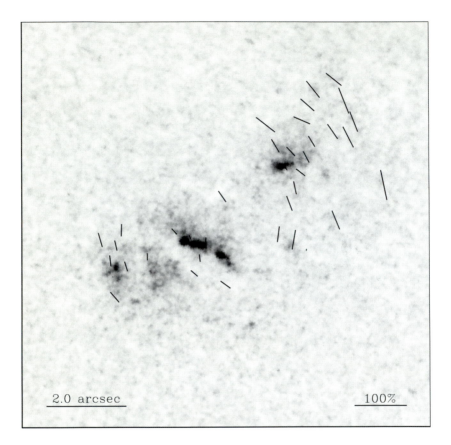

Plate 8   Near-UV (3200 Å) image of the radio galaxy 3C 321 (from Hurt et al. 1997; see also an R-band image in Plate 10). The short straight lines show the directions of the local polarization $\vec{E}$-vector; their lengths are proportional to the local polarization fraction, as shown by the scale bar. They clearly indicate that the polarization is due to scattering of light emanating from a small source, probably behind a dust lane (which is more clearly visible in Plate 10).

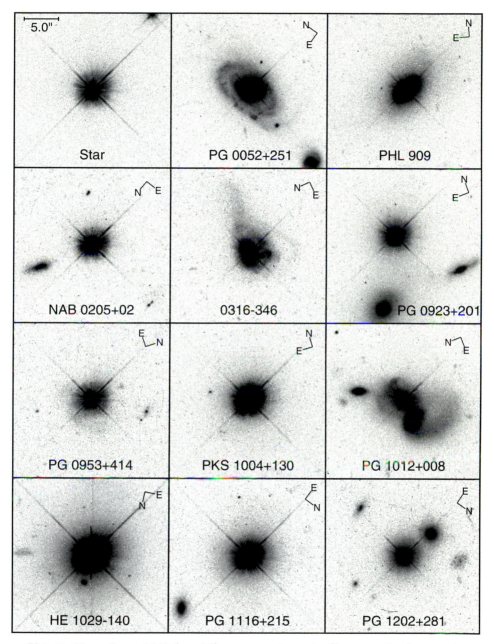

Plate 9   Eleven low-redshift ($z < 0.3$) quasars and one star (for comparison) as seen in red light by the *HST* (Bahcall et al. 1997). Clear spiral structure can be seen in some host galaxies (such as the host to PG 0052+251), but others appear to be ellipticals (e.g., PHL 909). In one case (PG 1012+008), the host is clearly undergoing a violent collision with another galaxy.

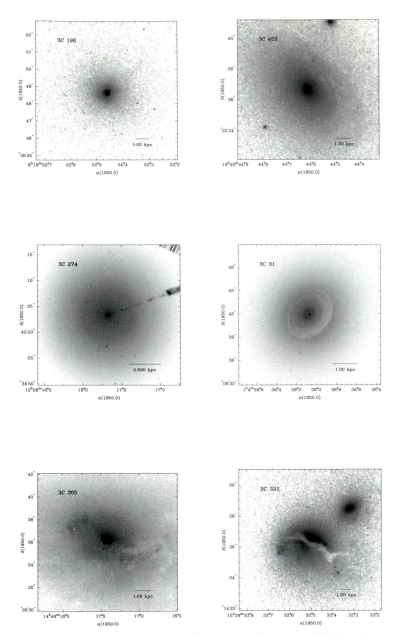

Plate 10 A sampler of R-band images of the hosts of low-redshift 3C radio galaxies, taken from Martel et al. (1997). From upper-left to lower-right, they are 3C 198 ($z = 0.082$), 3C 403 ($z = 0.059$), 3C 274 ($z = 0.004$), 3C 31 ($z = 0.017$), 3C 305 ($z = 0.041$), and 3C 321 ($z = 0.096$). The distance scale given for each is computed assuming $H_o = 75$ km s^{-1} Mpc^{-1}. Most of these are obviously elliptical galaxies, but several just as unmistakably show unusual features. 3C 274 is also known as M 87; its jet (apparent all the way from radio frequencies through X-rays) is clearly visible. 3C 305 has been disturbed in some fashion. 3C 321, for which a near-UV and polarimetric image is shown in Plate 8, has a dark dust lane across its center, and a nearby companion.

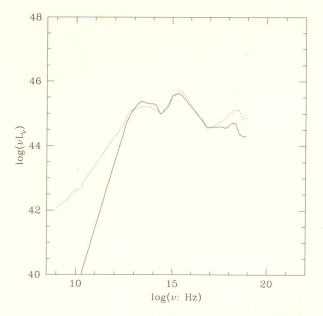

Fig. 1.3   Composite spectra for two different samples of high-redshift, high-luminosity AGNs, one (solid curve) relatively weak in the radio band, and the other relatively strong (dotted curve). The straight line segments in the far-IR/sub-mm band and the EUV are both interpolations across spectral regions where no real observations exist. Although the curves end at photon energies of a few tens of keV, this is not because AGNs are weak there (see, e.g., figs. 1.2 and 1.4).; it is because good data in the hard X-ray band and beyond exist for comparatively few objects. As these curves are composites, the luminosity scale (given in units of erg s^{-1}) should not be taken literally. Data are from Elvis et al. (1994).

examples of this subclass (e.g., Markarian 421; see fig. 1.4) with as much flux at 1 TeV as anywhere else in the electromagnetic spectrum. However, as always, there are exceptions. There are also AGNs in which the bulk of the light arrives at Earth within a frequency span of only one decade in the infrared (NGC 1068, another very nearby AGN, is a good example of this variety; see fig. 1.5).

1.2.4 Emission lines

AGN emission lines have received a great deal of attention for two reasons. First, they are often very prominent (equivalent widths are often ~ 100 Å). This makes AGN spectra stand in great contrast to the spectra of most stars and galaxies, where lines are generally relatively weak and

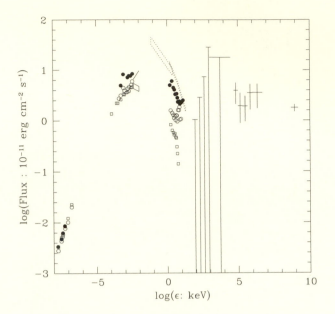

Fig. 1.4   Markarian 421 is a BL Lac object (see table 1.2). Its luminosity is spread almost equally per logarithmic frequency interval over 13 decades in frequency! Where different levels are shown at the same energy, the contrast denotes variability. The data for this plot were collected by Zdziarski and Krolik (1993).

predominantly in absorption. Figure 1.6 presents a composite spectrum that dramatically illustrates just how prominent the emission lines are. Second, because we know a great deal about atomic physics, it is easy and productive to study them (Chap. 10).

The emission lines that we see are remarkably stereotyped from one object to the next. When we have the wavelength coverage to look for them, if there are any lines at all, we almost always see Lyα, the Balmer lines, the CIV 1549 doublet, [OIII] 5007, and several others that are generally weaker. The Fe Kα X-ray line near 6.4 keV is also frequently seen.

However, there is an interesting split in the line width distribution. In some objects, many of the lines have broad wings extending out several thousand km s^{-1} from line center, whereas in others the lines are never broader than a few hundred km s^{-1}. Interestingly, the permitted and semi-forbidden lines are seen in both of these classes; in fact, when the broad wings appear, there is often a narrow core as well. The forbidden lines, on the other hand, are only seen with narrow profiles. In another

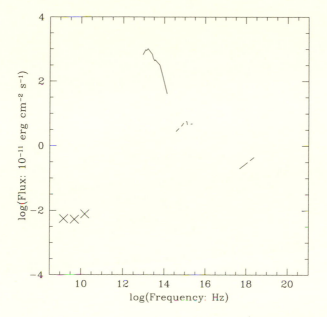

Fig. 1.5   NGC 1068 is another very nearby AGN. Unlike NGC 4151, its spectrum has a strong peak in the infrared, most likely due to very optically thick dust obscuration that reradiates the nucleus's luminosity in that band (§12.4.4). The data are from: Wilson and Ulvestad (1982), the ×'s; Rieke and Low (1975), the solid line; and Pier et al. (1994), the dashed lines. They have been analyzed in such a way as to show only nuclear radiation.

interesting correlation, those objects with only narrow lines are often quite weak from the near-infrared through the X-ray band; most of their light is generally emitted in the mid-infrared (as in the example shown in fig. 1.5).

1.2.5   Variability

It is often loosely remarked that variability is a hallmark of AGNs. This is only partially true. In the optical band, most AGNs, unlike normal galaxies, can be seen to vary, but the typical amplitude over human timescales (e.g., a few years) is often only 10% or so. Incomplete evidence suggests that the variability amplitude on the most easily observed timescales increases toward shorter wavelengths, with factors of two often seen in the X-rays. Figure 1.7 shows two cases in point, once again the frequently observed AGN NGC 4151, and the almost equally popular NGC 5548.

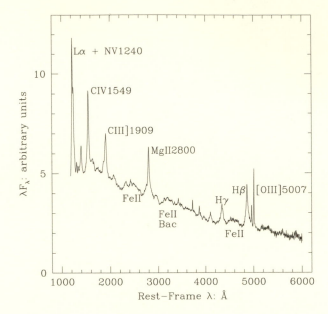

Fig. 1.6 A composite optical/ultraviolet AGN spectrum as compiled by Francis et al. (1991) from a large quasar survey. The very large contrast between emission lines and continuum, and the lines' substantial breadth, are both immediately apparent. Some of the more prominent emission lines are labeled; "FeII" is a shorthand for blends of many FeII multiplets, and "Bac" refers to the Balmer continuum.

It cannot be emphasized too strongly (fig. 1.7 underlines this point) that any statement about variability is strongly dependent on the timescale in question. A well known theorem in Fourier analysis states that the variance of a function of time $F(t)$ sampled at intervals Δt over a duration t_{tot} is equal to the integral of the power spectrum over the range of frequencies to which such measurements are sensitive:

$$\text{Var}(F) = \int_0^{t_{\text{tot}}} dt \; [F(t) - \langle F \rangle]^2 = \int_{1/t_{\text{tot}}}^{1/(2\Delta t)} df \, |\hat{F}(f)|^2. \qquad (1.2)$$

The form of this relation makes it very clear that the measured variance depends on the sampling unless it is good enough that essentially all the variability power is contained within the observed frequency range. Unlike stars, whose variability is often dominated by periodic components (consider eclipsing binaries or Cepheid variables), AGNs for the most part vary with no special timescales; that is, their Fourier spectra are broad-band, just as their photon spectra are. In consequence, the amplitude of variability for AGNs is a slippery thing to measure.

14

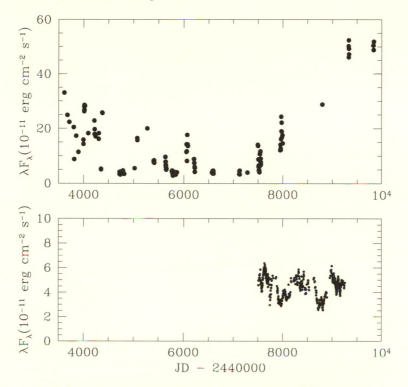

Fig. 1.7 (Top) A long-term UV (1455 Å) lightcurve for NGC 4151 (from Ulrich et al. 1997); (bottom) a somewhat shorter optical (5100 Å) lightcurve for NGC 5548 (data from Peterson et al. 1991, 1992, and 1994). In the UV, fluctuations of factors of several are common and can occur on timescales ranging from weeks to years. In the optical band, the fluctuations tend to be rather smaller.

A small subset of AGNs vary much more strongly, even in the optical band. In some cases, fluctuations of a factor of two have been seen from night to night, and cumulative changes of factors of 100 have occurred over year timescales (see, e.g., fig. 1.8). Strong variability is also very strongly correlated with three other properties: strong polarization, compact radio structure, and strong high-energy γ-ray emission.

1.2.6 Polarization

Most stars are intrinsically unpolarized, but the light we receive is generally linearly polarized by $\sim 0.5\%$ due to interstellar dust transmission polarization. The same is true for most galaxies. Most AGNs are also weakly polarized, but just enough more strongly for their polarization distribution to be statistically distinguishable from that of stars: typically they are lin-

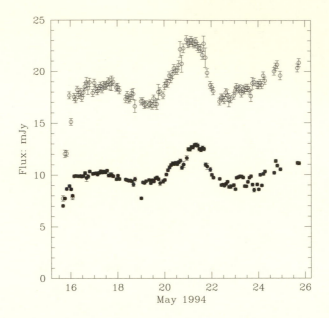

Fig. 1.8   Two ultraviolet lightcurves for PKS 2155-304. The open circles represent continuum flux at 2800 Å; the filled circles show the continuum flux at 1400 Å. In this case, changes of several tens of percent occurred within 1 day. Data courtesy of E. Pian, originally published in Pian et al. (1997).

early polarized, with fractional polarization $\simeq 0.5$–2%. However, again as for variability, a minority—including the same minority that have strong optical variability, as well as some of the AGNs with only narrow emission lines—are much more strongly polarized, often $\sim 10\%$ in linear polarization. Those objects that are strongly polarized and strongly variable in total flux are also highly variable in both the magnitude and the direction of the polarization. This variability has its bounds, however; circular polarization has never been detected, and the limits are reasonably tight.

Technical caveats are in order for discussing polarization as well. When the luminosity of the nucleus does not completely dominate the luminosity of the host galaxy, the polarization we measure depends on the aperture used, for starlight can substantially dilute the polarization of the nucleus. This effect is most noticeable in the AGNs with only narrow emission lines. In those cases, the (linear) polarization of the true nucleus can be as high as tens of percent but can only be seen after virtually all starlight has been eliminated (Chap. 12).

In addition, whether or not we detect polarization depends on the wavelength observed. Radio emission is usually linearly polarized at the few to few tens of percent level, but the observed value can be artificially suppressed by insufficient angular resolution, for the angle of polarization often varies from place to place (§9.1.4). Some AGNs whose spectra show exceedingly broad ultraviolet absorption troughs (see Chap. 11) are strongly polarized in the absorption features and more weakly polarized in the continuum. Others become strongly linearly polarized at wavelengths just shortward of the Lyman edge. Whether the X-ray emission is polarized is a question that awaits more sensitive instruments than those available in the 1990s.

1.2.7   Radio emission

Strong radio emission is the last of the distinguishing marks of AGNs. Historically, however, it was effectively the first. Some of the earliest radio astronomical observations discovered that many bright radio sources come in the form of double lobes with a galaxy located halfway between them (e.g., Cygnus A, shown in Plate 2). These radio lobes were the first evidence recognized as indicating nonstellar activity in external galaxies. After the compilation of the 3C catalog (see Chap. 2) in the late 1950s, efforts to identify its members led to Maarten Schmidt's realization in 1963 that the bright optical point source associated with one of these radio sources, 3C 273, possessed the (then) shockingly large redshift of 0.158. This realization cracked open the field of active galaxies; suddenly the unidentifiable optical point sources associated with numerous radio sources became objects with interpretable optical spectra, and known distances and luminosities.

Because radio astronomical techniques are extremely powerful, and a large fraction of all bright radio sources are AGNs, many of the *known* AGNs are strong radio emitters, and a great deal is known about the phenomenology of that emission. For example, only in the radio band is milliarcsecond imaging (by Very Long Baseline Interferometry, or VLBI) currently available, and AGN radio emission is mostly resolved on this angular scale.

However, this depth of knowledge is a bit misleading. Even in AGNs where the radio band is relatively strong, it never accounts for more than $\sim 1\%$ of the bolometric luminosity; in addition, less biased surveys have shown that the great majority of AGNs emit a much smaller fraction of their total power in the radio.

1.3 AGN Nomenclature

Because subclasses of AGNs exist that all share the same choices from this menu, numerous subvarieties have been named. Whether the taxonomical effort that created this "zoo" has helped clarify our view of this subject, or led to further confusion, is a matter of some debate. Nonetheless, to understand the conversation, one must learn the language.

Table 1.2 presents the lineaments of AGN zoology. The names themselves reveal how roundabout scientific progress can be. Some are descriptive: *radio-loud* and *radio-quiet* are fairly self-explanatory terms, and, as we shall discuss in Chapter 9, the distribution of the fraction of the bolometric luminosity that is radiated in the radio band is quite bimodal. *OVV* is an acronym for "Optically Violently Variable," a term that is equally direct: this class is marked by exceptionally rapid and large amplitude variability in the optical band. Other designations refer to the names of the first people to identify the class: Carl Seyfert pointed out the first six *Seyfert galaxies* (thirty years later they were subdivided into two principal types, according to whether their emission lines did or did not have broad wings: see §10.1.3); the "FR" in *FR1* and *FR2* stands for Fanaroff and Riley, who pointed out an interesting distinction in both luminosity and morphology among the radio galaxies; this will be discussed in detail in Chapter 9. Some class titles are deliberate coinings. *Quasar* was originally the pronounced form of "QSRS," an acronym for "quasi-stellar radio source," but over the years its usage evolved so that it now denotes nothing about the radio luminosity of an object. Today it is often used as a synonym for "generic AGN"; very low luminosity AGNs are sometimes called "micro-quasars," for example. And some have truly quirky histories. Variable stars are given names having two letters and an abbreviated form for the constellation in which they are found. The prototype of the AGN variety now called *BL Lac objects* was originally thought to be a variable star in the constellation Lacerta and was therefore given the name "BL Lac." Quirkier still is the origin of the term *blazar* (not found in the table because, 20 years after its introduction, its use is still somewhat nonstandard). This coinage is meant to unite the OVV and BL Lac classes and was invented as a joke by Ed Spiegel, the after-dinner speaker at the first conference organized on BL Lac objects. The name stuck both because the two varieties do resemble each other (compare their two lines in the table, deliberately put immediately adjacent to each other) and because its connotation of "blazing" is very appropriate to these objects whose power output varies so dramatically.

Other classes have been defined in the past, but are now rarely used. *N galaxies* are elliptical galaxies with bright optical nuclei, generally first noticed because of their large radio power. Most would now be called *broad line radio galaxies*. "QSO," a term still occasionally used, is an acronym for "quasi-stellar object" that has now been largely replaced by *radio-quiet quasar*. Osterbrock and his associates defined several Seyfert types intermediate between 1 and 2 (see §10.1.3 for a more detailed discussion); these classes can still sometimes be seen in the literature.

The column headings are abbreviations for items in the menu: "Point-like" refers to whether an optical point source can be seen. "Broad-band" means that there is comparable luminosity in the infrared, optical, and X-ray bands. "Broad lines" and "narrow lines" indicate the existence in the optical and ultraviolet spectra of lines several thousand km s^{-1}, or several hundred km s^{-1} in width, respectively. "Radio" means that the fraction of the luminosity emitted in the radio is relatively large, perhaps $\sim 10^{-3}$ of the bolometric luminosity (defining radio as $\nu \sim 10$ GHz). To be considered significantly variable, the members of the class should vary by an order of magnitude or more in the optical band over a timescale comparable to a human lifespan. To receive a "yes" in the "Polarization" column the optical light should be at least a few percent linearly polarized.

The astute reader will notice that these classifications divide up into groups: radio-loud versus radio-quiet, strongly variable versus all others, and narrow emission lines only versus broad emission lines as well. These groupings suggest an arrangement in a three-dimensional parameter space, illustrated in figure 1.9. The existence of these groupings has led to a great deal of effort to explain these multiple categories in terms of a smaller number of variables. These efforts are discussed in detail in Chapter 12. For now, it is best simply to take note of the connections.

This same hypothetical reader will notice that type 1 Seyfert galaxies have identical table entries to those of radio-quiet quasars. He or she might then ask why they are listed separately. That is a good question. In fact, there are a number of objects that have been classified "type 1 Seyfert" by some observers, and "radio-quiet quasar" by others. In practice, the only distinction is whether a host galaxy is visible. When it is, the AGN is called a Seyfert galaxy, whereas when none is visible, it is called a quasar. Since the distinction depends on the resolution and background level of the instrument, and does not depend primarily on the object itself, it is clearly not a very useful one. Its only objective correlate is with luminosity: Seyfert galaxies are on average two orders of magnitude less powerful than quasars. It is for this reason, of course, that the host galaxy is visible, for we can only see host galaxies when the luminosity from the AGN does not overwhelm the starlight.

Table 1.2: The AGN Bestiary

Beast	Pointlike	Broad-band	Broad Lines	Narrow Lines	Radio	Variable	Polarized
Radio-loud quasars	Yes	Yes	Yes	Yes	Yes	Some	Some
Radio-quiet quasars	Yes	Yes	Yes	Yes	Weak	Weak	Weak
Broad line radio galaxies (FR2 only)	Yes	Yes	Yes	Yes	Yes	Weak	Weak
Narrow line radio galaxies (FR1 and FR2)	No	No	No	Yes	Yes	No	No
OVV quasars	Yes	Yes	Yes	Yes	Yes	Yes	Yes
BL Lac objects	Yes	Yes	No	No	Yes	Yes	Yes
Seyferts type 1	Yes	Yes	Yes	Yes	Weak	Some	Weak
Seyferts type 2	No	Yes	No	Yes	Weak	No	Some
LINERs	No	No	No	Yes	No	No	No

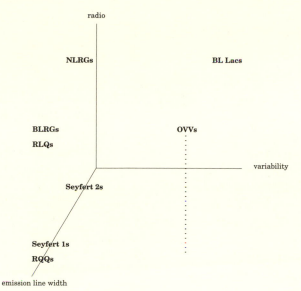

Fig. 1.9 The principal subvarieties of AGNs schematically arranged according to relative power in the radio band, emission line width, and variability. All combinations are possible except that there are no highly variable radio-quiet objects.

The last entry in the table, *LINERs*, denotes a category at the very margin of activity. As of this writing, it is still debatable whether these galaxies are truly active or not, or whether it is even a well defined category in any physical sense. Dubbed "LINER" for "Low-Ionization Nuclear Emission Region" by Tim Heckman in the early 1980s, these galaxies show strong emission lines, like Seyfert galaxies, but the relative strengths of those lines from low-ionization stages are rather greater than in Seyfert galaxies (see §10.1.3 for a quantitative definition).

It is also important to recognize that these categories may not exhaust the true list. It is entirely possible that there are varieties of AGNs that we do not yet recognize. Some people believe, for example, that the ultraluminous infrared galaxies discovered by the *Infrared Astronomical Satellite* (*IRAS*) may house AGNs (§13.6). Similarly, others suspect that there may be a class of *type 2 quasars*, by analogy with the division of the Seyfert class (§12.6.5). At present, limits on the existence of such objects are very weak. Surprises are always very much a possibility in the field of AGNs.

2   How to Find AGNs

As we have just seen, the definition of what constitutes an active galactic nucleus is a purely observational construct, and a somewhat waffly one at that. Therefore, what different people mean by the term tends to be based on their favorite means of finding AGNs. This anarchy does not lead to catastrophe because, fortunately for us, there is substantial overlap between the different search techniques in use. However, it does mean that one must be very careful when working at the quantitative level to make explicit one's usually tacit prejudices of definition. Because much hard work is necessary to accumulate AGN samples, the people responsible often develop very strong emotional attachments to them and make strong claims for their unique benefits. In evaluating these claims, one should always remember that *all surveys are biased*. This is not a pejorative remark; it simply reflects the fact that no single definition of AGNs has yet been found to be all-inclusive.

On the other hand, one should not lapse into complete relativism. There are certain qualities that distinguish good surveys from poor. Perhaps the most important of these is the ability of the observers to actually follow their ground rules. In that case, one can at least be confident that the actual selection criteria correspond to the ones stated. It also helps for these criteria to be framed simply, so that their implications can be (comparatively) easily recognized. Surveys also differ in their efficiency (the rate at which confirmed candidates can be found relative to the expenditure of human time, telescope time, and other resources), or in their degree of interest (do they explore new properties or reveal new classes of objects?).

Before describing the principal means different people have used to find, count, and describe AGNs, it is worthwhile to make another comment. If one is to compile a statistical sample, the large number of objects required to make the statistics meaningful generally means that one is resource-limited, that is, the information collected on each object will be the bare minimum required to fill its table entries in the sample. Because more detailed studies inevitably require more time both at the telescope and in analysis, in any given category there are generally only a few favorite objects about which a great deal is known. Consequently, the several brightest cases of each type are observed over and over again, with the hope that the detailed properties discovered in them will prove to be representative of the general class. Because they have little choice, many people proceed on this basis, but you should be warned that this is a dangerous expectation.

2.1   Optical Color

Most searches for AGNs are based on their extremely broad-band continua (see §1.2.3). In the optical band, that fact has the consequence of creating color distinctions between AGNs and ordinary stars or galaxies. These distinctions are the basis for the single most popular AGN survey technique. More surveys for AGNs have been conducted using optical color criteria than by any other method. This is partly because so many people are trained in the techniques of filter photometry, and partly because these methods are comparatively efficient—a few images suffice to supply a very large number of candidate objects.

To conduct such a survey, one first takes an image of the sky in some region through at least two different optical filters. In any given field, most of the objects in the image will not be AGNs. Stars and galaxies always outnumber AGNs, no matter where the field is or what its magnitude limit is. To cull the AGNs from the background, one must apply a filter to the object list.

The essence of the color selection method is that one can find AGNs because their optical spectra have different shapes from those of stars and galaxies, and hence the ratios of their fluxes in different bands (their "colors") can be distinguished. The distinction comes from several sources. First, the continua of AGNs tend to be broad and smooth, while stellar (and hence galactic) spectra are thermal and therefore strongly curved (see figs. 1.1 and 1.2). Thus, the color method is based directly on searching for objects with the sort of broad-band continua remarked on in §1.2.3. The breadth of typical AGN spectra means that they have more flux in both the UV and IR relative to the V band than do stars. Second, when the redshift is great enough to bring Lyα into the visible band, this emission line can be strong enough to substantially increase the flux in the band into which it falls. Third, when the redshift is somewhat greater yet, intervening Lyα and Lyman continuum absorption can remove enough flux to depress the band(s) in which this absorption falls.

Of course, if the colors of the AGN plus its host are to be different enough from normal galactic colors to be distinguished, it immediately follows that the optical luminosity of the AGN (in our frame) must be at least comparable to, or preferably much greater than, the luminosity of the host. Consequently, color surveys are automatically biased against low-luminosity AGNs. They also often carry a second, implicit bias: surveys hoping to find nearby AGNs often require all candidates to be extended in order to eliminate stars quickly, while surveys hoping to find more distant AGNs often require all candidates to be points in order to eliminate faint galaxies. The former criterion misses AGNs in which the nucleus is much more luminous than its host; the latter is biased in exactly the opposite way.

To see how surveys of this sort work in practice, consider first a simple cut in the single-color space of $U - B$ (the peak in the transmission coefficient for a standard "U" filter comes at about 3600 Å, while the greatest transmission for a "B" filter is near 4300 Å). The hottest (therefore UV-brightest) main sequence stars have $U - B \simeq 0.4$–0.5 mag. In contrast, typical (low-redshift) AGN spectra have $U - B \simeq -1$ mag. Thus, nearly all stars and galaxies immediately fall out if one takes only the "UV-excess" objects into the survey. The major contaminants in an UV-excess survey are hot subdwarfs (sdB stars) and white dwarfs, with both (and especially the subdwarfs) becoming relatively less of a problem as the flux limit is pushed to lower levels. Of course, stars of all varieties are fewer when the survey is done at high Galactic latitude. Conversely, the density of stars is so great in the plane of the Galaxy proper that it is frustratingly inefficient to look for AGNs within $\simeq 10°$ of the Galactic plane.

One of the earliest color surveys was done by Markarian (1967; the final list is in Markarian, Lipovetsky, and Stepanian 1981). In this case, in order to eliminate subdwarfs, he required his objects to be extended. The survey covers a large part of the northern sky ($\simeq 10,000$ square degrees) and extends down to about $m_B = 15.5$ mag. Roughly 1500 galaxies are on the Markarian list, about 10% of them Seyfert galaxies. Although the sample definition has never been well calibrated, this survey was an important pioneering effort, for it turned up many interesting objects.

The most thoroughly studied color survey is the Palomar Bright Quasar Survey of Schmidt and Green (1983). This survey is known by two alternate acronyms, BQS and PG (for Palomar Green). Selecting only the pointlike objects with $U - B < -0.44$ mag, they found $\simeq 1700$ candidates brighter than $m_B \simeq 16$ mag in 11,000 square degrees of sky. Follow-up spectroscopy confirmed that 114 are AGNs, for an AGN efficiency of $\simeq 7\%$.[*]

To find fainter AGNs by the same technique, one takes longer exposures on a smaller solid angle. Boyle et al. (1990), for example, were able to find quasars down to a limit of $m_B = 20.9$ mag, but over a field of only 245 square degrees. Their color cut was very similar to that of the BQS, $U - B < -0.35$ mag.

Unfortunately, we do not know whether *all* quasars have such smooth "blue" optical spectra. The only way to find out whether we are missing any by selecting only on the basis of UV excess is to search using other criteria. To this end, there have also been a number of searches using multiple colors. For example, Warren, Hewett, and Osmer (1991) took images in five filters: U, B, V, R, and I (the V band is centered on 5400 Å; R on 6400 Å; and I on 8500 Å). Stars occupy a well defined "sausage" in four-dimensional color space (five flux measurements yield four colors, as it is only flux ratios that count), so that Warren et al. could define their candidates to be any object with a stellar image that lay a certain number of standard deviations outside this sausage in any direction. Indeed, by generalizing the color criterion, Warren et al. were able to find many AGNs that would not have been found by a pure UV excess criterion. A still larger survey of this variety is planned as part of the Sloan Digital Sky Survey (SDSS). Again there will be five color images, but the limiting magnitude for the images will be $m_B \simeq 22$ mag, and one-quarter of the sky will be searched. With this size survey, $\sim 10^6$ pointlike quasar candidates should be found, and $\sim 10^5$ (most of the brightest $\sim 10\%$) confirmed by spectroscopy.

One final difficulty must be noted that applies to all optical surveys: there can be substantial intervening absorption between us and the AGNs we seek. Some of this absorption is local—dust extinction in the plane of the Milky Way is quite substantial—but very distant intervening absorption becomes increasingly important the farther out we look. On average, gas clouds optically thick at the Lyman edge are found on any given line of sight with a density per unit redshift roughly $0.25(1 + z)^{1.5}$ (Stengler-Larrea et al. 1995); if one with $z \geq 4$ happens to lie between us and an AGN, it would blank out the spectrum in the B band and blueward. Large numbers of neutral H clouds with smaller column densities can also be found. Their numbers increase rapidly with redshift, leading to a significant depression of AGN continua at observed wavelengths shorter than $1216(1 + z)$ Å when

---

[*]Viewed from a different perspective, the BQS survey had a much greater efficiency for finding either hot subdwarfs ($\simeq 40\%$) or white dwarfs ($\simeq 20\%$). Wisely not letting good data go to waste, Schmidt and Green became experts in the statistics of subdwarfs and white dwarfs as a by-product of performing an AGN survey.

$z \geq 2$. Thus, this intervening absorption artificially diminishes B-band flux when the quasar redshift is greater than about 2.5. In addition, the density of galaxies on the sky is great enough that, if they contain significant dust, the light from many very high redshift quasars may suffer considerable extinction. Taken together, all these mechanisms can have the effect of making it increasingly difficult for us to find AGNs by their (observed frame) optical light as their redshifts increase.

2.2 Optical Emission Lines

Another distinguishing characteristic of (some classes of) AGNs is strong optical and ultraviolet emission lines (as discussed in §1.2.4). These can carry enough flux that they are apparent in even crude, low-dispersion spectra. Because very few stars have emission lines, this contrast in properties creates another opportunity for a potentially efficient AGN search technique. The main limit is that conventional spectroscopic methods yield only one spectrum per observation. Given the density of faint objects on the sky ($\sim 10^4$ per square degree at a flux limit of $m_B \simeq 22$ mag), one clearly requires a much more rapid device to make this approach practical.

The solution adopted by many different groups is to place a dispersing element across the telescope's light path before the light hits an imaging detector. These devices are variously called "objective prisms," "grisms," and "grenses," but their basic principles are all the same. The image obtained when any one of them is used consists of a large number of strips, each one a low-dispersion spectrum of the object at that location. It is relatively easy to search these strips for emission bumps that identify AGN candidates (in fact, grisms were used in the Markarian survey to make it easier to find continuum color anomalies). Higher quality spectra are then taken for all the AGN candidates to confirm their identifications and more accurately measure such properties as redshift, emission line fluxes, and continuum shape. Searches of this sort are especially useful for finding high redshift AGNs ($z > 2$) because the ultraviolet rest-frame emission lines are very prominent, whereas, as we have just discussed in the previous section, the colors of high redshift AGNs are hard to define a priori.

A number of surveys of this sort have been very important historically. The very first AGNs to be identified as interesting objects, although they were not thought of as "AGNs" until much later, were the six emission line galaxies singled out by Seyfert in 1943 because they had strong, high-ionization emission lines that were concentrated in their nuclei. More recently, emission line surveys have done the most to expand our knowledge of high-redshift quasars; the first survey able to discover large numbers of quasars with $z > 2$ was done by Hoag and Smith (1977). In the earliest such surveys, though, there were problems with establishing a quantitative definition for the selection criteria. The search for emission bumps was done by eye, and it was found post hoc that different people had substantially different subjective selection criteria.

In modern versions of this survey technique, as much as possible is automated to avoid subjectivity and nonuniformity. To construct the "Large Bright Quasar Sample," for example, 800 square degrees of sky were searched down to $m_B = 18.8$ mag (Foltz et al. 1987, 1989; Hewett et al. 1991; Chaffee et al. 1991; Morris et al. 1991; Hewett, Foltz, and Chaffee 1995). Each spectrum produced on the image by the objective prism was measured, and candidates were selected on the basis of extreme blue color, strong emission or absorption features, or strong continuum breaks. This survey found $\simeq 1000$ quasars. Similar techniques were employed in the survey of Schmidt, Schneider, and Gunn (1986a,b; Schneider, Schmidt, and Gunn 1994). In their work, the candidates were selected on the basis of a threshold in emission line equivalent width and signal/noise ratio. This survey proved especially successful in discovering very high redshift quasars. Despite the great improvement in this method provided by automation, it still rises or falls, of course, on the validity of its fundamental assumption: that quasars can be distinguished on the basis of sharp spectral features such as emission lines.

There is another version of this search technique that is, in some ways, the least biased method of finding AGNs yet invented. We began this section by noting that because of the rarity of AGNs, any practical search technique must employ an efficient discriminator. But this requirement evaporates if the main point of the survey is to study normal stars or galaxies (imagine, for example, that the BQS survey had been motivated primarily by searching for hot dwarf stars). In that case, effort expended on the non-AGNs isn't wasted, and any AGNs found come "for free." The only special provisos are that the survey be large enough to uncover a statistically interesting number of AGNs, and that the selection criteria should not in some way eliminate AGNs. Moreover, because surveys focused on normal stars or galaxies will be keyed to their properties, they are (almost) independent of the properties of any AGNs they turn up and are therefore largely unbiased with respect to AGN variety or characteristics. The principal bias such surveys impose is between AGNs with and without visible hosts. That is, galaxy surveys ignore AGNs with such great contrast relative to their hosts that they appear pointlike, whereas stellar surveys ignore AGNs in which the hosts are visible. A secondary bias that arises in galaxy surveys is that they are also likely to miss very low luminosity AGNs because such weak AGNs do not contribute enough light to whole-galaxy spectra to be recognized.

Two such surveys have been especially useful. One is the CfA redshift survey. Primarily designed to study large-scale structure in the Universe, this sample contains all galaxies within selected regions of the sky brighter than a certain visual magnitude limit (originally the limit was $m_V = 14$ mag, but an extended survey stretched the limit to $m_V = 15.5$ mag, and further stretching is possible). Because the essential quantity desired for mapping is the redshift, spectra were taken of all the galaxies in the sample. Measuring a redshift demands a higher quality spectrum than merely looking for large equivalent width emission lines, but not such high quality that the job consumes prohibitive amounts of telescope time. Spectra of this quality are, however, quite good enough to enable a quantitative check for large equivalent width emission lines. On this basis Huchra and Burg (1992) found roughly 50 Seyfert galaxies, divided approximately equally between types 1 and 2. As the total sample contained $\simeq 2000$ galaxies, they found that the local Seyfert fraction is $\simeq 2.5\%$. There are, however, two biases in this survey, which act in opposite directions. On the one hand, an especially bright type 1 nucleus could increase the total flux from a galaxy sufficiently to bring it into the sample even though the host flux by itself would not be great enough. This effect overestimates the Seyfert fraction. On the other hand, low-luminosity Seyfert nuclei of both types are missed because their lines do not have great enough equivalent width relative to the stellar continuum of the host galaxy.

The second such sample is the Revised Shapley-Ames catalog of bright galaxies (Sandage and Tammann 1981). This sample is based on photographic plates and so has a fuzzy magnitude limit around $m_{\mathrm{pg}} = 13$ mag. Of its $\simeq 1300$ galaxies, $\simeq 50$ are Seyfert galaxies of luminosity comparable to those found in the CfA survey, and another $\simeq 20$ are radio galaxies. However, using very small apertures centered on the nuclei of a large (486 galaxies) complete subsample of this catalog, Ho, Filippenko, and Sargent (1997a) showed that very low luminosity AGN activity is extremely common: roughly 40% of the RSA galaxies contain LINERs, and 10% have Seyfert nuclei.

The Sloan Digital Sky Survey, mentioned above in the color context, also falls into this category. In this survey, galaxies will be selected for spectroscopy on a basis wholly independent of their nuclear properties, and therefore AGNs in hosts bright enough to select, but not so bright as to swamp the AGN, should be very fairly sampled.

2.3   Radio Flux

2.3.1   The 3C catalog and the confusion limit

Historically, the first method by which AGNs as such were found was the identification of radio sources (radio emission is the distinguishing mark of AGNs discussed in §1.2.7). In the early 1950s, radio telescopes with large enough total apertures to detect first dozens, and then hundreds, of sources were built by groups at both Cambridge University (U.K.) and Sydney University (Australia). However, it took almost a decade to assemble the first useful large catalog, the 3C (for 3rd Cambridge) catalog, a collection of almost 500 sources brighter than 9 Jy at 178 MHz. A few years after the initial publication of the 3C catalog, a revised version (called 3CR) was compiled with more accurate fluxes (Bennett 1962). This list contains 328 sources with $\delta > -5°$ and the same flux limit as in the original version of the 3C catalog.

The name of this catalog calls for a digression: why is the first radio source catalog called the third? The reason is instructive, and its lessons are by no means limited to radio surveys. It is solely a matter of his- torical accident that these lessons were first learned as a result of doing surveys in that band; they apply with equal force all the way across the electromagnetic spectrum.

Suppose that you use a telescope whose primary beam has an area of 1 square degree, and that in the entire sky there are 1000 sources bright enough for you to detect. Then in each telescope beam, the expectation value of finding a source is $\simeq 0.024$, and the probability of finding two in any particular beam is 3×10^{-4}. Because there are 41,000 beam areas over the sky, one expects there will be $\simeq 12$ beams in which two objects are found. Thus, you will measure the correct flux for $\simeq 990$ of the 1000 objects you find.

The situation changes dramatically, however, if the number of sources per beam bright enough to detect increases much beyond this level. If instead there are 10,000 sources bright enough to detect, the expectation value per beam is now 0.24, and one expects multiple sources in 2.5% of the beams, or 1020 cases. Now 10% of all the beams with sources have misleading flux levels, for wherever several sources can all be found within a single beam, their fluxes have been mistakenly combined. Moreover, the list which you are attempting to create is missing all but one of the sources at each location where several sources are merged into one.

In fact, the situation is even worse than this. Typically, the number of sources per unit solid angle increases fairly rapidly with decreasing flux (measuring this rate of increase is a matter of considerable interest; see Chap. 3). That means that if the probability of finding two individually detectable sources in a single beam is interestingly large, the probability of finding two sources, each slightly weaker than the sample's flux limit, is even greater. If their mean flux is greater than half the flux limit, their total flux exceeds the threshold for inclusion in the catalog. Thus, by this means inadequate angular resolution can create large numbers of spurious sources.

Defining the angular resolution becomes a bit tricky when the telescope beam is synthesized interferometrically. The beam shape of a single round dish is quite simple and concentrated: outside the central maximum, the Airy rings decline rapidly in amplitude over an angular scale of order the ratio of the wavelength to the telescope diameter (see §13.1.1). However, depending on its configuration, the beam shape of an interferometer can have very complicated side lobes, sometimes spreading all across the field of view. When this is the case, a single very bright source can create numerous spurious weak sources if these side lobes are not carefully removed. Thus, although one measure of the angular resolution is the width of the central peak in the beam pattern, if there is a large dynamic range between the brightest and faintest sources, the character of the side lobes also figures into what one means by the angular resolution. For this reason, it is wise to place surveys looking for faint sources in regions of the sky where there are no bright sources.

This whole set of problems is known as the *confusion limit* and was not well understood at the time of the first and second Cambridge surveys. Unfortunately for the astronomers compiling them, their surveys were, in fact, badly confusion limited. This fact was brought pointedly to their attention in a classic paper by members of the Sydney group (Mills and Slee 1957). Ultimately, the problem was solved by an upgrade of the Cambridge radio telescope to twice its former operating frequency, and the resulting fourfold decrease in the beam size permitted the measurement of reliable fluxes and positions on the third go-around.

One might think that the lesson of the confusion limit is to build survey telescopes with as few sources per beam as possible. Although there is some truth in this, it cannot be pushed to an extreme. For one thing, it is obvious that such an approach is very inefficient: why waste time observing all those empty fields? However, there is another, subtler problem. All flux measurements are subject to noise. If a large positive noise fluctuation occurs in a location on the sky where no real source exists, one might mistakenly identify it as a real object. Such large fluctuations are rare, but if nearly all antenna beams are empty, the chances are good that spurious sources will appear in at least some of them.

To illustrate this point quantitatively, let us return to the numerical example posed for our discussion of the confusion limit. When there are 1000 real sources distributed over 41,000 beams (mean number of sources per beam = 0.024), there are 40,000 empty beams. If the system noise is Gaussian, and we set the detection threshold at 3σ, the probability per beam of finding a spurious source is 0.00135, so the expectation value for the total number of spurious sources is $\simeq 540$; that is, the catalog is inflated by 50% over its real size! Increasing the detection threshold to, say, 5σ, makes a big improvement: the probability per beam of finding a spurious source falls to $\simeq 3 \times 10^{-7}$. Thus, the optimum balance between angular resolution and limiting flux is set by a trade-off between the conflicting goals of avoiding source confusion, maximizing observational efficiency, and rejecting spurious sources (a particularly complete discussion of these issues can be found in Murdoch, Crawford, and Jauncey 1973).

2.3.2 Optical identifications and the discovery of quasars

Listing radio sources and their fluxes has much intrinsic interest, but astronomical objects almost always yield the most detailed information when studied in the visible band. The familiar complexities of atomic spectroscopy give us a tremendous boost in diagnosing the circumstances inside distant objects. Consequently, as soon as the first sources were discovered, a few perceptive astronomers launched optical observational programs to find what optical sources might be identified with them and to study their properties.

The first step in this effort was to achieve enough angular resolution with radio observations that the optical objects in the radio error boxes were few enough to make it possible to choose which of them was responsible for the radio emission. The Sydney group was the first to achieve high enough angular resolution to permit identification of a few "radio stars" with optical counterparts and found (somewhat shakily) that both M 87 and NGC 5128 were the homes of powerful radio sources (Centaurus A is centered on the latter galaxy). Just a few years later, positions for a few more bright radio sources were refined by an order of magnitude, allowing Baade and Minkowski (1954) to measure the redshift of the radio galaxy Cygnus A, a (then) astonishing 0.056. By 1960, Minkowski was able to identify the radio source 3C 295 with a galaxy at $z = 0.46$ (Minkowski 1960).

Other 3C radio sources (e.g., 3C 48 and 3C 273) proved to be more difficult to identify. In these cases, unlike, for example, 3C 295, the optical image of the most likely candidate in the radio error box was pointlike, and so didn't appear to be a galaxy at all. Moreover, their optical spectra were dominated by emission lines, a fact that greatly puzzled astronomers used to the absorption line spectra of normal galaxies. Without guidance about what sort of object they might be, optical astronomers had few clues about what sorts of lines to look for in the spectra, and the measured wavelengths were not even close to anything familiar. The mystery was only cleared up in 1963 when Maarten Schmidt realized that the spectra could be understood if they were adjusted for what was still considered (despite the example of 3C 295) an extraordinarily large redshift—0.158 for 3C 273 (Schmidt 1963).

The fact that the first quasars were discovered through the identification of radio sources led to another instructive mistake. The very term *quasar*, now used for any high-luminosity AGN with broad optical and ultraviolet emission lines (and sometimes for AGNs of any variety), was coined as a contraction of "quasi-stellar radio source," and for almost a decade it was believed that the typical high-luminosity AGN (although to use this terminology is an anachronism) was "radio-loud." Only after about 1970 was it realized that only about 10% of all quasars radiate as much as 0.1% of their total luminosity in the radio range. This historical misunderstanding reemphasizes that surveys find what they are looking for, and that there is no single best way to find all examples of a phenomenon as diverse as AGNs.

Although the effort consumed nearly 30 years, the 3CR catalog is now virtually completely identified (one or two objects still remain outstanding). Its full redshift range extends past $z = 3$. Roughly three-fourths of its members are radio galaxies, with most of the rest quasars, but there also a few BL Lac objects and Seyfert galaxies (and also a small number of galactic supernova remnants, like the Crab Nebula).

2.3.3 Subsequent surveys

Other, deeper surveys followed the 3CR. The 4C survey (Pilkington and Scott 1965; Cower, Scott, and Wills 1967) covered nearly all the northern sky to a limiting flux of 2 Jy at 178 MHz. A comparable survey covering the southern sky was performed at the Parkes radio telescope in Australia (Bolton, Gardner, and Mackey 1964; Price and Milne 1965). In this case the observing frequency was somewhat higher (408 MHz), and the sensitivity a bit greater ($\simeq 3$ Jy limiting flux). Other surveys at this frequency were also completed in the early 1970's at the Molonglo (Robertson 1977) and Bologna (Colla et al. 1970; Bergamini et al. 1973) radio observatories.

As high-frequency receiver technology improved, higher radio frequencies could be explored, while the greatly improved angular resolution made possible by interferometry permitted much deeper surveys at all frequencies, high and low. Because the number of sources per sterradian increases rapidly at lower flux levels (see also §3.1), the sheer weight of numbers makes deep surveys over the whole sky cumbersome. Consequently, the surveys (up until recently) have tended to divide into (relatively) shallow all-sky surveys and deep "pencil-beam" surveys that cover only small fields of view.

Complete high-frequency sky coverage has been provided by a pair of surveys, one in the north and one in the south. The entire northern sky was surveyed at 1.4 and 5 GHz by the NRAO at Green Bank down to a limiting flux of 0.8 Jy at the higher frequency (Pauliny-Toth and Kellermann 1972); the Parkes Radio Observatory in Australia cataloged southern sky sources to a similar flux limit (0.7 Jy) at 2.7 GHz. The Parkes survey was published in many segments, of which the last was Wall, Wright, and Bolton (1976). Objects in this survey are denoted "PKS HHMMDDD", where "HHMM" is the right ascension of the source in hours and minutes and "DDD" is the declination to the nearest tenth of a degree. The brightest sources from the combined Green Bank, Parkes, and Bonn surveys can be found in the "2 Jansky" survey of Wall and Peacock (1985), which lists all those sources brighter at 2.7 GHZ than 2 Jy (of course) over 9.8 sr of sky. The remaining 2.8 sr covers the Galactic plane. Virtually all of these sources have optical identifications. Recently, a much more ambitious high-frequency large-area survey has begun at the VLA. The FIRST survey (for Faint Images of the Radio Sky at Twenty Centimeters) is imaging at 1.4 GHz the same 10,000 square degrees of the northern Galactic cap planned for the SDSS. The goal is to identify all sources brighter than 1 mJy (Becker, White, and Helfand 1995).

Until the FIRST survey is completed, deeper surveys have been limited to small solid angles. A great many have been made, of which we mention here only a few. Typically, as their sensitivity grows (i.e., the minimum flux for detection falls), their sky coverage shrinks. For example, the 5C survey is a series of pencil beams (13 so far: the latest is published in Benn 1995), each ~ 10 square degrees in size, in which sources are found at 408 MHz down to 10 mJy and at 1.4 GHz down to 2 mJy. On the other hand, Windhorst et al. (1985), also working at 1.4 GHz, covered 3 square degrees but identified sources as faint as 0.3 mJy. At higher frequencies, surveys have gone deeper still over even smaller areas: all the way down to $\simeq 15$ μJy at both 5 GHz (Fomalont et al. 1991) and 8.4 GHz (Windhorst et al. 1993). The former survey covered a region 16' in diameter, the latter two fields each 7' on a side. Although the great majority of the brighter sources are AGNs, at fluxes below ~ 1 mJy the nature of the host galaxies becomes difficult to determine because they are optically very faint. It is possible that galaxies without active nuclei (but perhaps unusual stellar populations—speculation centers on these being "starburst" galaxies; see also §13.6) predominate (Windhorst et al. 1993). For this reason, at very faint flux levels, searching for AGNs on the basis of radio emission alone no longer suffices; this search key must be combined with other information to discriminate between AGNs and other sorts of galaxies.

Although the GHz band is only a single order of magnitude higher than the frequencies at which the early surveys were made, a rather different population of objects is found there. In samples selected on the basis of flux in the few hundred MHz range, the radio spectra are most often fairly smooth power laws of the form $F_\nu \propto \nu^{-\alpha}$, with $0.5 < \alpha < 1$. Consequently, in terms of flux per unit frequency, these objects are often rather faint at frequencies much greater than a few GHz. Perhaps it should not be too surprising that in samples selected on the basis of flux at a few GHz, the spectra, especially above ~ 1 GHz, are often quite different. At these relatively high frequencies, objects discovered in GHz-band surveys usually have spectra that are nearly flat in F_ν, even if at lower frequencies they rise more steeply. Of course, it is just this spectral contrast that makes them the dominant population when objects are chosen for brightness in the 1–10 GHz band. In addition, although the low-frequency selected sources are almost always very large in physical extent (lengthscales of hundreds of kiloparsecs are common), high-frequency selected objects tend to be much more compact and are often unresolved unless very high angular resolution techniques are employed. The physics behind these effects will be discussed in Chapter 9.

2.4 X-rays

As we emphasized at the start (§1.2.3), AGNs, to a degree unknown in almost any other sort of astronomical object, emit their power over a very broad range of photon frequencies. For no ordinary galaxy or star does the ratio of X-ray to optical flux (in the sense of $dF/d\log\nu$) approach unity, as it does for many AGNs. For this reason, X-ray surveys are another very efficient way of finding AGNs.

Just as in the history of radio astronomy, AGNs were some of the earliest sources discovered. Using sounding rockets, Friedman and Byram first reported a marginal detection of the quasar 3C 273 in 1967; Bowyer et al. (1970) put this detection on a firmer footing and also detected the radio galaxy Centaurus A. Two more AGNs (the Seyfert galaxies NGC 4151 and NGC 1275) were found in the early 1970s by *Uhuru*, the first X-ray astronomy satellite. However, AGNs were first understood to be generically strong X-ray emitters as a result of the *Ariel V* survey, which found 13 Seyfert galaxies with 2–10 keV X-ray fluxes greater than a few times 10^{-11} erg cm^{-2} s^{-1} (Elvis et al. 1978).

Two surveys are now the main source of statistical information on AGNs in the X-ray band, the HEAO-1 All-Sky Survey (Piccinotti et al. 1982) and the merger of the *Einstein* Medium Sensitivity Survey (EMSS; see Gioia et al. 1990; Stocke et al. 1991) with the *Rosat* deep survey (Boyle et al. 1993). The former relied on "hard" X-rays, the 2–10 keV band; the latter was sensitive to "soft" X-rays, 0.3–3.5 keV. In both cases, intrinsic technological limits determined the character of the survey.

The fundamental distinction between "soft" and "hard" is enforced by a basic property of normal solids. Typical atomic spacings in solids are ~ 1 Å; interpreted as a wavelength, this distance corresponds to photons of energy ~ 10 keV. Consequently, it is very difficult to make reflecting materials in the X-ray band. At best, grazing incidence optics allow focussing telescopes to be constructed with reasonable throughput up to several keV. For this reason, it is relatively easy to make deep imaging surveys with high angular resolution—but small fields of view—in the soft X-ray band. In contrast, at higher energies the positions of sources are determined by collimators. Angular resolution better than a few arcminutes is very difficult to achieve, but it is possible to cover the entire sky with comparative ease.

These contrasts have a dramatic effect on the kinds of surveys done at soft and hard X-ray energies. Soft X-ray surveys cover small areas very deeply; hard X-ray surveys cover large areas but to much brighter limiting fluxes. For example, the EMSS covered only 780 square degrees but achieved a limiting sensitivity (in some parts of its field) of as little as 6×10^{-14} erg cm^{-2} s^{-1}. Roughly 30% of the $\simeq 1400$ sources discovered were AGNs, mostly low-redshift quasars (this demonstrates the high efficiency of X-ray surveys; only 7% of the BQS objects selected by color were AGNs). The *Rosat* deep survey was similar in design to the EMSS and was sensitive to a similar range of photon energies but had a variable flux limit ranging from 3×10^{-15} erg cm^{-2} s^{-1} at its deepest up to 6.5×10^{-14} erg cm^{-2} s^{-1}. That is, at its most sensitive it was able to find sources 20 times fainter than the faintest EMSS sources, while in its least well covered regions, the minimum flux was about the same as in the best EMSS fields. The trade-off, of course, is that only 0.06 square degrees were examined at the faintest limit, and 1.4 square degrees at the brightest limiting flux. In this survey, 107 quasars were found.

On the other hand, the Piccinotti survey, although covering much more of the sky (8.2 sr), was restricted to much brighter sources; its minimum flux was 3.1×10^{-11} erg cm^{-2} s^{-1} in the 2–10 keV band. Fainter sources were masked by the relatively high background included in this survey's large error boxes, and at fluxes not too much lower it would have been severely confusion-limited in any case. Of the 85 sources in the entire sample, 60 were extragalactic, roughly half galaxy clusters, and half AGNs. Almost all the AGNs were type 1 Seyfert galaxies.

A still larger X-ray survey of AGNs can be derived from the *Rosat* all-sky survey (nonuniform coverage of the entire sky with a typical flux limit $\simeq 5 \times 10^{-13}$ erg cm^{-2} s^{-1}) and accumulated *Rosat* pointed observations in which serendipitously discovered AGNs appear. Each of these survey techniques is likely to find a total of $\sim 10^5$ AGNs.

2.5 Infrared

Yet another way to find AGNs by contrasting their broad-band continua with the narrower spectrum emitted by galaxies is to look in the infrared. A comparison of figures 1.1 and 1.2 shows, however, that the advantage is not as great as in the X-ray band because many normal galaxies also emit a substantial part of their power in the infrared.

Another difficulty that limits the utility of infrared surveys for AGNs is that, by comparison with techniques in other bands, infrared work is still primitive. The only comprehensive survey available so far was done by the *IRAS* satellite, whose angular resolution was a few arcminutes and whose limiting sensitivity (in each of four bands centered at 12, 25, 60, and 100 μ)

was only $\simeq$ 0.3–3 Jy (depending on both wavelength and location on the sky). Nonetheless, the *IRAS* survey covered virtually the entire sky and revealed many important things.

Like the CfA survey, there were many goals for the *IRAS* survey other than discovering AGNs. Consequently, our discussion of it will be very limited. For our purposes, two results are most significant. First, AGNs can be efficiently discovered by selecting for "warm" infrared colors (de Grijp et al. 1992). Whereas the infrared flux from most galaxies is predominantly in the range 60–100 μ, the flux (again in the sense $dF/d\log\nu$) from AGNs in the range 12–25 μ is generally at least comparable to the flux at longer wavelengths. Because most (but not all; see Chap. 9) infrared emission is thought to be due to thermally radiating dust grains, spectral shapes are described in terms of equivalent temperatures. In this language, normal galaxies show dust color temperatures $\sim$ 30 K, while in AGN the color temperature (whether or not the actual radiation mechanism is thermal) is more commonly $\simeq$ 100–300 K.

Second, *IRAS* discovered a whole new class of galaxies whose bolometric luminosity is dominated by emission in the infrared (Soifer et al. 1986, 1987). Although rare with respect to all galaxies, in the local Universe these *ultra-luminous IRAS galaxies* are roughly as numerous as AGNs of comparable luminosity. It is currently a matter of considerable controversy which of these galaxies derive their extraordinary IR luminosity from a large population of young stars whose light heats surrounding dust clouds, and which owe their infrared power to a hidden active nucleus (see also §13.6). It may be that in some galaxies both mechanisms are active.

2.6 γ-rays

The last survey key to be developed was high-energy γ-ray emission. Searches for γ-ray sources are technically very difficult, both because it is hard to construct detectors capable of pulling very low flux levels out of the noise and because achieving even an angular resolution of $\sim 1°$ is not easy. Nonetheless, when a source is found, it is guaranteed to be interesting, for ordinary stars produce virtually no γ-rays at all.

The first such survey was performed by the Energetic Gamma-Ray Experiment Telescope (*EGRET*) on board the *Compton Gamma-Ray Observatory*. Covering the entire sky (although distinguishing point sources was difficult in the Galactic plane), it was able to detect AGNs with fluxes as low as $\sim 10^{-11}$ erg cm^{-2} s^{-1} in the energy range 30 MeV–3 GeV. Locations derived from such a survey are rough: the directions of 100 MeV photons were measured with an accuracy only $\simeq 5°$. Nonetheless, this survey discovered numerous AGNs. Of its 129 point sources, 40 are confirmed AGNs and another 11 are possible AGNs (Thompson et al. 1995).

2.7 Merged Catalogs and Their Flux Limits

To give an impression of the relative power of surveys conducted in different continuum bands and using different search strategies, the limiting fluxes (in νF_ν) of a variety of well known surveys are shown in figure 2.1. Not surprisingly, those covering the whole sky, or a large part of it, are much shallower than those focused on a small region. It is also interesting, however, to compare the flux limits, not as a function of survey area, but as a function of photon frequency. This comparison illuminates the peculiar mix of fundamental, technological, and social factors that determine the limits of survey sensitivity.

For all-sky surveys, the track of the limiting sensitivities as a function of frequency is not too different from the spectrum of a typical radio-loud and highly variable AGNs (cf. fig. 1.4), with all the bands from the infrared upward in frequency at roughly comparable levels, and the radio flux about four orders of magnitude lower. The picture is different for surveys in which the observing strategy was to look for very faint sources in a small area of sky. No "pencil-beam" surveys have yet been carried out in either the infrared or the high-energy γ-ray regimes. In addition, the character of the surveys changes. The radio limits are five orders of magnitude fainter than the optical limits, but it is hard to say precisely what fraction of the radio sources are bona fide AGNs. The relative positions of the flux limits in the optical and X-ray reverse, with the X-ray limit about 2.5 orders of magnitude deeper in the small-angle surveys.

The reasons for these contrasts can be understood in terms of the analysis presented in §2.3.1, where we discussed how both angular resolution and imaging noise control the limiting flux achievable with a given experimental apparatus. The constraints on each of these factors differ from band to band.

If all we saw in the sky were AGNs, avoiding source confusion would require an angular resolution of only $\simeq 30''$ because even the deepest AGN surveys find only ~ 100 AGNs per square degree (§3.1). Resolution at this level is very easy to achieve in either the radio or optical bands and

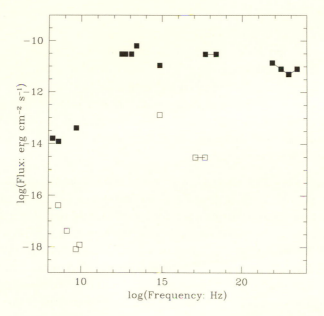

Fig. 2.1   The limiting fluxes for a variety of AGN surveys. Those marked with filled squares are shallow, large-area surveys; those with open squares are deeper, small-area surveys. In the radio, the large-area surveys are the 3C (at 178 MHz), Parkes (at 408 MHz), and Green Bank (at 5 GHz). Some representative radio pencil-beam surveys are the 5C (at 408 MHz), Windhorst et al. (1985) at 1.4 GHz, Fomalont et al. (1991) at 5 GHz, and Windhorst et al. (1993) at 8.4 GHz. Only the *IRAS* survey is available in the infrared. In the optical, the BQS survey represents what has been done over large areas of the sky, whereas the work by Boyle et al. (1990) shows how deep a small survey can go. In the X-ray band, the *Rosat* survey is a good example of a deep survey, the Piccinotti sample illustrates the depth of a shallower all-sky survey. As in the infrared, in the γ-ray band there has only been one survey, the one done by *EGRET*.

is not too difficult in soft X-rays using grazing incidence optics. In the radio band (down to ~ 1 mJy) and in the X-ray band (through most of the range of fluxes probed so far), the number of AGNs really is comparable to the total number of sources. However, in visible light, as well as at radio frequencies for flux levels below 1 mJy, galaxies far outnumber AGNs. In order to separate AGNs from galaxies, much greater resolution is needed. At an optical flux limit of $m_B = 22$ mag ($\simeq 5 \times 10^{-14}$ erg cm^{-2} s^{-1}), the necessary resolution is more like $10''$. This, too, is well within the capabilities of ordinary ground-based telescopes at reasonable sites. Thus, the flux limits for the deep surveys of figure 2.1 are not strongly affected by confusion.

Imaging noise is much more of a constraint. Its nature varies substantially with wavelength as a result of the very different detector technologies employed. In the radio regime, noise is generated by a variety of mechanisms. At low frequencies, plasma dispersion in the interstellar medium, the Solar System, or the ionosphere causes sources to scintillate. At high frequencies (tens of GHz and up), there are fluctuations in atmospheric attenuation and thermal emission. At all frequencies, there is thermal noise in the antenna-receiver system.

Although virtually all modern surveys have employed interferometers, the minimum flux level at which point sources can be detected is independent of the interferometer configuration because, by definition, they are unresolved. It is simply several (exactly how many depends on the trade-offs discussed in §2.3.1, but 5 is typical) times the *rms* fluctuations in the noise level. To translate the noise level into flux units, one needs to multiply by the beam size of the telescope that would be formed if all the elements of the interferometer were merged into a single telescope. Because this solid angle is $\simeq (\lambda/D)^2$ when the diameter of the "merged telescope" is D, the *rms* noise "flux" is

$$
\begin{aligned}
N_\nu &\simeq \frac{2k_{\mathrm{B}}T_s}{\lambda^2} \left(\frac{\lambda}{D}\right)^2 \frac{1}{(\Delta\nu\,\Delta t)^{1/2}} \\
&\simeq 3 \left(\frac{T_s}{20\ \mathrm{K}}\right) \left(\frac{\Delta\nu}{100\ \mathrm{MHz}}\right)^{-1/2} \left(\frac{\Delta t}{10^4\ \mathrm{s}}\right)^{-1/2} \mu\mathrm{Jy},
\end{aligned}
\tag{2.1}
$$

where the effective system temperature is T_s, the bandwidth is $\Delta\nu$, and the integration time is Δt. In the numerical evaluation of equation 2.1, we have used $D = 130\ m$, the number appropriate to the VLA. The sensitivity improves $\propto (\Delta t)^{1/2}$ because the energy in noise fluctuations grows only $\propto (\Delta t)^{1/2}$ while the energy collected from a source grows $\propto \Delta t$.

In the optical regime, thermal noise is negligible because we are dealing with photon energies far above $k_{\mathrm{B}}T$. On the other hand, other problems do crop up at low flux levels: sky brightness, photon-counting fluctuations, and "uninteresting" sources. Even at a good dark site, the night sky has a signficant surface brightness—integrated over the visible band, about 5×10^{-14} erg cm^{-2} s^{-1} in a square arcsecond. Because the sky brightness is very smooth on small angular scales and can be measured very accurately, it can be subtracted. However, no measurement is ever perfect (and sometimes the sky itself is time variable), so that the accuracy of this subtraction places a floor under the faintest source that can be detected in a broad-band measurement. For spectroscopy, on the other hand, where the source signal is spread out over many separate elements, the limit in modern photon-counting detectors tends to be simple Poisson fluctuations in the number of photons arriving. With current technology, this poses a lower limit at $\sim 10^{-15}$ erg cm^{-2} s^{-1}.

However, for AGN surveys, the real limit turns out to be sorting out AGNs from "uninteresting" objects, which, at very faint fluxes, are mostly inactive galaxies. As the flux limit falls below $\sim 10^{-14}$ erg cm^{-2} s^{-1}, these become progressively more and more numerous relative to the AGNs (see §3.1.3). Thus, the rarity of AGNs, and the complexities of distinguishing them from other galaxies, turn out to be the real limits on optical survey sensitivity. As we have already noted in §2.3.3, these same problems also arise in radio work at very faint flux levels.

In the X-ray band, yet a different situation prevails. Here thermal noise is even more negligible than in the optical, the sky is very dark, and in modern instruments there is hardly any detector noise (when high-resolution imaging telescopes are used). The sensitivity limit in this band is far simpler: the photon fluxes (at fixed energy flux) are far smaller, so the principal problem is collecting enough photons. This is essentially an economic limit, as it represents a product of collecting area and observing time.

Thus, the comparative sensitivity of different search techniques turns out to hinge primarily on imaging noise rather than angular resolution. Which method provides the most sensitive means of finding AGNs depends on a complex technological/social trade-off.

Ultimately, the amalgam of all search techniques produces a list of all known AGNs discovered by any means. Although this list is far from being an unbiased sample of all AGNs in the Universe, it is often useful to consult such a list in order to look up basic data on objects already identified. For many years, two different groups have worked hard to provide this service. Hewitt and Burbidge (1993) list the position, optical magnitude, color, redshift, variability character, optical polarization, X-ray flux, radio flux, and infrared flux for 7225 quasars and 90 BL Lac objects. References to the observational literature are also given. A parallel effort has been made by Veron-Cetty and Veron (their most recent edition was published in 1996, and also lists Seyfert galaxies and LINERs, as well as quasars and BL Lac objects). Both of these general catalogs, as well as many of the specific surveys, can be accessed electronically through a variety of gateways.

3 Evolution

We have just seen that there are many ways of finding AGNs, each with its own special biases. Our job now is to describe the results in a way that will tell us how the population of AGNs has changed with cosmic time. There are a number of ways to do so, which we will discuss in order of increasing specificity. Not surprisingly, increasing specificity also demands increasing data quality.

All these methods must, of course, be employed in the context of an expanding Universe. In such a Universe, several different sorts of distance and volume may be distinguished. These are described in Appendix F. Explicit evaluation of these quantities, and derived ones such as luminosity, depends on three parameters: the Hubble constant H_o; Ω_M, the ratio of the mean matter density to the cosmological critical density; and the cosmological constant Λ (these, too, are described in greater detail in Appendix F). Frequently the latter two parameters are combined into the deceleration parameter $q_o \equiv \Omega_M/2 - \Omega_\Lambda$, where Ω_Λ is the magnitude of the cosmological constant normalized to current values of the Hubble constant and cosmological scale factor. As of this writing, it seems that H_o most likely falls somewhere in the range 50 to 80 km s^{-1} Mpc^{-1}. We will therefore write it as $H_o = 100h$ km s^{-1} Mpc^{-1}, and scale h to 0.75. Because Ω_M and Λ are much less well determined, but the combination $\Omega_M = 1$ and $\Lambda = 0$ makes many algebraic expressions particularly simple, we will use those values for numerical examples.

3.1 Source Counts

The simplest statistical description of evolution is the distribution of source counts, that is, how many sources there are as a function of their brightness. In fact, it is so simple that what it really tests, strictly speaking, is not time evolution, but geometry. Its connection to cosmological evolution comes about because age correlates to distance through the finite speed of light, and because our Universe continually changes shape and size as it evolves.

3.1.1 Formalism

The historical roots of this method go back to the beginning of the twentieth century and the debate over the shape of our Galaxy. At that time the only way to measure distances was through parallax, and that could be done for only a handful of the nearest stars. Consequently, the only data that bore on the shape of the Galaxy were the distribution of stars on the sky and their apparent magnitudes. The beauty of the source counts method is that it works even on such simple, and easily obtainable, data.

To see how it works, suppose that you have measured the apparent magnitudes of all the stars brighter than some limiting magnitude $m_{\max}$ that are found in a small patch of solid angle $\Delta\Omega$ on the sky. First, you transform all the data into fluxes:

$$F_i = 10^{0.4(m_o - m_i)} F_o, \tag{3.1}$$

where the flux of the ith star is F_i, and m_o is a fiducial magnitude corresponding to flux F_o. Next, tabulate how many sources there are brighter than any given flux down to the limiting flux of the survey.

To understand the significance of this function $N(> F)$, consider how the number of sources would scale with limiting flux if they were distributed spherically symmetrically and with *luminosity function dn/dL* (i.e., density per unit luminosity) independent of distance. For each luminosity L, the limiting luminosity distance (see Appendix F for a definition) is $D_L(L) = (L/4\pi F)^{1/2}$. The number of such sources we can see with flux greater than F is then

$$N(> F) = \int dL\, V[D_L(L, F)] \frac{dn}{dL}, \tag{3.2}$$

where $V[D_L(L, F)]$ is the volume within which a source of luminosity L would be seen with flux at least F. In a Euclidean universe, $V(D_L) = (\Delta\Omega/3) D_L^3$, so

$$N_{\mathrm{Eucl}}(> F) = \frac{\Delta\Omega}{3(4\pi)^{3/2}} F^{-3/2} \int dL\, L^{3/2} \frac{dn}{dL}. \tag{3.3}$$

Thus, independent of the shape of the luminosity function, the integral counts for a spherically symmetric, homogeneous population in a Euclidean universe scale as $F^{-3/2}$. A corollary instantly follows: the source counts in a homogeneous Euclidean universe cannot be used to infer *any* information about the shape of the luminosity function.

If one of these assumptions is broken (e.g., the source density is not constant with either distance or direction, or the geometry is not Euclidean), then the curve can diverge from that prediction. For example, if we were studying star counts over the entire sky, we would find

$N(> F) \propto F^{-3/2}$ for the very brightest stars, but as F falls, the counts would rise more slowly due to both the planar geometry of our Galaxy and the diminution of the flux from distant stars by intervening dust.

In a cosmological setting there are analogous effects. Although the isotropy of the microwave background gives us strong reason to believe that the Universe is very nearly spherically symmetric on the largest scales, the relation between enclosed volume and luminosity distance is not as simple as in Euclidean geometry. The redshifts that distinguish the luminosity distance from the coordinate distance (see Appendix F) act in a way that is reminiscent of the way dust extinction dims distant stars in the Milky Way. Similarly, in the relation between volume and luminosity distance the curvature factor acts in a way analogous to the geometrical factors that influence Galactic star counts (see fig. 3.1). Volume in an expanding Universe out to a given luminosity distance increases much more slowly than in a Euclidean universe because the cosmological redshift has the effect of strongly diminishing the volume accessible out to a given luminosity distance. Because this effect depends only on the fact of expansion, its magnitude is rather insensitive to specific values of Ω_M and Λ.

To evaluate these effects quantitatively, one simply uses the expression for the appropriate volume measure as a function of luminosity distance (Appendix F). Often the most convenient is the *co-moving volume* V_{co}, the volume evaluated factoring out the cosmological expansion, because density measured with respect to this volume is constant if there is no change in the number of objects. It is the relation between V_{co} and D_L that is shown in figure 3.1. If there has been no evolution in the source luminosity function, the function $N(> F)$ then increases more slowly with decreasing F than the Euclidean model would predict, for any cosmological model. Without evolution in the luminosity function, when $V_{\mathrm{co}} \propto D_L^x$, $N(> F) \propto F^{-x/2}$; since x is fairly small by $z \simeq 1$ in almost any cosmology, if there is no evolution, the counts distribution flattens out as soon as the typical redshift becomes order unity. Larger values of Ω_M (for $\Lambda = 0$) flatten the source count curve more than smaller values, but the contrast is comparatively small unless either Ω_M or the typical redshift is fairly large.

The character of this argument is altered only slightly if we wish to allow for evolution. Then the integral for $N(> F)$ becomes

$$N(> F) = \int dL \int_0^{r_1(D_L)} dr \, \Delta\Omega \frac{r^2}{(1 - kr^2)^{1/2}} \frac{dn}{dL}(r), \qquad (3.4)$$

where r is the co-moving radial coordinate and k defines the sense of curvature (again see Appendix F for more detail). Objects with no pecu-

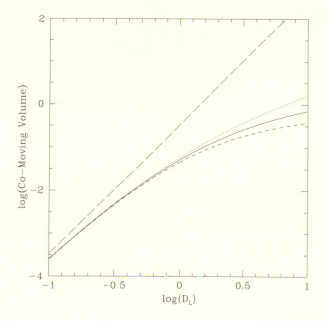

Fig. 3.1   The dependence of co-moving volume on luminosity distance (normalized to R_o, the current scale of the Universe) in four different cosmological models, all having zero cosmological constant. The long-dashed curve is the normal Euclidean volume (per unit solid angle)-radius relation $V = D_L^3/3$; the dotted curve is for an open Universe ($\Omega_M = 0.1$); the solid curve corresponds to a flat Universe ($\Omega_M = 1$); and the short-dashed curve is for a closed Universe ($\Omega_M = 2$). Note that a particular value of D_L corresponds to different values of z in different cosmological models; e.g., $D_L = 10$ is reached at $z = 6.8$ in a flat cosmology, whereas it is reached at only $z = 3.9$ in the open Universe shown here.

liar motion retain the same co-moving coordinates for all time; the factor $(1 - kr^2)^{-1/2}$ in equation 3.4 corrects the differential volume element for spatial curvature. Note that the complications due to the expansion of the Universe make the shape of the source counts dependent on the shape of the luminosity function even if there is no evolution. More luminous objects can be seen to greater luminosity distance, and the shape of the source counts changes with D_L.

3.1.2   AGN source counts in the real world

So far in this discussion it has been assumed that all data entering the source counts are perfect. Real surveys are prey to a variety of problems, of course. Several can be especially pernicious when one wishes to interpret the results of source counts.

Confusion, as discussed in Chapter 2, can often lead to spurious features in the function $N(>F)$. When the source count slope is steep (as is commonly found), the net effect of confusion is to create extra sources at fluxes a few times the sample threshold by merging real sources, many of which have fluxes near the threshold. A corollary effect is that all but the brightest source in the merger are missing from the sample, depressing the source counts at fluxes just above threshold.

Another problem that is exacerbated by a steep source count slope is called *population bias*. Random measurement errors can throw sources back and forth across the sample flux limit. When the slope is steep, there are many more sources just below the limit than just above, so the net effect is to increase the numbers of sources artificially.

A very clever partial solution to both this problem and the previous one was invented by Scheuer (1957). When using this technique, one relinquishes the attempt to find individual sources in order to gain the ability to describe their mean properties down to very low flux levels. Scheuer's idea was that, provided the sources are uncorrelated on the sky, the effects of both confusion and noise can be modeled if the source count distribution is known. This fact allows one to work backward: from the guessed source count distribution one *predicts* the statistical character of the actual data down to fluctuations of order the noise level. The parameters of the guessed distribution can then be modified to maximize the fit between the predicted and measured fluctuation distributions. This approach is known as the "$P(D)$" method in radio astronomy because the data whose probability distribution is predicted are the actual correlated electric field deflections recorded by the interferometer. However, it is clear that the basic principle extends to direct imaging methods as well.

Real variability can act in a fashion much like measurement error. The difference in this case is that the statistics of the fluctuations may not be Gaussian, so it is hard to compute the impact of these fluctuations on the source counts.

Gravitational lensing creates a related problem. If the light rays from a distant quasar pass through an intervening gravitational potential with a significant shear, it is possible for the rays to be focused in much the same way as if they had passed through a giant glass lens. The light we see is then amplified relative to what we would have seen without the lensing. Although the probability per quasar that its light is substantially amplified is small, when the counts distribution is steep this effect can in principle significantly perturb the observed counts distribution.

Finally, genuinely attaining completeness according to the stated rules of the sample can be problematic. Depending on the devices employed, there may be problems with maintaining a consistent flux limit. Sometimes calibration is difficult; sometimes backgrounds can pose problems; sometimes there are subtle changes in sensitivity correlated with the time of observation or the position of the field in the sky. All these can lead to the actual definition of the sample being different from the nominal one.

Having voiced these cautions, we are now ready to examine what the actual source count distributions look like for AGNs in the radio, optical, and soft X-ray bands. Interestingly, the distributions look quite different in each of the three spectral regions, and only a part of the reason is the insistence of the observers in each band that their convention about how to plot the data is best.

3.1.3 Radio source counts

Although radio astronomy is, of course, a much younger discipline than optical astronomy, extragalactic source counts were measured first in the radio. At least one reason contributing to this precociousness of radio astronomers is a surprising fact about the radio sky. Ordinarily one might expect that the apparently brightest objects are the nearest, as it is easy to implicitly assume that all sources have roughly similar luminosities. In a crude way, this is certainly true of the optical sky. Source counts in the optical are therefore completely dominated by Galactic stars until relatively faint levels are reached. The radio sky is spectacularly different. Stellar radio sources are mostly very dim. On the other hand, radio galaxies are extremely luminous, and their luminosity distribution is very broad. In addition, there were many more highly luminous sources at redshifts of a few than there are today. The result of all these factors is that many of the apparently brightest objects in any radio survey are quite far away. Put another way, if we had radio-sensitive eyes, many of the most familiar (because brightest) objects in the sky would lie halfway across the Universe from us.

Thus it was that radio astronomers recognized very early on (by the mid-1950s) that theirs was an essentially extragalactic subject, and that source counts taken from their surveys could provide fundamental information about the evolution (or lack of evolution) of the Universe. Indeed, a primary scientific goal of the first surveys was to test the then popular steady state model of cosmology.

The current state of the art is shown in figures 3.2 (showing the source counts at an observing frequency of 408 MHz) and 3.3 (for the counts as measured at 4.85 GHz). In radio work, the standard convention is to plot the differential number counts dN/dF normalized to the Euclidean slope by multiplying by $F^{5/2}$. Note that radio astronomers conventionally use S in place of F. This is why these diagrams are called "$\log N$–$\log S$" plots by radio astronomers.

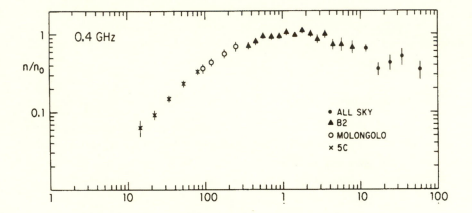

Fig. 3.2 Normalized differential source counts at 408 MHz assembled by Kellermann and Wall (1987). The unit n_o is 1125 sources times $Jy^{3/2}$ in the entire sky. The "all sky" survey is that of Robertson (1973); "B2" refers to the Bologna survey (Colla et al. 1970); "Molongolo" points are from the Molongolo Deep Survey (Robertson 1977); and the points identified as coming from the 5C catalog are from Pearson (1978).

Qualitatively, the source count curves at the two frequencies are fairly similar. In both cases, the slope over the brightest one to two decades in flux is slightly steeper than Euclidean, rolls over to match Euclidean for approximately the next decade, and then becomes shallower than Euclidean for the next two to three decades. Thus, as one looks back from our time into the past, the number of radio sources at first rises, but that growth eventually decreases in rate.

The high frequency counts span an extremely large range: a factor of 10^6 in flux. It should therefore not be too surprising that the population

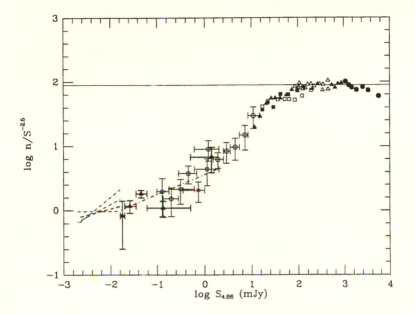

Fig. 3.3   Normalized differential source counts at 4.85 GHz from Fomalont et al. (1991) in units of sources times $Jy^{3/2}$ in the entire sky. Points without error bars are taken from the compilation of Kellermann and Wall (1987); the points with error bars are a mix of data obtained by Fomalont et al. , and other data they compiled from the literature. The horizontal solid line is a Euclidean source count distribution. The dot-dashed line is a fit to the source count distribution at the faint end. The dashed lines at very low fluxes are a variety of fits to the fluctuation distribution.

surveyed changes as a function of flux. As already remarked in §2.3.3, although the brightest radio sources are overwhelmingly genuine AGNs, the fainter ones (below ~ 1 mJy at 5 GHz) are a mixed bag, and AGNs may even be a minority (Windhorst et al. 1993). Therefore, if there were a clean way to discriminate between AGNs and other kinds of sources when $F < 1$ mJy, the number of AGNs per unit solid angle would rise even more slowly than the figure shows as the flux falls below that level.

3.1.4 Optical source counts

Before examining the optical counts, it is important to point out the peculiar units that are conventional in optical work. In most reports of optical source counts, the quantity that is plotted is the logarithm base 10 of the differential surface density in number per square degree per some fraction Δ of a magnitude. These units are chosen because in optical astronomy the measured quantity is only rarely a physical flux; much more commonly the measured quantity is a magnitude, that is, the logarithm of the flux ratio as compared to some standard star. The relation between the counts as conventionally displayed and their representation in more straightforward units is then

$$\log_{10} \frac{dN}{dF} = f(m) - \log_{10} F - \log_{10}(\Delta \ln m_o), \qquad (3.5)$$

where $f(m)$ is the count distribution in the conventional units and $m_o = 2.512\ldots$ is the logarithmic base of the magnitude system. Thus, the logarithmic derivative of the differential counts with respect to flux is one less than the slope of $\log_{10}(dN/dm)$ against m divided by $\log_{10} m_o = 0.4$.

Hartwick and Schade (1990) merged a large number of different surveys to form a consensus view of the measurements (see fig. 3.4). From $m_B \simeq 14$ to about $m_B \simeq 19$ mag, the integral slope (i.e., $d \ln N / d \ln F$) is $\simeq -2.2$, but the curve flattens out at fainter flux levels, rolling over to a slope of roughly -0.5 for $m_B > 20$ mag. At $m_B \simeq 22$ mag, the surface density of optically selected quasars reaches $\simeq 100\text{--}200$ deg^{-2}. The initial steep rise in the counts indicates dramatic evolution with lookback time. In addition, this steep rise extends over a greater range of fluxes than the gentler "super-Euclidean" slope in the radio—a span of $\simeq 2.5$ decades, as opposed to barely more than one.

3.1.5 X-ray source counts

In the X-ray community, integral source counts are standard. In the band around 1 keV (Boyle et al. 1993), the shape of the curve is more similar to the radio counts than to the optical, but the bright end is even closer to Euclidean (and the dynamic range of the entire curve is only about two orders of magnitude). The slope of the integral counts is only slightly steeper than -1.5 at the bright end (fluxes between 3×10^{-14} and 3×10^{-13} erg cm^{-2} s^{-1}), and gradually becomes even flatter as the flux limit falls below 10^{-14} erg cm^{-2} s^{-1}.

If we had no redshift information, the X-ray counts might be compatible with a homogeneous distribution. However, the redshift distribution for the X-ray data is in fact quite broad, with the greatest number of sources in the range $z \simeq 1\text{--}2$ once the flux falls below $\sim 10^{-13}$ erg cm^{-2} s^{-1}. Such a broad redshift distribution would lead to an $N(> F)$ that is *shallower* than $F^{-3/2}$ if the co-moving density were constant. Consequently, even this "pseudo-Euclidean" source count function indicates significant evolution.

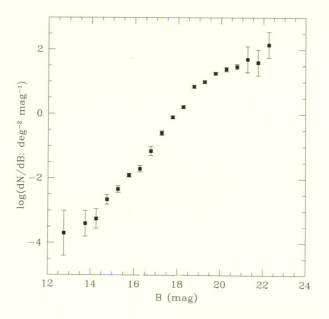

Fig. 3.4 Quasars per square degree per magnitude from a compilation by Hartwick and Schade (1990) of a number of different quasar surveys.

3.1.6 In which band do the counts go the "deepest"?

As the previous three subsections have shown, very thorough surveys spanning a wide range of sensitivities have now been performed in all three of the major observing bands: radio, optical, and X-ray. One might now naturally ask, "From which of these bands do we find the most complete sample of AGNs?" As we have striven to emphasize, the diversity of AGNs ensures that no one search technique can ever be best at finding all varieties. However, it is fair to ask the question, "Which currently finds the greatest density on the sky?"

Comparing the *Rosat*/EMSS counts data (fig. 3.5) with the Hartwick and Schade compilation of optical samples (fig. 3.4), it is clear that these two bands are similar—in both we can find about 100 AGNs per square degree. In other words, at this time the technological trade-offs discussed in §2.7 just about balance any intrinsic flux ratio in the sources. The radio data are not strictly comparable because it is not yet possible to clearly distinguish how many sources are AGNs at faint flux levels. However, if we suppose that most are still AGNs when the flux at 5 GHz is ~ 1 mJy,

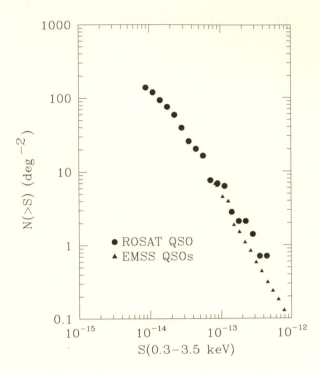

Fig. 3.5   Source counts from the combined *Rosat*/EMSS (i.e., soft X-ray) surveys (Boyle et al. 1993).

these surveys find $\simeq$ 25 per square degree. In that sense, 5 GHz radio work finds only about a quarter as many AGNs as the optical and X-ray surveys. Lower frequency samples (e.g., the 408 MHz surveys shown in fig. 3.2) are still shallower—at $\sim$ 10 mJy, they find only about one AGN per square degree. The relative scarcity of radio sources we are able to detect is due to a combination of technical issues and the comparative dimness of AGNs in the radio band (see Chap. 9).

3.2   The V/V_{max} Test

There is another way to look at source count data that, given some assumptions about geometry, reveals immediately whether the population in question has a density that depends on distance. In its simplest form, it requires only source count data; however, the same cosmological factors that complicated the interpretation of the source count distribution require the additional use of redshift information in this method of analysis.

In a flux-limited sample, all of the members are at least slightly brighter than the flux limit. Consequently, all of them could have been detected even if they had been at least (slightly) farther away than they actually are. More quantitatively, because flux falls as the inverse square of luminosity distance, the maximum luminosity distance at which a particular member of a sample could have been seen is $R_{max,i} = (F_i/F_{min})^{1/2} R_i$, where F_{min} is the minimum flux permitted in the sample, and R_i is the actual distance to object i.

The next step embodies the key idea of the method. Suppose that the density of objects in the direction of our survey is independent of distance. Then object i has an equal probability per unit volume of being located at any distance between us and $R_{max,i}$. Consequently, the mean value of the ratio between the volume within which sample members are found, and the volume within which they could have been found, should be 1/2.

In Euclidean geometry, we can test whether this is true by defining the volumes $V_i = (1/3)R_i^3\Delta\Omega$ and $V_{max,i} = (1/3)R_{max,i}^3\Delta\Omega$, and computing the mean value of their ratio

$$\langle V_i/V_{max,i}\rangle = \langle R_i^3/R_{max,i}^3\rangle = \langle(F_{min}/F_i)^{3/2}\rangle. \qquad (3.6)$$

Note that the result requires only the measured fluxes and the flux limit. It does not depend on the distribution of luminosity in the sample because the argument is completely independent of absolute magnitude. Moreover, if the flux limit varied during the experiment, it is a simple generalization to use whatever flux limit obtains for each point, so that F_{min} becomes $F_{min,i}$.

If $\langle(F_{min}/F_i)^{3/2}\rangle$ indeed turns out to be 0.5, then we have confirmed that the objects are distributed with uniform density in the radial direction. If, however, the mean value of this ratio is different from 0.5, there must be a gradient in the density as a function of radius. Larger values mean that the density increases outward; smaller values mean that the density decreases outward. If, in addition, the survey is uniform around the sky, the value of $\langle V/V_{max}\rangle$ also conveys information about the angular symmetry of the distribution. Smaller values indicate a distribution flatter than spherical, with $\langle V/V_{max}\rangle = 0.4$ for a planar distribution, for example.

In the context of cosmologically distributed objects, the non-Euclidean geometry of the Universe creates a relationship between the ratio L/F and the enclosed volume that is no longer a simple power law, and therefore the absolute luminosity of the sources does not simply scale out. Consequently, to apply the V/V_{max} test properly in a cosmological context requires additional information: either some a priori knowledge of the absolute luminosity, or measured redshifts. If one blindly goes ahead and attempts to apply the method assuming Euclidean geometry (as is sometimes done), the interpretation becomes, of course, ambiguous. Because the enclosed volume does not increase as fast with D_L as it would in Euclidean geometry, cosmological effects (absent evolution) cause a larger proportion of the sample to have fluxes well above the flux limit. Thus, if $\langle (F_{min}/F_i)^{3/2} \rangle = 0.5$ for a sample that reaches out to significant redshift, it means that the density of sources was actually *greater* at large distance than at small.

The correct way to apply the V/V_{max} test for cosmologically distributed objects is therefore to also acquire redshifts for each object. When this test was first applied to AGNs (on a sample of radio-loud quasars: Schmidt 1968), a very surprising result was found: $\langle V/V_{max} \rangle \simeq 0.7$; that is, their effective density increases sharply with increasing distance! Because readily obtainable samples often contain many objects with redshifts up to around 3, this simplest of tests immediately shows that quasars were in some sense far more common when the Universe was a fraction of its present age than they are today. Since Schmidt's original work, the same test has been applied to other AGN samples, e.g., radio galaxies containing no (apparent) optical quasar, and similar results have been found.

3.3   Luminosity Function as a Function of Redshift

Let us now consider the maximum information we could draw from samples with both flux and redshift information. Granted choices of the cosmological parameters H_o, Ω_M, and Λ, we can determine the absolute luminosity of all our sources. Now we can ask much more detailed questions of our data: "How does the density of sources evolve? And, what was the luminosity distribution at each epoch?"

The basic quantity we wish to obtain is $dn(z)/dL$, the luminosity function as a function of redshift. In most cases, the density n is most conveniently described with respect to co-moving coordinates. To obtain this quantity from a survey, we start with a scatter plot in the L–z plane. If our sample had a fixed flux limit, then objects would be found only in the lower right half of the diagram; that is, we can only detect those objects either close enough or luminous enough to pass our flux threshold. If z were simply proportional to distance, and the Universe were exactly Euclidean, this boundary would be a straight line with slope $d\log L_{min}/d\log z = 2$. However, if our sample extends to cosmological distances, in general the flux threshold boundary will be curved. Figure 3.6, with data taken from the combined *Rosat* and *Einstein* Medium Sensitivity Surveys, illustrates these effects. The flux limits of the two surveys are easily visible as the envelopes of the two point distributions. These envelopes are not perfectly sharp because different fields of view had different limiting fluxes.

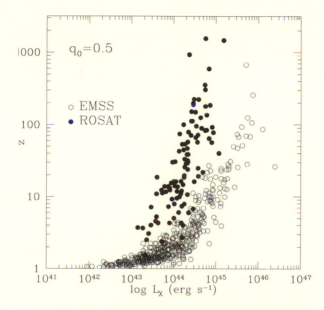

Fig. 3.6 The *L–z* diagram for the combined *Rosat*/EMSS surveys (Boyle et al. 1993).

It is now (in principle) straightforward to find the density of objects near redshift z with luminosity approximately L. First one defines a survey that (one hopes) is complete within certain well defined limits of flux- and AGN-defining properties. After identifying all the sources and measuring their redshifts, one constructs a luminosity-redshift scatterplot like the one in figure 3.6. At this point, there are several alternate paths to take. Here we adopt the conceptually simplest; more sophisticated approaches are described in Appendix D.

In this simplest method, we begin by ruling off the scatter plot into boxes. Suppose that in the box whose limits are $z \pm \Delta z$ and $L \pm \Delta L$ there are N sources. For each of these sources (which we label by the index i), one first defines the co-moving volume within which it was found:

$$V_{\text{co},i} = \Delta\Omega_i \left[V'_{\text{co}}(z + \Delta z) - V'_{\text{co}}(z - \Delta z) \right]$$

$$\simeq \Delta\Omega_i \frac{r^2(z)}{1+z} \frac{\Delta z}{\sqrt{1 + \Omega_M z + \Omega_\Lambda \left[1/(1+z)^2 - 1 \right]}}, \tag{3.7}$$

where V'_{co} is the co-moving volume per unit solid angle, and the second expression is appropriate when $\Delta z/z \ll 1$. When the survey limit cuts across the box, the limits on redshift must be adjusted to reflect the range within which that particular object could have been found. The resulting volume is called the *accessible volume*. The luminosity function at that redshift is then

$$\frac{dn}{dL} = \sum_i^N \frac{1}{2\Delta L V_{\text{co},i}}. \tag{3.8}$$

When the survey limit does not cut across the box, $2\Delta L$ is the full luminosity width of the box; when it does, it is the range of luminosities within which object i could have been found.

The only ambiguity in this procedure is that the co-moving radius associated with a given redshift depends on the cosmological parameters Ω_M and Ω_Λ. This dependence enters both into the inferred luminosity for a given flux and redshift, and the magnitude of the co-moving volume.

In practice, life is rarely so easy. Many subtleties must be thought through if one is to obtain meaningful results.

First, there are all the same problems we have already remarked upon in the context of source counts: confusion, population bias, variability, calibration errors, and backgrounds.

But there are additional problems specific to the task of measuring luminosity functions. Unless the sample is quite large, because the source counts rise so steeply toward fainter fluxes, there may be enough objects to avoid small number statistics in only those boxes very near the flux threshold curve. When that is the case, one can find the distribution in luminosity at a single redshift only over a very limited range of luminosities.

Another related problem is called the *Scott effect*. Very luminous objects can be seen to great distance and therefore turn up in disproportionate numbers in flux-limited catalogs. Proper attention to the accessible volume solves this problem.

Because we are interested in how the luminosity function evolves over a significant span in redshift, we must also assume something about the intrinsic spectral shape of AGNs, so that the luminosity can be compared for the same *rest-frame* wavelength interval. When quasars at $z = 3$ are being compared with quasars at $z = 0.3$, the uncertainty induced by unknown spectral shape can be considerable. Dispersion in the spectral shape over the sample can also lead to distortion in the luminosity function. Imagine, for example, that the fiducial rest-frame wavelength is in the visible band. To find the luminosity of high-redshift quasars at this rest-frame wavelength requires extrapolating their continua to wavelengths that we would observe in the infrared. The luminosity we infer for high-redshift quasars with spectra "bluer" than the guessed form will then be systematically too high, and vice versa for those with true spectra "redder" than the norm. Because high-luminosity objects tend to be rare, a few "blue" quasars could therefore substantially distort the shape of the luminosity function.

Another problem arises from a combination of measurement error and the sample selection method. We have already encountered a special case of this problem in the context of source count distributions, where it is called population bias. More generally, whenever the measurement error in the quantity whose distribution we are estimating is correlated with the observable on which the sample is selected, *Malmquist bias* may occur (for a detailed discussion, see Landy and Szalay 1992).* For example, consider the effect of flux errors when, as here, we are attempting to estimate a luminosity function (this particular case of Malmquist bias is also sometimes known as *Eddington bias*). If the source count distribution rises toward lower fluxes (as usually happens), there will be more objects in the sample whose fluxes are recorded as greater than they should be than the other way around. This means that systematically extra objects are placed in the higher luminosity bins.

Finally, if the sample is not truly complete to the actual flux limit, then one is subject to an uncontrollable error that grows with the fractional incompleteness. Because the presence or absence of objects in a sample is likely to be correlated with some particular property, this error is probably a systematic error, rather than a random one. In a jocular way, one might say that "All samples are complete; the only problem is figuring out their real definition".

---

*There is tremendous confusion in the astrophysical literature about the definition of Malmquist bias. The definition given here is a consensus view; dramatically different usages can be found.

All these problems can be dealt with, and corrected for (except the last, which can be cured only by either completing the sample according to its nominal terms, or quantitatively defining its actual selection criteria), but their neglect can lead to serious problems.

Having said all this, we present figures 3.7 and 3.8, which show how the luminosity function for optical color-selected AGNs (i.e., radio-quiet quasars) changed over time. The first figure shows the evolution of the AGN luminosity function from $z = 0$ to 2. By combining data from a large number of surveys, the first problem mentioned above (the difficulty of obtaining data at all redshifts over the same range of luminosities) has been somewhat mitigated. Rather than plotting dn/dL, we show $L dn/dL = dn/d\log L$. These units have been chosen for two reasons: because this matches the customary presentation in terms of number density per unit magnitude; and for reasons very similar to those behind the choice of νL_ν for spectral plots—these units immediately show where most of the numbers (or energy) is located.

Several qualitative points stand out clearly. First, high-luminosity AGNs were at least two orders of magnitude more numerous at $z = 2$ than they are today. Second, AGNs exist over an extremely wide range of luminosity: from the weakest Seyfert galaxies in the CfA sample to the most luminous quasars, the luminosity increases by a factor of 10^4. Even this span is merely a lower limit to the true dynamic range, for there are AGNs several orders of magnitude fainter than the CfA Seyferts, and we may yet find still more powerful quasars. Third, the shape of the luminosity function seems to be consistently one in which the high-luminosity end is fairly steep (in crude terms, $dn/dL \propto L^{-3}$), but the slope is shallower on the low-luminosity side. Fourth, despite the shallower slope at low luminosity, the total number density is very sensitive to the low-luminosity cutoff.

Figure 3.8 shows what we know of the luminosity function over the range $z = 2$ to 4. Two things are immediately obvious: the statistics are still rather fragmentary, and the luminosity range we are able to detect is quite limited; and the dramatic rise from $z = 0$ back to 2 does not continue beyond $z = 3$. If anything, there may have been slightly *fewer* high-luminosity quasars at $z = 4$ than at $z = 3$.

A comparable analysis can be performed for radio-loud objects. In this case it is useful to establish a division into two spectral classes: *flat* and *steep*. In this context, flat or steep means $d\ln F_\nu/d\ln\nu$ greater than or less than -0.5. For fluxes measured at a few GHz, this division correlates well (but not perfectly) with a division according to whether most of the

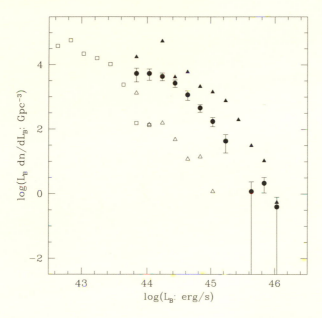

Fig. 3.7   The evolution of the AGN optical luminosity function from $z = 0$ to 2. The data were compiled from a large number of surveys by Hartwick and Schade (1990). Co-moving density is shown assuming $h = 0.75$, $q_o = 0.5$, and $\Lambda = 0$. In order to compare B-band luminosities, the continuum was assumed to have the shape $F_\nu \propto \nu^{-0.5}$ from 4000 to 1200 Å. Note that L_B is only $\simeq (1/4)\nu L_\nu$ at the central frequency of the B band, and the bolometric luminosity is at least several times greater still. The open squares are the local Seyfert galaxy luminosity function from Huchra and Burg (1992), using the total luminosity of the host plus nucleus; the open triangles are quasars with $0.16 \leq z < 0.4$; the filled circles are quasars with $1.0 \leq z < 1.2$; and the filled triangles are quasars with $1.9 \leq z < 2.2$. Error bars are shown only for the $z \simeq 1$ quasars to avoid clutter; the sizes of the error bars for the other points are similar.

radio flux comes from a fairly small source at the nucleus of the host galaxy, or from enormous radio lobes far larger than the host. Most members of the latter class (i.e., "steep" spectrum, extended sources) have only weak optical continuum emission at the nucleus, while most members of the former class would be classified as fairly ordinary quasars if the decision were based solely on optical data.

Both classes evolve strongly with redshift between $z = 0$ and 2.5, but the density of steep-spectrum objects seems to have declined more sharply since $z \simeq 2$ than did the density of the flat-spectrum objects (see fig. 3.9). Comparing both to the optical luminosity function, we see that radio-loud AGNs have always been a small minority (see also §9.1.1).

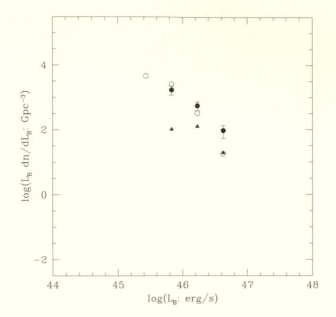

Fig. 3.8   The evolution of the AGN optical luminosity function from $z = 2$ to 4. The data are taken from Warren et al. (1994). Co-moving density is shown assuming $h = 0.75$, $q_o = 0.5$, and $\Lambda = 0$. Because these authors believed extrapolation to the B band involved too great an uncertainty for these high-redshift quasars, the absolute continuum luminosity is instead defined in terms of νL_ν at 1200 Å in the rest frame. The open circles are quasars with $2.0 \leq z < 2.2$; the filled circles are quasars with $3.0 \leq z < 3.5$; and the filled triangles are quasars with $3.5 \leq z < 4.5$. Error bars are shown only for the $z \simeq 3$ quasars to avoid clutter; the sizes of the error bars for the other points are similar.

The third band in which the survey data are complete enough to study the evolution of the AGN luminosity function is the soft X-ray band, i.e., $\simeq$ 0.5–2 keV, the range covered by grazing incidence X-ray telescopes. As always, quantitative details of the evolution depend on the choice of cosmology, but the qualitative character does not. Here, too, evolution has been dramatic. Figure 3.10 shows how the X-ray luminosity function evolved if one chooses $\Lambda = 0$ and $q_o = 0.5$. Just as in the optical and radio bands, the co-moving density of luminous quasars (in the soft X-ray band that means luminosities $\sim 10^{45}$ erg s^{-1}) increased by three orders of magnitude between $z = 0$ and $\simeq 2.5$.

One subvariety of AGNs may have a significantly different pattern of evolution. BL Lac objects (see §1.3) have flat radio spectra but are

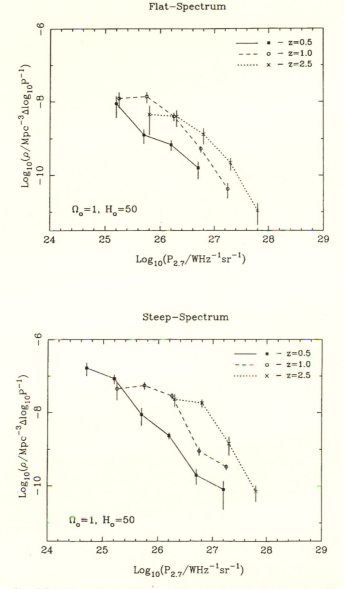

Fig. 3.9 The evolution of the AGN radio luminosity function (Dunlop and Peacock 1990). (Top) The space density of flat-spectrum objects; (bottom) the space density of steep-spectrum objects. Note that the units of space density and the Hubble constant used in this figure are different from those used in the previous two, but the cosmologies are the same (both $\Lambda = 0$ and $q_o = 0.5$). To compare the densities, numbers in this figure should be multiplied by $(0.75/0.5)^3 \times \log 2.512 \times 10^9 = 1.35 \times 10^9$; luminosities in this figure correspond to luminosities in the previous one when multiplied by $(0.5/0.75)^2 = 0.44$.

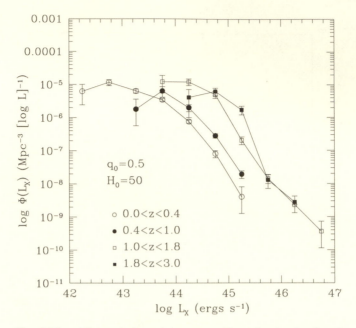

Fig. 3.10   The evolution of the quasar X-ray luminosity function (Boyle et al. 1993). Both the units of the luminosity function and the cosmological parameters assumed in this plot are the same as in fig. 3.9.

also strongly variable in all bands, usually polarized, and have very weak, often undetectable, optical emission lines. Radio and X-ray surveys have generally been the most efficient ways to find them—looking for emission lines is obviously unhelpful, and optical color searches fail because of the substantial dispersion in the shapes of their optical continua; many are so "red" that they are difficult to distinguish from stars by color tests. In fact, the range of continuum shapes is broad enough that radio and X-ray surveys discover different objects. Not surprisingly, the former find objects most of whose power is radiated at relatively low frequencies (typically in the infrared), while the latter mostly uncover objects whose spectra peak in the ultraviolet or soft X-ray bands.

For both the radio- and X-ray-selected subsets, two problems stand in the way of tracing the evolution of their luminosity functions: their comparative rarity makes it difficult to accumulate sufficient statistics over a range of redshifts, and their lack of emission lines often makes it hard even to measure their redshifts.

The first step in describing their evolution is therefore the relatively crude $\langle V/V_{max} \rangle$ statistic. Interestingly, although $\langle V/V_{max} \rangle$ for most other classes of AGNs is consistently greater than 0.5, for most BL Lac samples it is close enough to 0.5 to be consistent with that value to within the errors (Padovani and Urry 1995). It is possible, therefore, that this subclass of AGNs has not evolved as rapidly as the others.

Bolstered by this indication of weak evolution, we can construct a luminosity function under the assumption of *no* evolution. That is, even though the redshift distributions for each luminosity level found in the sample are different (in the usual behavior of flux-limited samples, the highest luminosities are usually found at the greatest distance), we hope that each is fairly sampled, as would be the case if the co-moving luminosity function actually is independent of redshift. Figure 3.11 shows the luminosity functions derived in this fashion from two different surveys, one at 5 GHz, the other at $\simeq 1$ keV. Comparing the radio-selected luminosity function to the ones shown in figure 3.9, we see that BL Lacs are rare indeed, less than 1% of all radio sources (flat or steep spectrum) at any specific radio luminosity. To compare properly the number densities of the radio- and X-ray-selected varieties, we should do so at fixed bolometric luminosity, which is not readily available. A rougher comparison may be made on the basis of mean ratios of X-ray luminosity to radio luminosity; these show that the number densities are not grossly different.

3.4 Parameterizing Evolution

The luminosity function curves contain all the information we have about the evolution of the quasar luminosity function. However, it is often clumsy to work with such a detailed description, and a simpler, parameterized description is sometimes preferable.

Two simple parameterizations immediately come to mind. One is the integrated density as a function of redshift; the other is the mean luminosity as a function of redshift. The study of the former is known as *density evolution*; the study of the latter, naturally enough, is called *luminosity evolution*.

To compute the density as a function of redshift, one would like to integrate the luminosity function at fixed redshift over the luminosity. However, as demonstrated by figures 3.7 and 3.8, the density integral is very sensitive to its low-luminosity end point. Consequently, when most people speak of density evolution in this sense, it is for a subclass of AGNs with absolute luminosity brighter than some minimum.

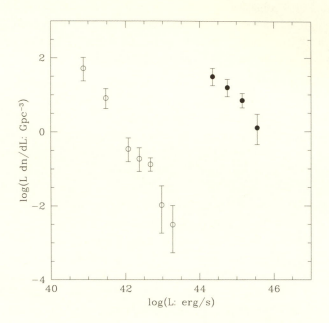

Fig. 3.11   Two "no evolution" estimates of the BL Lac luminosity function. The filled circles make use of the Einstein Medium Sensitivity Survey (Wolter et al. 1994) and are plotted as a function of their 0.3–3.5 keV X-ray luminosity; the open circles are from the "1 Jy" survey at 5 GHz (Stickel et al. 1991) and are plotted with reference to νL_ν at 5 GHz. Here $h = 0.75$ and $\Lambda = 0$, but, unlike the other luminosity functions shown in this section, these assume $q_o = 0$. Fortunately, the limited redshift ranges probed in the samples make these choices less critical. Data courtesy of Paolo Padovani, adapted from Padovani and Urry (1995).

Consider, then, the density evolution of only bright quasars, for example, those with $M_B < -23$ mag, or approximately 3 mag brighter than the characteristic luminosity of field galaxies, $L_* \simeq 1 \times 10^{43}(h/0.75)^{-2}$ erg s^{-1}. Using the data of figure 3.7, we find that the density of these AGNs increased from 120 Gpc^{-3} at redshifts of a few tenths to 3.2×10^4 Gpc^{-3} at $z = 2$. It is important to remember that these are, however, merely the proverbial tip of the iceberg. The Huchra and Burg (1992) analysis of the Center for Astrophysics redshift survey Seyfert galaxies found a space density for them of 8×10^4 Gpc^{-3} if one extends the luminosity range down to total (nucleus + host) $M_B = -18$ mag.

To put these numbers in perspective, it is useful to compare them to the density of bright field galaxies ($\simeq 2 \times 10^6 (h/0.75)^{-3}$ Gpc^{-3} for luminosities within one e-fold of the characteristic luminosity of field galaxies, L_*). As already discussed in §2.2, about 2.5% of contemporary galaxies house AGNs of at least modest luminosity (i.e., luminosity at least comparable to L_*), and this number rises to at least 10% when the minimum luminosity for detection is depressed. On the other hand, at $z \simeq 2$, almost as many galaxies contained luminous quasars as today house modest luminosity AGNs. How many more faint AGNs there were is an interesting matter for speculation.

The differential luminosity functions show that, just as for galaxies, there is a characteristic luminosity for quasars—but it changes rapidly in time. Although different surveys give somewhat different quantitative dependences for this characteristic luminosity, a scaling $\propto (1 + z)^3$ or faster from $z = 0$ to 2.5 seems likely. From $z = 2.5$ to 3.5 the characteristic luminosity appears to continue to climb, but more slowly (Hewett, Foltz, and Chaffee 1993). Beyond that redshift (fig. 3.8), the data are not yet able to define the characteristic luminosity well enough for its evolution to be well described.

These parameterizations have their place in providing a crude overview of AGN evolution, but they should not be taken too seriously. For far too long in the early days of AGN statistics, the temptation to search for a simple analytic fit to the data misled many workers into a sterile debate about which parameterization—density evolution or luminosity evolution—better described the history of the AGN population. The proponents of density evolution imagined that the shape of the luminosity function changed only slightly over time, but that its amplitude fell steadily from $z \simeq 2.5$ until the present; the advocates of luminosity evolution claimed that the apparent increase in numbers was due to a shape-invariant luminosity function sliding up from low luminosity to high as we look back toward $z = 2.5$. Now that the data have improved, it is apparent that neither camp was altogether correct. In some respects, luminosity evolution has turned out to be closer to the truth, but it is still not an adequate description of the actual history of evolution. It is now clear that *both* the shape of the luminosity function and its normalization changed over the history of the Universe.

It is amusing to note that if the real luminosity function had turned out to be a power law of fixed shape, both camps could have been right! If a power law moves in a log–log plot, one cannot distinguish whether it moves horizontally, vertically, or diagonally.

3.5   Can One Solve the Population Equation?

The debate over density evolution versus luminosity evolution was not purely over which parameterization better fit the data. At stake was a more important issue—what physics caused the evolution in AGN statistics that we observe? If pure density evolution had been correct, one might guess that the reason for the evolution was that at $z = 2.5$ many galaxies were active, and as time went on, more and more simply "turned off" abruptly, dropping from substantial luminosity to none in a very short time. Contrarily, if pure luminosity evolution were correct, one might invent an analogous picture in which some fraction of galaxies became active at redshifts of a few, but their characteristic luminosity declined smoothly, though rapidly, from $z = 2.5$ to the present.

However, both these interpretations, while attractively simple, are by no means unique interpretations of the data. There are two key issues: Were only certain special galaxies ever active, or did nearly all galaxies participate at one time or another? And can a formerly active galaxy ever be revived? Because only a small fraction of all galaxies are active at any given time, it is quite possible to mock up *any* evolutionary pattern either by using a small number of galaxies but changing their parameters, or by assigning an activity probability to all galaxies and letting the parameters of that probability distribution change with time. Likewise, it is just as easy to reproduce the observed evolution with unique epochs of activity or recurring episodes. In other words, even if we had perfect measurements of the evolution of the AGN luminosity function, we would still not have enough information to determine the real history of the phenomenon. The only tool we have that might resolve this fundamental ambiguity is the study of AGN remnants, the subject of the next chapter.

By way of introduction to that chapter, here we will simply remark on another way in which the evolution of the AGN luminosity function can be interpreted. Suppose that during a period of duration T, each galaxy has a probability p of being active. This probability can be estimated from the ratio of the AGN density to the density of bright galaxies. The mean duration of activity per galaxy at that epoch is then pT. If certain galaxies were never active, then the mean duration of activity for those which were active at some time is greater than pT. Thus, the minimum time spent by a present-day galaxy as a quasar (if it spends any time at all in that state) is $\sim 10^{-3}/H_o$; if we broaden our definition of activity to include the lower luminosity range in which Seyfert galaxies are found, the minimum active time per active galaxy at present is at least $0.1/H_o \simeq 1 \times 10^9 (h/0.75)^{-1}$ yr. Because the fraction of galaxies that were active decreased with increasing age of the Universe considerably faster than T^{-1} at redshifts smaller than $\simeq 3$, the total time spent in an active state by any galaxy that was ever active is rather larger than $\sim 0.1/H_o$. By this means, then, we are able to place a lower limit on the lifetime of activity. In the following chapter we will see how that also places a lower bound on the mass of AGN remnants.

Finally, it is also important to register a further caution: most work on the evolution of the quasar luminosity function has used only *optical* data. The rather different $N(F)$ curves found in X-ray and radio surveys suggest that spectral evolution may also be significant. Given that the B band probably carries less than ~ 0.1 of the *total* luminosity in a quasar, this is an important consideration.

4 Global Energetics and Black Holes

4.1 The Mean Energy Density of AGN Light

The fundamental problem in AGN studies is to determine the source(s) of their energy output. As we have just seen, their luminosities can be enormous, the duration of activity in any one object cannot be too short, and, although a minority population at any given time, they are not exceedingly rare. Consequently, the total amount of energy liberated by AGNs in the history of the Universe is quite large.

In fact, it is possible to estimate directly the total amount of energy radiated in photons by AGNs integrated from the beginning of the Universe up to the present. The only information required is the function $N(> F)$. This argument, invented by Sołtan (1982) and elaborated by Phinney (1983) and Chokshi and Turner (1992), begins with the reasonable assumption that photons emerging from distant sources, once created, are only rarely destroyed. It is true that the flux of high-redshift quasars at observed wavelengths $\lambda < 1216(1 + z)$ Å can be diminished by absorption and scattering in intervening clouds of neutral hydrogen; it is possible that there can also be extinction by intervening dust. However, even if these diminutions are significant, the photons we receive are then a lower bound on the number originally radiated in our direction.

If none of the photons has been destroyed, then they are here today, and if the Universe is truly homogeneous, the sample we see locally should be a good indicator of their mean density. Moreover, if the Universe is isotropic, to find the local density we need only measure those photons coming from a small solid angle and then scale that number to the full 4π. Thus, the local specific energy density of AGN photons is simply

$$U_\nu = \frac{4\pi}{c} \int dF_\nu \, \frac{dN}{dF_\nu} F_\nu, \qquad (4.1)$$

where the units of N are AGNs per sterradian and the counts are measured at frequency ν with respect to specific flux F_ν. This expression, then, immediately gives the contribution of AGNs to the cosmological radiation background at frequency ν at the present epoch.

To find the total amount of energy radiated by AGNs, we need to integrate over ν. This can be done in either of two ways: by adding together the analogous expressions for a number of different bands in which the counts are measured, or by using only the counts at a single frequency, and making a guess about an appropriate bolometric correction. Note that because we are talking about photon energy densities we directly measure, U_ν is entirely independent of the usual quibbles about H_o, Ω_M, and Ω_Λ.

In practice, the second alternative is the one most commonly chosen because there is usually one band in which the counts go deepest (§3.1.6). For example, using the results displayed in Chapter 3, the integral in equation 4.1 can be easily evaluated in B-band flux. The bolometric correction relative to the B band is uncertain at the factor of a few level, for we have truly broad-band spectra of rather few AGNs. In fact, extrapolating from the better measured bands, we have reason to believe that significant flux may be awaiting discovery in the mid-infrared (wavelengths of tens of microns), EUV (shortward of the Lyman limit), and hard X-ray (photon energy of order 100 keV) bands. However, if we use the data we have to guess at the bolometric correction, something of order 10 is probably called for.

If the AGNs that dominate the integral were local, equation 4.1 would give a good estimate of the total amount of energy radiated at frequency ν, and it would be relatively simple to estimate the bolometric correction. However, if some produced their energy at a significant redshift, two corrections must be made. First, $\int d\nu\, U_\nu$ is a quantity that, relative to co-moving volume, redshifts by one factor of $1 + z$ as the photon energies diminish along with the cosmological expansion. Second, because the frequency we observe as ν was $\nu(1 + z)$ when radiated, the bolometric correction for *observed* frequency ν is likely to depend on redshift. Given the strong cosmological evolution of AGNs (see Chap. 3), these two corrections combine to produce an adjustment that is probably at least a factor of a few. We designate this factor $A(\langle z \rangle)$.

With this correction factor in hand, we arrive at the local density of energy output due to AGNs: $2 \times 10^{-16}(F_{\rm bol}/10F_B)[A(\langle z \rangle)/3]$ erg cm^{-3}. To get a feel for the magnitude of this number, it is worth comparing it to a few other distributed energy densities. Starlight (mostly due to stars in the Galaxy) has a local density about 10^4 times greater. The energy density of the cosmic microwave background is 4.2×10^{-13} erg cm^{-3}, $\sim 10^3$ times greater. The number of normal galaxies at $m_B = 20$ mag is only about 10 times greater than the number of quasars, so the integrated light of galaxies brighter than that limit is larger than the integrated quasar light by a similar factor. Of course, the total light due to galaxies includes fainter galaxies, so the ratio between the summed output of all galaxies over the history of the Universe to the output of all quasars is considerably greater than 10.

If instead of comparing total energy densities, we compared specific energy density at a particular frequency, the contribution due to AGNs becomes the dominant one in a number of bands. We can estimate this contribution at frequencies for which we don't actually measure counts, or even at nonzero redshifts, provided we have sufficient information about the redshift distribution of AGNs and their mean spectral shapes. Suppose that we wish to use survey data at frequency ν to find the AGN contribution to the local background at frequency ν' (e.g., ν might be 1 keV, appropriate to a soft X-ray survey, and ν' might be 30 keV, where the energy density of the X-ray background peaks). If all the sources discovered at ν had the same redshift and the same spectrum, then we could simply use equation 4.1, but multiplying F_ν by a factor $S(\nu'; \nu)$, which accounts for the change in specific flux moving from ν to ν'. Empirically, we find that the dispersion in continuum shape is not very large, so use of a single spectral shape is a reasonable approximation. However, at any given flux level there is a wide range in observed redshift. Consequently, we replace equation 4.1 with a new, somewhat more complicated, expression

$$U_\nu = \frac{4\pi}{c} \int dF_\nu \int dz \, \frac{\partial^2 N}{\partial F_\nu \partial z} F_\nu S \left[\nu'(1+z); \nu(1+z) \right] \qquad (4.2)$$

that involves the redshift distribution at each flux level and a redshift-dependent spectral adjustment.

More generally, knowledge of the function $\partial^2 N / \partial F_\nu \partial z$ is equivalent, given values of H_o, Ω_M, and Ω_Λ, to knowledge of the luminosity function as a function of z (§3.3). In terms of this quantity, the AGN contribution to the energy density at redshift z is

$$U_\nu = \frac{4\pi}{c} \int dL_* \int_z^\infty dz' \, \frac{dr}{dz'} \frac{r^2(z')}{\sqrt{1 - kr^2(z')}} \frac{dn(z')}{dL_x} \frac{L_x}{4\pi D_L^2(z', z)}$$
$$\times \left[\frac{1+z'}{\nu_*(1+z)} \right] S \left[\nu \left(\frac{1+z'}{1+z} \right) ; \nu_x \left(\frac{1+z'}{1+z} \right) \right],$$
$$(4.3)$$

where L_x is the luminosity at frequency ν_x as measured at redshift z', $r(z)$ is given by equation F.19 (or, for $\Lambda = 0$, eq. F.20) appropriately adjusted so that it reflects the redshift of z' as seen at z, dn/dL_x is the luminosity function with respect to co-moving volume, and $D_L(z', z)$ is the luminosity distance from z' to z. If the expression for the luminosity function as a function of redshift used in equation 4.3 is genuinely consistent with the observed distribution of fluxes and redshifts, the dependence on cosmological parameters that appears in the integrand disappears.

Making use of a similar expression (and also adjusting for intervening absorption), Haardt and Madau (1996) found that the mean intensity of Lyman edge radiation due to quasars rose to as much as $\sim 5 \times 10^{-22}$ erg cm^{-2} s^{-1} Hz^{-1} ster^{-1} at $z \simeq 2.5$ and has fallen by perhaps a factor of 50 since then. It is quite possible that since they first "turned on," quasars have been the most important source of ionizing radiation in the Universe.

Similarly, it now seems likely that AGNs are the dominant source of the hard X-ray background. From the data shown in figure 3.5, it is apparent that the present-day energy density of quasar X-rays is at least $\simeq 6 \times 10^{-18}$ erg cm^{-3} at ~ 1 keV. To go from this energy density at 1 keV to a spectrum spanning a range of harder photon energies takes only two steps. From the counts and redshifts in a well defined sample (e.g., the combined *Rosat* and *Einstein* surveys), one builds a model of the evolution of the quasar luminosity function at the characterisic energy of the sample, and then one folds the model luminosity function through a mean X-ray spectral shape as in equation 4.3. When this procedure is followed (e.g., Zdziarski et al. 1995; Madau, Ghisellini, and Fabian 1994), one finds that the predicted intensity from ~ 1 to ~ 100 keV is in fact quite a good fit to the observed spectrum, both in amplitude and shape (fig. 4.1).

Thus, although the total photon energy production by AGNs is clearly not cosmologically important, neither is it trivial, and in certain bands AGNs dominate the observed background radiation. As we are about to see, the total energy produced is in itself a powerful constraint on the nature of the central engine in AGNs.

4.2   Energy Production Efficiency and Remnant Masses

Any mechanism you can think of for generating energy follows the same basic pattern: fuel is brought in, it changes form in some way that releases energy, and the spent fuel (the "ash") accumulates somewhere. In our own daily lives, most of the energy we control is derived from chemical reactions, generally oxidation. At the order of magnitude level, 1 eV per atom is produced in chemical reactions, so the energy efficiency per unit rest mass is $\sim 10^{-10}$, where we have taken the average reacting atom to have an atomic mass of $\simeq 10 m_p$. In some cases the "ash" is vented in gaseous form (e.g., the H_2O and CO_2 produced by burning hydrocarbons),

71

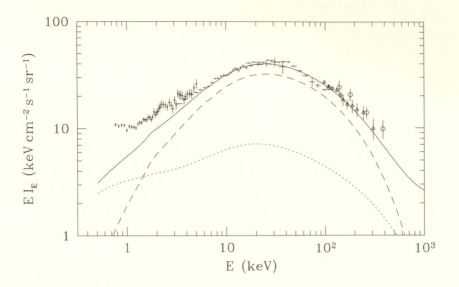

Fig. 4.1   Match between the measured X-ray background and the intensity predicted as described in the text. The points with error bars are data from Gendreau (1995), Gruber (1992), Kinzer et al. (1997), and Fukada et al. (1975); the curves represent a model described in Zdziarski et al. (1995). In this model, the co-moving volume emissivity of AGNs is taken to increase $\propto (1+z)^2$ from $z = 0$ to 3, in a rough approximation to the results of Boyle et al. (1993) when $q_o \simeq 0 = \Lambda$. For reasons explained more fully in §12.6.4, type 2 Seyferts (dashed curve) are assumed to be five times as numerous as the type 1 variety (dotted curve), but suffer local absorption by a column density of 10^{23} cm^{-2} (see §§8.7.1, 12.6.3). The solid curve is the sum of the contributions from the two types. It is quite a good fit from 3 to 300 keV, but other, similar models with slightly different quantitative choices do equally well.

but that doesn't change the basic character of this argument. If, as happens when wood is burnt, large amounts of the fuel mass do not go through a chemical reaction, the effective efficiency may be substantially smaller than 10^{-10}.

In stars, the fuel is the interstellar gas, mostly hydrogen, assembled to make the star, and then "burnt" to helium by nuclear reactions. In this case, the energy released per unit fuel mass is about 8 MeV/m_p, for a dimensionless efficiency $\sim 8 \times 10^{-3}$. In an old stellar population, essentially all the burnt fuel is retained in the white dwarf that is eventually formed, so it is easy to locate the "ash" (the unburned fuel is nearly all "vented").

In a young stellar population, the situation is a little more complicated but not essentially different. Massive stars produce most of the light, through a mix of nucleosynthesis and gravitational collapse. Most of their integrated energy output in photons comes from nuclear reactions and is radiated during the stellar lifetime. However, the majority of the processed mass is dispersed into the interstellar medium as a result of supernova explosions, with only a few tens of percent retained in a neutron star or black hole remnant. In addition, the energy released by the supernova is primarily produced by gravitational collapse, which has an intrinsically much higher efficiency than nuclear reactions (see below). But this does not end up raising the net photon efficiency of stellar processes very much because the great majority of the energy released by stellar gravitational collapse is radiated in neutrinos, not photons. Thus, in the end the net efficiency, measured in terms of photon energy per remnant mass, of even a young stellar population is still not much more than $\sim 10^{-3}$.

The most fuel-efficient way we know to release energy is accretion into a relativistically deep gravitational potential. Accretion onto either a neutron star or a black hole can result in radiation of ~ 0.1 of the rest-mass energy (see Chap. 5 for a more thorough discussion). In our own Galaxy, this takes place in numerous binary X-ray sources whose relativistic component has a mass $\sim 1\text{--}10 M_\odot$. As we are about to show, in AGNs the central mass is much larger.

We can now use these efficiency estimates to transform the mean energy density of AGN radiation into a mean mass density of AGN remnants, assuming that the spent fuel is kept in localized remnants. The result is $2 \times 10^{-36} (F_{\mathrm{bol}}/10 F_B)[A(\langle z \rangle)/3](\eta/0.1)^{-1}$ gm cm^{-3}, where we have normalized to an efficiency η appropriate to accretion onto a relativistically deep potential. To the degree that intervening absorption has diminished the quasar energy density we measure, or that our samples are either incomplete or fail to extend to faint enough levels, this is a lower limit.

This number can be made more meaningful by comparing to the space density of galaxies with luminosities within 1 mag of the characteristic luminosity of field galaxies, L_*, $2 \times 10^6 (h/0.75)^3$ Gpc^{-3}. If the remnant mass were divided equally among these galaxies, the mean mass per galaxy would be

$$\langle M_{\mathrm{rem}} \rangle \simeq 1.6 \times 10^7 (F_{\mathrm{bol}}/10 F_B)(\eta/0.1)^{-1}(h/0.75)^{-3}[A(\langle z \rangle)/3] \, M_\odot. \quad (4.4)$$

That is an interesting fraction of an average galaxy mass even if the efficiency is high; if the energy were generated through nuclear reactions, a considerable part of each galaxy would have had to be used as AGN fuel. The mass budget becomes even more constraining if only a minority of galaxies ever housed AGNs.

This argument (first clearly put by Lynden-Bell 1969) is the fundamental reason why nearly everyone in the AGN field believes the energy generation is due to accretion onto a relativistically deep gravitational potential, probably a black hole. We have enough difficulties trying to invent mechanisms that can assemble even the minimum quantity of fuel required; if the energy release were through stellar processes, the problem would be far harder. Stellar mechanisms entail the further drawback that we should be able to find massive remnants in at least some nearby galaxies. Even if a considerable part of the "ash" is thrown off into the interstellar medium, a large enough fraction would likely remain in compact remnants for their gravity to seriously affect the orbits of stars in the host galaxy.

4.3   Rees's Flow Chart

The efficiency arguments of the previous subsection lead us to consider accretion into a relativistically deep potential as the power source for AGNs, but that alone is not enough to require black holes. After all, neutron stars are objects that have relativistically deep potentials but are not black holes. It could be that the "central engines" of AGNs are just clusters of neutron stars.

Perhaps the most comprehensive and convincing case that a single very massive black hole can be found in the center of each AGN has been made by Rees (1977). In order to make this case, he added one more observed fact: the short variability timescales that are seen in many AGNs.

Before presenting Rees's argument, it is worth pausing to remark that important elements in it were guessed at within a year of Schmidt's identification of the redshift of 3C 273 and the consequent recognition that quasars emit extraordinary luminosities. Both Salpeter (1964) and Zel'dovich (1964) understood that the simplest way to generate very large amounts of power in a very small volume is to accrete matter onto a large black hole.

4.3.1   Causality bounds

The significance of the short variability timescales is that they place an upper bound on the size of the radiating region. Imagine first that a stationary system of size l suddenly increases its emissivity at all points simultaneously. This step in emissivity translates to a ramp in flux that we receive whose width is l/c, for we only receive the news from the far side at a delay of l/c with respect to the near side. Thus, a stationary source cannot fluctuate in a way that involves its entire volume on any timescale shorter than l/c.

In fact, real fluctuation timescales are likely to be rather longer than l/c because simultaneous fluctuations are hard to arrange. Presumably an event occurs at some place that triggers the fluctuation. Then "news" of this event must be conveyed throughout the source with a signal speed $v_s \leq c$. If, for example, the control point is located at the center of the source, and $v_s \ll c$, then we will first see the change in light from the center, and only after a delay $l/(2v_s)$ will we see the change in light from the near side. The light travel time delay will then spread out the fluctuation further.

One could in principle avoid this limit by imagining that the fluctuation control point is on the far side of the source, and the signal travels from the far side toward the near side. This would allow a ramp up on a rather shorter timescale, $(1 - v_s/c)(l/c) = [(1 + \beta_s)\gamma_s^2]^{-1}(l/c)$, where γ_s and β_s are the usual relativistic factors: $\beta_s = v_s/c$ and $\gamma_s = (1 - \beta_s^2)^{-1/2}$.

This would be all right in one source, but how could we explain such a special geometry in an entire sample? The only way in which we might reasonably expect to be in such a favored direction for all the objects is if the whole source is moving relativistically toward us and therefore beams its light preferentially in our direction. If all the sources were relativistically beamed, then it would be very hard for us to see any of them from the side or back because they would be much fainter in those directions. In fact, as we shall discuss in greater detail in Chapters 9 and 12, there is good evidence for just this sort of relativistic beaming in a number of AGNs. However, at this point it suffices to say that relativistically beamed objects are probably a minority, and even when beaming does occur, the Lorentz factors we infer are not great enough to loosen these causality bounds on size scales by more than one or two orders of magnitude.

4.3.2   Variability timescales

To apply Rees's argument, we must now look to the observations to see on what timescales AGNs actually vary. This is another subject to which we shall return later (§§7.6.4, 8.1.4, 9.1.5), but for the present purpose we need only qualitative statements.

Although the observational evidence is still quite sketchy, three facts do seem relatively well established at the qualitative level: (1) with only rare exception, one sees variability on a range of timescales, with no single scale dominating (i.e., periodicities are almost never seen), and the fluctuations retain significant autocorrelation out to fairly large lags (in the optical, this means ~ 1 yr or longer); (2) lower luminosity objects vary more rapidly than do higher luminosity objects; and (3) within an individual object, higher frequency continuum bands show faster variability than lower frequency bands. These comments do not apply to the small subset of AGNs that show very large amplitude variability. If we wish to find the timescale on which a source changes substantially, then it is important to bear in mind that this is *not* necessarily the timescale of most rapid observable change. When the power spectra are steeper than f^{-1} (as is often the case in AGNs), most of the variance is accumulated at the longest timescales, not the shortest. Therefore, to produce a fluctuation that is order unity with respect to the mean, one looks to the *low*-frequency end of the power-law segment of the fluctuation power density spectrum.

The specific timescale on which large fluctuations occur varies considerably from object to object and band to band. Some low-luminosity ($\sim 10^{43}$ erg s^{-1}) AGN have changed their 2–10 keV X-ray flux by factors of two in as little as a few hours, while some high-luminosity ($\sim 10^{47}$ erg s^{-1}) AGN vary by as little as 10% over spans of years. We therefore infer maximum lengthscales of anywhere from 10^{-4} to 10 pc.

4.3.3 The flow chart

The nature of AGN variability immediately directs one's attention to single black hole models. On the one hand, the broad autocorrelation functions argue against a cluster of independently radiating objects; except during those times when one member of the cluster dominates the emission, the flux time history from a collection of independent objects should resemble white noise, whose autocorrelation function is a simple delta function. On the other hand, the lack of periodicity argues against a single object with a surface that might support structure; any rotation would then induce a periodicity (cf. the "spinar" model proposed by Morrison 1969).

Still, one might argue that a cluster of neutron stars and stellar mass-scale black holes could radiate in a collective fashion (as actually proposed for other reasons by Arons, Kulsrud, and Ostriker 1975). It is a useful exercise to see how such models can be limited by physical arguments. Our aim will be to show that this model either is internally inconsistent or else naturally evolves toward collapse into a single supermassive black hole.

In order to make this argument applicable to the "typical" case, we will scale to fiducial numbers in the middle of the possible range. Suppose, then, that a particular AGN has a luminosity of 10^{45} erg s^{-1}and lives for 10^8 yr, so that the total mass accreted is $\sim 1.5 \times 10^7 (\eta/0.1)^{-1} M_\odot$. We will also suppose its luminosity in some band changes by order unity in one week. Note that such rapid variability, while not unheard of in the X-ray band, would be very unusual in the optical or ultraviolet in an object of this luminosity not classed as either an OVV or a BL Lac object. However, making that assumption, we would conclude that it must be no larger than 2×10^{16} cm, and the orbital speeds of the cluster members will be at least 3000 km s^{-1}. Two-body gravitational encounters would then cause relaxation on a characteristic timescale $\simeq 3 \times 10^5$ yr. Studies aimed at globular cluster dynamics have shown that a collisionless system evolving in this fashion develops a singular density core after a few hundred relaxation times. The variability time for this example was chosen so that the time to core collapse would coincide with its assumed lifetime, $\sim 10^8$ yr.

What is the nature of this core? First, it is necessary to realize that only a small part of the total mass is contained within the core. The cluster structure outside the core is close to isothermal, so the density $\rho \propto r^{-2}$ (actually closer to $r^{-2.2}$), and the mass interior to radius r is $M(r) \propto r^{0.8}$. If this cluster contained main sequence stars, the physical sizes of the stars would come to dominate the evolution: stellar collisions would destroy the stars because the random velocities are larger than the stellar surface escape speeds, and encounters closer than a few stellar radii would cause enough energy loss through viscous dissipation of tidal distortions to form binaries. However, in a cluster of neutron stars and black holes, the physical size only becomes significant when the periastron distances are a few gravitational radii. Thus, for this sort of cluster, the core ultimately evolves through the merger of neutron stars to form black holes, and the merger of black holes to form larger black holes. Because this reduces the number of stars, while conserving the mass, the evolution accelerates. Within a relatively short time, a relativistic instability may (depending on the total angular momentum of the system) be triggered that causes the entire core to collapse into a single massive black hole (Shapiro and Teukolsky 1993).

We conclude, therefore, that if the cluster is not too large (less than a light-week for intermediate luminosities), a dense cluster of neutron stars and black holes evolves into a massive black hole in any case. We do not know for certain that all AGNs must be this small, so this argument is not absolutely watertight. However, as variability studies improve, it is possible that this argument might be refined to the point where it really would apply in general.

Rees's famous flow chart diagram (fig. 4.2) hinges on this sort of argument but also attempts to show how a cluster of ordinary stars might evolve through the intermediate stage of a neutron star/black hole cluster, and thence into a supermassive black hole. Dense clusters of ordinary stars may form in the centers of galaxies (see §14.5), and thus the inevitability of massive black hole formation even in star clusters becomes an important possible pathway toward construction of the central engines in AGNs.

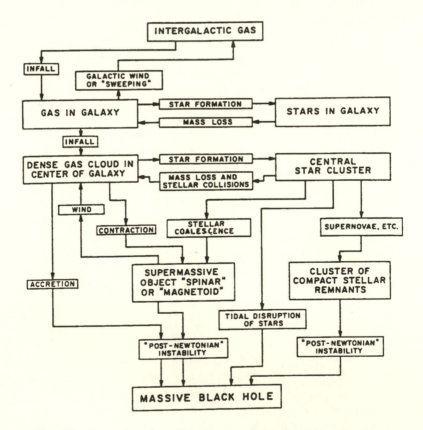

Fig. 4.2   Rees's flow chart (Begelman et al. 1984).

It is important to keep one's sense of perspective when making such broad-brush arguments. At an AGN conference held at the Institute of

Astronomy, Cambridge in the summer of 1977, Dick McCray invented a different diagram (fig. 4.3) that, while jocular, still conveys some home truths about the theory of AGNs that are as valid today as they were then. His diagram is entitled "Astronomers Accreting onto a Fashionable Idea."

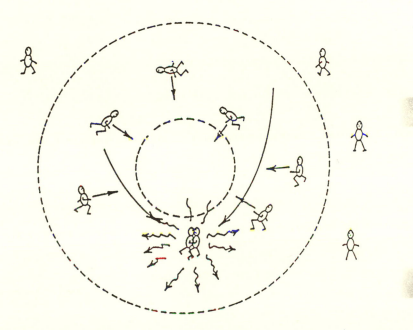

Fig. 4.3   McCray's cartoon (McCray 1979).

The best description of the events shown in this cartoon is McCray's own words:

"I cannot resist pointing to an example of the accretion process that is familiar to all of us: that is, the response of astrophysicists to a fashionable new idea. The system is characterized by two radii. Beyond the accretion radius, r_A, astrophysicists are sufficiently busy with other concerns not to be significantly influenced by this fashionable idea. But others, within r_A, begin a headlong plunge toward it. There is little communication among individuals as they follow random ballistic trajectories, depending on their initial conditions. In their rush to be the first, they almost invariably miss the central point, and fly off on some tangent. With a sufficient number of

astrophysicists in the vicinity of the idea, communication must finally occur, but it usually does so in violent collisions. Such collisions may result in some illumination, which may, in turn, modify the behavior of the others. Indeed, the illumination may be sufficient to counteract the attraction of the idea and stop the flow altogether. The only lasting effect in that case is that some individuals may have crossed the rationality horizon, r_S, beyond which the fashionable idea has become an article of faith. These unfortunate souls never escape; examples of this latter phenomenon are familiar to all of us." (McCray 1979)

4.4 Evidence for Massive Black Holes in Galaxies

In the second section of this chapter, we estimated that the mean remnant mass per galaxy today might be $\sim 1 \times 10^7 \, M_\odot$. If every galaxy participated equally in the AGN phenomenon, we would expect to see a mass of this magnitude in the center of every galaxy today. On the other hand, if only a minority of galaxies participated, that same minority might have considerably greater remnant masses. Thus, by observations of the masses in galactic centers today, we can map out the probability distribution $P_{\rm rem}(M)$ for how much mass each galaxy's nucleus accreted over its lifetime (this is the tool alluded to in §3.5). By definition, $\int_0^\infty dM \, P_{\rm rem}(M) = 1$; the argument leading to the mean mass per galaxy fixes $\int_0^\infty dM \, M P_{\rm rem}(M) = \langle M_{\rm rem} \rangle$. Future searches may help us further constrain the shape of $P_{\rm rem}(M)$.

It is important to remember, however, that although the discovery of large mass concentrations in galactic nuclei does bear on the question of which galaxies formerly housed AGNs, it does not directly implicate black holes as such. As discussed in §4.3, in principle an extremely compact cluster of objects with relativistic gravitational potentials would be equally consistent with the detection of very large mass concentrations at the centers of galaxies. Our confidence that a large mass in a galactic nucleus is a black hole rather than a dense stellar cluster grows as upper bounds on its size tighten.

The only way we might hope to detect a massive AGN remnant would be through its gravity. Roughly speaking, the "sphere of influence" of a large mass at the center of a galaxy is defined by the distance at which its potential significantly affects the orbital motions of stars or interstellar gas:

$$r_* = \frac{GM}{\sigma_*^2} \simeq 4 M_7 \sigma_{*,100}^{-2} \text{ pc}, \qquad (4.5)$$

where we have normalized to a central mass of $10^7 M_\odot$ and *rms* orbital speeds σ_* of 100 km s^{-1}. Thus, even fairly large remnants have a major influence out to only comparatively small distances. Even in nearby galaxies, the associated angular scale is small: $\sim 0.1''(D/10 \text{ Mpc})^{-1}$.

4.4.1 Expected response of stars and interstellar gas to a large nuclear mass

Because interstellar gas is a fluid, its response to a central mass is much simpler than the collisionless response of stars. The velocity distribution function for a gas is very constrained: only under most unusual circumstances can it be anything other than isotropic in the center of mass frame. Moreover, the random speeds of gas molecules are frequently small compared to their bulk speed. When that is the case, everything about the gas's motion can be characterized by a single function of position. The only potential complication is that gas can be subject to a variety of non-gravitational forces—pressure gradients, radiation pressure, and so on—to which stars are entirely impervious. For this reason, before interpreting gas kinematics as due to gravity, it is important to find reasons to exclude the presence of other forces.

By contrast, stellar velocity distribution functions can take on a very wide range of shapes, and the absence of dissipative processes means that any bulk streaming motion is generally small in magnitude compared to the random velocities. The much greater freedom possessed by stellar distribution functions as compared to gaseous ones makes it possible for the observable properties of stars in a purely stellar potential to mimic the observable properties of stars in a black hole-dominated potential. Consequently, if all forces but gravity are truly irrelevant to the gas dynamics, the interpretation of stellar data is much subtler than the interpretation of data regarding gaseous orbits.

We therefore begin by considering the signature a large remnant mass writes on surrounding gas. This signature is, indeed, often very simple. Low-density gas easily radiates away the energy of random motions, but finds it very hard to rid itself of angular momentum (see Chap. 7), so generically interstellar gas structures tend to be rather flat. This is why the random speeds in gaseous sytems are often much smaller than the bulk speed. At any radius r large compared to the outer boundary of the central mass, the rotation speed of an undisturbed flat gas disk is simply $(GM/r)^{1/2}$. In addition, to the extent that the gas's velocity dispersion is small compared to its bulk speed, and the bulk speed appears to vary smoothly with position, our confidence in a purely gravitational origin for the bulk motion is bolstered. When these confirming signs that the gas's motions are dominated by gravity are seen, observations of the rotational velocity of gas as a function of radius can provide a very clean measure of the total mass inside its innermost radius.

The impact of a large mass on surrounding stars is more complex because we must follow the stellar distribution function, rather than just the mean velocity, as a function of position. Moreover, the stellar distribution function can be expected to change as a function of time, both because of processes intrinsic to the stellar system and because the gravitational potential changes as the black hole gains mass. To distinguish important time dependence from trivial, we begin by identifying three significant timescales: the orbital timescale, the black hole growth timescale, and the stellar two-body relaxation timescale.

The orbital timescale at the edge of the black hole's sphere of influence is, by astronomical standards, relatively short:

$$t_{\rm orb} \sim 4 \times 10^4 M_7 \sigma_{*,100}^{-3} {\rm yr}. \qquad (4.6)$$

We can estimate the rate at which AGN remnant mass is accumulated by making use of the radiative efficiency arguments of §4.2. Scaling to a typical AGN luminosity gives a rate of growth for the central mass

$$\dot{M} \sim \frac{L}{c^2 \eta} \simeq 0.2 L_{45} \left(\frac{\eta}{0.1}\right)^{-1} M_\odot \; {\rm yr}^{-1}. \qquad (4.7)$$

At this rate, an average galaxy would take a time

$$t_{\rm growth} \sim 1 \times 10^8 L_{45}^{-1} \left(\frac{\eta}{0.1}\right) \left(\frac{F_{\rm bol}}{10 F_B}\right) \left(\frac{h}{0.75}\right)^{-3} \left(\frac{\langle 1+z \rangle}{3}\right) \; {\rm yr} \qquad (4.8)$$

to accumulate its total remnant mass. This is considerably longer than the orbital time at the edge of the central mass's sphere of influence:

$$\frac{t_{\rm growth}}{t_{\rm orb}} \sim 10^3 L_{45}^{-1} \sigma_{*,100}^3 \left(\frac{\eta}{0.1}\right). \qquad (4.9)$$

The two-body relaxation time is

$$t_{\rm relax} \simeq 0.1 \left(\frac{N}{\ln N}\right) t_{\rm orb}, \qquad (4.10)$$

where N is the total number of stars in the system and $t_{\rm orb}$ is the orbital period. This quantity is, to order of magnitude, the time required for a star to change its energy substantially by encounters with other stars; in other words, it is also the time required for the system to approach a Maxwellian distribution function (see Binney and Tremaine 1987 for a far more thorough discussion). The growth time for remnant masses is generally rather smaller than $t_{\rm relax}$:

$$\frac{t_{\rm growth}}{t_{\rm relax}} \sim 0.02 L_{45}^{-1} \sigma_{*,100}^3 \left(\frac{\eta}{0.1}\right) \left(\frac{\langle m \rangle}{M_\odot}\right) \left(\frac{M_*}{10^7 M_\odot}\right)^{-1}, \qquad (4.11)$$

where m is the stellar mass and M_* is the total mass of stars.

Two useful approximations may therefore be made: two-body encounters may be ignored, and the change in the potential occurs adiabatically. Thanks to the first approximation, we can adopt the collisionless Boltzmann equation as our fundamental rule for how the distribution function evolves. If the opposite limit applied (i.e., $t_{\rm relax} \ll t_{\rm growth}$), and no stars were lost from the system, the stellar distribution function would become (if it were not already) nearly isotropic and would also develop a nearly Gaussian dependence on the energy. However, in our context the system does not have time to relax, and a wide variety of distribution functions are possible.

The second approximation points us to a set of conserved quantities, each star's action variables

$$J_i \equiv \oint dq_i\, p_i, \qquad (4.12)$$

where the (q_i, p_i) are pairs of canonical coordinates and their conjugate momenta (see, e.g., Goldstein 1980 or Landau and Lifshitz 1960). The integral is taken over a single orbit. Note that these actions only exist for *bound* orbits, so we will make the (not very restrictive) assumption that the star's energy $E < 0$. They are constant for each star, provided only that the potential changes slowly relative to the orbital timescale; this is the significance of estimating the ratio $t_{\rm orb}/t_{\rm growth}$. This constancy makes the action description particularly useful because the distribution function defined in terms of the actions does not change in time.

In a spherically symmetric potential $\Phi(r)$, there are two nontrivial actions, the radial action (proportional to the mean radial kinetic energy)

$$J_r \equiv \oint dr\, \dot{r} = \oint dt\, m\dot{r}^2, \qquad (4.13)$$

and the magnitude of the angular momentum

$$J_\phi \equiv \oint d\phi\, mr^2 \dot{\phi}. \qquad (4.14)$$

Where the potential is Keplerian,

$$J_\phi = m \left[\frac{(GM)^2}{2|E|(1 - e^2)} \right]^{1/2} \tag{4.15}$$

and

$$J_r = -J_\phi + \frac{GMm^{3/2}}{\sqrt{2|E|}}; \tag{4.16}$$

here e is the orbital eccentricity. These relations can be inverted for E and e in terms of the actions:

$$|E| = \frac{(GM)^2}{2(J_r + J_\phi)^2} \tag{4.17}$$

and

$$e = \sqrt{1 - \frac{1}{(1 + J_r/J_\phi)^2}}. \tag{4.18}$$

Analogous relations exist for different forms of the potential. In practice, they are most easily found by computing

$$J_r(E, J_\phi) = \oint dr \, m^{-1/2} \sqrt{2[E - \Phi(r)] - J_\phi^2/mr^2}. \tag{4.19}$$

To find the stellar density, we begin by recalling that the complete phase space volume element $d^3q \, d^3p$ is the same independent of the choice of canonical variables (q_i, p_i). In particular, $d^3r \, d^3p = d^3\theta \, d^3J$, where the θ_i are the angle variables conjugate to the actions J_i. Therefore, the stellar distribution function

$$f_*(\vec{r}, \vec{v}) = f_*(\vec{J}, \vec{\theta}) = f_*(\vec{J}), \tag{4.20}$$

where the last equality follows because orbital phase-averaging removes any dependence of the distribution function on the angle variables. This equality of the distribution functions allows us to substitute $f_*(\vec{J})$ for $f_*(\vec{x}, \vec{v})$ in the density integral:

$$\rho_*(\vec{r}) = \int d^3v \, f_*(\vec{r}, \vec{v}) = \int dv_r \int dv_t \, 2\pi v_t f(J_r, J_\phi), \tag{4.21}$$

where v_r is the radial velocity and v_t is the magnitude of the tangential velocity.

The next step is to relate v_r and v_t to J_r and J_ϕ. Independent of the form of the potential, the tangential velocity is always

$$v_t = \frac{J_\phi}{mr}, \tag{4.22}$$

but the radial velocity does depend on the details of $\Phi(r)$:

$$v_r = m^{-1/2}\sqrt{2[E - \Phi(r)] - J_\phi^2/r^2}, \tag{4.23}$$

where E should be thought of as a function of J_r and J_ϕ whose form depends on the form of $\Phi(r)$. Changing integration variables to J_r and J_ϕ transforms the integral in equation 4.21 to

$$\rho_*(r) = \int_0^{J_{\phi,max}} dJ_\phi \int_{J_{r,min}}^\infty dJ_r \frac{2\pi J_\phi}{m^{3/2}r^2} \frac{\partial E}{\partial J_r} \frac{f(J_r, J_\phi)}{\sqrt{2[E(J_r, J_\phi) - \Phi(r)] - J_\phi^2/r^2}}, \tag{4.24}$$

where the bounds of integration are defined by the classical turning points of the motion so that

$$E(J_{r,min}, J_\phi) = \Phi(r) + \frac{J_\phi^2}{2mr^2} \tag{4.25}$$

and

$$\lim_{J_r \to \infty} E(J_r, J_{\phi,max}) = \Phi(r). \tag{4.26}$$

By normalizing the actions to $r\sqrt{|\Phi|}$ and the energy to $|\Phi|$, the integral can be put into dimensionless form:

$$\rho_*(r) = 2\pi \left(\frac{|\Phi|}{m}\right)^{3/2} \int_0^{x_{\phi,max}(r)} dx_\phi \int_{x_{r,min}(r,x_\phi)}^\infty dx_r \frac{\partial \hat{E}}{\partial x_r} \frac{f_*(x_r, x_\phi)}{\sqrt{2(1 - |\hat{E}|) - x_\phi^2}}, \tag{4.27}$$

where x_r and x_ϕ are the dimensionless actions and $\hat{E}$ is the dimensionless energy. In a Keplerian potential, the expressions simplify considerably:

$$x_{\phi,max} = \sqrt{2},$$
$$x_{r,min} = (2 - x_\phi^2)^{-1/2} - x_\phi, \tag{4.28}$$

and

$$\hat{E} = -\frac{1}{2(x_r + x_\phi)^2}, \tag{4.29}$$

85

so that

$$\frac{\partial \hat{E}}{\partial x_r} = (x_r + x_\phi)^{-3}. \tag{4.30}$$

In a region where the potential is close to Keplerian, dependence on r enters the right-hand side of equation 4.27 in only two places: the multiplicative factor $|\Phi|^{3/2}$, and $f_*(x_r, x_\phi)$, which retains a hidden dependence on r through the normalization of the actions in units of $r\sqrt{|\Phi|}$. If f_* is a slowly varying function of the actions, $\rho_* \propto |\Phi|^{3/2} \propto r^{-3/2}$ (Young 1980; Peebles 1972). Thus, sufficiently close to a black hole that its mass dominates the potential, the stellar density distribution should always have the same $r^{-3/2}$ radial dependence.

As the black hole mass grows, the isodensity contours in the Keplerian region remain fixed to the isopotential contours; that is, the stellar distribution is simply squeezed closer to the black hole while exactly preserving its shape. On the outside, of course, regions that were formerly dominated by the stellar potential gradually fall within the influence of the black hole. In these regions, there is a tendency for the stellar orbits to become more circular, as radial excursions become progressively more difficult in the steepening potential.

4.4.2 Observational tests

In the current state of technology, directly resolving angular scales in external galaxies small enough to search for the gravitational effects of black holes requires one of two approaches: the use of space-borne telescopes, in order to avoid atmospheric "seeing"; or radio VLBI techniques. Occasionally, however, when the effects are sufficiently dramatic, they can provide strong evidence for a central black hole even in the absence of bona fide imaging.

4.4.2.1 gas dynamical tests

As we have already remarked, where there is interstellar gas at once bright enough in some atomic line to establish a rotation curve, and "cool" enough dynamically to settle into a disk, it is relatively easy to establish the signature of a Keplerian potential. Happily, examples have been found, and the inferred mass within a few parsecs can be quite large.

One well established case is an $H\alpha$-emitting disk in the radio galaxy M 87. To resolve it requires the fractional arcsecond imaging capability of the *Hubble Space Telescope* (*HST*). Fitting its rotation curve to a Keplerian model yields an estimate of the mass within 18 pc of the galactic nucleus: $2.4 \times 10^9 \, M_\odot$ (Ford et al. 1994; Harms et al. 1994).

Radio VLBI work is possible only on very high brightness temperature sources. Although such sources are rare, there is at least one case, the LINER NGC 4258, in which they provide very strong evidence for the existence of an extremely large mass confined within a very small volume. In that galaxy, there are a great many very small H_2O masers scattered over a region ~ 0.1 pc in extent. Remarkably, they are very closely confined to a single (slightly warped) plane, and their line-of-sight velocities are fitted extremely well by a simple model of circular orbits in a Keplerian potential. The inferred enclosed mass is $3.6 \times 10^7 \, M_\odot$ (Miyoshi et al. 1995). Because the residuals from the fit are exceedingly small (considerably less than 1%) and the corresponding mass density is several orders of magnitude larger than the density of any known stellar cluster, the maser spots in NGC 4258 present very strong evidence for the existence of a supermassive object at the center of a galaxy.

Finally, there are numerous Seyfert galaxies (see, e.g., fig. 8.12) in which the Fe Kα X-ray line is so broad that if its width is interpreted as Doppler broadening, the emitting gas must be moving relativistically. Although an integrated line profile cannot be used to directly infer a central mass the way a rotation curve can, and alternative explanations not requiring orbital motion around a black hole can be invented, such broad profiles are certainly consistent with the idea that there are relativistically deep gravitational potentials inside AGNs.

4.4.2.2 stellar dynamical tests

Working backwards from observations of the stellar density and velocities is, not surprisingly, more complex. In the previous section we assumed we knew the distribution function in complete detail and predicted the response to a growing central mass. In the real world, we begin with a partial description of the stellar distribution function (i.e., the stellar density and a few moments of the distribution function) and must work backwards toward the enclosed mass. To make this connection, we again begin with the collisionless Boltzmann equation. It is most convenient to express it in spherical coordinates:

$$
v_r \frac{\partial f}{\partial r} + \left(\frac{v_\theta^2 + v_\phi^2}{r} - \frac{\partial \Phi}{\partial r} \right) \frac{\partial f}{\partial v_r} + \frac{1}{r} \left(v_\phi^2 \cot\theta - v_r v_\theta - \frac{\partial \Phi}{\partial \theta} \right) \frac{\partial f}{\partial v_\theta}
$$

$$
- \frac{1}{r} \left[v_\phi \left(v_r + v_\theta \cot\theta \right) + \frac{1}{\sin\theta} \frac{\partial \Phi}{\partial \phi} \right] \frac{\partial f}{\partial v_\phi} = 0, \tag{4.31}
$$

assuming a steady state and spherical symmetry. Taking the moment of this equation with respect to v_r yields

$$r\frac{\partial \Phi}{\partial r} = V^2 - \sigma_r^2 \left[\frac{d\ln \rho_*}{d\ln r} + \frac{d\ln \sigma_r^2}{d\ln r} + 2 - \frac{\sigma_t^2}{\sigma_r^2} \right], \qquad (4.32)$$

where V is the rotational speed $\langle v_\phi \rangle$, the $\sigma_{r,\theta,\phi}^2$ are the velocity dispersions in those directions, and $\sigma_t^2 = \sigma_\theta^2 + \sigma_\phi^2$.

When the potential is spherically symmetric, $r(\partial \Phi / \partial r) = GM(r)/r$, where $M(r)$ is the mass enclosed within r. Viewed in this way, measurements of the rotational and random velocities of stars as functions of r permit us to map out $M(r)$. $M(r)$ must, of course, decline inwards; if it is flat inside some critical radius, we may then believe we have measured the mass of the central object. Even if we do not find the point at which $M(r)$ levels off, if the mass/light ratio becomes sufficiently large inside some radius, that in itself is of interest.

A glance at the form of equation 4.32 immediately shows why this program is easier to implement on dynamically "cold," rotating, fluidlike systems than on dynamically "hot," slowly rotating, collisionless systems. In the former case, only $V(r)$ is needed; in the latter, three more quantities must be measured.

Moreover, those three extra quantities are not easily come by. The quantities we actually measure are all projections onto the sky plane: the surface brightness of the stars, the component of their velocity dispersion parallel to the line of sight, and the component of their bulk velocity along the line of sight. If we may assume that the density distribution is spherically symmetric, the only difficulty in transforming stellar surface brightness as a function of cylindrical radius into stellar mass density as a function of spherical radius is a mass/light ratio that is a priori unknown (and that may not be uniform from place to place). However, the velocity dispersions present more of a problem. There is no way to transform a single function (the line-of-sight velocity dispersion) uniquely into the two functions that are required (σ_r and σ_t^2/σ_r^2). All we can do is make guesses about the shape of the stellar velocity ellipsoid.

Unfortunately, there are forms of the velocity ellipsoid that can "spoof" the effects of a central point mass (Duncan and Wheeler 1980; Binney and Mamon 1982). Consider, for example, a case in which the rotational velocity is zero. The parameter σ_t^2/σ_r^2 can range all the way from 0 (purely radial orbits) to 2 (isotropic motions) to ∞ (purely tangential orbits). For a fixed measured velocity dispersion, the inferred mass rises steadily as the assumed orbital distribution changes from tangential to radial. Thus, if one interprets the velocity dispersion data in terms of, for example, an isotropic orbit distribution, while near the center of the galaxy the real distribution is dominated by stars on radial orbits, one could be misled into substantially overestimating the mass close to the galaxy's center. To what degree such anisotropic distribution functions can exist in Nature is somewhat uncertain; the most extreme forms are unstable (Merritt 1987), but these complications cast a shadow on many attempts to infer central masses from measurements of stellar densities and velocities.

Despite these difficulties, there are a number of cases in which the stellar surface brightness and velocity dispersion give strong reason to believe that there is a massive ($\sim 10^7$–10^9 $M_\odot$), dark (i.e., M/L greater than a few tens) object in the center of the host galaxy (Kormendy and Richstone 1995). The best cases are, of course, those in which the stellar characteristics are most "hydrodynamic," that is, in which the rotational speed is relatively large. In addition, surveys looking for candidate black holes (e.g., Kormendy 1992) find that more than 30% of normal elliptical galaxies have large enough implied mass/light ratios to suggest dark mass in the range 10^8–10^9 $M_\odot$.

5   Black Hole Physics

In the last chapter we developed the arguments suggesting that black hole accretion really is the power source for AGN. We must now start to understand the physics of how this takes place. Black holes are intrinsically general relativistic creatures, but a complete treatment of those aspects of general relativity necessary to understand the dynamics of material near them is far beyond the scope of this book. Readers with little background in this subject should read the *very* limited introduction given in Appendix A. For a more complete presentation, there are numerous excellent texts on general relativity (e.g., Weinberg 1972; Misner, Thorne, and Wheeler 1973).

In this chapter, we give only a very brief introduction to the general principles of the subject, sufficient to permit a qualitative understanding of the most astrophysically relevant properties of black holes: the basics of particle orbits, the maximum efficiency of accretion, how to relate photon intensities near the black hole to those seen by distant observers, and the special character of electromagnetic fields near black holes.

5.1   Particle Mechanics in the Vicinity of Black Holes

5.1.1   General properties of relativistic particle mechanics

If we wished to describe the complete history of a particle's motion, we would supply its three-dimensional position $\vec{\xi}$ as a function of time t. It is, of course, one of the principal achievements of special relativity to show that this is an ambiguous description—measurements of both distance and space depend on the observer's state of motion. From the relativistic point of view, it is preferable to put everything in terms of quantities whose transformation properties from frame to frame are well defined—Lorentz scalars, four-vectors, tensors, and so on. Expressions written this way are said to be *covariant*. In this language, the kinematic system mentioned at the beginning of the paragraph—a sequence of three-dimensional position vectors each associated with a particular time (and all as measured with respect to some particular reference frame)—becomes a sequence of four-vectors $\xi^\mu \equiv (t, \vec{\xi})$.

It is often convenient to label these four-vectors by the *proper time s* associated with each one. The change of proper time from one moment to the next is defined by

$$ds^2 = g_{\mu\nu}d\xi^\mu d\xi^\nu, \tag{5.1}$$

where $g_{\mu\nu}$ is the *metric*. In the world of special relativity (the world of high speeds and weak gravity),

$$g_{\mu\nu} = \eta_{\mu\nu} = \begin{pmatrix} 1 & 0 & 0 & 0 \\ 0 & -1 & 0 & 0 \\ 0 & 0 & -1 & 0 \\ 0 & 0 & 0 & -1 \end{pmatrix}. \tag{5.2}$$

Here, and for the rest of this chapter, we set $c = 1$ (as soon as we need to introduce it, we will also set Newton's $G = 1$). As the form of the definition in equation 5.1 makes plain, proper time is a Lorentz invariant, which accounts for much of its utility. It is also very useful because it can be interpeted as the elapsed time in the particle's rest frame. In that frame, $\vec{\xi} \equiv 0$, so an evaluation of equation 5.1 always yields an identity between proper time and coordinate time in that frame.

The simplest formulation of classical particle dynamics is in terms of Newton's Second Law:

$$m\frac{d^2\vec{\xi}}{dt^2} = \vec{F}, \tag{5.3}$$

where m is the particle's (rest) mass and the relevant forces are given by $\vec{F}$. Suppose (to simplify still further) that $\vec{F} = 0$. In the particle's rest frame, the four-vector version of Newton's Second Law,

$$\frac{d^2\xi^\mu}{ds^2} = 0, \tag{5.4}$$

is trivially true because coordinate time in that frame is identical to proper time. However, this equation is now manifestly in covariant form and so must also be correct in *any* inertial frame.

One of the key ideas of general relativity is the indistinguishability of inertial and gravitational mass. This fact has the consequence that, for any gravitational field, we can always find *local* coordinate systems in which there is no apparent gravity. These are called *freely falling* or *inertial* frames because they accelerate with the acceleration of gravity. Objects observed relative to them show no apparent response to gravity because the frame itself is falling in exactly the same way. We can therefore reinterpret equation 5.4 as a *general* statement that governs motion equally well in the presence or absence of gravity. The only restrictions to its validity are that any nongravitational forces are negligible, and it must be applied in an inertial frame.

However, we cannot restrict ourselves to only thinking about physics in freely falling frames. We often prefer to work in some other coordinate frame, a frame, for example, in which some important object (like a massive black hole at the center of an AGN) is at rest, but other objects (e.g., accreting matter) are not. In that case we need to learn how to transform the particle's coordinates from ξ^μ, as they are measured in the freely falling frame, to x^μ, as they would be found in the new frame. Following the usual chain-rule prescription, we write

$$
\begin{aligned}
\frac{d^2\xi^\mu}{ds^2} &= \frac{d}{ds}\left(\frac{d\xi^\mu}{dx^\nu}\frac{dx^\nu}{ds}\right) \\
&= \frac{d\xi^\mu}{dx^\nu}\frac{d^2x^\nu}{ds^2} + \frac{\partial^2\xi^\mu}{\partial x^\nu \partial x^\kappa}\frac{dx^\nu}{ds}\frac{dx^\kappa}{ds}.
\end{aligned}
\tag{5.5}
$$

Multiplication by $dx^\kappa/d\xi^\mu$ reduces the right-hand side of equation 5.5 to the form

$$
0 = \frac{d^2x^\kappa}{ds^2} + \Gamma^\kappa_{\mu\nu}\frac{dx^\mu}{ds}\frac{dx^\nu}{ds},
\tag{5.6}
$$

where we have made use of the identity $dx^\kappa/dx^\nu = \delta^\kappa_\nu$, the Kronecker δ. The quantity $\Gamma^\kappa_{\mu\nu}$ is called the *affine connection* and describes the transformation required from the freely falling ξ frame to the different frame in which the four-vector is x. Although in general this extra term incorporates all the peculiarities of four-dimensional space-time curvature that can be present in general relativity, it also has a much simpler physical interpretation. It can be seen as the generalization of the extra terms that enter the equation of motion when it is expressed in any system of coordinates in which there is either curvature, for example, spherical coordinates, or acceleration, for example, a rotating frame in which centrifugal and Coriolis force act. Further tensor manipulations give the relation between the affine connection and the metric:

$$
\Gamma^\sigma_{\kappa\mu} = \frac{1}{2}g^{\nu\sigma}\left[\frac{\partial g_{\mu\nu}}{\partial x^\kappa} + \frac{\partial g_{\kappa\nu}}{\partial x^\mu} - \frac{\partial g_{\mu\kappa}}{\partial x^\nu}\right].
\tag{5.7}
$$

Just as for classical mechanics, relativistic particle mechanics can also be derived more elegantly using the technique of Lagrangians. In nonrelativistic mechanics, the Lagrangian $\mathcal{L}$ is the kinetic energy of a system minus its potential energy, and its integral with respect to time along whatever trajectory the system follows is called the action. The fundamental principle of the Lagrangian method is that the correct path is found by searching for the one that produces an extremal value of the action. This path can be found explicitly by solving the Euler-Lagrange equations

$$
\frac{d}{dt}\frac{\partial\mathcal{L}}{\partial\dot{x}_i} = \frac{\partial\mathcal{L}}{\partial x_i},
\tag{5.8}
$$

where $i = 1, 2, \ldots$ labels the system's degrees of freedom, and the coordinate of each is x_i.

To discover the relativistic generalization of the Lagrangian method, we begin by noting that the path that is ultimately chosen exists independent of the frame from which we view it. If it is determined by a minimization, the quantity minimized must then be a Lorentz invariant. The question is, "Which?" The most obvious candidate is the proper time. If we wished to tie our calculation as tightly as possible to the classical limit, we would make the Lagrangian the proper time times the rest mass. However, because all gravitational trajectories are independent of the mass of the particle, we work instead with the Lagrangian per unit mass, that is, the proper time alone. Let us now test whether finding an extremum for the proper time yields the correct equations of motion.

The proper time to go from A^μ to B^μ is

$$S = \int_A^B d\lambda \left[g_{\mu\nu} \frac{dx^\mu}{d\lambda} \frac{dx^\nu}{d\lambda} \right]^{1/2}, \tag{5.9}$$

where λ is a parameter describing the path. We have tremendous freedom in choosing λ; all that is really necessary is that $\lambda_A \neq \lambda_B$ if $A \neq B$, and that λ change monotonically from A to B. It is often convenient to choose λ to be a Lorentz invariant, but it is not strictly necessary.

To locate an extremum of S we vary the paths from A to B with the proviso that A and B are fixed, and search for that trajectory such that $\delta S = 0$. That is, we compute the variation

$$\delta S = - \int_A^B \frac{d\lambda}{2} \left[g_{\mu\nu} \frac{dx^\mu}{d\lambda} \frac{dx^\nu}{d\lambda} \right]^{-1/2} \left[\frac{\partial g_{\mu\nu}}{\partial x^\alpha} \frac{dx^\mu}{d\lambda} \frac{dx^\nu}{d\lambda} \delta x^\alpha + 2 g_{\mu\nu} \delta \left(\frac{dx^\mu}{d\lambda} \right) \frac{dx^\nu}{d\lambda} \right] \tag{5.10}$$

and then find $\delta x^\alpha(\lambda)$ and $\delta(dx^\mu/d\lambda)$ such that $\delta S = 0$. The factor of two in the second term on the right-hand side of equation 5.10 comes from the fact that $g_{\mu\nu}$ is symmetric with respect to interchange of the μ and ν indices. The multiplicative factor in the integrand is just $ds/d\lambda$, so we have

$$\delta S = - \int_A^B ds \left[\frac{1}{2} \frac{\partial g_{\mu\nu}}{\partial x^\alpha} \frac{dx^\mu}{ds} \frac{dx^\nu}{ds} \delta x^\alpha + g_{\mu\nu} \delta \left(\frac{dx^\mu}{ds} \right) \frac{dx^\nu}{ds} \right]. \tag{5.11}$$

The second term in the integral can be integrated by parts. Because there is no variation of the end points, the variation in S becomes

$$\delta S = - \int_A^B ds \left[\frac{1}{2} \frac{\partial g_{\mu\nu}}{\partial x^\alpha} \frac{dx^\mu}{ds} \frac{dx^\nu}{ds} - \frac{\partial g_{\alpha\nu}}{\partial x^\mu} \frac{dx^\mu}{ds} \frac{dx^\nu}{ds} - g_{\alpha\nu} \frac{d^2 x^\nu}{ds^2} \right] \delta x^\alpha. \tag{5.12}$$

Setting $\delta S = 0$ and using equation 5.7 transforms equation 5.12 into

$$-\int_A^B ds \, g_{\alpha\nu} \left[\frac{d^2 x^\nu}{ds^2} + \Gamma^\nu_{\mu\beta} \frac{dx^\mu}{ds} \frac{dx^\beta}{ds} \right] \delta x^\alpha = 0. \qquad (5.13)$$

Because $\delta x^\alpha(s)$ is a function with arbitrary values at each s, the integrand must be equal to zero everywhere. Successfully reproducing equation 5.6 demonstrates that the proper time is, indeed, the correct action density.

In the nonrelativistic limit, the Lagrangian is the coefficient of dt in the differential action. Comparison with equation 5.9 then makes it clear that the general relativistic Lagrangian is

$$\mathcal{L} = \left[g_{\mu\nu} \frac{dx^\mu}{d\lambda} \frac{dx^\nu}{d\lambda} \right]^{1/2}. \qquad (5.14)$$

By direct analogy with the nonrelativistic case, the actual trajectory is found by solving the Euler-Lagrange equations, which are now written as

$$\frac{d}{d\lambda} \left(\frac{\partial \mathcal{L}}{\partial \dot{x}^\mu} \right) = \frac{\partial \mathcal{L}}{\partial x^\mu}, \qquad (5.15)$$

where $\dot{x}^\mu \equiv dx^\mu/d\lambda$. The derivatives of the Lagrangian with respect to $\dot{x}^\mu$ are the canonical momenta

$$p_\mu \equiv \frac{\partial \mathcal{L}}{\partial \dot{x}^\mu}. \qquad (5.16)$$

Note that because they are found by differentiating a scalar, they are covariant vectors. Their associated contravariant forms are given by $p^\mu = g^{\mu\nu} p_\nu$.

When we follow a particle with nonzero rest mass, we might as well choose $\lambda = s$. It is then obvious that the Lagrangian is constant along worldlines (the partial derivatives that go into the Euler-Lagrange equations are *across* the worldline, so they are not necessarily zero). However, this choice of λ would not be satisfactory for zero rest-mass particles, for which the accumulated proper time is always zero. For them, it is important to find a different parameter λ. However, even in this case, it is possible to choose a λ that maintains the constancy of $\mathcal{L}$ (these special worldline parameters are called *affine parameters*). When we do so, the same dynamical equations are derived if we define an effective Lagrangian

$$\mathcal{L}_{\text{eff}} = \frac{1}{2} g_{\mu\nu} \dot{x}^\mu \dot{x}^\nu. \qquad (5.17)$$

5.1.2 Particle motion in a spherically symmetric potential

We are now ready to solve for particle orbits in a specific potential. Suppose the metric is spherically symmetric. Using the proper time to illustrate its components, we can write it as

$$ds^2 = g_{tt}dt^2 - g_{rr}dr^2 - g_{\theta\theta}d\theta^2 - g_{\phi\phi}\sin^2\theta d\phi^2, \qquad (5.18)$$

where all the elements of the metric are dependent on only r and t and we adopt a sign convention in which all the coefficients of differentials in equation 5.18 are positive. We also assume that the region in which there is significant gravity is bounded, so that $g_{\mu\nu} \to \eta_{\mu\nu}$ at infinity. The corresponding effective Lagrangian is

$$\mathcal{L}_{\text{eff}} = \frac{1}{2}\left(g_{tt}\dot{t}^2 - g_{rr}\dot{r}^2 - g_{\theta\theta}\dot{\theta}^2 - g_{\phi\phi}\sin^2\theta\dot{\phi}^2\right). \qquad (5.19)$$

Each coordinate's time derivative is taken with respect to the parameter λ.

Because the effective Lagrangian has no explicit dependence on ϕ, the associated canonical momentum p_ϕ is conserved. This is, of course, the angular momentum component parallel to the polar axis:

$$p_\phi = -g_{\phi\phi}\sin^2\theta\dot{\phi} \equiv L.$$

Note, however, that L is only equal to the *locally measured* angular momentum about the polar axis when the local measurement is made at infinity.

In addition, because we are now dealing with a set of *four* Euler-Lagrange equations, there is another conserved quantity if the metric is stationary with respect to t:

$$p_t = \frac{\partial \mathcal{L}_{\text{eff}}}{\partial \dot{t}} = g_{tt}\dot{t}. \qquad (5.20)$$

Because p_t is the energy-component of the conserved four-momentum, this quantity can be interpreted as the *energy at infinity* E_∞. To find the *locally measured* value of the energy, we must project the t-component of the momentum four-vector onto the unit four-vector in the t-direction, $\hat{t}^\mu$. By definition, $\hat{t}^\mu \cdot \hat{t}^\mu = \hat{t}^\mu g_{\mu\nu}\hat{t}^\nu$, so in components, $\hat{t}^\mu = (1/\sqrt{g_{tt}},0,0,0)$. That means the locally measured energy is

$$p_\mu\hat{t}^\mu = \frac{p_t}{\sqrt{g^{tt}}} = \frac{E_\infty}{\sqrt{g_{tt}}}. \qquad (5.21)$$

Equation 5.21 approaches an identity at infinity as the metric becomes closer and closer to the flat-space metric.

The equation of motion for θ is

$$\frac{d}{d\lambda}\left(-g_{\theta\theta}\dot{\theta}\right) = g_{\phi\phi}\sin\theta\cos\theta\dot{\phi}^2 = \frac{\cos\theta}{\sin^3\theta}\frac{L^2}{g_{\phi\phi}}. \tag{5.22}$$

If we consider motions initially in the equatorial plane ($\theta = \pi/2$ and $\dot{\theta}(0) = 0$), then they stay in the plane. Given the symmetry of the problem we have set—we have the freedom to define the polar direction as the axis perpendicular to the initial velocity and initial acceleration—this should come as no surprise. This fact also allows us to simply set $\theta = \pi/2$ and ignore θ hereafter.

The equation of motion for r could be obtained from the Euler-Lagrange equation in r, but it is simpler to arrive at it by making use of the fact that the invariant magnitude squared of the four-velocity ($u^2 \equiv g_{\mu\nu}\dot{x}^\mu\dot{x}^\nu$) is 1 for particles with nonzero rest-mass and 0 for zero rest-mass particles. Then

$$\dot{r}^2 = \frac{1}{g_{rr}}\left[\frac{E_\infty^2}{g_{tt}} - u^2 - \frac{L^2}{g_{\phi\phi}}\right], \tag{5.23}$$

where the components entering the four-velocity have been written in terms of conserved quantities.

5.1.3 Motion in the Schwarzschild metric

Of course, the most interesting spherically symmetric metric for our purposes is that corresponding to a time-steady nonrotating black hole. It is called the Schwarzschild metric in honor of Karl Schwarzschild, who found this solution of the Einstein field equations (Schwarzschild 1916) immediately after their publication. He performed the work while hospitalized with the autoimmune disease from which he was soon to die.

Having set $G = c = 1$, we see that masses are equivalent to lengths (and therefore times) via the relations

$$l_o = 1.47\frac{M}{M_\odot}\,\text{km} = t_o = 4.9 \times 10^{-6}\frac{M}{M_\odot}\,\text{s}, \tag{5.24}$$

so that the units of length and time, l_o and t_o, are proportional to the mass M. For this reason, we call the natural unit of length near black holes the *gravitational radius* $r_g \equiv GM/c^2$. Expressed in these units, the Schwarzschild metric describing space-time around a point mass of mass M is

$$g_{\mu\nu} = \begin{pmatrix} (1 - 2M/r) & 0 & 0 & 0 \\ 0 & -(1 - 2M/r)^{-1} & 0 & 0 \\ 0 & 0 & -r^2 & 0 \\ 0 & 0 & 0 & -r^2\sin^2\theta \end{pmatrix}. \tag{5.25}$$

Although the g_{rr} element of this metric formally diverges at $r = 2M$ (the *Schwarzschild radius*), no locally measured physical quantity diverges at this point. Indeed, other coordinate systems can be defined in which the metric for space surrounding a point mass is perfectly regular everywhere, including at $r = 2M$ (see, e.g., the discussion in Shapiro and Teukolsky 1983, §12.6). True divergences are encountered at $r = 0$, but, as we shall see, these are screened from outside view.

Using this metric, we find that the radial motion equation for particles of nonzero rest mass is

$$\dot{r}^2 = E_\infty^2 - \left(1 - \frac{2M}{r}\right)\left(1 + \frac{L^2}{r^2}\right). \tag{5.26}$$

Similarly, photon radial motions are described by

$$\dot{r}^2 = E_\infty^2 - \frac{L^2}{r^2}\left(1 - \frac{2M}{r}\right). \tag{5.27}$$

Although this would not be a correct physical interpretation, we may analyze both equations as if they were classical equations of motion for particles of energy $E_* = E_\infty^2/2$ in an effective potential

$$V(r) = \begin{cases} (1/2)(1 - 2M/r)[1 + (L/r)^2] & \text{nonzero rest mass} \\ (1/2)(1 - 2M/r)(L/r)^2 & \text{zero rest mass .} \end{cases} \tag{5.28}$$

It is instructive to compare the nonzero rest-mass effective potential to the Newtonian one: $V_N(r) = -M/r + j^2/(2r^2)$, where j is the angular momentum per unit mass. The relativistic version can be written as $V(r) = V_N(r) + 1/2 - Mj^2/r^3$; that is, it contains the Newtonian effective potential but adds two terms. One is simply a constant, which may be seen as reflecting the rest mass, but as a constant, has no effect on the motion. All the relativistic effects end up being confined within the last term $\propto r^{-3}$.

5.1.3.1 gravitational redshift

To get a feel for the meaning of these equations of motion, let us consider several simple cases. First consider the relation between coordinate time and proper time for a stationary observer. Because the differential proper time $ds = \sqrt{g_{tt}}dt$, we have

$$ds = \sqrt{1 - \frac{2M}{r}}\,dt. \tag{5.29}$$

As $r \to 2M$ from outside, the ratio between the proper time seen by an observer near the black hole and the proper time (which in the limit of large

r coincides with coordinate time) measured by an observer at large distance becomes smaller and smaller. That is, events that take a finite amount of time as measured near the black hole appear to take divergently long times at large distance. This effect is called the *gravitational redshift*. An atom (miraculously) stationary just outside $r = 2M$ radiates photons of the normal quantum mechanical frequency in its frame, but their oscillations appear to be extremely slow as viewed by an outside observer.

The function $\sqrt{g_{tt}}$ is also called the *lapse function* (usually written as α) because it describes the ratio between the elapse of apparent time at a particular location and the elapse of coordinate time, usually defined in the weak-field, flat-space region at infinity.

It also has yet another physical interpretation. The gravitational acceleration $\vec{g} = -\nabla \ln \alpha$, so that $\ln \alpha$ becomes, in the weak-field limit, the ordinary Newtonian gravitational potential. We shall return to this alternate interpretation later (§5.1.4.3).

5.1.3.2 photon trajectories

Next consider photon trajectories. Let $b = L/E_\infty$; that is, it is the impact parameter. Then the radial equation of motion (eq. 5.27) becomes

$$\dot{r}^2 = E_\infty^2 \left[1 - \frac{b^2}{r^2} \left(1 - \frac{2M}{r} \right) \right]. \qquad (5.30)$$

Clearly, we can always rescale our worldline parameter to remove the dependence on E_∞. What is left depends only on b, and can be interpreted as a quasi-effective potential relation in which the photon "energy" has unit value. The implications of this effective potential are revealed in figure 5.1.

As can be seen from this figure, the shape of the effective potential is independent of b, but its amplitude is $\propto b^2$; in other words, it becomes more of a barrier as b grows. Its peak is always at $r = 3M$, where the value of V_{eff} is $b^2/(27M^2)$. Thus, if a photon starts out at large radius and is directed toward the black hole with $b > 3\sqrt{3}M$, it always encounters a turning point at some $r \geq 3M$. Photons with smaller impact parameters, however, can be captured.

On the other hand, to determine which photons radiated from near the black hole can escape, we must also consider their directions. Let us describe the initial direction by the angle Ψ between the photon's initial path and the outward radial direction. Because all photons everywhere travel with unit velocity, $v^\phi = \sin \Psi$ and $v^r = \cos \Psi$. To relate v^ϕ to b,

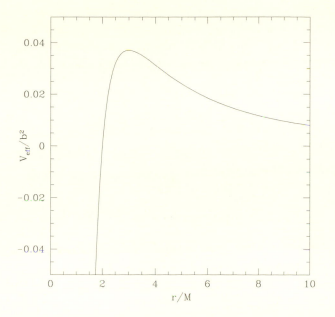

Fig. 5.1 The photon effective potential in the Schwarzschild metric, scaled by b^2.

we write v^ϕ as the ratio between the differential proper length the photon moves in the azimuthal direction and the differential proper time $d\tau$ that motion took, where both measurements are taken by a stationary observer at the place where the photon was emitted:

$$v^\phi = \frac{\sqrt{g_{\phi\phi}}}{\sqrt{g_{tt}}}\frac{d\phi}{dt} = \frac{r}{\sqrt{1-2M/r}}\frac{d\phi}{dt}. \tag{5.31}$$

From the definition of L we have $d\phi/ds = L/r^2 = bE_\infty/r^2$, where ds is the differential worldline parameter. The definition of E_∞ is $E_\infty = g_{tt}dt/ds$, so $d\phi/dt = (d\phi/ds)(ds/dt) = (bE_\infty/r^2)(g_{tt}/E_\infty)$ and

$$\sin\Psi = v^\phi = \frac{b}{r}\sqrt{1-\frac{2M}{r}}. \tag{5.32}$$

For inward-directed photons emitted outside $r = 3M$, $b < 3\sqrt{3}M$ spells capture, so the criterion on $\sin\Psi$ for escape translates to

$$\sin\Psi > \frac{3\sqrt{3}M}{r}\sqrt{1-\frac{2M}{r}}. \tag{5.33}$$

For outward-directed photons emitted from $r > 3M$, escape is automatic. Inside $r = 3M$, photons must be directed sufficiently close to purely radially outward, that is, have impact parameters sufficiently *small*, to clear the hump in the effective potential. Thus, for outward-directed photons radiated inside $3M$, the escape criterion is

$$\sin \Psi < \frac{3\sqrt{3}M}{r}\sqrt{1 - \frac{2M}{r}}. \tag{5.34}$$

Finally, photons directed inward from any $r < 3M$ are surely captured. These results are summarized in table 5.1.

Table 5.1: The Criteria for Photon Escape

Direction	$r < 3M$	$r > 3M$
Inward	Certain capture	$\sin \Psi > \frac{3\sqrt{3}M}{r}\sqrt{1 - \frac{2M}{r}}$
Outward	$\sin \Psi < \frac{3\sqrt{3}M}{r}\sqrt{1 - \frac{2M}{r}}$	Certain escape

When $r \leq 2M$, there is *no* direction, even straight outward, for which photons escape. This is the meaning of the *event horizon*, the surface from within which no photons ever leave. Note that it coincides with the surface of infinite redshift. It is for this reason that the true singularity at $r = 0$ is guarded from external observation.

5.1.3.3 particle trajectories

We now return to consideration of the dynamics of nonzero rest-mass particles. Let us begin with the very simplest case, a particle initially at rest at infinity. It therefore has no angular momentum, and $E_\infty = 1$ (i.e., its energy is purely due to its own rest mass). From equation 5.26, we see that the rate at which it plunges into the black hole *as measured with respect to its own proper time* is

$$\dot{r} = -\sqrt{E_\infty^2 - \left(1 - \frac{2M}{r}\right)} = -\sqrt{\frac{2M}{r}}. \tag{5.35}$$

Thus, relative to its own internal clock, the time required to fall all the way in depends primarily on how far away it started:

$$\Delta s = \frac{2/3}{\sqrt{2M}} \left[r_o^{3/2} - (2M)^{3/2} \right]. \tag{5.36}$$

On the other hand, the Schwarzschild coordinate time, which is the same as the time measured by a distant observer, is quite different. For this example, proper time is related to coordinate time by

$$ds^2 = \left(1 - \frac{2M}{r} \right) dt^2 - \frac{dr^2}{1 - 2M/r}. \tag{5.37}$$

Solving for dt in terms of ds (and using eq. 5.35 for dr/ds) gives

$$\frac{dr}{dt} = \sqrt{\frac{2M}{r}} \left(1 - \frac{2M}{r} \right). \tag{5.38}$$

In other words, as the particle approaches $r = 2M$, its *apparent* speed becomes slower and slower. As viewed by a distant observer, infinite time is actually required for the particle to reach the event horizon.

Next consider orbits with nonzero angular momentum. It is immediately obvious that for any value of L, $V_{\text{eff}} = 0$ at $r = 2M$ and tends toward $1/2$ at large r. However, as L increases, the shape of V_{eff} changes from a curve that rises monotonically to one that has both a maximum and a minimum before reaching its asymptotic limit (see fig. 5.2). The positions of these extrema are given by

$$r_m = \frac{1}{2} \left(\frac{L}{M} \right)^2 \left[1 \pm \sqrt{1 - 12(M/L)^2} \right]. \tag{5.39}$$

From this expression, it is clear that the extrema appear when $L \geq \sqrt{12}M$. The minus sign gives the position of the maximum, the plus sign the position of the minimum. When $L \gg M$, the dynamics are in the classical limit, and the connection between radius and angular momentum for a circular orbit is the familiar Newtonian one.

Just as for classical motion in an effective potential, circular orbits occur when the energy exactly matches the value of the effective potential at a place where the effective potential is a minimum. Consequently, there are *no* stable orbits for $L < \sqrt{12}M$, and the smallest radius for which a stable orbit exists, called the *marginally stable orbit* is $r_{ms} = 6M$. Any

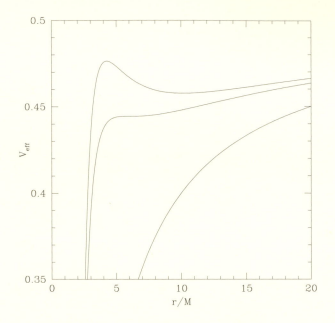

Fig. 5.2 The effective potential for nonzero rest-mass particles in the Schwarzschild metric. The lowest curve is for $L = 0$, the middle curve has the critical value $L = \sqrt{12}$, and the top curve has an angular momentum at infinity slightly greater than critical, $L = 3.8$.

particle attempting to move in a circle inside that radius must fail, and fall inward.

This result has an immediate consequence of fundamental importance for accretion models. If accretion is to be maximally efficient, then of course the particles must somehow lose as much energy as possible by radiation before disappearing into the black hole. The least energy orbit for any given angular momentum is always a circle, so maximally efficient accretion proceeds by a succession of (nearly) circular orbits as the particles lose their angular momentum, and with it, their energy. But because there is a smallest stable circular orbit, we expect that energy loss will be very inefficient for particles of smaller angular momentum than that associated with this orbit. The reason is that if no stable circular orbit exists, they will fall at nearly c and reach $r \leq 2M$ before they have time to radiate. As Bardeen, Press, and Teukolsky (1972) showed, the acceleration involved in the gravitational plunge is not enough to cause significant electromagnetic radiation, as measured on the scale of the particle rest mass. In addition, of course, any photons emitted that close to the event horizon have a significant probability of being captured. Therefore, the maximum accretion efficiency for material falling into a Schwarzschild black hole is given by the binding energy at the smallest stable orbit. We find this binding energy by setting $E_*(6M) = V_{\text{eff}}(6M) = 4/9$. The energy at infinity associated with this pseudo-energy is $E_\infty = \sqrt{2E_*} = \sqrt{8/3}$. The binding energy is the difference between the rest mass of the particle and its energy at infinity: $E_b = 1 - \sqrt{8/3} = 0.057$. Thus, the maximum radiative efficiency of accretion for particles falling into a Schwarzschild black hole is 5.7% in rest mass units. Note that because we have computed the energy at infinity, we have already accounted for the gravitational redshift of escaping photons.

Amusingly, some of these results can be obtained from a Newtonian analysis using a pseudo-potential cooked up to mimic general relativistic dynamics (Paczyński and Wiita 1980). Suppose that we replace the real potential with $V_{\text{fake}} = -M/(r - 2M)$, and examine the motion of particles with angular momentum L. Under the rules of Newtonian mechanics, we could treat their motion as if it occurred in an effective potential

$$V_* = -\frac{M}{r - 2M} + \frac{L^2}{2r^2}. \tag{5.40}$$

Circular orbits with angular momentum L would take place at radii such that $dV_*(L)/dr = 0$, but they become unstable inside the point where $d^2V_*(L)/dr^2 = 0$. Computing these two derivatives and requiring both to be zero yields the result that the marginally stable orbit with respect to V_{fake} also occurs at $r = 6M$. Moreover, the binding energy of a circular orbit at that radius is $1/16$, only 10% greater than the correct figure. These coincidences make it possible to do "Newtonian" calculations to describe approximately what in reality is a strongly relativistic situation.

5.1.4 Motion in the Kerr metric

Life gets more complicated when the black hole rotates. In that case the metric is no longer spherically symmetric; in fact, in the most "natural" coordinates (we will explain momentarily what that means), they are not even diagonal.

Suppose, then, that we have a black hole whose angular momentum $J = aM$, and let us choose a coordinate system that goes over smoothly at infinity to ordinary flat-space spherical coordinates. In this system, called

Boyer-Lindquist coordinates, the differential proper time is

$$ds^2 = \left(1 - \frac{2Mr}{\Sigma}\right)dt^2 + \frac{4aMr\sin^2\theta}{\Sigma}dtd\phi - \frac{\Sigma}{\Delta}dr^2$$
$$- \Sigma d\theta^2 - \left(r^2 + a^2 + \frac{2Mra^2\sin^2\theta}{\Sigma}\right)\sin^2\theta d\phi^2,$$

(5.41)

where

$$\Sigma \equiv r^2 + a^2\cos^2\theta$$
$$\Delta \equiv r^2 - 2Mr + a^2,$$

(5.42)

and the polar direction of the coordinates coincides with the direction of the angular momentum. Note that this reduces to the Schwarzschild metric in the limit $a \to 0$. Here a is bounded above by M because any additional angular momentum would increase the energy of the black hole and, therefore, its mass.

We may now ask, "How large is a likely to be?" If the mass of the black hole is accumulated in a fashion that is nearly spherically symmetric, a might stay small compared to M. Stellar collapse to a black hole might occur this symmetrically, so the initial black hole formed in a galactic nucleus might indeed have $a \ll M$. However, subsequent accretion (as we shall argue in Chap. 7) is much more likely to be through a thin disk. When this is so, and the axis of the disk remains stable over long times, the angular momentum of the black hole builds up rapidly. In the initial stages, when the Schwarzschild metric is still a good approximation, the angular momentum per unit mass with which material arrives is just the angular momentum per unit mass of the last stable orbit, $\sqrt{12}M$. As a rises toward M, the angular momentum per unit mass of the last stable orbit slowly falls toward M. Thus, the added angular momentum per unit mass is always $\sim M$. Therefore, once the black hole mass has been doubled by thin disk accretion, its angular momentum is *automatically* $\sim M^2$, and $a \sim M$. Further accretion simply pushes a closer and closer to M. However, in realistic disks, although a is likely to approach very close to M, it is unlikely to quite reach it. Some of the photons carrying away the energy released by accretion travel on negative angular momentum orbits. These are more likely to be captured by the black hole than those on positive angular momentum orbits. As a result, depending on the exact angular distribution of the radiated photons, a is limited to $\simeq 0.998M$ (Thorne 1974).

5.1.4.1 the ergosphere

A curious thing happens very close to a rotating black hole—the very notion of a stationary observer becomes suspect! Consider, for example, a particle following a circular orbit with fixed r and θ, and angular azimuthal frequency Ω. The magnitude of its four-velocity is

$$u^\mu u_\mu = u^\mu g_{\mu\nu} u^\nu = \left(u^t\right)^2 \left[g_{tt} + 2g_{t\phi}\Omega + \Omega^2 g_{\phi\phi}\right]. \qquad (5.43)$$

Here we have made use of the fact that $\Omega = d\phi/dt = u^\phi/u^t$. Because $u^\mu u_\mu = p^\mu p_\mu/m^2$, we would expect that the expression in the square brackets should be positive, but this can only be if Ω lies between its two roots, or

$$\frac{1}{g_{\phi\phi}}\left[-g_{t\phi} - \sqrt{g_{t\phi}^2 - g_{tt}g_{\phi\phi}}\right] < \Omega < \frac{1}{g_{\phi\phi}}\left[-g_{t\phi} + \sqrt{g_{t\phi}^2 - g_{tt}g_{\phi\phi}}\right]. \qquad (5.44)$$

When $a = 0$, these restrictions simplify to just requiring particles to move slower than c. However, when $a \neq 0$, the limits are more interesting. In particular, the lower limit on Ω becomes greater than zero when $g_{tt} = 0$, or

$$r \leq M + \sqrt{M^2 - a^2\cos^2\theta}. \qquad (5.45)$$

This location is called the *static limit* because observers with $\Omega = 0$ are forbidden inside this surface. In other words, this close to a rotating black hole, there is no such thing as a stationary observer. Every physically realizable frame of motion must rotate. In a loose way, we can think of the rotating black hole as "dragging" frames of motion along as it spins.

Inside the static limit, a different form for the metric allows a more physically transparent picture. The issue, of course, is that no observer can stand still, so the usual definition of proper time, that is, the lapse of time for an observer who doesn't move, can't be used. To solve this problem, Bardeen et al. (1972) invented a new form, called the *locally nonrotating frame* metric, in which the necessary rotation is automatically taken care of by building it into the motion of the observer. We can then define a new azimuthal angle ϕ' that is appropriate to an observer in a circular orbit of frequency Ω by setting $d\phi' = d\phi - \Omega dt$. For the special value $\Omega_o = -g_{t\phi}/g_{\phi\phi}$, $L = 0$ (hence the acronym "ZAMO," for "Zero Angular Momentum Observer"), and

$$ds^2 = \frac{\Sigma\Delta}{A}dt^2 - \frac{A}{\Sigma}\sin^2\theta\,(d\phi')^2 - \frac{\Sigma}{\Delta}dr^2 - \Sigma d\theta^2. \qquad (5.46)$$

The new function A is

$$A = (r^2 + a^2)^2 - a^2\Delta\sin^2\theta. \qquad (5.47)$$

The transformation between Boyer-Lindquist coordinates and locally nonrotating frame coordinates is accomplished by "completing the square" in the metric. A modest amount of algebra will confirm that the two metrics agree. Locally nonrotating coordinates are much more physically transparent than Boyer-Lindquist coordinates because they automatically display the true *locally* radial direction, that is, the local set of basis vectors incorporates the effects of frame-dragging. Note, however, that ZAMOs are not quite perfect solutions to the complications of the Kerr metric—their frames are not inertial, so no particle without an engine can follow a ZAMO orbit.

In these coordinates, the relation between proper time and coordinate time for "stationary" observers is well defined:

$$ds = \sqrt{\frac{\Sigma\Delta}{A}}\,dt. \tag{5.48}$$

Because Σ and A are both always greater than zero, the gravitational redshift diverges where $\Delta = 0$, or

$$r = M \pm \sqrt{M^2 - a^2}. \tag{5.49}$$

The outermost of these two surfaces also defines the event horizon. Thus, when $a \to 0$, the event horizon moves toward the Schwarzschild position of $r = 2M$, whereas when $a \to M$, the event horizon moves in to $r = M$. Interestingly, its radial coordinate r_h remains independent of polar angle even when the metric possesses only axial symmetry.

Comparing the event horizion to the static limit, we see that the oblate spheroid that is the static limit wraps around the spherical event horizon, with the two surfaces touching at $\theta = 0, \pi$. The region between them is called the *ergosphere*, and has some remarkable properties. First, of course, as we have already discussed, there can be no static observers in it. But more surprisingly, it is possible to have negative energy orbits, that is, particles can have trajectories that are so strongly bound that their total energy, including rest mass, can be negative. In the discussion leading up to equation 5.44, we assumed that the energy had to be positive; inside the ergosphere, this is not necessarily true, although it is true that orbits extending to infinity must have positive energy. Negative energy orbits occur when the angular momentum of the orbit is large and directed opposite the angular momentum of the black hole. Penrose (1969) suggested that the existence of negative energy orbits would permit a new way to extract energy from a black hole: a pair of particles could be created in the ergosphere, with one on an orbit with such large negative energy that the other could have enough positive energy to be unbound. As the negative energy particle enters the black hole (and therefore decreases its mass), the positive energy particle would carry energy off to infinity. Bardeen et al. (1972) showed, however, that this process is unlikely to be realized astrophysically because positive and negative energy orbits are separated by large velocity differences, and it would therefore be very difficult to create such a pair.

5.1.4.2 particle orbits

Just as for the Schwarzschild metric, particle orbits in the Kerr metric are most easily analyzed using the effective Lagrangian $\mathcal{L}_{\text{eff}} = (1/2)g_{\mu\nu}\dot{x}^{\mu}\dot{x}^{\nu}$. Once again, there are two constants of the motion, E_{∞} and L, that arise directly from symmetries of the Lagrangian. However, it turns out that there is another constant of the motion, uninteresting in the Schwarzschild limit, which seems to arise "accidentally" (Misner et al. 1973). This constant, called Carter's constant, is

$$Q = (p_{\theta})^2 + \cos^2\theta \left[a^2 \left(m^2 - E_{\infty}^2 \right) + \left(\frac{L}{\sin\theta} \right)^2 \right]. \tag{5.50}$$

Using these three constants and the fixed magnitude of the four-velocity, it is possible to find all four equations of motion. Because the Kerr metric has only cylindrical symmetry, orbits in different planes can have quite different behaviors (in fact, only when the orbit is in the equatorial plane does it stay in a single plane). In addition, the more complicated forms of the metric elements lead to rather lengthier algebra. Consequently, it will suffice for our purposes to merely quote results. A fuller sketch can be found in Shapiro and Teukolsky (1983) or in greater detail in Bardeen et al. (1972).

Several new points emerge in the consideration of circular orbits in the equatorial plane of a Kerr black hole. First, the innermost circular orbit (the *photon orbit*) actually lies well outside the event horizon. Only photons can follow the circular orbit at the limiting radius, whence the name:

$$r_{\text{ph}} = 2M \left\{ 1 + \cos \left[\frac{2}{3} \cos^{-1} \left(\mp a/M \right) \right] \right\}. \tag{5.51}$$

The upper sign refers to prograde orbits, the lower sign to retrograde. Note that $r_{\text{ph}}(a = 0) = 3M$, that is, this orbit lies right at the peak of the photon effective potential. At the other extreme in a,

$$r_{\text{ph}}(a = M) = \begin{cases} M & \text{prograde} \\ 4M & \text{retrograde} . \end{cases} \tag{5.52}$$

These orbits have a rather curious character, for they are unbound (positive energy) and therefore unstable (the fact that they fall at the peak of the effective potential makes this obvious in the $a \to 0$ limit). They do not even exist outside the marginally bound radius

$$r_{\text{mb}} = 2M \mp a + 2\sqrt{M(M \mp a)}. \tag{5.53}$$

As a function of a, r_{mb} has the limiting values $r_{\text{mb}}(a = 0) = 4M$ and

$$r_{\text{mb}}(a = M) = \begin{cases} M & \text{prograde} \\ 5.83M & \text{retrograde} \end{cases} \tag{5.54}$$

Because the marginally bound radius is the radius at which a circular orbit has zero binding energy, this is also the minimum periastron of all parabolic orbits.

Even bound circular orbits are not necessarily stable. The innermost radius at which they are stable is

$$r_{\text{ms}} = M \left\{ 3 + Z_2 \mp [(3 - Z_1)(3 + Z_1 + 2Z_2)]^{1/2} \right\}, \tag{5.55}$$

where

$$Z_1 = 1 + \left(1 - \frac{a^2}{M^2}\right)^{1/3} \left[\left(1 + \frac{a}{M}\right)^{1/3} + \left(1 - \frac{a}{M}\right)^{1/3}\right] \tag{5.56}$$

and

$$Z_2 = \left(3\frac{a^2}{M^2} + Z_1^2\right)^{1/2}. \tag{5.57}$$

For $a = 0$, as we have already seen, $r_{\text{ms}} = 6M$; for a maximally rotating black hole,

$$r_{\text{ms}}(a = M) = \begin{cases} M & \text{prograde} \\ 9M & \text{retrograde} \end{cases} \tag{5.58}$$

Finally, just as for Schwarzschild black holes, the most significant quantity required is the binding energy of the marginally stable circular orbit, the quantity that determines the maximal efficiency of energy release as a result of accretion. As a increases from 0 to M, this quantity increases from $1 - \sqrt{8}/3$ (as we already determined for Schwarzschild black holes) to $1 - 1/\sqrt{3}$, assuming prograde orbits. The increased efficiency (from 5.7% to 42%) results from the decrease in the size of the marginally stable orbit from $6M$ to M. If a is limited by photon capture to 0.998, the efficiency diminishes substantially, to $\simeq 30\%$ (Thorne 1974).

The alert reader may notice that in the limit $a \to M$, all the special radii we have identified—the event horizon, r_{ph}, r_{mb}, and r_{ms}—move toward $r = M$. Do they really all coincide in that limit? This apparent coincidence is another example of the delicacy required to interpret relativistic coordinates properly. All physical measurements require appropriate use of the metric and cannot be found by using unmodified coordinates. When the metric is employed in the correct way to find proper lengths, in fact these radii are all separated from one another by nonzero distances.

5.1.4.3 Lense-Thirring precession

Thorne and collaborators (Thorne, Price, and MacDonald 1986) have worked out a weak-field (post-Newtonian) expansion of relativistic gravity in which both the gravitational field and the particle motions follow equations having the form of classical electromagnetism. In this formulation, the gravitational field is described by four quantities, a scalar field Φ and a vector field (the "gravito-magnetic vector potential") $\vec{A}$. Just as in electromagnetism, the gravito-electric field is $\vec{g} = -\nabla \Phi$ and the gravito-magnetic field is $\vec{H} = \nabla \times \vec{A}$. The potentials are related to the metric by

$$\Phi = \frac{1 + g_{tt}}{2},$$
$$A_i = -g_{ti}, \tag{5.59}$$

where the spatial components are labeled by $i = 1, 2, 3$. Recall that $\alpha \equiv \sqrt{g_{tt}}$, so that we would have recovered the first of these expressions through the identification of $\ln \alpha$ with the gravitational potential we remarked upon earlier. When the metric is spherically symmetric (e.g., the Schwarzschild metric), $\vec{A} = 0$, so there is no gravito-magnetic field.

This formalism possesses several advantages. Physical insight, borrowed by analogy from electromagnetism, is certainly the foremost of them. However, it also simplifies the process of finding the gravitational fields from their sources. Instead of solving the full Einstein field equations to first order in the field strengths and speeds, one needs only to solve four pseudo-Maxwell equations:

$$\nabla \cdot \vec{g} = -4\pi\rho,$$
$$\nabla \cdot \vec{H} = 0,$$
$$\nabla \times \vec{g} = 0, \tag{5.60}$$
$$\nabla \times \vec{H} = -4\left(4\pi\rho\vec{v} - \frac{\partial \vec{g}}{\partial t}\right).$$

The only differences from the true Maxwell equations are several sign changes (because gravity is purely attractive), a change of the order unity coefficient in the "gravito-Ampére's law" (because gravity is a tensor force), and the missing $-\partial \vec{H}/\partial t$ term from the $\nabla \times \vec{g}$ equation. When the sources are time steady and motionless, the gravitational field reduces to the Newtonian limit.

The equations of motion are also found very simply. Particle motions follow an analogue of the Lorentz equation:

$$\frac{\partial \vec{v}}{\partial t} = \vec{g} + \vec{v} \times \vec{H}. \tag{5.61}$$

A spinning black hole acts as a source of gravito-magnetic field much as a spinning charge creates an ordinary magnetic field. The most important component is, of course, a dipole, whose effective dipole moment is simply aM. The gravito-magnetic field created by a Kerr black hole can then be approximated by

$$\vec{H} = \frac{aM}{r^3} \left[3(\hat{z} \cdot \hat{r})\hat{r} - \hat{z}\right], \tag{5.62}$$

where $\hat{z}$ and $\hat{r}$ are unit vectors parallel to the spin axis and the radial direction, respectively.

A magnetic dipole $\vec{\mu}$ embedded in a magnetic field $\vec{B}$ feels a torque $\vec{N} = \vec{\mu} \times \vec{B}$. In just the same fashion, a gravito-magnetic dipole (a circulating mass) feels a torque when it is exposed to a gravito-magnetic field:

$$\vec{N} = 2aM\hat{z} \times \frac{\vec{L}}{r^3}, \tag{5.63}$$

where $\vec{L}$ is the integrated angular momentum of a test mass in a circular orbit of radius r. This torque will cause the ring to precess unless $\vec{L}$ is aligned with the black hole spin.

If there is an entire disk of material orbiting at some inclination relative to the black hole's equator, the changing precession frequency as a function of radius will lead to friction between adjacent rings. Simple estimates suggest that the associated dissipation would then cause the matter to settle into the equatorial plane of the black hole within that radius where the inflow timescale is comparable to the precession period (Bardeen and Petterson 1975). However, the argument for this criterion is fairly crude, and it is possible that the true physics may be more complicated.

What is not controversial is that the innermost parts of such a disk (even if "innermost" is only vaguely defined) should align with the spin of the black hole, even if they do not trivially align (as would occur if the spin of the black hole arose through accretion of material with constant angular momentum direction). Consequently, spinning black holes have a powerfully stabilizing effect on any characteristic directions in AGNs. It is often suggested that the extraordinary alignment of radio jets over extremely long timescales (vide the infamous NGC 6251; see fig. 9.4) is due to this sort of stabilization.

5.2   Electromagnetic Fields

5.2.1   Introductory comments

In Chapter 7 we will discuss in detail the physical conditions of the accreting material. Here it suffices to say that this gas is almost certainly an ionized plasma with such high electrical conductivity that magnetic field lines are "frozen" into the fluid, that is, there is virtually no slippage of the plasma across the field (see Appendix B). For this reason, we can expect significant magnetic flux to be brought into the vicinity of these black holes, and it behooves us to explore what happens to it.

The discussion presented here will stay at a heuristic level. However, Thorne and collaborators (see Thorne et al. 1986) have demonstrated that these intuitive arguments can be developed on a rigorous basis. In fact, there is a formulation of the general relativistic physics of black holes (the "membrane paradigm") that allows real meaning to be attached to such apparently loose concepts as the resistance, viscosity, and temperature of a black hole. Those who wish to pursue this in greater depth are strongly urged to read the Thorne et al. book, which presents these arguments in a beautifully clear and insightful way.

Here we will merely quote a few key results from that analysis. Before citing these results, it is necessary to make a few preliminary remarks. First, it makes sense to measure electric and magnetic fields only in locally nonrotating frames; however, for the purposes of visualization, and the connection of events near the black hole to events at large distances, it is more convenient to describe things in terms of coordinates like the Boyer-Lindquist system. Therefore, we will be forced to flip back and forth between the two points of view.

Second, Maxwell's equations in this context are most conveniently expressed in integral form. Under these circumstances, they are

$$\int_S d\vec{S} \cdot \vec{E} = \int_V dV \, 4\pi \rho_e,$$

$$\int_S d\vec{S} \cdot \vec{B} = 0,$$

$$\int_C d\vec{l} \cdot \left(\vec{E} + \vec{v} \times \vec{B} \right) \alpha = -\frac{d}{dt} \int_S d\vec{S} \cdot \vec{B},$$

$$\int_C d\vec{l} \cdot \left(\vec{B} - \vec{v} \times \vec{E} \right) \alpha = \frac{d}{dt} \int_S d\vec{S} \cdot \vec{E} + 4\pi \int_S d\vec{S} \cdot \left(\vec{j} - \rho_e \vec{v} \right) \alpha.$$

(5.64)

The integration volumes, surfaces, and curves can in principle move in co-ordinate time t. When they do, appropriate derivatives must be computed. This is not an issue for the first two equations because they contain no explicit time dependence. However, for the latter two equations (the Faraday equation and the Ampére equation) such questions arise in most applications. In fact, one may think of the lapse functions appearing in the latter two equations as providing the appropriate conversion between coordinate time and proper time for locally nonrotating observers.

Next, we point out that the Faraday equation has a piece $\propto \vec{v} \times \vec{B}$. The velocity here is meant to be as measured by an observer in a locally nonrotating frame; if the integration curve is fixed in Boyer-Lindquist co-ordinate space, it moves with respect to locally nonrotating frames, and there is a significant effective electric field due to the motion with respect to the physical magnetic field. An analogous thing happens in the Ampére equation (but, of course, with opposite sign). In addition, in that equation one must be careful to separate out physical current from the mere motion of charge density caused by frame-dragging.

5.2.2   Surface resistance

Returning to the thread, we already know (although it hasn't been said explicitly) the boundary conditions for electromagnetic waves approaching a black hole: An event horizon is a perfect outgoing surface, in the sense that all EM waves approaching it from the outside are absorbed with zero reflection. Thus, an event horizon acts like free space with respect to the radiation of EM waves. Free space, for example, an open wave guide, has an impedance of 4π stat-ohm = 377 ohm, and so does an event horizon around a black hole. This statement can be shown to be precise (Thorne et al. 1986).

Another heuristic approach to deriving the resistance R_H of an event horizon comes from imagining what happens to the magnetic flux which is brought to it by accretion. The currents that supported this flux disappear inside the hole and are causally disconnected from the outside world, so these fields must decay. Because structures on the scale of the event horizon can only be destroyed by the propagation of a signal, and signal speeds are always limited by c, the characteristic decay time $t_d \sim M$. Faraday's law states that $\nabla \times \vec{E} = -\partial \vec{B}/\partial t$, so we may make the order of magnitude estimate $t_d \sim M|\vec{B}|/|\vec{E}|$, where we have also taken the characteristic lengthscale on which $\vec{E}$ varies to be $\sim M$. From Ohm's law we can estimate $|\vec{E}| \sim |\vec{J}|R_H M$, making $t_d \sim |\vec{B}|/(R_H|\vec{J}|)$. But we know from Ampére's law that $|\vec{B}| \sim 4\pi M|\vec{J}|$, so that finally we have $R_H \sim 4\pi$. Remarkably, this heuristic order of magnitude argument reproduces the exact result that follows from the argument of the previous paragraph.

5.2.3 The rotating black hole battery

We have also already seen that there is no such thing as a truly static observer once one gets close to a rotating black hole. Consequently, if there is any magnetic field in the vicinity of the black hole, there is at least some electric field in *every* observer's frame (Blandford and Znajek 1977). This is the reason why electromagnetic effects are generally more important around spinning black holes than in the vicinity of nonrotating black holes.

Imagine a very simple situation first, as shown in figure 5.3. Suppose that there are no bulk charges or currents outside the black hole, and that the magnetic field is constant in coordinate time. To find the voltage around a circuit, we use the relativistic Faraday equation,

$$\int_C d\vec{l} \cdot \vec{E}\alpha = -\int_C d\vec{l} \cdot \left(\alpha \vec{v} \times \vec{B}\right). \tag{5.65}$$

If the circuit in question follows the path shown by the dashed line in figure 5.3, only the portion near the black hole horizon contributes to the integral on the right-hand side of equation 5.65 because everywhere else either $\vec{v}$ is parallel to $\vec{B}$ or $\vec{B} = 0$. If the position of the path is stationary in Boyer-Lindquist coordinates, then $\vec{v}$ is exactly the frame-dragging velocity just outside the black hole. Thus, a rotating black hole embedded in a magnetic field appears to have a "battery" in its surface, and the magnitude of its EMF is $\sim r_h|\vec{B}| \sim 10^7 (M/M_\odot)(B/1 \text{ G})$ V. It is legitimate to think of the origin of this EMF as due to the frame-dragging motion across the imposed magnetic field, or as an interaction of the electromagnetic and gravitomagnetic fields. It is the latter, of course, that causes frame-dragging.

Fig. 5.3   A rotating black hole immersed in a (coordinate)-time-steady magnetic field. The Faraday equation path of integration is the dashed line.

Although frame-dragging is an essentially relativistic phenomenon, once enforced rotation is recognized, the rest of the physics is almost mundane. Ordinary generators work in basically the same way—conducting wires are forced to move through a magnetic field, creating a net EMF around the circuit.

Now we will add some new elements to the picture, specifically a resistive region at large radius from the black hole, where the currents must cross field lines, and some plasma so that there are charges available to move. By Ohm's law, the current I flowing through this loop is simply $(R_H + R_L)^{-1} \int_C d\vec{l} \cdot \alpha \vec{v} \times \vec{B}$, where R_L is the resistance of the "astrophysical load" at infinity. Both in the load and in the surface of the black hole, Ohmic heating dissipates energy at the usual rates $I^2 R_L$ and $I^2 R_H$. Not surprisingly, the energy for this comes from the rotational energy of the black hole, but in the case of the dissipational heating in the event horizon, the heat stays in the black hole, increasing its entropy.

114

We can estimate the magnitude of this power dissipation by supposing $R_L \sim R_H$. When the two are equal, the power dissipated in the load is maximal. If that is the case, $I^2 R_L \sim (r_h B)^2 / R_H \sim 3 \times 10^{19} (B/1 \text{ G})^2 (M/M_\odot)^2$ erg s^{-1}. Scaling this to AGN parameters, and estimating the magnetic field strength by supposing it to be comparable to the gas pressure near the black hole (see Chap. 7), we find $P \sim 3 \times 10^{45} (B/10^4 \text{ G})^2 (M/10^9 M_\odot)^2$ erg s^{-1}, a very respectable quantity.

Because all this energy is coming from the stored rotational energy of the black hole, we must ask how much rotational energy is available. The answer to this question is given by the black hole "area theorem" (Hawking and Ellis 1973), which states that the surface area of a black hole cannot decrease. In a beautiful analogy, this theorem can be identified with the Second Law of Thermodynamics, for we may interpret the black hole surface area as the black hole's entropy (Bekenstein 1973; Hawking 1974).

The area is simply

$$A = \int d\theta d\phi \sqrt{g_{\theta\theta} g_{\phi\phi}} = 8\pi M r_h = 8\pi M^2 \left[1 + \sqrt{1 - a/M}\right], \qquad (5.66)$$

so it may vary between $8\pi M^2$ (in the case of maximally rotating black holes) and $16\pi M^2$ (in the case of Schwarzschild black holes). Motivated by this relation, the *irreducible mass* is defined as $M_i \equiv (A/16\pi)^{1/2}$; no matter what happens, the black hole cannot do anything to reduce its energy, and therefore its mass, below M_i. Thus, the fraction of the black hole mass that is available to do work on outside material is $1 - M_i/M = 1 - [(1 + \sqrt{1 - a/M})/2]^{1/2}$, which for a maximally rotating black hole is $1 - 1/\sqrt{2} = 0.293$.

5.3   Photon Propagation

As astronomers, the only way we have to learn about what happens near black holes is to observe photons (or perhaps neutrinos) that make their way from the vicinity of a black hole to Earth. We can then hope to use the spectrum and other properties of these photons to infer something about what is going on in the source region near the black hole. Of course, there are dramatic differences between the (nearly) flat-space in which we live and the highly curved space-time around a black hole, so in interpreting the spectra of radiation produced near a black hole, we must be sure to allow for those contrasts in reference frame (Cunningham and Bardeen 1973; Cunningham 1975). Some of these contrasts may be thought of as special relativistic (the boosting and beaming due just to the fact that the atoms radiating the photons are moving relativistically), but there are also purely general relativistic effects (e.g., gravitational redshift and the curvature in the photons' trajectories).

Imagine that we ride along with the emitting material. A newly born photon's character can be completely specified by its energy $E^{(e)}$ (we will label emission frame quantities with superscript (e), observer frame quantities with superscript (o)), and (in a Kerr metric) two invariant constants of motion, $l = L/E_\infty$ and $q = \sqrt{Q}/E_\infty$. The rate at which photons are created varies from place to place, but possibly with some symmetry (e.g., disk emission may be independent of azimuthal angle or time). Thus, (at least) four variables—$E^{(e)}$, l, q, and one or more coordinates—are required to describe the photon output completely.

On the other hand, in most circumstances we are able to measure the dependence of the radiation on only two quantities—the observed energy $E^{(o)} = E_\infty$ and (possibly) the angle $\theta^{(o)}$ between our line-of-sight and some axis of symmetry. Consequently, the spectrum we see is usually an integration over at least two of the independent variables.

Specifically, the flux that we measure at energy $E^{(o)}$ is the product of the intensity we observe with the solid angle we see the source as subtending:

$$F_{\mathrm{E}}^{(o)} = \int d\Omega^{(o)} \, I_{\mathrm{E}}^{(o)}. \tag{5.67}$$

Because we know how the initial conditions of photons lead to their final state, we can always change the two angle variables in the integral of equation 5.67 to two of the other variables, provided we have a model for how the emitted intensity depends on the other independent parameters. Because we commonly do know something about how the emission might depend on position within the source, and on $E^{(e)}$, that is exactly how we will proceed.

First, however, we need to transform the intensity and the solid angle from our frame back to the emission frame. It is easy to transform intensities because the quantity I_{E}/E^3 is a relativistic invariant. That it is invariant follows directly from its physical interpretation—$I_{\mathrm{E}}/E^3 = 8\pi\mathcal{N}/(hc)^3$, where $\mathcal{N}$ is the photon occupation number per quantum state. Therefore,

$$I_{\mathrm{E}}^{(o)} = \delta^3 I_{\mathrm{E}}^{(e)}, \tag{5.68}$$

where $\delta = E^{(o)}/E^{(e)}$ is the Doppler shift between the frame in which the photon is emitted and the frame in which it is observed.

We know from special relativity that

$$\delta = \frac{1}{\gamma(1 - \vec{\beta} \cdot \hat{n}^{(o)})}, \tag{5.69}$$

for a photon seen traveling in direction $\hat{n}^{(o)}$ after leaving a source with Lorentz factor γ and velocity β (both of these also as seen by the observer). The factor in the denominator of equation 5.69 can be rewritten in terms of four-vectors: $\gamma(1 - \vec{\beta} \cdot \hat{n}^{(o)}) = p_\mu u^\mu$, where p_μ is the photon four-momentum and u^μ is the source four-velocity. Thus, the general relativistic Doppler shift factor between a given source and a distant observer in flat-space can be written as

$$\delta = \frac{E_\infty}{p_\mu u^\mu}. \tag{5.70}$$

To make this program more concrete, we will give specific expressions for a special case in which the emitter follows a circular orbit of radius r in the equatorial plane of a rotating black hole. The Doppler shift for this special case is

$$\delta = \left[\left(\frac{\Delta \Sigma}{A} \right) (1 - \beta_e^2) \right]^{1/2} (1 - \Omega l)^{-1}, \tag{5.71}$$

where the orbital frequency Ω is

$$\Omega = \frac{M^{1/2}}{r^{3/2} + aM^{1/2}}, \tag{5.72}$$

and the orbital speed (in the locally nonrotating frame) is

$$\beta_e = (\Omega - \Omega_o) \frac{A \sin \theta^{(e)}}{\Sigma \Delta^{1/2}}. \tag{5.73}$$

Here $\theta^{(e)}$ is the polar angle of the emitted photon relative to the black hole rotation axis.

If we know something about the physics of the emission process, we can generally make some statement about how the intensity depends on direction relative to the normal to the emission surface $\hat{n}^{(e)}$. In special relativity, this angle is given by

$$\cos \theta^{(e)} = \frac{\vec{p} \cdot \hat{n}}{p_t}. \tag{5.74}$$

To find the general relativistic analogue, we need to define a four-vector extension n^μ of $\hat{n}$. This four-vector must, of course, have unit magnitude

117

and must also reduce to $\hat{n}^{(e)}$ in the emitter rest frame. With n^μ in hand, equation 5.74 becomes

$$\cos\theta^{(e)} = -\frac{p_\mu n^\mu}{p_\mu u^\mu}. \tag{5.75}$$

Returning to our favored special case, $n^t = n^r = n^\phi = 0$, and $n^\theta = \Sigma^{-1/2}$ in the associated locally nonrotating frame. Because $\cos\theta = 0$ in the black hole's equatorial plane,

$$p_\mu n^\mu = -\frac{p_\theta}{r} = -\frac{qE_\infty}{r}. \tag{5.76}$$

Combining equations 5.70 and 5.76 then gives

$$\cos\theta^{(e)} = \frac{q\delta}{r} \tag{5.77}$$

because the emission surface normal is parallel to the black hole rotation axis. Thus, for this special case, we have

$$I_{\mathrm{E}}^{(o)} = \delta^3 I_{\mathrm{E}}^{(e)} \left(E^{(o)}/\delta, q\delta/r, r \right). \tag{5.78}$$

Next we turn to the transformation of solid angle. We begin by defining a two-dimensional coordinate system on the sky plane, (ψ, ζ), where ψ is measured from the projected black hole rotation axis, and ζ is distance in the perpendicular direction. In terms of these coordinates, $d\Omega^{(o)} = d\psi\, d\zeta/D^2$, where D is the distance from the source to the observer. Changing to our preferred integration variables r (and other coordinates, as required) and δ (the latter quantity is basically equivalent to a description in terms of $E^{(e)}$) means that we must compute the Jacobian

$$\frac{\partial(\psi,\zeta)}{\partial(r,\delta)} = \frac{\partial(\psi,\zeta)}{\partial(l,q)}\frac{\partial(l,q)}{\partial(r,\delta)}. \tag{5.79}$$

The Jacobian of ψ and ζ with respect to l and q is readily computed because

$$\psi = -\frac{l}{\sin\theta^{(o)}} \tag{5.80}$$

and

$$\zeta = \frac{p_\theta}{E_\infty} = \left(q + \cos^2\theta^{(o)} - l^2\cot^2\theta^{(o)} \right)^{1/2}. \tag{5.81}$$

The second Jacobian comes partly from equation 5.71, and partly from a numerical solution for the photon trajectories. The latter source is necessary in order to find q as a function of δ and r (unless $I_{\mathrm{E}}^{(e)}$ is independent of $\theta^{(e)}$, $q(\delta, r)$ is also necessary to complete the description of the emitted intensity). Note that $q(\delta, r)$ can be a multiple-valued function. In our favored special case, for example, it is double-valued, as is readily seen by considering those photons departing from locations in the plane containing the rotation axis and our line of sight.

118

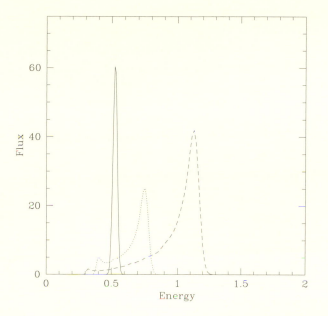

Fig. 5.4 The flux per unit energy from an emitter in a circular orbit at $r = 3M$ around a black hole with $a = 0.998M$, as viewed from $\theta^{(o)} = 0$ (solid curve), $\theta^{(o)} = \pi/3$ (dotted curve), and $\theta^{(o)} = \pi/6$ (dashed curve). The intensity in the source frame is isotropic, and its spectrum is Gaussian with central energy 1 and characteristic width 0.03. As our view moves from the axis toward the equator, the line shifts blueward, becomes more asymmetric, and (in terms of integrated flux) brightens. Figure courtesy of E. Agol.

Combining all these quantities, we arrive at the final result:

$$F_{\mathrm{E}}^{(o)}(E, \theta^{(o)}) = \frac{1}{D^2} \int d\delta \, dr \, \frac{\partial(\psi, \zeta)}{\partial(l, q)} \frac{\partial(l, q)}{\partial(r, \delta)} \delta^3 I_{\mathrm{E}}^{(e)} \left[E^{(o)}/\delta, q(\delta, r)\delta/r \right],$$

(5.82)

where r may be regarded as either the real radial coordinate or a stand-in for additional coordinates, if necessary.

The magnitude of these effects (for the special case of a circular orbit in the equatorial plane) is illustrated in figure 5.4. Particularly when a is close to unity, so that r_{ms} may be not much greater than M, there can be substantial Doppler shifts, both larger and smaller than unity. In addition, the large velocity of the emitting matter can strongly focus the radiation into the orbital plane.

6   Spherical Accretion

Now that we possess some basic understanding of black hole physics, it is time to see how much energy can, in practice, be liberated as material falls into a black hole. It is immediately obvious that if matter falls in as discrete lumps, its prospects for radiating significant amounts of energy are dim. Swallowing stars whole, therefore, is not a very good way for black holes to generate light.

Unfortunately, if a very massive black hole is embedded in a stellar cluster, just this would be expected. As a rough indicator of when stars would plunge straight into a black hole without first being disrupted by tidal forces, consider the familiar Roche criterion. If we follow the usual Newtonian prescription, this criterion boils down to a comparison between the smoothed out mass density of the disruptor (here a massive black hole) and the internal mean density of the disruptee (here a star). That is, a self-gravitating object of mass M_* and radius R_* is pulled apart by tidal forces when it ventures closer to a black hole of mass M than

$$r_t \simeq R_* \left(\frac{M}{M_*} \right)^{1/3} . \tag{6.1}$$

Scaling to the Solar density and to the gravitational radius of the black hole, we find that

$$\frac{r_t}{r_g} \simeq 2 M_8^{-2/3} \left(\frac{\rho_*}{\rho_\odot} \right)^{-1/3} . \tag{6.2}$$

In other words, the tidal radius for Solar-type stars lies *within* the event horizon of the black hole when $M > 10^8 \, M_\odot$. The criterion changes quantitatively, but not qualitatively, for main sequence stars of different mass. Roughly speaking, the mean density of main sequence stars is $\simeq 1 (M_*/M_\odot)^{-2}$ gm cm^{-3}, so the tidal radius for higher mass stars is slightly larger, and the tidal radius for lower mass stars slightly smaller. Only for black holes considerably smaller than $10^8 M_\odot$, or stars considerably less dense than main sequence stars (e.g., red giants), might we expect that tidal forces could effectively dissolve a star well outside the black hole event horizon.

We have already estimated (§4.2) that the *minimum* mean AGN mass is $\sim 10^7 \, M_\odot$; later in this chapter we will show that, depending on the luminosity, the minimum mass could be as large as $\sim 10^9 \, M_\odot$. Given the above estimate, although tidal disruption might be effective in some AGNs, it does seem that we need to seek alternative mechanisms.

Supplying the material at lower density, so that the black hole's gravity dominates any self-gravity well outside the event horizon, is more promising. Depending on just how this material is delivered (e.g., mass loss from nearby stars, radial inflow from farther out in the interstellar medium; see Chap. 14 for further discussion), one might imagine that the nature of the accretion flow could be complicated, asymmetric, and time-varying. We begin our discussion, instead, with the simplest possible physics: smooth time-steady accretion that is perfectly spherically symmetric.

This model, although highly idealized, can still provide insight in several different contexts relevant to AGNs. It defines several characteristic scales—most importantly, the Eddington luminosity—whose significance is independent of accretion geometry. It also provides a first illustration of several phenomena that arise time after time in AGNs: forces due to radiation pressure (see also §§9.3.2, 10.7, and 11.4.3), the interplay between dynamics and thermodynamics (the implicit subject of much of Chap. 7, as well as §§10.7 and 11.4.1), and thermal instability (see also §§7.4.2 and 10.7). Moreover, there are several situations in which the spherical approximation may not be so bad.

The most obvious is the edge of the "sphere of influence" defined in Chapter 4. In some cases, the scale height of the host's interstellar medium may be large enough relative to r_* for the spherical approximation to be reasonable (as in the case of the hot gas surrounding elliptical galaxies; see §14.4.1). Two other situations also arise in which quasi-spherical flow may be a realistic description. Both will be discussed at greater length in Chapter 7 (§7.3.5). In one of these, the accreting gas becomes so hot that its temperature (or at least the temperature of its ions) approaches the *virial temperature*, that is, the temperature at which the mean thermal energy of the particles is comparable to their mean gravitational binding energy. In the other, the pressure of trapped photons becomes so great that radiation pressure can be a significant support against gravity.

6.1 Bondi Accretion

Imagine, then, that at large distances from the black hole, the gas moves very slowly (relative to the hole) and has some specified density and temperature. To describe the gas's motion toward the black hole, we employ the standard equations of fluid mechanics. Throughout this discussion, we will write them in Eulerian form.

First there is the continuity equation, i.e., the equation enforcing mass conservation:

$$\frac{\partial \rho}{\partial t} + \nabla \cdot (\rho \vec{v}) = 0. \tag{6.3}$$

The right-hand side is zero whenever there are no sources or sinks of matter other than at the origin and the boundary at infinity. Here ρ is the gas density and $\vec{v}$ its velocity. In spherical symmetry, if the flow is time steady, this equation has an immediate integral for the total mass accretion rate

$$\dot{M} = 4\pi r^2 \rho v, \tag{6.4}$$

where, of course, the condition of spherical symmetry forces $\vec{v} = v\hat{r}$.

Next is the force equation

$$\rho \frac{\partial \vec{v}}{\partial t} + \rho \vec{v} \cdot \nabla \vec{v} = -\nabla p + \rho \vec{g}, \tag{6.5}$$

where p is the gas pressure, $\vec{g}$ is the acceleration due to gravity, and we have (temporarily) ignored any effects due to radiation pressure. When the flow is time steady, integrating this equation along stream-lines gives the Bernoulli law. For purely radial flow onto a point mass M this becomes

$$\frac{v^2}{2} + \int_{p_\infty}^{p} \frac{dp'}{\rho} - \frac{GM}{r} = \text{const.} \tag{6.6}$$

The constant is clearly zero by virtue of our boundary conditions. Note that because the velocity enters only in the square, this equation describes inflow and outflow equally well.

Finally there is the energy equation, or the equation of state. Treating the simplest possible case first (Bondi 1952), we will examine what happens when the equation of state is a simple adiabatic law $p \propto \rho^\gamma$.

As is often wise policy with sets of differential equations, we can concentrate the dependence on dimensional quantities into a few parameters. Let $c_s = \sqrt{\gamma p_\infty / \rho_\infty}$ (the speed of sound in the gas at infinity) be the unit of speed, let $r_s \equiv GM/c_s^2$ be the unit of length, and let ρ_∞ be the unit of density. Then the mass conservation law and the Bernoulli law can be written

$$x^2 yz = \lambda,$$
$$\frac{1}{2}y^2 + \frac{1}{\gamma - 1}\left(z^{\gamma-1} - 1\right) = \frac{1}{x}, \tag{6.7}$$

where x is the dimensionless radial variable, y is the dimensionless speed, and z is the dimensionless density. Clearly, the only parameters in the problem are γ and λ, where the latter is the dimensionless mass accretion rate.

It is possible to determine analytically the eigenvalue λ with a clever trick having both mathematical and physical significance. Let $u = y/z^{(\gamma-1)/2}$. This is not just an arbitrary algebraic combination of variables: it is exactly the local Mach number, i.e., the ratio of bulk speed to the local sound speed. After a little rearrangement, equations 6.7 combine to form:

$$u^{4/(\gamma+1)} \left[\frac{1}{2} + \frac{1}{(\gamma-1)u^2} \right] = \lambda^{-2(\gamma-1)/(\gamma+1)} \left[\frac{1}{\gamma-1} x^{4(\gamma-1)/(\gamma+1)} + x^{-(5-3\gamma)/(\gamma+1)} \right] \tag{6.8}$$

or

$$g(u) = \lambda^{-2(\gamma-1)/(\gamma+1)} f(x). \tag{6.9}$$

First, let us assume that $1 < \gamma < 5/3$. Then both f and g have unique minima, occurring respectively at $u_m = 1$ and $x_m = (5 - 3\gamma)/4$. The location of the minimum of g is physically significant: $u = 1$ at that point. Therefore, wherever that minimum is achieved, the Mach number is unity. Places where $u = 1$ are called *sonic points*. On the other hand,

$$x_m = \begin{cases} 0.5 & \gamma = 1 \\ \to 0 & \gamma \to 5/3 . \end{cases} \tag{6.10}$$

That is, the closer γ approaches to $5/3$, the smaller the radius at which the minimum of f is found.

A solution that spans the range from $u = 0$ to $u \gg 1$ can be found in two steps: First match the right-hand side to the left-hand side at the respective minima, so that $u(x_m) = 1$, making x_m the sonic point. Next, because the scales of x and u relative to each other are completely arbitrary, we are free to define a mapping between them such that the curves stay aligned everywhere else.

The first step is accomplished by finding λ such that

$$g(1) = \lambda^{-2(\gamma-1)/(\gamma+1)} f(x_m),$$

that is,

$$\lambda = \left(\frac{1}{2} \right)^{(\gamma+1)/[2(\gamma-1)]} \left(\frac{5 - 3\gamma}{4} \right)^{-(5-3\gamma)/[2(\gamma-1)]} . \tag{6.11}$$

123

The eigenvalue λ (which, as we have just seen, depends only on γ) specifies the dimensionless accretion rate, so the physical accretion rate is

$$\dot{M} = 4\pi\lambda(\gamma)\frac{(GM)^2\rho_\infty}{c_s^3}. \tag{6.12}$$

This expression for the accretion rate is significant in two regards. First, it sets a characteristic scale for the accretion rate in terms of the central mass and the conditions at infinity. For an AGN, "infinity" must be placed roughly at r_* because outside that point the gravitational potential is no longer dominated by the central mass. At that point, then, we can estimate the gas density necessary to feed an AGN of a given luminosity and mass:

$$\rho_* \sim 5 \times 10^{-27} L_{45} T_4^{3/2} \left(\frac{\eta}{0.1}\right)^{-1} M_8^{-2}\mathrm{gm~cm}^{-3}. \tag{6.13}$$

Thus, the density of interstellar gas required to support accretion is quite modest compared to typical interstellar densities $\sim 10^{-24}$ gm cm^{-3}. Even if the gas were kept considerably warmer than the fiducial temperature of 10^4 K to which we have scaled, the density required would not be prohibitively large.

Second, it demonstrates that the accretion rate (and the global character of the accretion flow) depend on the equation of state. As γ increases from unity to 5/3, λ falls from 1.12 to 0.25. At the upper limit for γ (the value corresponding to particles with no internal degrees of freedom), the sonic point moves closer and closer to the origin and the accretion rate falls. On the other hand, an equation of state that is so soft as to be nearly isothermal finds a sonic point at relatively large distance and permits a larger accretion rate.

Real astrophysical plasmas without heating or cooling have an equation of state that is generally best described as adiabatic with $\gamma = 5/3$. However, heating and cooling are often important, so a proper treatment requires integrating a genuine energy equation rather than simply using an adiabatic equation of state (§6.3). Nonetheless, for toy models one can sometimes mock up the effects of heating and cooling by the simple formal device of altering the effective adiabatic index. It is in this light that the preceding section has its greatest significance. Because the sonic point approaches $r = 0$ for a strictly adiabatic gas with $\gamma \to 5/3$ from below, we can already see that to achieve a smooth transonic accretion flow requires at least some net radiative cooling. In §6.3, this point will be made more quantitatively.

6.2   The Eddington Limit

Our description of the flow so far has neglected the fact that the central object is generating a very large amount of radiation. This radiation exerts an outward force on the gas:

$$\vec{F}_{\rm rad} = \frac{1}{c} \int d\nu\, \vec{\mathcal{F}}_\nu \kappa(\nu)\rho, \tag{6.14}$$

where $\vec{\mathcal{F}}_\nu$ is the radiation flux per unit frequency ν and $\kappa(\nu)$ is the total opacity of this material per unit mass (scattering plus absorption) as a function of frequency ν. The radiation flux is defined in the usual way as

$$\vec{\mathcal{F}}_\nu \equiv \int d\Omega\, \hat{n}(\Omega) I_\nu(\Omega), \tag{6.15}$$

where $I_\nu(\Omega)$ is the intensity at ν traveling in the direction of the unit vector $\hat{n}(\Omega)$. The intensity obeys the transfer equation

$$\frac{1}{c}\frac{\partial I_\nu}{\partial t} + \nabla\cdot(\hat{n}I_\nu) = \int d\Omega' \int d\nu'\, \kappa^s_{\nu',\hat{n}'}(\nu,\hat{n}) I_{\nu'}(\hat{n}') - (\kappa^s_\nu + \kappa^a_\nu) I_\nu(\hat{n}). \tag{6.16}$$

where κ^a is the absorption opacity, $\kappa^s_{\nu',\hat{n}'}(\nu,\hat{n})$ is the partial scattering opacity for transforming photons with initial frequency ν' and direction $\hat{n}'$ into outgoing photons with frequency ν and direction $\hat{n}$, and we assume there is no true emission outside the central source.

In general, the actual opacity of the gas is a function of its physical state and depends substantially on frequency. The transfer equation can then be very messy, and the flux becomes a complicated function of frequency and distance. However, if the gas is completely ionized (as may be a good approximation in many regimes), the opacity is very simply described for all photon energies below ~ 100 keV: it is just the Thomson opacity. Thomson opacity has unit albedo and is exactly elastic in the limit of $h\nu/(m_e c^2) \to 0$, so it conserves flux frequency by frequency. When the only source of radiation is very small compared to the radii of interest, and the electron density distribution is spherically symmetric, the flux from a source with luminosity per unit frequency L_ν is $\mathcal{F}_\nu = L_\nu/(4\pi r^2)$, and the force is just

$$F_{\rm rad} = \sigma_{\rm T} n_e \int d\nu\, \frac{L_\nu}{4\pi c r^2} \tag{6.17}$$

in the radial direction. If the mass per unit electron μ_e is fixed (as it is for any fully ionized gas), the radiative acceleration is

$$g_{\rm rad} = \frac{\sigma_{\rm T}}{\mu_e} \frac{L}{4\pi c r^2}, \tag{6.18}$$

where σ_T is the Thomson cross section.

Thus, the familiar inverse-square law of the diminution of radiative flux has the corollary that radiative acceleration due to a central source embedded in a purely scattering atmosphere likewise falls off as distance squared. That is, it depends on distance exactly the same way gravity does.

There is a dramatic consequence to the fact that both the radiative force and gravity depend on distance in exactly the same way: if one exceeds the other at any one place, it will also do so (and by the same ratio) everywhere else:

$$\frac{g_{\rm rad}}{g} = \frac{\sigma_T L}{4\pi c \mu_e GM}. \tag{6.19}$$

This relation can be rewritten in an instructive way:

$$\frac{g_{\rm rad}}{g} = L/L_{\rm E}, \tag{6.20}$$

where

$$L_{\rm E} = \frac{4\pi c GM \mu_e}{\sigma_T} = 1.51 \times 10^{38} \frac{M}{M_\odot} \ {\rm erg\ s}^{-1}. \tag{6.21}$$

That is, for any given mass, there is a critical luminosity, the *Eddington luminosity* $L_{\rm E}$, beyond which the radiation force must overpower gravity.

It immediately follows that the large luminosities we observe from AGNs—anywhere from 10^{43} to 10^{47} erg s^{-1}—demand minimum central masses in the range 10^5–$10^9 M_\odot$. This provides a second argument to bolster the "ash" argument (§4.2) on behalf of AGNs requiring masses in this range.

Furthermore, if the accretion flow produces light with fixed efficiency, the central mass sets a characteristic scale for the mass accretion rate, the *Eddington accretion rate*:

$$\dot{M}_{\rm E} = \frac{L_{\rm E}}{c^2 \eta} = 3 M_8 \left(\frac{\eta}{0.1}\right)^{-1} M_\odot \ {\rm yr}^{-1}, \tag{6.22}$$

where η is the radiative efficiency in rest-mass units, and in light of the results of Chapter 5, we expect $\eta \sim 0.1$. That the central mass should play any part in controlling the accretion rate may seem a surprising result. We have just shown that in the case of adiabatic spherical accretion, the accretion rate is determined entirely by ρ_∞ / c_s^3 and the equation of state, which are completely independent of what goes on inside the accreting object. How is it that the central mass can affect what happens so far away? And what happens when $\dot{M} > \dot{M}_E$?

The answer to the first question is simple: in our analysis of spherical adiabatic accretion, we made no allowance for radiation forces. It is their inclusion that creates a link between the accretion rate and the central mass.

Answering the second question is more difficult, in part because it has several answers. First, suppose we retain the assumption of perfect spherical symmetry. In this case, we must be careful about how the radiation efficiency is determined. Although it is true that the *maximum* efficiency for accretion onto a black hole is determined by the binding energy of the last stable orbit, there is no guarantee that this maximum efficiency is achieved. If there is little dissipation to transform organized kinetic energy into heat, or if radiative losses are slow compared to the infall speed, it is possible for the material to fall into the black hole with almost its initial energy, and therefore having radiated very little (§§6.3, 7.3.5). Even if substantial radiation is created, it is possible (if the gas is optically thick enough) for the radiation to be trapped in the flow and carried back inward (§§6.3.2, 7.3.5). If that happens, the effective outward flux is diminished. It is also possible for energy to be lost in some form that interacts only weakly with the outer accretion flow, e.g., neutrinos. Finally, it is possible to destroy the symmetry assumption by a break-up into a two-dimensional pattern, or a two-phase flow that is spherically symmetric only in a statistical sense (§6.3.3). We will discuss some of these issues more quantitatively in the next few pages.

Before we do so, however, we digress to discuss how the Eddington limit may regulate the growth of mass in the central object. Because the accreted "ash" from radiation production is all retained, the black hole's mass grows as

$$\frac{dM}{dt} = \frac{L}{c^2\eta}, \tag{6.23}$$

Writing the luminosity in terms of the Eddington luminosity transforms equation 6.23 to

$$\frac{dM}{dt} = \frac{L}{L_E} \frac{4\pi G \mu_e}{\eta c \sigma_T} M; \tag{6.24}$$

that is, the mass grows exponentially with characteristic growth time

$$t_{\text{growth}} = \frac{\eta c \sigma_T}{4\pi G \mu_e} \frac{L_E}{L}$$
$$= 3.7 \times 10^8 \eta \frac{L_E}{L} \text{ yr.} \tag{6.25}$$

127

Thus we see that Eddington-limited accretion can make the central mass grow exponentially on timescales as short as $\sim 10^8$ yr, even when the radiative efficiency is the maximum value possible for a Kerr black hole. Further consequences of this characteristic timescale are explored in Chapter 14.

6.3   Nonadiabatic Accretion

If we are to take a more realistic view of the accretion flow's equation of state, we would do well to begin by estimating the characteristic physical conditions in the flow. Far from the black hole, we might imagine that the gas has the properties of ordinary interstellar matter—density within a few orders of magnitude of 1 atom cm^{-3}, temperature anywhere from a few K to $\sim 10^6$ K, and, with respect to electron scattering, quite transparent.

Closer to the black hole, but far enough away to make the Newtonian approximation reasonable, let us suppose that the gas is falling inward with nearly the free-fall speed. If we scale to an accretion rate close to the Eddington rate for $M \sim 10^8 \, M_\odot$, we find that the electron density is

$$n_e = \frac{\dot{M}}{4\pi r^2 \mu_e v} = 3.8 \times 10^{11} \left(\frac{\dot{M}}{M_\odot \, \mathrm{yr}^{-1}} \right) M_8^{-2} x^{-3/2} \ \mathrm{cm}^{-3}, \qquad (6.26)$$

where x is the radius in units of gravitational radii, GM/c^2. Clearly, the gas is compressed to densities much greater than are commonly seen in the interstellar medium. The Thomson optical depth from the outside in to radius x is

$$\tau_{\mathrm{T}} \simeq 4 \left(\frac{\dot{M}}{M_\odot \, \mathrm{yr}^{-1}} \right) M_8^{-1} x^{-1/2}. \qquad (6.27)$$

The physical significance of the latter estimate is better put into focus when written in terms of the Eddington accretion rate:

$$\tau_{\mathrm{T}} = \frac{\dot{M}/\dot{M}_{\mathrm{E}}}{x^{1/2}}. \qquad (6.28)$$

In other words, any time the Eddington limit is approached, the Thomson optical depth becomes significant near the black hole.

The temperature is harder to estimate. If the density increases inward $\propto x^{-3/2}$, and the gas is adiabatic, its temperature increases $\propto x^{-(3/2)(\gamma-1)}$, that is, $\propto x^{-1}$ if $\gamma = 5/3$. At the very most, the temperature could be of order the virial temperature, that is, $k_{\mathrm{B}}T \sim GMm_p/r$ (k_{B} is the

Boltzmann constant). However, as demonstrated in Chapter 8, a variety of processes make it difficult to push the electron temperature above $\sim m_e c^2/k_B$, although it is possible that the ions thermally decouple from the electrons, and remain much hotter.

Combining these estimates, we can make a rough guess of the likely radiative efficiency of a spherical accretion flow. If the gas cools by a two-body collisional process, the power radiated per unit volume is $\sim n_e^2 \Lambda$, where (as will be demonstrated in §8.2) the *cooling function* Λ is at most $\sim \Lambda_o = \alpha_{fs} \sigma_T m_e c^3$ (α_{fs} is the fine-structure constant). Because the density increases so rapidly inward, most of the power output occurs just outside the black hole. The maximum total luminosity is then

$$L \sim \frac{\alpha_{fs}}{12\pi} \frac{\dot{M}^2 \sigma_T m_e c^3}{GM \mu_e^2} \sim 2 \times 10^{41} \left(\frac{\dot{M}}{M_\odot \ \text{yr}^{-1}}\right)^2 M_8^{-1} \ \text{erg s}^{-1}. \tag{6.29}$$

This characteristic value, too, is thrown into clearer relief when written in terms of the Eddington scale:

$$\frac{L}{L_E} \sim 1 \times 10^{-4} \left(\frac{\dot{M}}{\dot{M}_E}\right)^2. \tag{6.30}$$

The fact that $L/L_E \ll 1$ even when $\dot{M} \sim \dot{M}_E$ means that the efficiency of spherical accretion is generically very low. Although the amount of potential energy transformed into heat can be very great, inflow at the free-fall speed keeps the density so low that very little of that is converted into photons before the matter—and its energy content—stream irrevocably into the black hole.

6.3.1   Critical point structure

Our next step is to consider in greater detail the impact on the flow dynamics due to radiative heating and cooling. In place of the adiabatic equation of state, we must solve an explicit energy equation:

$$\left[\frac{\partial}{\partial t} + v\frac{\partial}{\partial r}\right]\left[\frac{v^2}{2} + \frac{\gamma}{\gamma-1}\frac{p}{\rho} - \left(1 - \frac{L}{L_E}\right)\frac{GM}{r}\right] = H - C, \tag{6.31}$$

where γ is the adiabatic index in the event of zero net radiative heating or cooling, $H - C$ is the radiative heat balance in units of erg s^{-1} gm^{-1}, and the gravitational potential has been adjusted to take into account radiation force (assuming nearly all hydrogen and helium atoms are ionized). Note that the force equation should also be adjusted to account for radiation force.

We now seek a steady-state solution. To simplify the equations, we retain the integral of the mass continuity equation used in the Bondi problem ($\dot{M} = 4\pi r^2 \rho v$) and rewrite the luminosity in terms of the accretion rate: $L = \dot{M} c^2 \eta$. Only in a time-steady state, of course, can the luminosity at any given time be identified with the instantaneous accretion rate. The energy equation then becomes

$$
\frac{1}{\gamma - 1}\left(\frac{\gamma k_B T}{\mu v^2} - 1\right)\frac{d}{dr}\left(\frac{k_B T}{\mu}\right) = -\frac{k_B T}{\mu v^2}\left(1 - \frac{L}{L_E}\right)\frac{\sigma_T L_E}{4\pi c \mu_e r^2}
$$
$$
+ \frac{2k_B T}{\mu r} + \frac{1}{v}\left(\frac{k_B T}{\mu v^2} - 1\right)(H - C),
$$
(6.32)

where μ is the mean mass per particle. When $|v| = \sqrt{\gamma k_B T/\mu}$ (i.e., at the sonic point), dT/dr becomes singular unless the right-hand side is also zero. Thus, for the flow to pass smoothly through a sonic point at some radius r_s, there is a special constraint:

$$
2r_s = \left(1 - \frac{L}{L_E}\right)\frac{GM}{c_s^2} - \frac{r_s^2}{c_s^3}(\gamma - 1)(H - C).
$$
(6.33)

Because the sonic point cannot occur at a negative radius, it is clear that the heating at the sonic point cannot be too great. In fact, when $\gamma = 5/3$, there must be net cooling at least somewhere in the flow in order to push the sonic point out from $r = 0$. The differential equation for velocity has a form very similar to the one for the temperature, including the same singular point, and the same constraint on the sonic radius. Note also that a spherically symmetric wind is subject to very similar constraints, for v appears only as v^2 everywhere except in the coefficient of $H - C$. Time-steady spherically symmetric transonic winds therefore also have a critical point where the flow becomes supersonic, but, not surprisingly, net heating pushes out the sonic point.

6.3.2   Optical depth and photon trapping

As we have already seen (eq. 6.28), the optical depth can be considerable when the accretion rate is close to or exceeds Eddington. As a result of this opacity, photons are not able to travel freely outward (Begelman 1978). Suppose that the Thomson depth from a given radius r to the outside is $\tau > 1$. Then it is a well known property of random walks that on average τ^2 scatterings occur before any particular photon finds its way to the outside edge. Because the density scales as $r^{-3/2}$, most of these scatters occur at radii within a factor of two of r, and the mean free path at r is $\simeq r/\tau$. Thus, it takes the average photon a time $\simeq \tau r/c$ to escape. This diffusion takes place relative to a frame moving with the fluid, and (well inside r_s) that frame is moving inward with a speed $\simeq c(r/r_g)^{-1/2}$. Therefore, the ratio between the time it takes the fluid to carry the photons inward one factor of two in radius and the time it takes the photons to diffuse outward a factor of two is $\simeq (r/r_g)^{1/2}/\tau \sim (r/r_g)(\dot{M}/\dot{M}_E)^{-1}$. When the accretion rate is comparable to (or possibly even larger than) Eddington, near the black hole it is quite possible for the advection time to be shorter than the diffusion time, so that most photons do not escape. The net effect of this *photon trapping* is, of course, to reduce the radiative efficiency (as seen from large distance). In this fashion, although the accretion rate can be super-Eddington, the escaping luminosity remains below L_E. The consequences of this effect are developed further in §7.3.5.

6.3.3 Thermal instability

Another problem besets spherical accretion: thermal instability. To see this problem in its simplest setting, we will select one heating mechanism (Compton heating) and one cooling mechanism (ordinary nonrelativistic electron-proton bremsstrahlung). Many different heating and cooling mechanisms may operate (more detailed discussions of these can be found in Chapters 8, 9, and especially 10), but it is useful for this discussion to introduce a pair of mechanisms that are both very often important and also have formal properties similar to many of the other mechanisms.

Let us first consider Compton heating. This is simply the transfer of energy from photons to electrons via Compton recoil. When the photon energy $\epsilon \ll m_e c^2$, the scattering cross section is independent of ϵ, and the average energy transfer per scattering is $\epsilon^2/m_e c^2$ (see also §8.2.2). The heating per unit mass due to the scattering of photons with energy ϵ is then

$$H_\epsilon = \frac{4\pi\sigma_T}{\mu_e} J_\epsilon \frac{\epsilon}{m_e c^2}, \tag{6.34}$$

where J_ϵ is the mean intensity at energy ϵ. Integrating over the entire spectrum gives the total heating rate

$$H = \frac{4\pi\sigma_T}{\mu_e} J \frac{4k_B T_C}{m_e c^2}, \tag{6.35}$$

where the mean intensity integrated over photon energy is J, and the *Compton temperature* is

$$T_C = \frac{1}{4k_B} \frac{\int d\epsilon\, J_\epsilon \epsilon}{\int d\epsilon\, J_\epsilon}. \tag{6.36}$$

Energetic electrons can return energy to the photons, but this only counterbalances the Compton heating when $T \geq T_C$ (this is why T_C is called the Compton temperature; see §8.3.1), and the T_C associated with AGN spectra is in general $\gg 10^6$ K (see §10.5.1).

Like many (but not all) processes by which a plasma can cool radiatively, bremsstrahlung requires the collision of two particles. Consequently, the cooling rate associated with it scales in proportion to density:

$$C_{\text{brems}} = n_e \Lambda_{\text{brems}}(T), \tag{6.37}$$

where (non-relativistic) $\Lambda_{\text{brems}} \propto T^{1/2}$. For scaling purposes, it is handy to note (as will be discussed further in §8.2.1.1) that

$$\Lambda_{\text{brems}} \sim \Lambda_o \left(\frac{k_B T}{m_e c^2} \right)^{1/2}. \tag{6.38}$$

To further simplify matters, we will restrict our attention to cases in which radiative equilibrium is achieved on a timescale short compared to the flow time. In other words, we will require that $t_{\text{cool}}/t_{\text{dyn}} \ll 1$, where

$$\begin{aligned}
\frac{t_{\text{cool}}}{t_{\text{dyn}}} &= \frac{3 k_B T v}{n_e \Lambda r} \\
&= 3 \left(\frac{k_B T}{m_e c^2} \right)^{1/2} \frac{\eta (v/v_{\text{ff}})^2}{\alpha_{\text{fs}} (\dot{M}/\dot{M}_E) \Lambda_o},
\end{aligned} \tag{6.39}$$

in which the rewritten version assumes the free-fall density scaling, and the free-fall speed is v_{ff}. Radiative equilibrium is most likely where the inflow is slow compared to free-fall, and the accretion rate is relatively high. When this criterion is satisfied, temperature balance between Compton heating and bremsstrahlung is fixed by the condition

$$n_e \Lambda_{\text{brems}}(T) = \frac{L \sigma_T}{4 \pi r^2} \frac{4 k_B T_C}{m_e c^2}. \tag{6.40}$$

Because the right-hand side of equation 6.40 is independent of temperature (for $T \ll T_C$), at fixed density increasing heating rate results in increasing temperature. If the appropriate boundary condition is fixed pressure, however, as is more generally the case, the left-hand side is $\propto p T^{-1/2}$, and increasing heating rate can only be balanced by *decreasing* temperature.

The paradoxical character of this result correctly signals a physical problem. Indeed, as Field (1965) first showed, this sort of thermal balance

132

is in fact linearly unstable. To isolate the basic physics of this instability, consider a very simple model: zero gravity, constant pressure and density in the equilibrium state, and a time-steady temperature balance. If the gas in question is actually part of an accretion flow, these assumptions are equivalent to two restrictions: the perturbations must vary on such short lengthscales that none of the equilibrium quantities vary significantly on that scale, and the frequency of the perturbations must be rapid compared to the time required for the equilibrium conditions to change significantly within a fluid element. The time development of small perturbations is then governed by the perturbed continuity equation

$$\frac{\partial \delta\rho}{\partial t} + \frac{\partial(\rho\delta v_x)}{\partial x} = 0, \tag{6.41}$$

the perturbed force equation

$$\rho\frac{\partial \delta v_x}{\partial t} = -\frac{\partial \delta p}{\partial x}, \tag{6.42}$$

and the perturbed energy equation

$$\frac{\partial}{\partial t}\left[\frac{\gamma}{\gamma-1}\frac{\delta p}{\rho} - \frac{\gamma}{\gamma-1}\frac{p}{\rho^2}\delta\rho\right] = \frac{\partial(H-C)}{\partial p}\delta p + \frac{\partial(H-C)}{\partial \rho}\delta\rho + \frac{\partial}{\partial x}\left(\kappa_{\rm th}\frac{\partial \delta T}{\partial x}\right). \tag{6.43}$$

Here δX is the perturbation in quantity X, and we (arbitrarily) select the x-axis as the direction of the perturbed velocity; $\kappa_{\rm th}$ is the thermal conductivity.

Suppose that all perturbed quantities vary $\propto \exp[i(kx - \omega t)]$. A little algebra suffices to arrive at the dispersion relation

$$\omega = \frac{i(\gamma-1)}{\gamma\omega^2/k^2 - c_s^2}\left[\frac{\gamma\omega^2}{k^2c_s^2}\frac{\partial(H-C)}{\partial \ln p} + \frac{\partial(H-C)}{\partial \ln \rho}\right] - i\frac{\gamma-1}{\gamma}k^2\kappa_{\rm th}. \tag{6.44}$$

For very long wavelengths ($k \ll \omega/c_s$), the phase speed is large compared to the sound speed, and the perturbations grow (or decay) at constant density. In this case, the dispersion relation reduces to

$$\omega \simeq i(\gamma-1)\left[\frac{1}{c_s^2}\frac{\partial(H-C)}{\partial \ln p} - \frac{1}{\gamma}k^2\kappa_{\rm th}\right]. \tag{6.45}$$

In other words, the mode is either purely growing or purely decaying. Growth occurs when the heating increases with increasing pressure at fixed density. For our example of Compton heating and bremsstrahlung cooling, quite the opposite occurs, so long-wavelength perturbations are damped.

133

In the short wavelength limit (i.e., $k \gg \omega/c_s$), the pressure, but not the density, is kept fixed, so the dispersion relation becomes

$$\omega \simeq -i(\gamma - 1) \left[\frac{1}{c_s^2} \frac{\partial(H - C)}{\partial \ln \rho} + \frac{1}{\gamma} k^2 \kappa_{\text{th}} \right]. \tag{6.46}$$

Neglecting heat conduction, there is growth when $H - C$ declines with increasing density; that is, the cooling rate increases with increasing density at fixed pressure. This is exactly what happens when Compton heating is balanced by bremsstrahlung, and so this equilibrium is generically unstable to short wavelength perturbations. Because $H - C \sim k_{\text{B}}T/(\mu t_{\text{cool}})$, the growth rate is $\sim t_{\text{cool}}^{-1}$. However, at sufficiently short wavelengths that the heat diffusion time is short compared to the cooling time, heat conduction damps the instability.

This criterion allows us to use a simple plot of how the equilibrium temperature depends on the heating rate to flag thermal instability. Suppose, as is often the case, that H is independent of either ρ or T, and that $C \propto \rho T^s$ (see e.g., §10.3.3). Then $\partial \ln(H - C)/\partial \ln \rho|_p = 1 - s$, and the equilibrium temperature $T_{\text{eq}} \propto (H/p)^{1/(s-1)}$. Equation 6.46 informs us that short-wavelength thermal instability occurs when $s < 1$, but this is equivalent to the condition that T_{eq} *falls as H/p rises*.

Instability of exactly this sort can occur within a spherical accretion flow (it can also occur in other contexts: see, e.g., §§7.4.2, 10.5.1 and 10.7). Because one of the conditions for exponential growth is $t_{\text{cool}} \ll t_{\text{dyn}}$, instability is favored when $\dot{M}/\dot{M}_{\text{E}}$ is relatively large and the temperature at infinity relatively low. High efficiency also favors the development of thermal instabilities, but the argument for this is more indirect (Ostriker et al. 1976; Cowie, Ostriker, and Stark 1978; Krolik and London 1983). At somewhat smaller values of η and L/L_{E}, instability can still be possible, but the growth time may no longer be rapid compared to the dynamical time. In that case, perturbations grow, but only algebraically in time. Whether growth is exponential or algebraic, the endpoint of the instability is not well understood. It may be intrinsically unsteady flow, it may be a breakup into two-phase flow, or it may be something else entirely. One way or another, time-steady, spherical accretion with high efficiency and high L/L_{E} probably does not exist in the real world.

7 Accretion Disks and the Optical/Ultraviolet Continuum

At the last stable orbit of even a $10^8 \, M_\odot$ black hole, the specific angular momentum is $\sim 1 \times 10^{24}$ cm^2 s^{-1}. By comparison, the specific angular momentum of matter orbiting in a typical galaxy is $\sim 6 \times 10^{28}$ cm^2 s^{-1}. Clearly, approach of accreting material toward a black hole requires eliminating all but a very small fraction of the material's initial angular momentum. Given how much easier it is to lose energy (by radiation of photons) than it is to lose angular momentum (by viscosity? by nonaxisymmetric gravitational forces? by radiation of MHD waves?), we expect that accretion onto an AGN will be limited largely by how rapidly angular momentum can be lost.

In any spherically symmetric potential, the orbit of minimum energy for fixed angular momentum is a circle. On this basis, we expect that the orbits of accreting material will be a succession of ever-so-slightly smaller circles. In addition, we can expect the matter to fall into a common plane because matter traveling along orbits inclined to one another will collide in the plane of intersection. The result of these collisions is to mix the angular momentum of the different gas streams. Consequently, at any given radius, all the matter will eventually acquire the same specific angular momentum and therefore orbit in a single plane. For these reasons, it seems very likely that the material accretes not spherically, but through a disk.

7.1 Fundamental Equations of Thin Disks

The arguments just given strongly indicate that we consider what happens to gaseous material all orbiting in a common plane. In fact, in many circumstances, we can make the approximation that this is almost a two-dimensional system because the vertical thickness h is often much smaller than the radius r. If $L/L_{\rm E}$ is close to unity, we already know that the force due to radiation pressure is comparable to that of gravity, and the scale height of matter could be significant compared to r. However, if $L/L_{\rm E} \ll 1$, radiation alone cannot support the matter against gravity at substantial altitudes above the disk plane, so we must also check to see what thermal motions may do. Viewed from a frame of reference moving with the orbital velocity v_{orb}, the gas atoms have random speeds $\sim c_s$, the sound speed. Because the vertical component of the gravitational acceleration is GMz/r^3 at altitudes $z \ll r$ above the equatorial plane, the typical height an atom can reach before falling back is

$$h \sim r \frac{c_s}{v_{\text{orb}}}. \tag{7.1}$$

In other words, the aspect ratio of the disk is roughly the ratio of the gas's sound speed to its orbital speed. So long as the temperature T satisfies the criterion

$$\frac{k_{\text{B}}T}{\mu c^2} \frac{r}{r_g} \ll 1, \tag{7.2}$$

the disk is indeed thin. Here μ is the mean mass per particle, and we scale the radius to $r_g = GM/c^2$. As we have seen in Chapter 5, the size of the smallest orbits is $\sim r_g$. Because radiative cooling is often strong and becomes particularly effective when $k_{\text{B}}T > m_e c^2$, this condition is fairly easy to meet.

In the simplest model for AGNs we can verify quite directly that the thin disk criterion is met. Suppose that the disk radiates by emission of a local blackbody spectrum. The $1/r$ dependence of the gravitational potential compels most of the energy release to be close to the last stable orbit around the black hole. Exactly how large this orbit is depends on the angular momentum of the black hole, but it is generally $\sim r_g$ (§§5.1.3.3, 5.1.4.2). Then we have

$$L \sim \pi r_g^2 \sigma T^4, \tag{7.3}$$

where σ is the Stefan-Boltzmann constant, which leads to

$$T \sim 10^6 L_{46}^{-1/4} \left(\frac{L}{L_{\text{E}}}\right)^{1/2} \text{K}. \tag{7.4}$$

Thus, we expect the inner region of such a system to be quite cool compared to the rest mass of a proton $m_p c^2 \sim 1$ GeV$\sim 10^{13}$ K. On this basis (again neglecting radiation pressure), we should expect the system to be quite geometrically thin.

It is also important to notice two other consequences of this estimate. First, if blackbody emission is anywhere near a correct description for the radiation mechanism, characteristic photon energies in the UV are to be expected; in view of the local maximum in νF_ν that often appears around 10 eV, this is very satisfying (see further discussion in §7.6). Second, for fixed L/L_{E}, the characteristic temperature *decreases* with increasing luminosity (or equivalently, increasing mass). This is because the radiating surface area grows with M^2, requiring a smaller surface brightness to radiate the same total luminosity. Note that stellar mass black holes can therefore be expected to have a characteristic temperature $\sim 10^7$ K, and indeed their spectra generally display thermal humps with just that temperature (Tanaka 1989).

When the assumption of geometrical thinness is justified, it makes sense to describe the disk as a two-dimensional system. If, in addition, the disk is azimuthally symmetric, instead of following the density as a function of cylindrical radius r, z, and ϕ, we need only concern ourselves with the surface density: $\Sigma(r) = \int dz\, \rho(z, r)$. Our first order of business is to ask how $\Sigma(r)$ evolves as a function of time.

By definition, inside the disk the gas velocity can have only two components: in the azimuthal direction (i.e., the orbital motion with speed v_{orb}) and possibly in the radial direction, with speed v_r. If there are no sources or sinks of matter within the disk proper, the surface density evolves according to the continuity equation:

$$\frac{\partial \Sigma}{\partial t} + \frac{1}{r} \frac{\partial}{\partial r} \left(r \Sigma v_r \right) = 0. \tag{7.5}$$

There is also a continuity equation for the axial component of the angular momentum (it is the analogue of the rectilinear continuity equation for linear momentum, the Navier-Stokes equation):

$$\frac{\partial}{\partial t} \left(r^2 \Sigma \Omega \right) + \frac{1}{r} \frac{\partial}{\partial r} \left(r^3 \Sigma \Omega v_r \right) = \frac{1}{2\pi r} \frac{\partial \mathcal{G}}{\partial r}, \tag{7.6}$$

where Ω is the orbital frequency and is determined by the underlying gravitational potential. If the central mass dominates the potential, Ω has the usual Keplerian dependence $\propto r^{-3/2}$ for $r \gg r_g$ (see eq. 5.72); but if the disk mass is significant, or there is some other substantial amount of mass in the region (a dense stellar cluster?), Ω may have a more complicated dependence on r.

Equations 7.5 and 7.6 define how matter and angular momentum flow through the disk, but they require boundary conditions for a complete specification of the problem. When accretion disks form in stellar binaries, mass is fed from the donor star in a well defined stream that joins the accretion disk at its outer radius. Mass leaves the disk primarily by accretion onto the central object from the inner ring of the disk. Angular momentum lost by accreting material is transferred through the disk to its outer edge, and thence to the donor star by tidal torques. In AGNs we likewise expect mass to leave the disk primarily by falling off the inner edge and being swallowed by the central object, but none of the rest of the story carries over. We do not know the source of mass, or where it enters the disk. There is no object suitable for absorbing angular momentum from the outer edge of the disk, so angular momentum must somehow be transported away from the disk in some other way. Conceivably this could be by pushing outward a small amount of mass; other mechanisms (e.g., electromagnetic waves) are also possible. These uncertainties about the boundary conditions mean that there may be interesting portions of the disk where the right-hand side of equation 7.5 is nonzero, and the right-hand side of equation 7.6 may substantially change its character.

If we put aside these concerns about boundary conditions, the quantity $\mathcal{G}$ can be interpreted as the torque between two neighboring annuli. The local rate of change of the surface density of angular momentum is the *gradient* of the torque (if the ring at $r - dr$ torques the ring at r by the same amount that the ring at r torques the one at $r + dr$, there is no net change in the angular momentum at r); the factor $2\pi r$ transforms the torque per annulus to the torque per unit length around an annulus.

7.2 What Is the Torque?

7.2.1 Viscosity?

A simple phenomenological way to describe the torque is to suppose that it is created by a local viscosity of unspecified character. Although this assumption is a plausible one, it is important to bear in mind that other possibilities do exist. It is possible to imagine (and a number of people have) schemes for producing torque that either are not local or, even if they are local, do not produce torque proportional to the local shear in the way that viscosity does. Nonetheless, we will begin by supposing that the torque is describable in the language of local viscosity.

A fundamental object in that language is the viscous stress tensor. Like any stress tensor (e.g., an ordinary pressure tensor), its (i, j) component gives the rate at which momenum in the ith direction is carried in direction j by viscosity:

$$T_{ij} = \rho \nu \left[\frac{\partial v_i}{\partial x_j} + \frac{\partial v_j}{\partial x_i} - \frac{2}{3} \left(\nabla \cdot \vec{v} \right) \delta_{ij} \right], \tag{7.7}$$

where ν is the kinematic shear viscosity, δ_{ij} is the usual Kronecker δ, and the x_j are the three coordinates (with appropriate scale factors if non-Cartesian).

What Is the Torque?

In an accretion disk, the only significant velocity component is the azimuthal one, v_ϕ, so it is advantageous to write the stress tensor in cylindrical coordinates. When that is done, the only interesting component is

$$T_{r\phi} = \rho\nu\left(\frac{\partial v_\phi}{\partial r} - \frac{v_\phi}{r}\right) = \rho\nu r\frac{\partial\Omega}{\partial r}. \tag{7.8}$$

The torque between two adjacent rings is then

$$
\begin{aligned}
\mathcal{G} &= \int d\phi\, r \int dz\, r T_{r\phi} \\
&= 2\pi r^3 \nu\Sigma\frac{\partial\Omega}{\partial r},
\end{aligned}
\tag{7.9}
$$

where both ν and Σ are azimuthal averages, and ν has been vertically averaged as well. If $\partial\Omega/\partial r < 0$, then $\mathcal{G} < 0$, and angular momentum is carried outward.

With this model expression for the torque, we can use equations 7.5 and 7.6 to find the radial velocity:

$$v_r = \frac{\frac{\partial}{\partial r}\left(r^3\nu\Sigma\partial\Omega/\partial r\right)}{r\Sigma\frac{\partial}{\partial r}\left(r^2\Omega\right)}. \tag{7.10}$$

From this result we see that the radial velocity is proportional to the shear (a property of the potential) and to the viscosity. In order of magnitude terms, $v_r \sim \nu/r$. Using the above expression for v_r, we can combine the mass conservation and angular momentum conservation equations into one:

$$\frac{\partial\Sigma}{\partial t} = -\frac{1}{r}\frac{\partial}{\partial r}\left[\frac{\frac{\partial}{\partial r}\left(r^3\nu\Sigma\partial\Omega/\partial r\right)}{\frac{\partial}{\partial r}\left(r^2\Omega\right)}\right]. \tag{7.11}$$

Thus the relaxation time for the surface density is $t_{\text{visc}} \sim r^2/\nu$.

To make quantitative statements about the surface density of a disk, it is clear that we must obtain some knowledge of the viscosity. Unfortunately, for reasons that will be explained shortly, no satisfactory model for disk viscosity has ever been advanced. As a result, for many years astrophysicists have been forced (by want of anything better) to simply parameterize their ignorance by guessing at the magnitude of the viscosity via dimensional analysis. Surprisingly, this simple device has proved to be quite powerful. In an extremely influential paper, Shakura and Sunyaev (1973) pointed out that the natural scale for *any* hydrodynamic stress is p, so that it makes sense to write the viscous stress as $-\alpha p$, where α might be as large as ~ 1. It is important to remember that the total pressure can include radiation pressure if the photons are closely coupled to the fluid. The effective viscosity implied by this parameterization is

$$\nu = \frac{\alpha c_s^2}{\gamma |\partial \Omega / \partial \log r|} \sim \alpha c_s^2 / \Omega \sim \alpha c_s h, \qquad (7.12)$$

where γ is the adiabatic index, and the final equality follows from equation 7.1.

Identifying an actual mechanism that produces a strong enough viscosity (i.e., an α that is not trivially small) has proven to be very difficult. Normal atomic scattering viscosity is hopelessly weak—in fact, if one can describe the disk as a fluid, the effective α of atomic viscosity is automatically tiny because the assumption of fluidlike behavior requires the atomic mean free path $\lambda_{\mathrm{mfp}} \ll h$.

One way that viscosity with a reasonable effective α might be realized is if some mechanism (to be determined) generated turbulence in the disk. In that case, the effective viscosity due to the turbulence would be $\nu_{turb} \sim \delta v_{turb} l_{turb}$, where δv_{turb} is the characteristic speed of the turbulent eddies, and l_{turb} is their size. Supersonic turbulence damps quickly as its energy is dissipated in shocks, so $\delta v_{turb} < c_s$; on the other hand, the size of the eddies can hardly be larger than the thickness of the disk, so $l_{turb} < h$. Therefore, we might expect that turbulence could produce an α that might approach unity but is unlikely to exceed it (Shakura and Sunyaev 1973).

However, there is no obvious reason why the disk should be turbulent. The Rayleigh stability criterion $d(r^2\Omega)/dr > 0$ is well satisfied for many potentials, including Keplerian. Under certain special circumstances, particularly when there is a hydrogen or helium ionization transition zone, disks can be convectively unstable. Since the interesting regions of AGN disks have temperatures $\sim 10^5$ K, HeII/HeIII ionization transitions may occur in some regions, but convection is unlikely to act throughout the disk. In addition, although considerable controversy existed on this point for a number of years (Ruden, Papaloizou, and Lin 1988; Ryu and Goodman 1992; Lin, Papaloizou, and Kley 1993; Kumar, Narayan, and Loeb 1995), it now appears that even when convection does exist, it tends to carry angular momentum *in* rather than *out* (Stone and Balbus 1996). Another possibility, again likely to be inappropriate to AGN disks, is that the disk is so cold that the electrons are degenerate. This would greatly stretch out their scattering lengths and correspondingly increase the viscosity coefficient (Paczyński 1978a). Photon viscosity has also been considered (e.g., Loeb and Laor 1992) but also seems inadequate in the AGN context.

A particularly interesting suggestion was made by Paczyński (1978b), who pointed out that if the viscosity due to other mechanisms is too low, matter will build up in the disk until it becomes self-gravitating in the vertical direction. The criterion for this to occur (see also §14.4.3) is equivalent to the familiar tidal limit

$$\Sigma \sim \frac{Mh}{r^3},$$
(7.13)

which implies that the disk mass

$$M_d \sim M\frac{h}{r}.$$
(7.14)

In such a case, gravitational collapse on scales $\sim h$ would create turbulence with an amplitude such that $\alpha \sim 1$.

7.2.2   Magnetized winds?

Others have suggested mechanisms, mostly based on ordered magnetic fields, in which the torque is not well described by a simple viscosity. Blandford (1976) and Lovelace (1976) both suggested that if the disk possesses an ordered poloidal field, angular momentum may be carried away to infinity by electromagnetic waves. Blandford and Payne (1982) pointed out that when an ordered poloidal field is present, any matter tied to the field lines can also remove angular momentum from the disk by being spun up to speeds greater than escape. Note that such a wind could help solve the angular momentum boundary condition problem discussed in §7.1; indeed, this idea was partially motivated by the desire to solve that problem.

To understand the magnetized wind mechanism at the order of magnitude level, imagine that the disk is threaded by a poloidal magnetic field strong enough that (at least near the disk surface) the Alfvén speed is much greater than the sound speed (see fig. 7.1). When that is so, the gas pressure at and above the disk surface has very little influence on the shape of the field lines; so long as the gas speed also remains small compared to the Alfvén speed, the fluid then has no choice but to slide along the field lines like a succession of beads on a wire. On the other hand, if the Alfvén speed deep inside the disk is small compared to the orbital speed (a likely situation when $c_s \ll v_{\rm orb}$ and the midplane density is much greater than the density at the surface), each field line is constrained to rotate with the orbital frequency of the point where it passes through the disk midplane. In this situation, the gas at the disk surface feels as if it is in an effective potential

$$\Phi_{\rm eff} = -\Omega^2(r_o)\left[\frac{1}{2}r^2 + \frac{r_o}{\sqrt{r^2 + (r - r_o)^2 \cot^2\theta}}\right],$$
(7.15)

where r_o is the cylindrical radius of the appropriate field line footpoint, θ is the angle between the field line and the rotational axis, and we have made use of the approximation $h \ll r$.

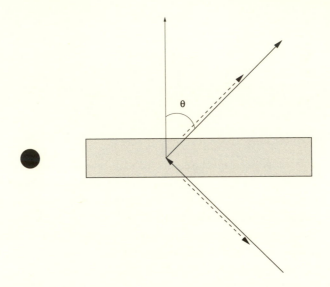

Fig. 7.1 A schematic view of (half) an accretion disk with a magnetized wind. Poloidal field lines (the heavy solid lines) run in from very large distances, pass through the disk and go back out again. Near the disk, their direction lies at an angle θ from the rotation axis (the light solid line). Low-pressure gas from the surface of the disk (the dashed lines) is flung outward along the field lines.

Because the gas in the disk is, after all, in stable circular orbit around the central mass, $\partial\Phi_{\mathrm{eff}}/\partial r = 0$ when $r = r_o$. When $\partial^2\Phi_{\mathrm{eff}}/\partial r^2 > 0$, this is a stable equilibrium, but when the second derivative is negative, it is unstable. Applying these criteria to the potential of equation 7.15, we find that the condition for instability is $\theta > 30°$. That is, if the field lines are swept back from the axis at even a rather modest angle, once a fluid element begins to move outward, the field forces it to rotate faster and faster because the component of gravity projected on the field line is insufficient to counter the centrifugal force felt in the rotating frame. This acceleration outward (and in the azimuthal direction) continues unabated until the fluid speed becomes comparable to the Alfvén speed.

The net result of this process is to force matter to leave the disk and fling it to infinity carrying a specific angular momentum $\sim r_A v_A$, where r_A is the radius at which the flow speed is comparable to to the local Alfvén speed v_A. Because this specific angular momentum can easily exceed $r_o v_{orb}(r_o)$, it is possible to vent a large amount of angular momentum by expelling a relatively small amount of mass. Whether this process can remove *enough* mass to account for the entire accretion torque depends sensitively on the upper structure of the disk. It also depends, of course, on the existence of field lines that are sufficiently swept back near the disk surface and reach radii very large compared to their starting points.

It is difficult to say a priori whether magnetized winds exist, or how strong they are, but it is possible in principle for them to be very efficient at generating torque on an accretion disk. Although it does not truly give the r–ϕ component of the stress tensor in terms of the gas pressure, we can define an effective "α"-parameter for a magnetized wind that does provide an estimate of its angular momentum-removal efficiency. To order of magnitude accuracy, α_{eff} can be estimated by taking the ratio of the torque on a given ring to the vertically integrated pressure there: $\alpha_{eff} \sim [r_A/h(r_o)](B^2[v_A/c_s(r_o)]^2$, where c_s is the sound speed in the disk. It is quite plausible (though hardly guaranteed) that the field strength at the Alfvén radius is not greatly weaker than the field in the disk proper, and that the field inside the disk has an energy density comparable to the disk pressure. If that is so, $\alpha_{eff} \sim r_A/h(r_o) \gg 1$.

7.2.3 Turbulent Maxwell stresses: the Velikhov-Chandrasekhar-Balbus-Hawley instability

As Shakura and Sunyaev (1973) presciently also pointed out, electromagnetic stresses can also transmit angular momentum through an accretion disk, and their contribution to the torque can be expressed in the language of the α formalism. The Maxwell stress tensor is

$$M_{ij} = \frac{1}{4\pi} \left[E_i E_j + B_i B_j - \frac{1}{2}(E^2 + B^2)\delta_{ij} \right], \qquad (7.16)$$

where $\vec{E}$ and $\vec{B}$ are the electric and magnetic fields. The electromagnetic torque between two rings can then be found in direct analogy to the viscous torque:

$$\mathcal{G}_m = \int d\phi\, r \int dz\, r M_{r\phi}. \qquad (7.17)$$

Plasmas generally have such high conductivity that electric fields are all shorted out, at least for slowly varying fields (see Appendix B). Taking only the magnetic part, the inter-ring torque reduces to

$$\mathcal{G}_m = \frac{1}{2}r^2 \int dz\, B_r B_\phi. \qquad (7.18)$$

If on average the field lines are trailing (e.g., $B_r > 0$, $B_\phi < 0$), $\mathcal{G}_m < 0$, and angular momentum is transported outward. Given the character of differential rotation in a disk around a point mass, this would seem the most likely configuration, but it is not absolutely required. If the magnetic field is so tangled that $B_r \sim B_\phi \sim B$, the electromagnetic stress can be incorporated into the α-formalism by a very simple device: just write it as $T_m = -\alpha_m p$ and let the total α be the sum of the magnetic part and the part due to turbulent viscosity.

It was only in the early 1990s, however, that a magnetic stress-based torque-creation mechanism was suggested that appears to be sufficiently generally applicable, and sufficiently powerful, to do the job. This mechanism is based on the recognition that although the Rayleigh stability criterion, $d(r^2\Omega)/dr > 0$ holds for *fluid* systems, it does not necessarily apply to *magneto*-hydrodynamic systems. Balbus and Hawley (1991), rediscovering an instability originally described by Velikhov (1959) and Chandrasekhar (1981), pointed out that a radial perturbation of an initially poloidal magnetic field is unstable, and the instability's growth rate can be as rapid as the orbital frequency. The basic free energy for this instability comes from the shear, but the magnetic field, rather than atomic (or turbulent eddy) diffusion, establishes the coupling between different radii. In that sense, it is fundamentally nonviscous and therefore conceptually quite different from any mechanism that can be written as an α-model.

A qualitative picture of this instability is surprisingly simple (see fig. 7.2). Because the mean position of a perturbed field line is its initial location, it continues to orbit with the frequency of its initial radius. However, where $\delta B_r > 0$, that is, where the field lines loop out to larger radius, this orbital frequency is larger than the one that produces a circular orbit. Similarly, where the field lines are drawn inward, the mean field line orbital frequency is too low for the material attached to that field line to stay at the perturbed radius. Consequently, the field loops stretch in the radial direction. Larger radial perturbations lead to still larger mismatch, and exponential growth results.

Although this mechanism is quite general, its essential physics can be seen in a stripped-down model. Suppose that in the equilibrium state of

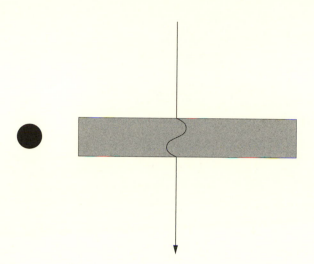

Fig. 7.2   Schematic view of an axisymmetric perturbation to a
weak poloidal magnetic field in an accretion disk. Because the perturbed loop rotates with the orbital frequency of its base, and that
frequency is greater than the orbital frequency where the loop has
been stretched out, but less than the orbital frequency where the
loop has been pulled in toward the black hole, material attached to
it moves farther and farther away from its starting point.

the disk there is a dynamically unimportant (i.e., $B^2/(8\pi) \ll p$, where p is
the total pressure) magnetic field that runs parallel to the disk axis $\hat{z}$. In
order to concentrate attention on the magnetic effects, suppose also that all
the gas in the disk has the same specific entropy, and that there is no radial
pressure gradient in equilibrium. We further restrict the perturbations by
setting $\delta v_z = 0$; this has the effect of eliminating sound waves and fixing
$\delta B_z = 0$, but not materially altering the growth of the instability. Now
consider what happens to perturbations with wavevector $\vec{k} = k\hat{z}$ and frequency ω. The equations governing the development of these perturbations
are the linearized MHD equations (Appendix B; also note that because the
equilibrium state has rotation, the perturbation equations are effectively
written in the equilibrium rotating frame):

$$-i\omega \frac{\delta\rho}{\rho} = 0, \qquad (7.19)$$

$$-i\omega\delta v_r - 2\Omega\delta v_\phi - \frac{ikB}{4\pi\rho}\delta B_r = 0, \tag{7.20}$$

$$-i\omega\delta v_\phi + \frac{1}{2}\Omega\delta v_r - \frac{ikB}{4\pi\rho}\delta B_\phi = 0, \tag{7.21}$$

$$-i\omega\delta B_r - ikB\delta v_r = 0, \tag{7.22}$$

$$-i\omega\delta B_\phi - \frac{d\Omega}{d\ln r}\delta B_r - ikB\delta v_\phi = 0, \tag{7.23}$$

and the adiabatic condition, which in this simplified context reduces to

$$\gamma\frac{\delta\rho}{\rho} = \frac{\delta p}{p} = 0. \tag{7.24}$$

Combining these equations results in the linear dispersion relation

$$\frac{\omega^2}{\Omega^2} = q^2 + \frac{1}{2}\left(1 \pm \sqrt{1 + 16q^2}\right), \tag{7.25}$$

where the reduced wavenumber $q = kv_A/\Omega$, for the Alfvén speed $v_A = B/\sqrt{4\pi\rho}$, and we have set $d\Omega/d\ln r = -(3/2)\Omega$, as is appropriate for a Keplerian disk. The upper sign corresponds to purely oscillating internal waves; the mode we are interested in is the one whose frequency is given by equation 7.25 with the lower sign. For small q, the growth rate is $\text{Im}(\omega) \simeq \sqrt{3}q\Omega$, but physically it is clear that k cannot be much smaller than $\sim 1/h$, so $q > v_A/c_s$. The maximum growth rate is achieved at $q = \sqrt{15/16}$, where it is $(3/4)\Omega$. For $q \geq \sqrt{3}$, the mode is stable. Thus we see that the perturbations can grow on the orbital timescale, provided $v_A < c_s$.

At first it might seem paradoxical that a magnetic instability is important only when the magnetic field is comparatively weak. The reason that this is so is that the magnetic field is used only as a binding element; as already remarked above, the free energy drawn upon by the instability is all in the orbital shear flow. Indeed, the mode becomes damped when the magnetic field becomes comparable in energy density to the gas pressure because at that point the energy cost of bending the field becomes comparable to the energy gain provided by radial motion of fluid elements. Yet another way to see this point is to observe that when the dispersion relation is written in terms of the reduced wavenumber q, the strength of the field becomes invisible. The only role the magnetic field strength plays is to define the characteristic wavenumber scale Ω/v_A of the instability. Stronger field just means that significant magnetic tension forces can be derived from longer wavelength perturbations.

With such a large growth rate, nonlinear amplitude can be reached in just a few orbital periods. MHD simulations indicate that $\langle \delta B^2 \rangle$ can quickly exceed the original unperturbed energy density in magnetic field. Consequently, the primary limit to growth is the point at which $\langle \delta B^2 \rangle$ becomes comparable to the gas pressure.

Because the entire instability is founded on magnetic forces pushing matter to rotate either faster or slower than the local orbital frequency, it is obvious that torque across radial rings is intrinsic to its nature. The guess by Shakura and Sunyaev that Maxwell stresses may be at least as important as Reynolds (i.e., fluid) stresses is thus vindicated. In the linear regime, it is easy to show that for $q \sim 1$, the Maxwell stress is always comparable to the Reynolds stress and has the correct sign. Numerical simulations of idealized problems indicate that, if anything, in the nonlinear regime the Maxwell stresses may be a few times *greater* than the Reynolds stresses (Stone et al. 1996; Brandenburg et al. 1996).

There is another regard in which torque via this mechanism is fundamentally different from viscous torque. Even if it turns out that the Maxwell stress is, in fact, $\sim -\alpha p$, there is no dissipation directly connected with it. Therefore, the normal formulation of the energy equation (eq. 7.32), in which there is local heat production in proportion to the local torque because the torque is proportional to the local viscosity, may not be correct. Nonlinear processes ultimately transfer energy from long wavelength motions to small, where dissipative processes work most efficiently, but it is entirely possible for energy to move from place to place within the disk in the course of the waves' propagation.

7.3 Time-Steady Disks

7.3.1 Radial structure: angular momentum and energy conservation

Granting ourselves the assumption that *something* creates a stress adequate to the job, let us now consider the simple case of time-steady disks. Like any other dynamical system, the structure of an accretion disk is determined by the conservation of mass, momentum, and energy. In this case, these equations become particularly simple: Because we are focusing on vertically integrated properties of azimuthally symmetric disks, the only interesting spatial variation is in the radial direction, and the only interesting momentum component is the vertical component of the angular momentum. Energy conservation can be separated from the other two equations under the assumption that any heat created by dissipation is radiated locally. Finally, because we are specializing to time-steady disks, conservation of mass and angular momentum amounts to the simple specification of the two associated conserved fluxes. Consequently, solving for the structure of a time-steady disk reduces to choosing a mass accretion rate, solving the local angular momentum conservation equation for the surface density required to produce that accretion rate, and separately finding the temperature at which the local dissipation rate is balanced by radiation.

That mass conservation in a time-steady disk is equivalent to conservation of the mass flux can be seen from the fact that the time-steady mass conservation equation (eq. 7.5) has an immediate integral, the mass inflow rate:

$$\dot{M} = -2\pi r \Sigma v_r. \tag{7.26}$$

Similarly, the angular momentum conservation equation (eq. 7.6) can be integrated to yield:

$$\frac{\dot{M}\Omega}{2\pi} + \frac{C}{r^2} = -\int dz\, T_{r\phi}, \tag{7.27}$$

which has been simplified by the use of equation 7.26. The constant C can be interpreted by looking for a radius r_m where the stress in the disk is zero. For example, if we were considering an accretion disk around a star, this might occur at the stellar surface, where the material has been forced to come into corotation with the star. Around a black hole, this location again lies at the inner edge of the disk, but for the opposite reason: we expect the stress there to go to zero as the material peels away from the last stable orbit and falls into the hole. In either case, at the point where the stress is zero, $C = -r_m^2\Omega(r_m)\dot{M}/(2\pi)$, that is, C is the angular momentum flow in the disk, a conserved flux in the same sense that the mass accretion rate $\dot{M}$ is. Note that the outward angular momentum flux is identical to the entire angular momentum of the accreting material in its innermost orbit times the mass accretion rate; at larger radii, the outward angular momentum flux is a progressively smaller fraction of the angular momentum possessed by the accreting matter.

It is convenient to rewrite equation 7.27 in a more compact form:

$$\frac{\dot{M}\Omega(r)}{2\pi} R(r) = -\int dz\, T_{r\phi}, \tag{7.28}$$

where the *reduction factor* is

$$R(r) = 1 - \left(\frac{r_m}{r}\right)^2 \frac{\Omega(r_m)}{\Omega(r)}. \tag{7.29}$$

In a Keplerian potential,

$$R(r) = 1 - \left(\frac{r_m}{r}\right)^{1/2}.$$ (7.30)

Given a physical model relating the stress to the local surface density and temperature, equation 7.28 combined with the local heat balance equation would determine the entire radial structure of the disk.

For example, if we adopt the α-parameterization, equation 7.28 becomes

$$\int dz\, p(z) = \frac{\dot{M}\Omega}{2\pi\alpha} R(r)$$
$$= \frac{2}{e\alpha} \frac{L}{L_{\rm E}} \frac{c^2}{\kappa_{\rm T}} x^{-3/2} R(x).$$ (7.31)

Here $\kappa_{\rm T}$ is the Thomson opacity per unit mass and $x = r/r_g$. The first version of equation 7.31 shows explicitly how the local temperature and surface density are constrained by the twin requirements of mass and angular momentum conservation; the second, dimensionless, version shows that the integrated pressure through the disk at any particular normalized radius depends only on $L/L_{\rm E}$ and α. There is *no* dependence on the absolute luminosity scale.

Our next step is to use the fact that energy is conserved. As the fluid loses angular momentum and gradually falls inward, it also loses potential energy. Part of that energy goes into orbital kinetic energy; part is carried to outer rings of the disk, accompanying the outward transport of angular momentum; part may be put into MHD waves; and part is transformed locally into heat if the torque is dissipative. If we make the assumption that the torque is essentially viscous (more precisely, that the dissipation is proportional to the local shear), the local rate of dissipation into heat is

$$Q = \nu\Sigma \left(r \frac{d\Omega}{dr}\right)^2.$$ (7.32)

To put this heating rate in terms of the mass accretion rate, we combine equations 7.8 and 7.28 to write $\nu\Sigma$ as

$$-\nu\Sigma = \frac{\dot{M}}{2\pi} \frac{d\ln r}{d\ln\Omega} R(r).$$ (7.33)

From this form for $\nu\Sigma$, we find the heat release per unit disk area:

$$Q = \frac{\dot{M}}{2\pi}\Omega^2 \left|\frac{d\ln\Omega}{d\ln r}\right| \left[1 - \left(\frac{r_m}{r}\right)^2 \left(\frac{\Omega_*}{\Omega}\right)\right].$$ (7.34)

In a Keplerian potential this is

$$Q = \frac{3GM\dot{M}}{4\pi r^3} R(r). \tag{7.35}$$

If we assume that all the dissipated heat is radiated (this assumption is not always justified; see §7.3.5) and integrate the radiated luminosity over a section of a Keplerian disk far outside r_m, we find

$$L(r_1, r_2) = \frac{3}{2} \left(\frac{GM\dot{M}}{r_1} - \frac{GM\dot{M}}{r_2} \right). \tag{7.36}$$

That is, the energy radiated in this region is 3/2 times the amount of gravitational potential energy lost by the accreting matter. Where does this extra energy come from? The answer is that potential energy released closer in is carried out with the angular momentum because the outer disk must be spun up and pushed outward as a corollary of the torque on the inner portions. There is no paradox in total, however, when we use the $R(r)$ factor in the inner region. Integrating over the whole disk using that factor results in a total luminosity of

$$L_{\text{tot}} = \frac{1}{2} \frac{GM\dot{M}}{r_m}, \tag{7.37}$$

where the factor 1/2 is only correct for a nonrelativistic disk (for a disk around a black hole, the correct coefficient is the one that corresponds to the binding energy at the marginally stable orbit). Now we must ask why the heat loss is *less* than the potential energy loss. The answer to this is that when the accretion flow moves inside the zero shear-stress ring it still carries kinetic energy, and this energy—half the total in a Kepler potential—is not available for radiation. In fact, the decrease in dissipation as the zero shear-stress ring is approached means that the maximum surface brightness is actually found slightly outside this radius.

7.3.2   Relativistic effects

So far in this discussion the only allowance we have made for relativistic effects is the boundary condition at the inner edge of the disk, where we have forced the stress to go to zero. A more careful approach incorporates relativistic dynamics in several places, including the conservation equations we have just derived.

Novikov and Thorne (1973) and Page and Thorne (1974) demonstrated that, when the disk is time steady and its rotation axis is parallel to the rotation axis of the black hole, all the relativistic effects relevant to disk dynamics can be encapsulated in three multiplicative functions, all of which approach unity as $x \to \infty$. Thus, by using these three functions, the radial structure of the disk can be found in a manner very similar to the procedure appropriate to Newtonian disks.

Specifically, assuming only that the disk is geometrically thin, the energy dissipation equation (eq. 7.35) can be written as

$$Q = \frac{3GM\dot{M}}{4\pi r^3} R_R(x). \tag{7.38}$$

That is, the ordinary reduction factor simply changes form to accommodate the relativistic corrections. Similarly, the angular momentum conservation equation (eq. 7.28) becomes

$$-\int dz\, T_{r\phi}(z) = \frac{\dot{M}\Omega_{\mathrm{K}}(r)}{2\pi} R_T(x), \tag{7.39}$$

where Ω_{K} is the Keplerian orbital frequency (as opposed to the true, general relativistic one). The third correction factor applies to the vertical component of gravity:

$$g_z = -\frac{GMz}{r^3} R_z(r). \tag{7.40}$$

Because we are concerned with relativistic effects, it is, of course, very important to be clear about the frame of reference and coordinate system with respect to which quantities are measured. Here the radius is the radial coordinate of the Boyer-Lindquist metric, equation 5.41, and it coincides with the radial coordinate of the locally nonrotating frame. M and $\dot{M}$ are the values as measured at infinity. Q, $T_{r\phi}$, and g_z are measured in the locally co-moving frame of the fluid; these forms assume that the fluid is following a circular orbit at radius r. It is especially important to distinguish Q *as it is seen by a distant observer* from Q as it is used in equation 7.38, its value in the fluid frame (see §§5.3 and 7.5.5).

In terms of a_*, the black hole angular momentum normalized to M, the three functions are

$$R_R(x) = \frac{C(x)}{B(x)}, \tag{7.41}$$

$$R_T(x) = \frac{C(x)}{A(x)}, \tag{7.42}$$

and (following a correction due to Abramowicz, Lanza, and Percival 1997),

$$R_z(x) = x^{-1}\left[L^2 - a_*^2\left(E_\infty - 1 \right) \right], \tag{7.43}$$

where L and E_∞ are the constants of motion for circular orbits (prograde and in the equatorial plane of the black hole if it is spinning; see also §5.1). The functions A, B, and C are

$$A(x) = 1 - \frac{2}{x} + \frac{a_*^2}{x^2}, \tag{7.44}$$

$$B(x) = 1 - \frac{3}{x} + \frac{2a_*}{x^{3/2}}, \tag{7.45}$$

and

$$C(x) = 1 - \frac{y_{\mathrm{ms}}}{y} - \frac{3a_*}{2y} \ln\left(\frac{y}{y_{\mathrm{ms}}}\right) - \frac{3(y_1 - a_*)^2}{yy_1(y_1 - y_2)(y_1 - y_3)} \ln\left(\frac{y - y_1}{y_{\mathrm{ms}} - y_1}\right)$$
$$- \frac{3(y_2 - a_*)^2}{yy_2(y_2 - y_1)(y_2 - y_3)} \ln\left(\frac{y - y_2}{y_{\mathrm{ms}} - y_2}\right)$$
$$- \frac{3(y_3 - a_*)^2}{yy_3(y_3 - y_1)(y_3 - y_2)} \ln\left(\frac{y - y_3}{y_{\mathrm{ms}} - y_3}\right), \tag{7.46}$$

with $y = x^{1/2}$. The special values of y in these equations are y_{ms}, the value of y at the marginally stable orbit, and $y_{1,2,3}$, the three roots of $y^3 - 3y + 2a_* = 0$. L and E_∞ can similarly be written in terms of functions going to unity at large radius:

$$L(x) = x^{1/2} \frac{1 - 2a_* x^{-3/2} + a_*^2 x^{-2}}{B^{1/2}(x)} \tag{7.47}$$

and

$$E_\infty(x) = \frac{1 - 2/x + a_* x^{-3/2}}{B^{1/2}(x)}. \tag{7.48}$$

The dependence of these three functions on radius is illustrated in figure 7.3. For fixed a/M, R_R and R_T are very similar to each other, but plotted as a function of x, both depend strongly on a/M. If they had been displayed as functions of x/x_{ms}, the dependence on a/M would appear much weaker. No matter how they are considered, however, these factors are significantly different from the Newtonian reduction factor even at distances quite far from the black hole. In fact, there is hardly any significant regime (in logarithmic terms) in which the Newtonian reduction factor is different enough from unity to be interesting and yet is also approximately correct. By contrast, $R_z \simeq 1$ except very near the black hole; there the vertical gravity can be somewhat stronger than would be predicted by Newtonian theory.

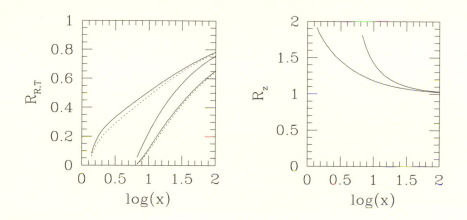

Fig. 7.3 The general relativistic reduction factors. (Left) R_R (solid curves) and R_T (dotted curves). The solid curve in the middle is the Keplerian $R(x)$, with $x_* = 6$. The curves extending to $x = 1.24$ are for a Kerr black hole with $a/M = 0.998$ (the effective maximum a_*; see §5.1.4); those stopping at $x = 6$ are for Schwarzschild black holes. (Right) R_z, again for a Schwarzschild black hole and a Kerr black hole with $a/M = 0.998$.

Because these reduction factors are so significant, the radius x_{max} where the maximum dissipation per unit area occurs is somewhat outside the marginally stable orbit, and the heating rate there falls considerably below the simple x^{-3} scaling suggested by equation 7.35. For example, when $a = 0$, $x_{ms} = 6$, but $x_{max} \simeq 9.5$ and $R_R(x_{max}) \simeq 0.50$. Similarly, in the opposite limit of $a = 0.998$, whereas $x_{ms} \simeq 1.3$, $x_{max} \simeq 1.56$, and $R_R(x_{max}) \simeq 0.15$.

7.3.3 LTE spectrum

The expressions we have just derived also demonstrate an important fact that, though obvious, is sometimes easy to lose amid the details of disk structure: If all the energy released by accretion is dissipated into heat within the disk, and if all the heat is radiated, the luminosity is independent of the viscosity, provided the accretion rate is fixed. In fact, if the local radiation is thermal, so that the surface brightness is independent of disk thickness, the run of surface temperature also depends only on the accretion rate and is independent of the viscosity. In terms of the accretion efficiency

η, it is

$$T_s = \left(\frac{Q}{2\sigma}\right)^{1/4} = \left[\frac{3GM\dot{M}}{8\pi\sigma r^3}R_R(r)\right]^{1/4}$$

$$= 6.8 \times 10^5 \eta^{-1/4} \left(\frac{L}{L_E}\right)^{1/2} L_{46}^{-1/4} R_R^{1/4}(x) x^{-3/4} \text{ K,} \quad (7.49)$$

independent of ν, as advertised. Far enough from the inner edge that $R_R(x) \simeq 1$, $T_s \propto x^{-3/4}$, but the temperature scale varies $\propto (\dot{M}/M^2)^{1/4} \propto (L/L_E)^{1/2}L^{-1/4}$. That is, we can expect the characteristic temperature to *decrease* with increasing luminosity (if L/L_E is constant), even though there is more light to radiate, because the larger central black hole spreads the radiation over an even larger area. Note that the fiducial temperature in equation 7.49 is merely the temperature in the outer disk extrapolated with constant logarithmic slope to $x = 1$; because R_R falls well below unity, the maximum temperature actually achieved is rather smaller than the fiducial value shown. For matter accreting onto a Kerr black hole with $a = 0.998$, the maximum temperature—as measured in the fluid frame—is only 0.44 times the fiducial value. At the opposite extreme in black hole spin (a Schwarzschild black hole), the maximum temperature is only 0.10 times the fiducial value.

In this limit, that is, when the disk is geometrically thin and in LTE, we can immediately derive the total radiated spectrum. Remarkably, its low-frequency shape is universal, and only its high-frequency cutoff depends on any specific characteristics of the disk. It is easy to derive because all LTE surfaces at the same temperature emit the same Planckian spectrum. All we need do is weight by the radiating surface area at each temperature and integrate in to the radius of maximum surface brightness:

$$L_\epsilon = 4\pi r_g^2 \int_{x_{\max}} dx\, x \frac{2\epsilon^3/(h^3 c^2)}{\exp[\epsilon/k_B T_s(x)] - 1}, \quad (7.50)$$

where L_ϵ is the luminosity per unit photon energy ϵ and k_B is Boltzmann's constant (note that the total luminosity counts *both* sides of the disk). Changing integration variable to $u = \epsilon/[k_B T_s(x)]$ and approximating $T_s(x)$ by $T_o x^{-3/4}$ (cf. eq. 7.49) transforms the integral to

$$L_\epsilon = \frac{32\pi}{3}\frac{r_g^2}{c^2 h^3}(k_B T_o)^{8/3}\epsilon^{1/3}\int_{u_{\max}} du\, \frac{u^{5/3}}{e^u - 1}, \quad (7.51)$$

where $u_{\max} = \epsilon x_{\max}^{3/4}/(k_B T_o)$. The only dependence on ϵ in the dimensionless integral is through its lower bound. When $u_{\max} \ll 1$, the integral effectively runs from 0 to ∞ and has a value of order unity (the integral is dominated by $u \sim 1$). When $u_{max} \gg 1$, the integrand is dominated by the lower bound because the argument of the exponential in the denominator is always larger than one. Thus, the generic LTE disk spectrum can be written approximately as

$$L_\epsilon \propto \epsilon^{1/3} \exp \left[-\frac{\epsilon}{k_B T_s(x_{max})} \right]. \qquad (7.52)$$

All disks of this sort should have $L_\epsilon \propto \epsilon^{1/3}$ at low frequencies (Lynden-Bell 1969); the only changes made by the specifics of the disk are the proportionality constant and the frequency of the exponential cutoff, which depends on $T_s(x_{max})$.

Another consequence of the LTE radiation assumption is that specific photon frequencies map to specific radii in the disk. If the local spectrum is a Planckian, very little light at energy ϵ is emitted anywhere $\epsilon \gg k_B T_s$, that is, outside the radius r_ϵ at which $k_B T_s(r_\epsilon) \simeq \epsilon$. On the other hand, at smaller radii, where $\epsilon \ll k_B T_s$, the blackbody spectrum is well approximated by the Rayleigh-Jeans expression $B_\epsilon = 2\epsilon^2/(h^3 c^2) k_B T_s$. Therefore, we can write the total radiated energy at energy ϵ as

$$L_\epsilon \simeq \int_{r_{max}}^{r_\epsilon} dr\, 2\pi r \frac{2\epsilon^2}{h^3 c^2} T_s(x_{max}) \left(\frac{r}{r_{max}} \right)^{-3/4}. \qquad (7.53)$$

The integrand depends on r as $r^{1/4}$, so the integral is dominated by its upper limit, r_ϵ. In other words, most of the light at energy ϵ is emitted near the radius r_ϵ.

One might now ask, "Under what circumstances is the disk thermal?" Thermodynamic equilibrium is the result of having both large optical depth and high density. Although equation 7.31 constrains the vertically integrated pressure (in the α-model), it does not by itself define the density and optical depth. A closer look at the vertical structure is required to gain this information.

7.3.4 Vertical structure

Two processes determine the most important aspects of the disk's vertical structure: the diffusion of heat, and force balance between gravity and a combination of gas and radiation pressure. We shall treat the latter process in either of two limits, gas pressure dominated or radiation pressure dominated.

7.3.4.1   heat diffusion

In principle, numerous mechanisms could distribute heat through the disk: radiative diffusion, heat conduction by gas atoms, convection, and turbulent heat conduction. Our ignorance of how dissipation is distributed through the disk makes categorical statements difficult, but it is likely that where the bulk of the radiation occurs, radiative diffusion is the most important. However, there are circumstances in which any of the others could also dominate.

Consider radiative diffusion first. If both photon scattering and the creation of new photons are isotropic, radiative diffusion is described by the transfer equation

$$\cos\theta \frac{\partial I}{\partial \tau} = J - I + \frac{\partial f}{\partial \tau}\mathcal{F}_{\text{rad}}(\tau_{\text{tot}}), \tag{7.54}$$

where θ is the angle between the photon travel direction and the disk normal, I is the intensity, J is the mean intensity, $\mathcal{F}_{\text{rad}}$ is the flux, and τ the optical depth, which we define as running from 0 at the midplane to τ_{tot} at the outer edge of the disk. The function $f(\tau)$ encapsulates our ignorance of the location of dissipation; the only a priori constraints we can put on it are that $f(0) = 0$ and $f(\tau_{\text{tot}}) = 1$. It is convenient to factor out $f(\tau)$ because we know $\mathcal{F}_{\text{rad}}(\tau_{\text{tot}})$; it is $(3/8\pi)R_R(r)GM\dot{M}/r^3$ (eq. 7.38; note that $\mathcal{F}_{\text{rad}}$ is the flux leaving *one* of the two sides of the disk). Multiplying equation 7.54 by $\cos\theta$ and integrating gives

$$\frac{\partial K}{\partial \tau} = -\mathcal{F}_{\text{rad}}(\tau) = -f(\tau)\mathcal{F}_{\text{rad}}(\tau_{\text{tot}}), \tag{7.55}$$

where K is the second moment of I with respect to $\cos\theta$. Conveniently, K/c is the radiation pressure. The second equality follows from the definition of $f(\tau)$. The solution for $K(\tau)$ is thus reduced formally to simply an integration of $f(\tau)$ with a boundary condition that $K(0) \gg K(\tau_{\text{tot}})$ if $\tau_{\text{tot}} \gg 1$. If we make the simple guess that $f(\tau) = \tau/\tau_{\text{tot}}$ (i.e., the dissipation per unit optical depth is constant), we find

$$K(\tau) = \frac{1}{2}\tau_{\text{tot}}\mathcal{F}_{\text{rad}}\left[1 - \left(\frac{\tau}{\tau_{\text{tot}}}\right)^2\right]. \tag{7.56}$$

In other words, the radiation pressure at the disk center is $\sim \tau_{\text{tot}}$ times its value near the outer edge.

Let us now turn to thermal conduction. Thermal conduction and viscosity are closely related. Whenever particles wander from place to place without a collision, they carry with them all their conserved quantities, including both energy and momentum. Consequently, the transport co-efficient for heat is in general comparable to the transport coefficient for momentum. Thus, for the same reason that atomic viscosity is generally negligible in accretion disks, so, too, is atomic heat conduction. On the other hand, if there is some dynamical instability that creates fluid turbulence, the efficiency of entropy transport by fluid eddies is automatically comparable to the efficiency of momentum transport by those same eddies. To order of magnitude, then, the importance of turbulent heat conduction relative to radiative diffusion is

$$\frac{\mathcal{F}_{\text{heat}}}{\mathcal{F}_{\text{rad}}} \sim \alpha \tau_{\text{tot}} \frac{c_s}{c} \frac{p_g}{p_{\text{rad}}}, \tag{7.57}$$

where p_g/p_{rad} is the ratio of gas to radiation pressure.

Last, we discuss the possible role of convection in vertical heat transport. Although the turbulence that mixes entropy in turbulent heat conduction is created by some other dynamical instability, the turbulence that mixes entropy in convection depends for its existence on having an entropy gradient directed parallel to gravity. Whether the entropy gradient has this sign depends strongly, of course, on the distribution of dissipation. When convection does occur, its importance relative to radiative diffusion can be gauged in a way very similar to the way we have just done for turbulent conduction.

7.3.4.2 gas-pressure-dominated vertical support

Suppose, first, that gas pressure provides most of the vertical support against gravity. In the thin disk approximation, the equation of hydrostatic equilibrium in the vertical direction is

$$\frac{\partial p}{\partial z} = \rho g_z = -\frac{GM\rho z}{r^3} R_z(r). \tag{7.58}$$

To get an order of magnitude feel for what happens, let us make the approximation that the temperature is constant with height. Then the hydrostatic equilibrium equation becomes

$$\frac{k_{\text{B}}T}{\mu} \frac{\partial \rho}{\partial z} = -\frac{GM\rho z}{r^3} R_z. \tag{7.59}$$

This simple differential equation has the solution

$$\rho(z) = \rho(0) \exp\left(-\frac{GM\mu R_z}{2r^3 k_{\text{B}}T} z^2\right). \tag{7.60}$$

Thus, this simplest limit gives a Gaussian roll-off in density with a scale height

$$h = \left(2k_{\mathrm{B}}Tr^3/GMR_z\mu\right)^{1/2} \sim \frac{c_s}{\Omega}$$

$$= 1.7 \times 10^9 T_5^{1/2} \left(\frac{L}{L_{\mathrm{E}}}\right)^{-1} L_{46} x^{3/2} R_z^{-1/2} \text{ cm.} \tag{7.61}$$

As we have already seen, if the disk is optically thick (a guess we shall verify shortly), the central radiation pressure (and therefore temperature) can be rather greater than the surface temperature: $T_c \sim \tau_{\mathrm{tot}}^{1/4} T_s$ if the LTE approximation applies. In evaluating the vertical thickness of the disk, it is of course the *maximum* scale height that determines the total, so the appropriate temperature to use in estimating h is T_c, not T_s.

In order to find that temperature correction, we approximate the vertically integrated pressure as

$$\int dz\, p(z) \simeq \frac{\Sigma k_{\mathrm{B}} T_c}{\mu}. \tag{7.62}$$

Using the fact that $\tau_{\mathrm{tot}} = \kappa\Sigma/2$, and supposing that the radiation is in LTE, we find

$$\Sigma = \kappa^{-1/5} \left[\frac{\mu \dot{M}\Omega R_T(r)}{2^{3/4}\pi\alpha k_{\mathrm{B}} T_s}\right]^{4/5}. \tag{7.63}$$

In §7.5.1 we discuss the various opacity mechanisms that are relevant to AGN disks; at these densities and temperatures Thomson opacity is generally the single most important opacity, although free-free and bound-free absorption by H and He (and in the highest density regions, a variety of processes involving other elements) can be competitive. When Thomson opacity dominates all others, yet there is enough absorptive opacity to justify the LTE approximation,

$$\Sigma = 2.1 \times 10^6 \alpha^{-4/5} \eta^{-3/5} \left(\frac{L}{L_{\mathrm{E}}}\right)^{2/5} L_{46}^{1/5} R_T^{4/5}(x) R_R^{-1/5}(x) x^{-3/5} \text{gm cm}^{-2}. \tag{7.64}$$

Given a Thomson opacity of 0.33 cm^2 gm^{-1}, gas-pressure-dominated disks should be very optically thick except for extreme values of L/L_{E} or x.

Such a large column density can boost T_c by a subtantial amount over T_s. Taking this effect into account, we find a scale height

$$h = 2.2 \times 10^{10} \alpha^{-1/10} \eta^{-1/5} \left(\frac{L}{L_{\mathrm{E}}}\right)^{-7/10} L_{46}^{9/10}$$

$$\times [R_R(x)R_T(x)]^{1/10} R_z^{-1/2}(x) x^{21/20} \text{cm.} \tag{7.65}$$

Because the thickness increases slightly more rapidly than the radius, the disk flares slowly toward the outside. This flare raises the interesting possibility of irradiation of the outer portion of the disk by photons created in the inner portion. We shall return to this thought at the end of this chapter and again in Chapter 8. From the surface density and the scale height, the central density follows:

$$\rho_c = 5.6 \times 10^{-5} \alpha^{-7/10} \eta^{-2/5} \left(\frac{L}{L_{\rm E}} \right)^{11/10} L_{46}^{-7/10}$$
$$\times R_T^{7/10}(x) R_R^{-3/10}(x) R_z^{1/2}(x) x^{-33/20} \text{ gm cm}^{-3}. \tag{7.66}$$

7.3.4.3 radiation-pressure-dominated vertical support

If the blackbody radiation assumption holds, the gas pressure at the disk midplane rises as one moves inward as $x^{-51/20}$. However, the radiation pressure rises $\propto x^{-18/5}$. The steeper dependence of radiation pressure on radius suggests that at sufficiently small radius radiation pressure might actually contribute more to vertical support than gas pressure does.

To show more clearly how radiation pressure acts, we rewrite the equation of hydrostatic equilibrium explicitly including it:

$$\frac{\partial p_g}{\partial z} - \frac{\kappa \rho}{c} \mathcal{F}_{\rm rad} = -\frac{GM\rho z}{r^3} R_z(r), \tag{7.67}$$

where κ is the Rosseland mean opacity per unit mass. As in the discussion of heat diffusion, it is convenient to write $\mathcal{F}_{\rm rad} = f\mathcal{F}_{\rm rad}(\tau_{\rm tot})$. If radiation pressure entirely dominates gas pressure, the hydrostatic equilibrium equation then becomes

$$\frac{3}{8\pi} \frac{\kappa \dot{M}}{c} R_R(r) f(z) = z R_z(r). \tag{7.68}$$

When the opacity is independent of z, it is clear that the only possible dissipation profile is $f(z) = z/h$ for $z \leq h$, and $f(z) = \rho(z) = 0$ for $z > h$. Moreover, the thickness of the disk depends only on $\dot{M}$ and κ and is almost (modulo the reduction factors) independent of r:

$$h = \frac{3}{8\pi} \frac{\kappa \dot{M}}{c} \frac{R_R(r)}{R_z(r)} = 1.5 \times 10^{13} \frac{L_{46}}{\eta} \frac{\kappa}{\kappa_{\rm T}} \frac{R_R(r)}{R_z(r)} \text{ cm.} \tag{7.69}$$

We normalize the opacity to the Thomson opacity because it so often dominates. It is especially important to emphasize that the disk thickness in this regime is completely independent of α. In fact, because ρ cancels out of the hydrostatic equilibrium equation in the radiation-pressure-dominated limit, $\rho(z)$ is also completely unspecified; only $f(z)$ is fixed. The only assumption required to achieve this solution is that the energy released by accretion is dissipated locally.

The fact that the balance between gravity and radiation pressure is independent of the local density is the analogue in disk geometry of the Eddington luminosity argument (§6.2). Both radiation force and gravity are proportional to density, so their ratio is independent of density. However, the relative magnitude of the radiation flux and the gravity depends on geometrical symmetry. In the spherical case, if all the flux is made in the center, the ratio of flux to gravity is constant because both fall with radius as r^{-2}. However, in disks, if all the dissipation occurs at the midplane, the ratio of flux to gravity falls upward, for the flux is constant while the gravity increases $\propto z$. It is for this reason that the flux must increase with altitude if radiation pressure is to support the disk against gravity.

The disk flare in the radiation-pressure-dominated regime can be found even more simply:

$$\frac{h}{r} = \frac{3}{2} \frac{\kappa}{\kappa_{\mathrm{T}}} \frac{L}{L_{\mathrm{E}}} \frac{R_R(x)}{\eta x R_z(x)}. \tag{7.70}$$

This expression shows clearly how the disk can become geometrically thick in its inner portions if L/L_{E} is close to unity. If there is any mechanism (e.g., free-free absorption in some circumstances) that significantly increases the opacity above Thomson, puffing is further encouraged.

To find the surface density in the radiation-pressure-dominated regime, we must assume something about the nature of the torque. Here we shall again adopt the α-model, but using the radiation pressure instead of the gas pressure. Note that this not an obvious step; it could well be that the viscosity scales in proportion to the gas pressure (or depends in some complicated way on both p_g and p_{rad}) even when most of the vertical support comes from radiation pressure.

Given this assumption, the function $p_{\mathrm{rad}}(z)$ can be found in (almost) the same fashion as in the discussion of heat diffusion. The only difference is that we now *know* $f(z)$. We can then leap directly to the solution

$$p_{\mathrm{rad}}(z) = \frac{\tau_{\mathrm{tot}}}{c} \mathcal{F}_{\mathrm{rad}}(\tau_{\mathrm{tot}}) \left[1 - \frac{1}{\tau_m} \int_0^z dz' \frac{z'}{h} \kappa(z')\rho(z') \right]. \tag{7.71}$$

With the simplifying assumption that $\kappa\rho$ is a constant in z,

$$p_{\mathrm{rad}}(z) = \frac{\tau_{\mathrm{tot}}}{c} \mathcal{F}_{\mathrm{rad}}(\tau_{\mathrm{tot}}) \left[1 - \left(\frac{\tau}{\tau_{\mathrm{tot}}} \right)^2 \right]. \tag{7.72}$$

This result points out an important fact about the α-model: strictly speaking, it applies only to vertically-integrated quantities and does not make a clear prediction about the local stress. We have just shown that when radiation pressure dominates, hydrostatic equilibrium requires $f(z) = z/h$ and constant dissipation per unit volume. However, if the dissipation per unit volume were $\alpha p |d\Omega/d\ln r|$ (as suggested by eq. 7.32), equation 7.72 would indicate that it would instead be proportional to $1 - (z/h)^2$.

Nonetheless, substituting equation 7.72 in equation 7.39 and using the hydrostatic equilibrium solution for h, we now find

$$\Sigma = \frac{8}{3\alpha} \frac{\eta}{L/L_{\mathrm{E}}} \left(\frac{\kappa}{\kappa_{\mathrm{T}}} \right)^{-2} \frac{x^{3/2} R_z R_T}{R_R^2(x)} \frac{\mu_e}{\sigma_{\mathrm{T}}}. \tag{7.73}$$

Thus we see that the radiation-pressure-dominated part of the disk is generally quite optically thick to Thomson scattering (except possibly in its innermost rings), justifying our assumption of large optical depth. Moreover, the surface density in this case depends on the accretion rate normalized to the Eddington value, but *not* on the absolute luminosity.

The density follows immediately:

$$\begin{aligned}
\rho &= \frac{16\pi}{9} \frac{\eta^2}{\alpha} \left(\frac{L}{L_{\mathrm{E}}} \right)^{-1} \left(\frac{\kappa}{\kappa_{\mathrm{T}}} \right)^{-3} \frac{x^{3/2} R_z^2(x) R_T(x)}{R_R^3(x)} \frac{\mu_e^2 c^3}{\sigma_T^2 L} \\
&= 2.8 \times 10^{-13} \frac{\eta^2}{\alpha} \frac{L_{\mathrm{E}}}{L} L_{46}^{-1} \left(\frac{\kappa}{\kappa_{\mathrm{T}}} \right)^{-3} x^{3/2} \frac{R_z^2 R_T}{R_R^3} \ \mathrm{gm \ cm}^{-3}.
\end{aligned} \tag{7.74}$$

When $L/L_{\mathrm{E}} \sim 1$ and $x \sim \alpha \sim 1$, the density is as small as the naive spherical accretion estimate (eq. 6.26) because the flow in the innermost part of the disk is then quasi-spherical.

The boundary between the radiation-pressure-supported region and the gas-pressure-supported region can now be found in retrospect: whichever produces the greater disk thickness dominates the support. From this criterion, we find that (assuming Thomson-dominated opacity and the maximal Kerr efficiency) the boundary occurs near

$$x_{rg} \simeq 1000 \alpha^{2/21} L_{46}^{2/21} \left(\frac{L}{L_{\mathrm{E}}} \right)^{2/3} \left(\frac{\kappa}{\kappa_{\mathrm{T}}} \right)^{20/21}. \tag{7.75}$$

In other words, the position of this boundary in terms of gravitational radii is almost completely insensitive to the absolute scale of the luminosity but does move outward with increasing L/L_E.

7.3.5 Advection-dominated disks

So far we have always assumed that all the energy dissipated at a particular radius is radiated very near that radius. That assumption is not necessarily justified, as has been shown (from different points of view) by Paczyński and Wiita (1980), Rees et al. (1982), Abramowicz et al. (1988) and Narayan and Yi (1994).

If the accretion rate is low enough, the local cooling rate is small because the density is comparatively low. When the cooling time is long compared to the inflow time, the disk is *advection-dominated*, that is, most of the dissipated heat is advected inward along with the flow instead of being radiated. As a result, the radiative efficiency can be several orders of magnitude smaller than the nominal 0.06–0.42 that one might expect from accretion onto a black hole.

If indeed most of the dissipated potential energy is retained in the accreting gas as heat, then the temperature of the ions must rise to roughly the virial temperature. Two consequences follow. First, treating the structure as a thin disk is no longer appropriate—the ion thermal speeds are automatically comparable to the orbital speed. Therefore, the dynamics of accretion become more similar to spherical accretion, and the results of Chapter 6 approximately apply. Second, because electrons radiate far more readily than do ions, their temperature is likely to be far lower; how much lower depends on the competition between ion-electron heat transfer and the specific electron cooling mechanisms. Unfortunately, we are at present unable to evaluate this balance with any confidence. Although we know a great deal about electron cooling mechanisms, we are much more ignorant about ion-electron heat transfer. The slowest possible rate is via Coulomb collisions, but it is quite possible that when $T_i > T_e$ wave-particle interactions may transfer energy at rates far higher (e.g., Begelman and Chiueh 1988).

Advection can also dominate in a high accretion rate regime if the optical depth is so high that the photon diffusion time is long compared to the inflow time. This latter case is closely analogous to super-Eddington spherical accretion (§6.3.2), where the optical depth is so large that photons cannot escape before being dragged into the black hole. The criterion for achieving a quasi-spherical structure, and therefore one in which the heat flux is no longer closely parallel to the rotation axis, is simply $h \sim r$. Disk models in which the efficiency is small enough and L/L_E large enough to puff up the inner portions were the subject of considerable interest in the mid-1980s because general relativistic effects lead to the creation of a funnel along the rotation axis at small radii. These funnels seemed to offer the promise of a mechanism to collimate the directed outflows seen in many AGNs (see §9.3). In addition, by breaking the assumption of spherical symmetry, it is possible for these disks to radiate luminosities somewhat greater than Eddington.

However, shortly after these models were proposed it was discovered by Papaloizou and Pringle (1984; see also Goldreich, Goodman, and Narayan 1986; Goodman, Narayan, and Goldreich 1987; Hawley 1987; Blaes 1985; and Blaes and Hawley 1988 for physical interpretation) that thick disks of this sort are unstable to perturbations resembling gravity waves that grow on a dynamical timescale. Although the matter is not firmly settled, it seems likely that an element not considered in the original calculations stabilizes these waves—the very mass inflow that makes disks interesting (Blaes 1987).

7.3.6   Inside the marginally stable orbit

When the accreting matter finally reaches the vicinity of the marginally stable orbit, the nature of disk accretion changes dramatically. Because all orbits within that radius have a substantial inward radial component, it is no longer a matter of gas patiently waiting while its angular momentum is removed by some slow transport mechanism. Instead, the gas plunges inward and passes through the event horizon in a time (in the fluid frame) only a few multiples of GM/c^3. Not only gravity pushes it in: the pressure just outside r_{ms} is greater than the pressure closer in, further accelerating the inflow. Falling density accompanies acceleration, causing the gas to cool as it does expansion work. The approximation of adiabatic flow should be very good in this regime, for, whether or not the flow is advection-dominated outside r_{ms}, it is certainly advection-dominated inside that point. Slow dissipative processes have no time to heat the gas, nor is there time for slow cooling processes to remove heat by creating photons. As a result, if the disk is geometrically thin, the inflow speed v^r becomes supersonic only a short distance inside r_{ms} (Chen and Taam 1993) and relativistic soon after.

Although the density here is much smaller than in the disk proper, there can still be enough material to make it optically thick. The surface density in the co-moving frame (assuming time-steady flow and axisymmetry) is

$$\Sigma = \frac{\dot{M}}{2\pi r|v^r|} \sim \frac{L/L_{\rm E}}{\eta x}\frac{\mu_e}{\sigma_{\rm T}}, \tag{7.76}$$

where $v^r \sim c$.

7.4 Time-Dependent Disks

One might easily imagine that accretion disks do not stay absolutely fixed in structure forever. There is no reason to suppose that the rate at which their fuel is supplied stays constant, nor have we even shown that they are stable. In this section we will examine some of the issues involved in studying time-dependent phenomena in disks.

7.4.1 Characteristic timescales

To begin, let us identify the several characteristic timescales relevant to disk dynamics. The shortest is the orbital timescale

$$t_{\rm dyn} \sim \frac{1}{\Omega}. \tag{7.77}$$

It is, of course, the timescale associated with response to gravitational forces in the radial direction. Next consider the timescale for response to gravitational forces in the vertical direction t_z. It can be estimated by

$$t_z \sim \frac{h}{c_s} \sim \frac{c_s/\Omega}{c_s} \sim \frac{1}{\Omega} \sim t_{\rm dyn}. \tag{7.78}$$

So we see that the dynamical times for motion in both the radial and vertical directions are comparable.

The next shortest timescale is generally the thermal timescale, that is, the time for a given patch of the disk to change its heat content. Given by the ratio of the disk thermal content to its heating rate, it is

$$t_h \sim \frac{\Sigma c_s^2}{\nu\Sigma\Omega^2\left(d\log\Omega/d\log r\right)^2} \sim \frac{c_s^2}{\nu\Omega^2} \sim \frac{c_s^2\Omega}{\alpha c_s^2\Omega^2} \sim \frac{1}{\alpha\Omega}, \tag{7.79}$$

provided that most of the heat is radiated locally. If the viscosity is near the maximal level expected in the event of fully developed turbulence, the thermal timescale is not much longer than the dynamical timescale.

Finally, consider the viscous timescale t_{visc}. This can also be thought of as the accretion timescale, or the timescale for the surface density to change. All these formulations are equivalent. We have already estimated $t_{\text{visc}} \sim r^2/\nu$. In terms of the α-formalism, this timescale can also be expressed as

$$t_{\text{visc}} \sim \frac{r^2 \Omega}{\alpha c_s^2} \sim \frac{1}{\alpha} \frac{r^2}{h^2} \frac{1}{\Omega}. \tag{7.80}$$

The viscous time is by far the slowest if the disk is geometrically thin.

All these timescales have the interesting property of scaling as power laws of the radius. The dynamical timescale changes $\propto r^{3/2}$, the thermal timescale $\propto \alpha^{-1} r^{3/2}$, and the viscous timescale $\propto \alpha^{-1} r^{7/2}$ (for a radiation-pressure-supported disk; when gas pressure dominates and the disk radiates like a blackbody, $t_{\text{visc}} \propto \alpha^{-1} r^{29/20}$). Thus, if there are fluctuations in the disk driven by a mechanism associated with any of these timescales, we can expect their characteristic frequencies to scale as power laws in radius. Because radiation at frequencies less than $k_B T(x_{\text{max}})/h$ is created over a sizable dynamic range in radius, we would then expect a power-law character in the observed Fourier power spectrum (Abramowicz et al. 1991; Krolik et al. 1991).

With this hierarchy of timescales in hand, we can also establish a hierarchy of approximations: If we are interested in changes in the surface density or accretion rate, we can expect local thermal balance to obtain and certainly expect dynamical balance in both directions; if we are interested in local temperature changes, the accretion rate can appear to be constant and out of equilibrium, but force balance should (usually) be a good approximation; if we are interested in either vertical or azimuthal velocity changes, it is possible for both the heat content and the surface density to be out of equilibrium, but also changing so slowly as to be effectively constant.

7.4.2 Stability

We are now prepared to test the equilibria for instability. For reasons of simplicity, we will confine our attention to modes that are azimuthally symmetric and perturb only vertically integrated quantities. The latter restriction is equivalent to requiring the radial wavelength to be $\gg h$. To apply the WKB approximation, we will also insist that the radial wavelength be $\ll r$.

Working within this set of assumptions, Piran (1978) developed a formalism, applicable to a wide variety of disk models, for deriving dispersion relations for modes involving both thermal and viscous effects. As this formalism makes plain, because changes in the heat content ultimately derive their energy from changes in the accretion rate, fluctuations in temperature and surface density can couple, despite the disparity in characteristic timescales. However, in the interest of physical insight, we will take a simpler approach and examine the mechanisms of instability in isolation without deriving full dispersion relations.

The reason why $t_{\rm visc}/t_h$ is $\sim (r/h)^2$ is just that the surface density changes by diffusion (cf. eq. 7.11) and the properties of time-steady disks change on the lengthscale r. For the same reason, perturbations with radial wavelength λ_r exhibit changes in surface density that are order $(h/\lambda_r)^2$ relative to changes in the thermal content. Thus, reasonably long ($h \ll \lambda_r \ll r$) wavelength perturbations act at constant surface density, while changing the local thermal content.

For modes of this sort, then, the focus of attention is on the local heat balance, and the criterion for instability is

$$\frac{\partial Q^+}{d\mathcal{E}}/_\Sigma > \frac{\partial Q^-}{\partial \mathcal{E}}/_\Sigma, \qquad (7.81)$$

where $Q^+ = \nu\Sigma\Omega^2(d\log\Omega/d\log r)^2$ is the local heating rate per unit area, and Q^- is the local cooling rate. $\mathcal{E}$ is the internal energy per particle in the gas. Because density and temperature are inversely proportional at constant pressure, this criterion is very similar to the one displayed in equation 6.46, derived in the context of spherical accretion: $\partial H/\partial \ln \rho/_p < \partial C/\partial \ln \rho/_p$. In disks, the relatively weak torque fixes Σ; in spherical accretion, the high speed of sound fixes p (at least on sufficiently short lengthscales).

If the viscosity is treated in the α-formalism,

$$Q^+ = \frac{1}{\gamma}\alpha c_s^2 \Sigma\Omega \left(\frac{d\log\Omega}{d\log r}\right)^2 \qquad (7.82)$$

so that

$$\frac{\partial \log Q^+}{\partial \log \mathcal{E}} = \frac{\partial \log \alpha}{\partial \log \mathcal{E}} + \frac{\partial \log c_s^2}{\partial \log \mathcal{E}}. \qquad (7.83)$$

When gas pressure dominates, $\mathcal{E} \propto T$, and $d\log c_s^2/d\log T = 1$, of course. If such a disk radiates like a blackbody (and we ignore any sensitivity to temperature in its opacity, since $T_s \simeq \tau_{\rm tot}^{1/4} T_c$),

$$\frac{\partial \log Q^-}{\partial \log T} = 4. \qquad (7.84)$$

Thus, a blackbody disk dominated by gas pressure is clearly thermally stable unless there is some peculiar sensitivity to temperature in α.

On the other hand, if the disk does not radiate like a blackbody, thermal instability is possible (Pringle, Rees, and Pacholczyk 1973). Suppose, for example, that the disk is gas pressure dominated and optically thin, and radiates by the bremsstrahlung mechanism (in the next section we will discuss specific radiation mechanisms in more detail). In this case,

$$
\begin{aligned}
Q^- &\propto \frac{T^{1/2}\Sigma^2}{h} \\
&\propto \frac{T^{1/2}\Sigma^2\Omega}{c_s} \, .
\end{aligned}
\tag{7.85}
$$

Because $c_s \propto T^{1/2}$, Q^- is now independent of T, so in this case the rapidity of dynamical equilibrium (embodied in the dependence of h on temperature) leads to a thermal instability of exactly the same variety as was discussed in the context of spherical accretion (§6.3.3).

If radiation pressure dominates, the situation changes (Shakura and Sunyaev 1976). We begin the analysis of time variability in this situation by redefining hydrostatic equilibrium in terms of the instantaneous radiation flux:

$$
\mathcal{F}_{\mathrm{rad}}(\tau_{\mathrm{tot}}) = Q^- = \frac{4cp_{\mathrm{rad}}}{\kappa\Sigma} = \frac{ch\Omega^2}{\kappa} \, .
\tag{7.86}
$$

Thus, the instantaneous disk thickness is

$$
h = \frac{4p_{\mathrm{rad}}}{\Omega^2\Sigma} \, .
\tag{7.87}
$$

In the α-model, the heating rate is

$$
\begin{aligned}
Q^+ &= \frac{d\log\Omega}{d\log r}\Omega\alpha \int dz\, p_{\mathrm{rad}} \\
&= 3\alpha\Omega h p_{\mathrm{rad}} \\
&= 12\alpha\frac{p_{\mathrm{rad}}^2}{\Omega\Sigma} \, .
\end{aligned}
\tag{7.88}
$$

Because the internal energy is, by definition, dominated by radiation pressure, $\partial/\partial\mathcal{E} = (1/3)\partial/\partial p_{\mathrm{rad}}$. Therefore, $\partial\log Q^+/\partial\log\mathcal{E} = 2$, while $\partial\log Q^-/\partial\log\mathcal{E} = 1$, and the equilibrium is unstable.

However, some caution should be exercised in applying this result. As we remarked above, it is not at all obvious, for example, that the heating rate always scales in proportion to the total pressure. Even within the α-model paradigm, it might be that the dissipation is proportional to the gas pressure alone, while if the stress is not $\sim -\alpha p$, all bets are off. Still another possibility, made particularly plausible by the coherent nature of the torque produced by magnetic stresses (§7.2.3), is that dissipation is spatially separated from torque. For example (see also §8.6.1), MHD waves may carry energy from the body of the disk out to its surface, dumping it into heat only at the disk edge. Any of these complications might entirely overturn the expectation of thermal instability in the simplest equilibria. On the other hand, if this equilibrium is genuinely unstable, the consequences are profound: it would mean that the most "natural" equilibrium we might guess for the state of the inner part of accretion disks in AGNs (and many other accreting black hole systems) is not the one found in Nature.

The longest of the characteristic timescales is that associated with viscosity and mass inflow. If there is a perturbation on this timescale, all forces are, of course, in equilibrium, but so is the heat content, that is, $Q^+ = Q^-$. Because the local disk properties are all determined by Σ and ν, the heat balance equation allows us (formally) to consider ν as a function of Σ. If we now examine the time-dependent equation for the surface density,

$$\frac{\partial \Sigma}{\partial t} = -\frac{1}{r} \frac{\partial}{\partial r} \left[\frac{\frac{\partial}{\partial r} \left(r^3 \nu \Sigma \partial \Omega / \partial r \right)}{\frac{\partial}{\partial r} \left(r^2 \Omega \right)} \right], \tag{7.89}$$

it is clear that what we care about is how the product $\nu \Sigma$, which is proportional to the *local* $\dot{M}$ (eq. 7.33), depends on Σ. If an increase in the surface density leads to an increase in the mass accretion rate, then there is stabilizing negative feedback; an opposite behavior leads to instability.

This is just what occurs in the α-model when radiation pressure supports the disk (Lightman and Eardley 1974). By definition,

$$\nu \Sigma = \frac{4 \alpha p_{\mathrm{rad}} h}{3 \Omega}. \tag{7.90}$$

Hydrostatic equilibrium (as expressed by eq. 7.69) and temperature balance (because $t_{\mathrm{visc}} \ll t_h$) then imply the relation

$$\nu \Sigma = \frac{64 c^2}{27 \alpha \kappa^2 \Omega \Sigma}. \tag{7.91}$$

In fact, this result was adumbrated by the equilibrium solution summarized in equation 7.73: under these assumptions, $\dot{M} \propto \Sigma^{-1}$.

Thus we have seen that the simplest formal solution for disk properties, the α-model, is unstable both thermally and viscously when radiation pressure dominates, and yet this is just what is expected for the innermost part of the disk, where the majority of the energy is liberated. These instabilities pose serious problems to anyone attempting to construct a clear picture of the nature of AGN accretion disks. It is unlikely that disks are created in an unstable equilibrium and then evolve away; it is rather more likely that, because of these instabilities, such equilibria are avoided from the start. Perhaps the stress does not scale as $-\alpha p$; perhaps advection-dominated regions (§7.3.5) are created where otherwise radiation pressure would be expected to dominate; or perhaps so much energy is dissipated outside the disk that internal radiation pressure is never the major support against the gravity (§8.5.1.1); or ...? At present, these questions are entirely open. As we shall see in the next section, their resolution has tremendous import for our expectations of the spectra emitted by accretion disks in AGNs.

7.5 Emitted Spectrum

Despite the qualms detailed in the previous subsection, it is possible to identify the key issues determining how disks radiate and, at least in model-dependent fashion, predict a number of the characteristics of their emitted spectra. From estimates of the column density, midplane volume density, and effective temperature (at the surface) of the material in the disk, we can evaluate the opacity of the disk and therefore improve our description of both its thermodynamics and its radiated spectrum. In this way, we can check just how close to reality the thermodynamic equilibrium approximation (§7.3.3) comes, as well as make more specific statements about spectral features to expect.

The organization of this section will be from the ground up: first, the most important physical processes to be considered; then the spectrum to be expected from disks with temperature constant (as a function of altitude) in their atmospheres; next angle-averaged spectra of disks with realistic internal temperature gradients; finally, discussions of angle dependence and polarization in the emitted flux.

7.5.1 Opacity mechanisms

Our procedure is to examine each of the likely opacity mechanisms and evaluate its impact on the emitted spectrum. There are three principal candidates: Compton scattering, free-free absorption, and bound-free absorption. The latter two are in fact quite closely related, as the quantum mechanical matrix element for the former is simply the analytic continuation of the matrix element for the latter. Both involve the absorption of a photon by an electron in the Coulomb field of a positive ion; the only difference is in the final state of the electron, whether it is bound or free.

From our previous estimates we have already seen that most disks are likely to be optically thick to Compton scattering in their inner, and most interesting, regions (cf. eqs. 7.64 and 7.73).

Consider free-free opacity next. This is a three-body process (a photon must encounter an electron near a nucleus in order to be absorbed), and so its opacity per unit mass is proportional to the density. Keeping the focus on the thermodynamics of disks, it is convenient to write the photon frequency in terms of the local temperature, that is, $\omega \equiv \epsilon/(k_B T)$. In that notation, the opacity posed by ions of net charge Z is

$$\kappa_{ff} = \frac{256\pi^{5/2}}{3^{3/2}} Z^2 \alpha_{fs} a_o^5 \left(\frac{I_H}{k_B T}\right)^{7/2} \frac{\rho g(\epsilon, T)}{\mu_e \mu_Z} \varepsilon^{-3} \left(1 - e^{-\varepsilon}\right), \qquad (7.92)$$

where μ_z is the mass per ion of charge Z, I_H is the ionization potential of H, and the subtracted exponential accounts for the stimulated reemission correction. The function $g(\epsilon, T)$ is the Gaunt factor, which embodies the corrections to the semiclassical Kramers formula caused by careful evaluation of the quantum mechanical matrix elements (Karzas and Latter 1961; summarized in Rybicki and Lightman 1979). It is generally of order unity (see eq. 8.2). In a plasma with normal cosmic abundances in which both H and He are fully ionized,

$$\kappa_{ff} = 4.90 \times 10^7 g(\varepsilon, T) \rho T_5^{-7/2} \varepsilon^{-3} \left(1 - e^{-\varepsilon}\right) \text{ cm}^2 \text{ gm}^{-1}. \qquad (7.93)$$

To put this expression in context, we evaluate the free-free optical thickness of a radiation-pressure-supported disk:

$$\tau_{ff} \simeq 1.1 \times 10^{-4} \frac{1 - \exp(-\varepsilon)}{\varepsilon^3} \frac{g(\varepsilon, T)}{T_5^{7/2}} \frac{\eta^3}{\alpha^2} \left(\frac{L}{L_E}\right)^{-2} L_{46}^{-1} x^3 \frac{R_z^3 R_T^2}{R_R^5}, \qquad (7.94)$$

where we have assumed that H and He are completely ionized and, as these estimates justify, that electron scattering dominates the total opacity. At this stage of the argument we do not yet know the temperature. If the radiation were thermalized, then we would be able to use the LTE description already derived; however, we are now trying to find those circumstances in which LTE breaks down. When LTE is not a good approximation, we can expect temperatures somewhat higher than those found using the LTE prescription because the efficiency of emission can never exceed the Planckian efficiency. For this reason, temperatures rather in excess of 10^5 K can be expected.

Thus, in the most interesting portions of accretion disks (the inner rings, where radiation pressure dominance is most likely), the photons of most interest locally (i.e., $\varepsilon \sim 1$) will not be stopped by free-free absorption alone. However, it is important to take note of the high powers in some of the dependences: $\tau_{\rm ff} \propto x^3 (L/L_{\rm E})^{-2} T_5^{-7/2}$; one need not go too far out, or require a terribly small $L/L_{\rm E}$, to reach a different conclusion.

Next consider bound-free opacity due to H (or any species with only a single electron) in the approximation that its ionization balance is in thermodynamic equilibrium. The cross section for ionization from a state with principal quantum number n is

$$\sigma_{\rm bf} = \frac{64\pi}{3\sqrt{3}} g \frac{\alpha_{\rm fs}}{Z^2} n a_o^2 \left(\frac{\epsilon}{\epsilon_n} \right)^{-3}, \tag{7.95}$$

where g is again a Gaunt factor close to unity, $\alpha_{\rm fs}$ is the fine-structure constant, a_o is the Bohr radius, Z is the number of protons in the nucleus, and $\epsilon_n = (1/2)(Z\alpha_{\rm fs})^2 m_e c^2 n^{-2}$ is the ionization potential. The ϵ^{-3} scaling is an approximation: at energies not far above threshold, the dependence of $\sigma_{\rm bf}$ on ϵ is shallower; when ϵ/ϵ_n is large enough that the Born approximation applies, $\sigma_{\rm bf} \propto \epsilon^{-7/2}$.

From the Saha equation, we know that in thermodynamic equilibrium the population with quantum number n is

$$n_o(n) = n_Z n_e n^2 \frac{(2\pi)^{3/2} \hbar^3}{(m_e k_{\rm B} T)^{3/2}} \exp \left(\frac{\epsilon_n}{k_{\rm B} T} \right), \tag{7.96}$$

where n_o is the density of H-like ions (or neutral H if $Z = 1$) and n_Z is the density of bare nuclei with Z protons. Thus, if the excited states' populations are in LTE, the opacity is

$$\kappa_{\rm bf} = 1.10 \times 10^7 g \frac{n_Z n_e}{n_H^2} Z^{-5} n^6 \left(\frac{\epsilon_n}{k_{\rm B} T} \right)^{3/2} \exp \left(\frac{\epsilon_n}{k_{\rm B} T} \right)$$
$$\rho \left(\frac{\epsilon}{\epsilon_n} \right)^{-3} \left(1 - e^{-\varepsilon} \right) \ {\rm cm^2 \ gm^{-1}}. \tag{7.97}$$

From this expression we see that bound-free opacity scales in much the same way as free-free opacity, as their similar physical natures require. In fact, it is convenient to evaluate the ratio of the bound-free opacity for H-like ions with Z protons to the free-free opacity including all ions (whose total density is n_i):

$$\frac{\kappa_{\text{bf}}}{\kappa_{\text{ff}}} = 1.7 \frac{g_{\text{bf}}}{g_{\text{ff}}} \frac{n_Z}{n_i} Z^4 n^{-3} T_5^{-1} \exp\left(\frac{1.58 Z^2}{T_5 n^2}\right), \tag{7.98}$$

where we suppose H and He are both fully ionized, and the expression applies only for photons above the threshold for state n. As can be seen, at temperatures of the sort expected in the inner regions of AGN accretion disks and frequencies above the ionization edge(s), the H (and possibly He) bound-free opacity can be expected to be comparable to free-free opacity. At lower temperatures, found either around more massive black holes, or farther out in the disk, bound-free opacity becomes progressively more important.

7.5.2 Thermalization

Absorption processes are significantly enhanced by the extra pathlength imposed on escaping photons by scattering. If the scattering optical depth is τ_s and the absorption optical depth is τ_{abs}, the total path length to escape is on average the number of photon flights $(\tau_s + \tau_{\text{abs}})^2$ times the pathlength per flight, $h/(\tau_s + \tau_{\text{abs}})$. As a result, the effective absorptive opacity is enhanced by the factor $\sqrt{1 + \tau_s/\tau_{\text{abs}}}$ (see eqs. 7.106–7.110 for a formal proof). In effect, the combination of scattering and absorption acts like an effective absorptive opacity that is the geometric mean of the total and absorptive opacities.

If the effective absorptive optical depth is larger than unity, there is a surface, the *thermalization photosphere*, below which the photons are near thermal equilibrium with the electrons, and the intensity at frequency ε approaches the Planckian value, that is, $I_\varepsilon \simeq B_\varepsilon(T)$. The column density Σ_{ph} above the thermalization photosphere is determined by $\tau_{\text{eff}}(\Sigma_{\text{ph}}) \simeq 1$, so

$$\Sigma_{\text{ph}} \simeq [\kappa_{\text{abs}}(\kappa_{\text{abs}} + \kappa_s)]^{-1/2}. \tag{7.99}$$

The density at the photosphere follows immediately if the scale height is known: $\rho_{\text{ph}} \sim \Sigma_{\text{ph}}/h$. When, as here, $\kappa_s \simeq \kappa_T > \kappa_{\text{abs}} \propto \rho$, we have

$$\rho_{\text{ph}} \simeq \frac{1}{h^{2/3} \left[(\kappa_{\text{abs}}/\rho)\kappa_T\right]^{1/3}}. \tag{7.100}$$

For example, if the dominant absorptive opacity is the free-free process,

$$\rho_{\text{ph}} \simeq 9 \times 10^{-12} h_{13}^{-2/3} T_5^{7/6} \frac{\varepsilon}{g^{1/3} (1 - e^{-\varepsilon})^{1/3}} \text{ gm cm}^{-3}. \tag{7.101}$$

When, as is often the case in accretion disks, $\tau_T \gg 1$, the additional pathlength due to scattering can substantially augment the disk's ability to come into thermal balance with the radiation. For example, the effective optical depth (counting only free-free opacity) for a radiation-pressure-supported disk is

$$\tau_{\text{eff}} \simeq 0.016 \left[g \varepsilon^{-3} (1 - e^{-\varepsilon}) \right]^{1/2} \frac{\eta^2}{\alpha^{3/2}} T_5^{-7/4} \left(\frac{L}{L_E} \right)^{-3/2} L_{46}^{-1/2} x^{9/4} \frac{R_z^2 R_T^{3/2}}{R_R^{7/2}}.$$
$$(7.102)$$

A glance at equations 7.64 and 7.66 shows that τ_{eff} in gas-pressure-supported disks is extremely large unless x is very very large or L/L_E is very small. Thus, we expect thermalization to be significant over much of the locally emitted spectrum except in radiation-pressure-supported disks when conditions are pushed to their extreme nominal values (i.e., $\alpha \sim L/L_E \sim x \sim 1$).

7.5.3   Emergent spectrum from a vertically isothermal disk

By definition, the final shape of the emergent spectrum is determined in the upper layer of the disk that comprises its photosphere (sometimes, if the disk is thin enough and hot enough, the photosphere may take in the entire thickness of the disk). As is obvious from §7.5.2, different frequencies find their photospheres at different locations. In general, therefore, we can expect the temperatures at their photospheres to be different.

In this subsection we will consider the simplest picture, in which the temperature is constant across the region containing the photospheres for the entire interesting range of frequencies. With less loss of generality, we will also simplify the argument by assuming that only the free-free process contributes to the absorption opacity. It is not hard to adjust the results to account for other sorts of opacity.

When the atmosphere does have constant temperature, it is convenient to divide its spectrum into four different regimes, separated by three critical frequencies. The lowest interesting frequency is ε_o, the frequency below which free-free opacity (or any other absorptive opacity) is actually greater than electron scattering opacity:

$$\varepsilon_o \simeq 5.2 \times 10^{-3} \eta \alpha^{-1/2} T_5^{-7/4} \left(\frac{L}{L_E} \right)^{-1/2} L_{46}^{-1/2} x^{3/4} \frac{R_z R_T^{1/2}}{R_R^{3/2}}. \qquad (7.103)$$

Here we have taken the Gaunt factor to be unity and found scaling appropriate to $\varepsilon_o < 1$. We see from this estimate that if α and L/L_E are not too small, in the interesting portions of AGN disks the opacity to frequencies near the thermal peak is generally not dominated by free-free absorption. However, at sufficiently low frequencies, free-free opacity becomes stronger and stronger. Provided $\tau_{ff}(\varepsilon) > 1$, the emitted spectrum at frequencies below ε_o is a Planckian at the disk temperature.

The next highest critical frequency is ε_t, the frequency above which even the additional pathlength incurred due to electron scattering is not enough to make the path optically thick to absorption:

$$\varepsilon_t \simeq 0.016 \frac{\eta^2}{\alpha^{3/2}} T_5^{-7/4} \left(\frac{L}{L_E} \right)^{-3/2} L_{46}^{-1/2} x^{9/4} \frac{R_z^2 R_T^{3/2}}{R_R^{7/2}}, \qquad (7.104)$$

where we have again taken the $\varepsilon \ll 1$ limit. When the fiducial factors are order unity, ε_t is only a few times bigger than ε_o, and so there is a rather limited range of frequencies in which Compton scattering aids free-free opacity in the effort to achieve thermodynamic equilibrium between radiation and matter. However, ε_t is more sensitive to most of these factors than ε_o is, so there is a general tendency for this range to open up as L/L_E decreases and x increases:

$$\frac{\varepsilon_t}{\varepsilon_o} \simeq 3.0 \frac{\eta}{\alpha} \left(\frac{L}{L_E} \right)^{-1} x^{3/2} \frac{R_z R_T}{R_R^2}. \qquad (7.105)$$

Thus, much of the disk's emission (in a radially integrated sense) falls into the range between ε_o and ε_t.

Between ε_o and ε_t, the intensity is Planckian if one goes sufficiently deep inside the disk. However, the varying position of the thermalization photosphere combined with the dominance of scattering over absorptive opacity means that there is a thicker scattering blanket over the thermalization photosphere at higher frequencies than at lower frequencies. Consequently, the emergent flux does *not* have a Planck spectrum, instead taking a shape called *modified blackbody*.

To see how this works, let us first write down the transfer equation in terms of the free-free optical depth (it is a simple matter to generalize to other kinds of absorptive opacity):

$$\mu \frac{\partial I_\varepsilon}{\partial \tau_{ff}} = \left(1 + \frac{\kappa_T}{\kappa_{ff}} \right) I_\varepsilon - \frac{\kappa_T}{\kappa_{ff}} J_\varepsilon - B_\varepsilon. \qquad (7.106)$$

In this expression we have adopted a sign convention such that optical depth increases as one looks deeper into the disk, and we suppose that in scattering events there is no angular correlation between incoming and outgoing photons. J_ε is the mean intensity. Integrating this equation with respect to μ gives

$$\frac{\partial F_\varepsilon}{\partial \tau_{ff}} = 2\left(1 + \frac{\kappa_T}{\kappa_{ff}}\right) J_\varepsilon - 2\frac{\kappa_T}{\kappa_{ff}} J_\varepsilon - 2B_\varepsilon = 2\left(J_\varepsilon - B_\varepsilon\right). \qquad (7.107)$$

We obtain a second equation for these moments of the intensity by multiplying the transfer equation by μ and integrating:

$$2\frac{\partial K_\varepsilon}{\partial \tau_{ff}} = \left(1 + \frac{\kappa_T}{\kappa_{ff}}\right) F_\varepsilon, \qquad (7.108)$$

where K_ε is the second moment of the intensity with respect to μ, divided by the total solid angle. In order to close this set of moment equations, we need a relation between K_ε and J_ε. A natural choice (called the *Eddington approximation*) is to set $K_\varepsilon = (1/3)J_\varepsilon$ because this is the correct relation when the photon distribution is isotropic. Thus, this approximation, although likely to be quite good deep inside an optically thick region, must fail near the surface. Let us, nonetheless, adopt it, so that the second moment equation (eq. 7.108) becomes

$$2\frac{\partial J_\varepsilon}{\partial \tau_{ff}} = 3\left(1 + \frac{\kappa_T}{\kappa_{ff}}\right) F_\varepsilon. \qquad (7.109)$$

Now that we have an expression for F_ε in terms of J_ε, we use it to substitute in the first moment equation (eq. 7.107) and find

$$\frac{1}{3(1 + \kappa_T/\kappa_{ff})} \frac{\partial^2 J_\varepsilon}{\partial \tau_{ff}^2} = J_\varepsilon - B_\varepsilon. \qquad (7.110)$$

It is obviously convenient to rescale the independent variable in this differential equation so that $\tau_* = \sqrt{3(1 + \kappa_T/\kappa_{ff})}\tau_{ff}$. This rescaling also embodies the physical point that the effective opacity is $\sim \sqrt{\kappa_{ff}(\kappa_{ff} + \kappa_T)}$. In fact, this is also the justification for the scaling argument made in §7.5.2. With this redefinition of the independent variable, equation 7.110 becomes

$$\frac{\partial^2 J_\varepsilon}{\partial \tau_*^2} - J_\varepsilon = -B_\varepsilon. \qquad (7.111)$$

The asymptotic solution of this equation is obvious: when $\tau_* \gg 1$, $J_\varepsilon \to B_\varepsilon$. That is, at a depth such that $\sqrt{\tau_{ff}\tau_T} \sim 1$, the intensity approaches thermal. In fact, the form of the complete solution (when T, and therefore B_ε, is independent of z) is equally obvious:

$$J_\varepsilon = C\exp(-\tau_*) + B_\varepsilon. \qquad (7.112)$$

In principle there might be an additional additive term $\propto \exp(+\tau_*)$, but because it diverges at large τ_* its coefficient must be zero. We have already applied one of the two boundary conditions required for our second-order differential equation (that $J_\varepsilon \to B_\varepsilon$ as $\tau_* \to \infty$); the second boundary condition is that $I_\varepsilon(\mu > 0) = 0$ at $\tau_* = 0$, that is, that there are no incoming photons at the exposed edge of the disk. To implement this we make the approximation that $F_\varepsilon = J_\varepsilon/\sqrt{3}$ at $\tau = 0$; this would be exact if, for example, the radiation is isotropic in the half-space. This extra condition defines C through equation 7.109, and we arrive at the final result for the outgoing intensity:

$$I_\varepsilon(\mu < 0) = \frac{2B_\varepsilon}{1 + \sqrt{1 + \kappa_T/\kappa_{\text{ff}}}}. \tag{7.113}$$

The physical content of this result is that a blanket of Thomson scattering matter lies on top of the thermalization photosphere. This blanket retards the escape of photons from that photosphere; it is thickest where the absorptive opacity is *least*.

In the limit that $\kappa_T \gg \kappa_{\text{ff}}$ (as generally applies in AGN accretion disks),

$$I_\varepsilon \simeq 2B_\varepsilon \sqrt{\frac{\kappa_{\text{ff}}}{\kappa_T}}. \tag{7.114}$$

That is, the intensity is reduced below Planckian by a factor which is the square root of the ratio between absorptive and scattering opacity. Using the explicit expression for free-free opacity given in equation 7.92, we arrive at the final result for the modified blackbody spectrum

$$I_\epsilon \propto \rho^{1/2} \begin{cases} \epsilon T^{1/4} & \epsilon \ll k_B T \\ \epsilon^{3/2} T^{-1/4} e^{-\epsilon} & \epsilon \gg k_B T. \end{cases} \tag{7.115}$$

On the Rayleigh-Jeans side it is less steep than a true blackbody spectrum by one power of frequency and depressed in amplitude by a factor $\propto \rho^{1/2} T^{-3/4}$; on the Wien side, the shape remains exponential in $k_B T$, but the amplitude is lower by a factor $\propto \rho^{1/2} \epsilon^{-3/2} T^{-1/4}$. In both cases, the decrease in free-free opacity accompanying increases in temperature suppresses the output by a factor that increases with increasing temperature.

These effects are illustrated in figure 7.4. At the lowest frequencies ($\varepsilon \leq \varepsilon_o$), the modified blackbody spectrum coincides with the Planck spectrum at the gas temperature T. At higher frequencies, Thomson scattering retards photon escape from the thermalization photosphere, depressing the spectrum, but I_ε is still $\propto e^{-\varepsilon}$ for $\varepsilon \gg 1$. The corresponding effective temperature T_{eff} is therefore rather smaller than T. For this reason, the modified blackbody spectrum can be significantly broader than a blackbody spectrum at T_{eff}.

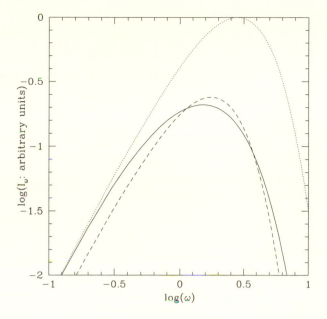

Fig. 7.4   A comparison of the modified blackbody spectrum (solid curve) with true blackbodies at the gas temperature (dotted curve) and effective temperature (dashed curve). Qualitative relations between these curves are generic; the specific curves plotted assume $\rho = 10^{-9}$ gm cm^{-3} and $T = 10^5$ K, so that $T_{\text{eff}} = 6.2 \times 10^4$ K.

In all our previous estimates of physical conditions in disks, we have supposed that the local output was similar enough to blackbody that it was proportional to T^4. However, where the disk spectrum is well approximated by the modified blackbody form, the total frequency-integrated flux scales with temperature as $T^{9/4}\rho^{1/2}$. Thus, the dependence of temperature on radius may be shallower than the simple blackbody prediction $T \propto r^{-3/4}$. In addition, because $T_{\text{eff}} < T$, T is elevated above the prediction of equation 7.49, which is predicated on local blackbody emission.

The third, and highest, critical frequency is ε_{IC}, the *inverse Compton frequency*. It is defined by $\kappa_{\text{ff}}(\varepsilon_{\text{IC}}) = (4k_{\text{B}}T/m_e c^2)\kappa_{\text{T}}$, so that

$$\frac{\varepsilon_{\text{IC}}}{(1 - e^{-\varepsilon_{\text{IC}}})^{1/3}} = 1.1 \frac{\eta^{2/3}}{\alpha^{1/3}} T_5^{-3/2} \left(\frac{L}{L_{\text{E}}}\right)^{-1/3} L_{46}^{-1/3} x^{1/2} \frac{R_z^{2/3} R_T^{1/3}}{R_R(r)}. \quad (7.116)$$

When the fiducial factors are order unity, so is ε_{IC}; it is generally rather greater than ε_t. ε_{IC} is a characteristic frequency for inverse Compton scattering because it is the frequency at which the small amount of energy lost (on average) by free-free absorption in traversing a given pathlength is offset by the energy gain due to inverse Compton scattering (§8.2.2).

Because inverse Compton scattering will be discussed at considerably greater length in Chapter 8, here we give only a very brief synopsis of its effects in accretion disks. At frequencies between ε_t and $\varepsilon_{\mathrm{IC}}$, the disk is not thick enough to achieve the boundary condition at large depth of $J_\varepsilon = B_\varepsilon$, but the free-free opacity is enough to offset inverse Compton scattering. In this range, the emitted spectrum is not far from optically thin thermal bremsstrahlung. In the range $\varepsilon_{\mathrm{IC}} < \varepsilon < 4$, photons gain energy by inverse Compton scattering, but when $\varepsilon > 4$, on average they lose energy to electrons by Compton recoil.

As will be explained in greater detail in §8.3.2.1, the importance of Comptonization can be gauged in terms of the *Compton y parameter*, $y \equiv (4kT/m_ec^2)\max(\tau_\mathrm{T}, \tau_T^2)$. Here, the relevant Thomson optical depth is only that part of the optical depth lying above the thermalization photosphere. Using our earlier estimate of the density and assuming that the only relevant absorptive opacity is the free-free process, we find

$$y \simeq 0.05 h_{13}^{2/3} T_5^{10/3} \frac{\varepsilon^2}{(1 - e^{-\varepsilon})^{2/3}}, \tag{7.117}$$

where h is the *local* density scale height near the photosphere. Thus, Comptonization is most effective for relatively high-frequency photons, particularly in regions of high temperature.

7.5.4 Emergent spectrum from a disk with vertical temperature gradients

It is now time to consider the possible influence of temperature gradients inside the disk. Anywhere $\tau_{\mathrm{eff}} > 1$ for $\varepsilon \sim 1$, such gradients must exist. The analogy implied by the term "thermalization photosphere" now becomes explicit: the structure of a disk atmosphere is very similar to the structure of a stellar atmosphere.

In stars, the outflowing energy is generated at such large optical depth that LTE is an excellent approximation at the base of the atmosphere: the distribution functions for every atomic property are thermal, and the photon intensity is very precisely Planckian at the matter temperature. The job of the atmosphere calculation is to follow the evolution of the photon spectrum as it approaches the surface of the star and gradually departs from Planckian. Two parameters suffice to determine the entire system:

the total heat flux (often stated in terms of the effective temperature T_{eff}), and the surface gravity. On the main sequence, the former ranges from $\simeq 2000$ to $40{,}000$ K, and the latter is

$$g_s \simeq 3 \times 10^4 \left(\frac{M}{M_\odot}\right)^{-1} \text{ cm s}^{-2}. \tag{7.118}$$

In white dwarfs, the effective temperature can be as high as $\sim 10^5$ K, and surface gravities rise to $\sim 10^8$ cm s^{-2}.

In many respects, AGN accretion disk atmospheres are similar, provided that most of the dissipation takes place deep inside. Disk optical depths can be great enough that, somewhere below the surface, LTE does become a good approximation. Unless the dissipation is concentrated toward the disk surface, the heat flux is very nearly constant through the photospheric region. As we have already seen (eq. 7.48), disk effective temperatures span the range from the hotter main sequence stars to hot white dwarfs. The temperature scale for any particular disk is $\propto [(L/L_{\text{E}})^2 L^{-1}]^{1/4}$. Surface gravities in AGN disks resemble those of main sequence stars if gas pressure dominates, but if radiation pressure is most important, the surface gravity can span a tremendous range, from values typical of red giants up to values more commonly seen in white dwarf atmospheres:

$$g_z \simeq \begin{cases} 1.6 \times 10^4 T_5^{1/2}(L/L_{\text{E}})L_{46}^{-1}x^{-3/2}R_z^{1/2} \text{ cm s}^{-2} & \text{gas pressure} \\ 1.4 \times 10^8 \eta^{-1}(L/L_{\text{E}})^2 L_{46}^{-1}x^{-3}R_R(x) \text{ cm s}^{-2} & \text{radiation pressure .} \end{cases} \tag{7.119}$$

Disks in the radiation-pressure-dominated regime also resemble hot stars in the sense that the radiation force nearly cancels gravity. Conveniently, in the radiation-pressure-dominated case the characteristic magnitudes for *both* T_{eff} and g_z depend only on the combination $(L/L_{\text{E}})^2 L^{-1}$.

However, there are also notable contrasts. Although it is possible that most of the heat is released at large optical depth in AGN disks, it is not necessarily so. We do not know the distribution of dissipation with altitude, and there may be significant heating due to external irradiation (§8.7.4). In fact, if the incident radiation is significant energetically, its absorption or scattering may also result in a significant *downward* force. Even if the incident radiation is not enough to produce much heat, it may still have a major effect on the ionization balance of many elements, leading to substantial alteration in the opacity.

For these reasons, one cannot simply carry over the results of stellar atmosphere calculations to AGN accretion disks. However, the structure of the problem is very similar, if one regards each ring of the disk as a separate star: One begins with a total amount of mass (in the stellar case) or surface density (for disks). With a rule for determining the rate of local heat production (nuclear reaction rates in stars, a viscosity prescription or external heating rate in disks), one solves the hydrostatic equilibrium equation, subject to the condition of local thermal balance. At this level, heat diffusion is adequately approximated by a Rosseland mean opacity.

Once one knows the pressure and temperature as functions of z, the next step is to solve the transfer equation for all relevant frequencies. Transfer solutions can easily become extremely complex and lengthy, so shortcuts are frequently taken. Whether these shortcuts are appropriate depends on the questions being asked.

For example, in the early days of stellar atmosphere calculations the problem was commonly made more tractable by the LTE approximation. That is, the distribution functions for all material properties (ionization, excited state population, etc.) were assumed to be thermal, but the photon intensity was allowed to deviate from the associated blackbody spectrum. This approximation is most nearly correct when the density in the photosphere is high, so that most transitions into and out of different atomic states are mediated by collisions (see further discussion of this issue in §10.3.5).

The Eddington approximation is also frequently used. A second look at equations 7.109 and 7.110 reveals that use of the Eddington approximation effectively transforms the transfer problem into a diffusion problem in which the transport coefficient is just the inverse of the opacity. Two happy consequences follow. Instead of following photons traveling along numerous different directions, one needs to track only one moment of the intensity for each photon energy. In addition, it is particularly easy to meld this kind of transfer solution with a Comptonization calculation (see §8.3.2.1). These advantages come at a cost, of course. The solution is likely to be inaccurate wherever the diffusion approximation breaks down (e.g., if there are emissivity gradients on scales not much longer than a photon mean free path). Moreover, one loses all angular information, of course (see §7.5.5).

Two main obstacles stand in the way of successfully completing this program for accretion disks. First, we do not know for sure how the local stress is determined and therefore how to find $\Sigma(r)$ given M and $\dot{M}$. Despite all its successes, setting $T_{r\phi} = -\alpha p$ is merely a stopgap, pending a deeper understanding of the physics; the fact that we can only guess at an appropriate value of α (and which sort of pressure to use in this expression) underscores the tentative character of this parameterization. Because the stress prescription influences both the total optical depth of the disk and its scale height, different choices lead to somewhat different expectations for the emergent spectrum (fig. 7.5). In addition, we must learn how the dissipation is distributed in altitude within the disk. If the stress is entirely viscous, the dissipation at a given spot is simply proportional to the local stress, but if nonviscous stress is important, simple local proportionality may not be correct.

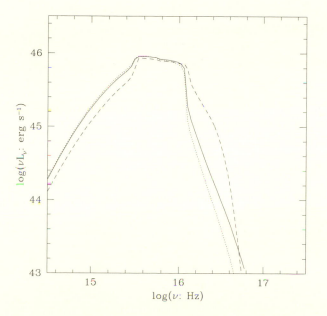

Fig. 7.5   The poleward spectrum of an accretion disk around a Schwarzschild black hole, as predicted by three different versions of the α-model, in which $T_{r\phi}$ is variously proportional to the total pressure including radiation (dashed curve), the gas pressure alone (dotted curve), and the geometric mean of radiation and gas pressure (solid curve). In this example, $M = 2.7 \times 10^8 \, M_\odot$, $\dot{M} = 0.3 \, \dot{M}_\mathrm{E}$, and $\alpha = 0.1$. The transfer solution was done in the LTE approximation. This figure and the following four are all courtesy of M. W. Sincell.

   Given all these uncertainties, there are few robust results. A few, however, do stand out. As expected on the basis of the fundamental thermal scaling laws, when the central mass and accretion rate are in the range corresponding to AGN conditions, predicted disk spectra peak in the ultraviolet and grow progressively "cooler" with increasing central mass for fixed luminosity (fig. 7.6).

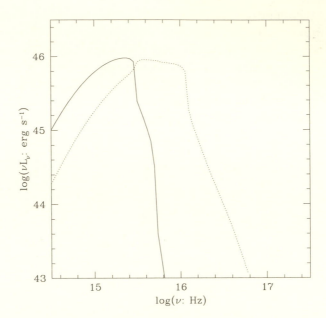

Fig. 7.6   Predicted poleward spectra for fixed luminosity (1.1×10^{46} erg s^{-1}), as a function of central mass. The curves shown have central masses of $2.7 \times 10^9\ M_\odot$ (solid line) and $2.7 \times 10^8\ M_\odot$ (dotted line). As in the previous figure, $a = 0$, and LTE ionization balance is assumed. Note that the Lyman edge is in absorption in the cooler disk, but in emission in the warmer one.

Another generic feature of these predicted spectra is that disks viewed more-or-less face-on (see §7.5.5 for further discussion of angle dependence) nearly always show a strong feature at the Lyman edge, whether in absorption or emission (Kolykhalov and Sunyaev 1984; Sun and Malkan 1989; Laor and Netzer 1989; Sincell and Krolik 1998). When the temperature declines sharply outward, the jump in opacity as the frequency increases across the Lyman edge leads to a *depression* in the spectrum just above the edge; on the other hand, if the temperature gradient is weaker, the same jump in opacity increases the effective emissivity and could produce an *emission* edge. Only a very careful choice of parameters can lead to a cancellation when one averages over the disk (as shown in fig. 7.7). When the disk is hot enough that there is significant emission there, a feature at the He II edge (55.4 eV) is also almost always present, for essentially the same reasons.

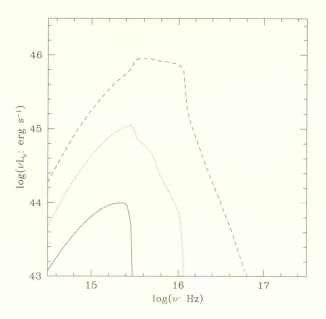

Fig. 7.7 Poleward spectrum as a function of $\dot{M}/\dot{M}_{\rm E}$ for fixed $M = 2.7 \times 10^8\, M_\odot$ for the same assumptions as in fig. 7.6. $\dot{M}/\dot{M}_{\rm E} = 0.3$ is shown by the dashed line, 0.03 by the dotted line, and 0.003 by the solid line. The hottest disks push the Lyman edge (at $\log \nu = 15.5$) into emisison, the cooler ones show it in absorption, but the sharp change in opacity there always produces some sort of feature.

7.5.5 Angle dependence

Three separate effects must be combined in order to predict the angular dependence of the radiation emerging from an accretion disk: the local limb-darkening, projection of the disk on the sky plane, and the impact of relativity on photon propagation: trajectory curvature, beaming, and frequency-shifting.

One of the classic results of radiation transfer theory is that if the flux F_ν at a given frequency is constant through the photospheric region, the emergent intensity is limb-darkened according to the relation

$$I_\nu \simeq \frac{3}{4} F_\nu \left(\mu + \frac{2}{3} \right) \tag{7.120}$$

(the exact solution is slightly steeper: Mihalas 1978). Because this form for the limb-darkening depends crucially on the assumption of conserved flux, the distribution of dissipation with height is critical to an evaluation of the angular distribution of the frequency-integrated flux. Even when all the dissipation takes place deep inside, a genuine transfer calculation is required to evaluate the frequency-dependent limb-darkening because energy can be moved from one frequency to another within the atmosphere while still conserving integrated flux. These effects are often very sensitive to details of the disk model.

Classically, if one knows I_ν as a function of viewing angle θ, the flux observed at polar angle θ is simply $F_\nu = \cos\theta I_\nu$. However, as already discussed in §5.3, general relativity introduces additional complexities: photon trajectories are bent, so that the photons that ultimately arrive at Earth were not necessarily originally directed toward us; and photons are shifted in frequency by both gravitational redshift and Doppler shifts of either sign due to orbital motion. The result is that the flux an observer receives is found from expressions having the form of equation 5.82. Because $r_{\rm ms}$ is closer to the event horizon when the black hole is spinning relatively quickly, relativistic effects are greatest when a/M is close to unity.

The single most important effect due to relativistic photon propagation is that the high-energy photons emitted by disks are squeezed toward the equatorial plane. This is because the highest energy photons are made exclusively in the innermost part of the disk, where relativistic effects are strongest. There Doppler boosting enhances the intensity of photons moving parallel to the orbital velocity, and the black hole's gravity strongly focuses their trajectories into the plane. Conversely, the spectrum as seen along the polar axis is relatively weak in high-energy photons.

In addition, when we observe photons with $\epsilon \sim k_{\rm B} T_s$, their energies in the emission frame can range over a factor of several. As a result, sharp spectral features in this energy range can be smeared out, particularly when $a/M \simeq 1$.

Figures 7.8 and 7.9 illustrate how these effects combine. As shown in figure 7.8, because the innermost ring of a disk around a Schwarzschild black hole is not that close to the event horizon, the general relativistic effects are relatively weak. In fact, taking the special relativistic limit (i.e., considering only special relativistic boosting and beaming without worrying about trajectory bending and gravitational redshift) is a good approximation for disks in this metric. However, when a/M is close to unity, the disk moves in so close to the event horizon that the general relativistic effects are extremely strong.

Figure 7.9 illustrates how much the light we see depends on angle when $a/M \lesssim 1$. As our point of view moves from the pole to the equator, we see

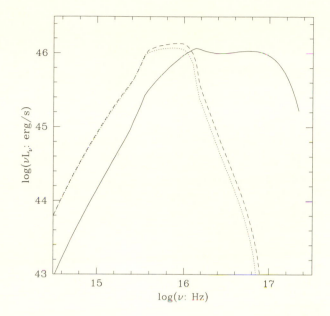

Fig. 7.8   All three curves pertain to a single set of accretion parameters: $\dot{M}/\dot{M}_E = 0.3$, $M = 3 \times 10^8\, M_\odot$, and $T_{r\phi} = -\alpha\sqrt{p_{\mathrm{rad}}p_g}$ viewed from the same angle ($\theta = 40°$), but differ in how relativity is treated. The spectrum shown by the dotted curve corresponds to a relativistic disk around a Schwarzschild black hole, but without any relativistic photon propagation effects. When the relativistic photon propagation effects are included, the result is the dashed curve. The solid curve is what the spectrum would look like if the black hole were spun up to $a/M = 0.998$. Greater black hole spin also permits greater radiative efficiency and hence greater integrated luminosity.

more and more of the high-frequency photons, and fewer and fewer of the low-frequency ones. In addition, when we view a disk from any angle other than face-on, the light we see from any single ring is subject to a broad range of Doppler shifts, depending on azimuthal angle. The result is that sharp features (like the Lyman edge) are made much smoother.

7.5.6 Polarization

Simple accretion disks are also expected to have one other characteristic signature: linear polarization. The basic idea is that the opacity in the atmosphere is generally dominated by electron scattering, and this is an electric dipole process. The induced dipole moment of the electron defines a favored direction, and this is imprinted on the distribution of outgoing photons. In quantitative terms, the cross section is proportional to $|\hat{e}_f^* \cdot \hat{e}_i|^2$,

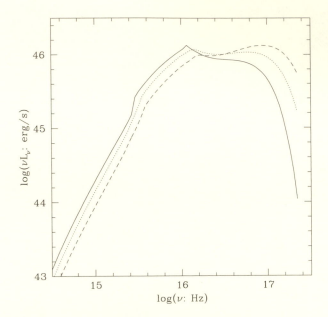

Fig. 7.9   The disk parameters are identical to those of fig. 7.8, but all three curves here assume full Kerr geometry with $a/M = 0.998$ and differ in viewing angle. The solid curve is the spectrum as seen pole-on ($\theta = 0°$), the dotted curve as seen from $\theta = 40°$, and the dashed curve from $\theta = 60°$.

where $\hat{e}_{i,f}$ are the initial and final polarization unit vectors (see the more detailed discussion in §12.4.1).

If our view is in the equatorial plane of the disk, photons rising out of the disk must scatter nearly 90° in order to enter our line-of-sight. Two linear polarization modes are possible for the photons we see: one parallel to the disk surface, the other parallel to the disk normal. However, there are relatively few emergent photons with polarization parallel to the disk normal because that is nearly parallel to their direction of motion. Therefore, a disproportionate fraction of the photons we see are polarized with $\vec{E}$-vectors parallel to the disk plane. Exact calculation of the polarization as a function of viewing angle in the limit of a plane parallel atmosphere with infinite scattering optical depth (Chandrasekhar 1960) shows that the polarization fraction varies from zero (when the disk is viewed pole-on, it is circularly symmetric on the sky, so there can be no preferred direction) to 12%, when it is viewed in the equatorial plane.

Realistic atmosphere calculations generally diminish the expected polarization somewhat for several reasons. First, general relativistic effects dilute polarization by swinging photons away from trajectories at small polar angles and pushing them toward larger polar angles (Connors, Piran, and Stark 1980). In addition, curvature of photon trajectories can lead to rotation of their polarization planes. Second, the Thomson depth above the photosphere may not be large enough for the infinite optical depth limit to apply (Angel 1969).

Third, the electron densities and magnetic field strengths in accretion disks could easily be great enough to create substantial Faraday depolarization. When electromagnetic waves travel through a magnetized plasma, the refractive index for the two circular polarization modes are different because the sign of the magnetic field projected on the wavevector breaks their symmetry. Electrons orbit in one sense around the field, positive ions in the other; the much greater sensitivity of the electrons to the wave fields creates a phase difference between the two modes. This phase difference, in turn, induces a rotation in the plane of any linear polarization. When the wave frequency $\epsilon/\hbar$ is much higher than both the electron plasma frequency $\omega_p = \sqrt{4\pi n_e e^2/m_e}$ and the electron cyclotron frequency $\omega_{\rm ce} = eB/(m_e c)$, the rotation angle is

$$\Delta\phi = \int dz \, \frac{2\pi c^3 \hbar^2}{m_e^2 c^2 \epsilon^2} n_e B_z, \tag{7.121}$$

where the photon travels in the z direction. Because the light we receive comes from many different layers within the disk, even if its density and magnetic field were uniform, there could be substantial depolarization; realistic variations in n_e and B_z can further depolarize the light.

This may be a significant effect in accretion disks. Suppose, for example, that, as predicted by the VCBH mechanism, the field strength is close to equipartition. In the inner part of an AGN accretion disk, where the effective temperature is $\sim 10^5$ K, equipartition of magnetic field energy density with radiation energy density implies a field $B \sim 4 \times 10^3 T_5^2$ G. Because the polarization of the outgoing flux is established above the thermalization photosphere, it is convenient to estimate the Faraday rotation in terms of the Thomson optical depth of the atmosphere:

$$\Delta\theta \simeq 36\tau_{\rm T} \left(\frac{\langle B_z \rangle}{10^3 \text{ G}}\right) \left(\frac{\lambda}{3000 \text{ Å}}\right)^2 \text{ radians}, \tag{7.122}$$

where the average on the vertical magnetic field is particularly sensitive to reversals of direction, and we have chosen a fiducial photon wavelength in the middle of the optical/ultraviolet band. Faraday rotation clearly has the potential to be an effect worth reckoning (Agol and Blaes 1996).

Fourth, atmospheric opacity is partly absorptive. This last effect can lead to creation of photons throughout the atmosphere, and those emitted from above the scattering photosphere can be seen from viewing angles in the equatorial plane. Unless some effect (e.g., Zeeman splitting) introduces an intrinsic polarization, these photons are not polarized. Through this mechanism, bound-free opacity might be great enough in disk atmospheres to reduce substantially the expected polarization at frequencies above the Lyman edge (Laor, Netzer, and Piran 1990).

However, absorptive opacity does not always diminish polarization (Blaes and Agol 1996). For example, if the top layer of the atmosphere has an electron scattering optical depth ~ 0.1, but there is considerable absorptive opacity in the layer beneath it, most of the photons seen by off-axis viewers come their way because they have undergone a single large-angle electron scattering event in the top layer. As a result, these observers may see a very large polarization. Therefore, predictions of the disk polarization spectrum must be done via true transfer solutions, and there are no easy "rules of thumb" to help one predict the outcome in advance.

7.6   Observational Tests

Over portions of the spectrum whose span is a factor of two or so in wavelength, it is generally possible to describe the optical/UV continuum with a powerlaw. If we define the local power-law index as $s \equiv d \ln F_\epsilon / d \ln \epsilon$, in the *observed* frame optical band there are examples spanning the range $-1.5 \leq s \leq 1$, with the peak of the distribution near $s \simeq -0.5$–0 (Francis et al. 1991). Radio-loud quasars (see §9.1 for a quantitative definition) tend to have somewhat steeper spectra, with s decreasing as the radio spectrum becomes flatter (Wills et al. 1995).

However, quasar surveys such as the one compiled by Francis et al. (the LBQS: see §2.2) mingle objects of many different redshifts, so that the portion of the spectrum we see at optical wavelengths may have been originally radiated anywhere from the optical to the far ultraviolet. The distribution of optical band spectral indices reported by Francis et al. is then a convolution of the intrinsic distribution with their survey's redshift distribution.

To get a better idea of the spectral shape in the emission rest frame, one may assemble a composite spectrum by stitching together the spectra of quasars of many different redshifts, placing the spectrum we observe from each at its emitted wavelength. Francis et al. did this for the LBQS survey; Zheng et al. (1997) did the same thing for a compilation of all quasar spectra available to them that were observed by *HST* in the (local) ultraviolet. By this means they were able to extend their composite spectrum far into the rest-frame EUV. Both composite spectra are shown in figure 7.10. Roughly speaking, $dF/d\ln\lambda$ peaks in the UV and declines with $s \simeq 2$ at wavelengths shorter than the Lyman edge (912 Å).

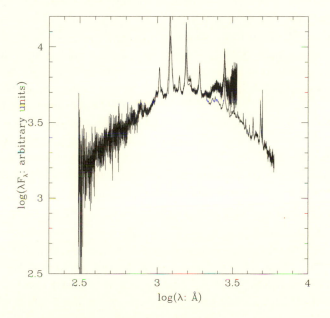

Fig. 7.10 The Francis et al. (1991) and Zheng et al. (1997) composite quasar spectral shapes, shifted so as to coincide in the vicinity of Lyα. In terms of $dF/d\ln\lambda$, there appears to be a broad, flat maximum in the vicinity of $\epsilon \simeq 10$ eV. The sharp peaks are emission lines (the subject of Chap. 10).

7.6.1 The big bump and disk model fits

The gentle peak in $dF/d\ln\lambda$ near 10 eV is often called the *Big Blue Bump*. That the flux per logarithmic wavelength should peak here is nicely in keeping with the most basic prediction of thermal disk models (eq. 7.4). However, this consistency is not a strong statement in support of the details of the accretion disk theory just presented.

As we have seen (§7.5.5), despite their differences in regard to details of spectral shape, all varieties of conventional accretion disk models predict that AGN spectra should peak in the ultraviolet. In fact, all that the existence of the big blue bump really signals is thermal emission from a surface area several tens of r_g^2. In principle, the radiating region need not even have a disk geometry—a lot of dense lumps could do just as well (Guilbert and Rees 1988).

At the same time, discrepancies with almost any conventional disk model appear once we look a little more closely. Although it is possible to fit adequately the UV continua of many AGNs with accretion disk models (e.g., Laor 1990), often this "success" is possible only because we do not have data spanning a great enough wavelength range (particularly into the EUV) to really test the models. In addition, as one looks beyond the ultraviolet toward both longer and shorter wavelengths, the flux predicted by simple accretion disk models generally falls farther and farther below the observed spectrum, indicating that additional components are required to explain both the infrared (see §§9.3.4.1 and 12.4.4) and soft X-ray (see §8.8.1) continua of AGNs. That is, straightforward accretion disk models generally predict spectra that are considerably too peaked in the ultraviolet to match what we see in real AGNs.

Moreover, not all AGNs have distinct big blue bumps. Radio-loud quasars and BL Lac objects in particular often don't conform to this picture, particularly when the radio structure is dominated by a flat-spectrum core (see §§9.1.2, 9.1.3 for a discussion of the morphological and spectral subclasses of radio-loud AGNs). The optical/ultraviolet spectra of these AGNs are often better described by smooth power laws, whose slopes (in the sense $d\ln F_\epsilon/d\ln\epsilon$) can be anywhere from 0.2 to 1.8—in fact, the spectral slopes of individual objects vary from time to time by as much as ~ 0.5 (Falomo et al. 1994). That their light is also frequently strongly polarized adds to our suspicion that it comes from a nonthermal source bearing little resemblance to an accretion disk. Synchrotron radiation, possibly augmented by inverse Compton scattering, is a more likely radiation mechanism for the optical/ultraviolet light in these objects (§9.3.4.1). How the energy is transferred from an accretion flow to a nonthermal radiator is an outstanding question (§9.3).

These facts are illustrated by the fits compiled by Sun and Malkan (1989) and Laor (1990). As figure 7.11 shows, disk models always need to be supplemented by an independent infrared component in order to explain the light longward of $\sim 1 \mu$. A smaller contribution from Balmer edge recombination emission and blended Fe II multiplets is also necessary to create the so-called *Little Blue Bump*; see Wills, Netzer, and Wills 1985; see also §10.1.1.1).

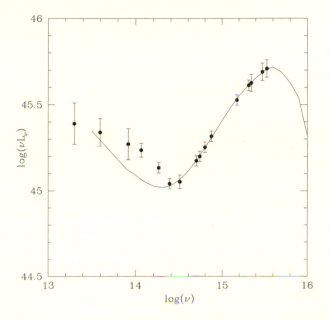

Fig. 7.11 The spectrum of the quasar Q1001+292, as fit by Laor (1990) with a combination of an accretion disk model and an arbitrary power law that accounts for the infrared emission. Although the accretion disk model fits the optical-ultraviolet spectrum reasonably well, its one distinctive feature—a sharp turndown in the far ultraviolet—occurs at frequencies just beyond where the data end.

7.6.2 Lyman edge features

The results of other tests of predicted accretion disk spectra are equally discouraging. Koratkar, Kinney, and Bohlin (1992) studied a sample of 49 quasars with redshifts high enough to bring 912 Å into the range 1200–3000 Å in our frame and found that fewer than 10% have Lyman edge features that could be due to an accretion disk. All of these possible edge features are in absorption. Another 22% have edge features at the redshift of the quasar, but with associated line absorption indicating an origin in lower density gas far from the accretion disk. Most quasar spectra look more like the one shown in figure 7.12: absolutely smooth across the Lyman edge.

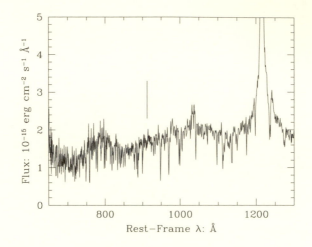

Fig. 7.12   Rest-frame FUV spectrum of the $z = 1.512$ quasar PG 0743-673 (data courtesy of A. Koratkar). The position of the Lyman edge is marked by the vertical line. There is clearly little that distinguishes this wavelength.

Several mechanisms may account for the lack of Lyman edge features, but we do not yet know which, if any, does the job. One way to eliminate an apparent edge is relativistic smearing (§7.5.5), although this is relatively ineffective for face-on views. Another is for the ionization balance of H to move away from LTE in the sense necessary to cancel the edge (§7.5.4; Störzer, Hauschildt, and Allard 1994). Comptonization in an overlying corona (§§8.3, 8.8.1) may also help.

7.6.3   Polarization

Most AGNs have measurable linear polarization, but only at about the 1% level (Berriman et al. 1990). A modest amount of absorptive opacity in the disk atmosphere may explain why the polarization at optical wavelengths isn't larger (Laor et al. 1990). However, *HST* observations by Koratkar et al. (1995) and Impey et al. (1995) show the polarization in some instances dramatically *rises* across the Lyman edge, in a fashion not predicted by any accretion disk model.

7.6.4   Variability

Yet another test of quite a different sort has also strongly suggested a problem with the simplest accretion disk models. Most AGNs are at least mildly variable. The few that have been monitored carefully in the optical or ultraviolet band typically show a roughly power-law power spectrum over the range of timescales from roughly a week to a year (see, e.g., the lightcurves shown in fig. 7.13, whose power spectrum is shown in fig. 7.14). In low-luminosity radio-quiet AGNs, the *rms* fluctuation amplitude on timescales ~ 1 yr might be typically tens of percent, with the amplitude often diminishing with increasing wavelength. Otherwise similar objects with luminosities two or three orders of magnitude greater appear to vary with an amplitude perhaps one order of magnitude smaller.

These fluctuations offer several possible avenues for exploitation as diagnostics. Discovery of characteristic timescales might permit the identification of the physical mechanisms driving the fluctuations by following arguments such as those presented in §7.4. Accretion rate variations, for example, should be far slower than variations in the local thermal content of regions in the disk. In addition, because thermal disks map frequency to location, it might be possible to track the propagation of a fluctuation through the disk by watching the sequence in which different frequency bands vary. Conversely, by comparing lightcurves in different continuum bands, it might be possible to relate the history of fluctuations at different locations.

The lightcurves shown in figure 7.13 are taken from one experiment of the last sort. Throughout the optical and ultraviolet, the fluctuations are reasonably well described by a Fourier power spectrum $|\hat{F}_c(f)|^2 \propto f^{-2}$ over the range $(240\text{d})^{-1}$ to $(30\text{d})^{-1}$. Because the variance during an experiment with sampling interval Δt and duration t_{tot} is

$$\text{Var}(F_c) = \int_{1/t_{\text{tot}}}^{1/(2\Delta t)} df \, |\hat{F}_c(f)|^2, \qquad (7.123)$$

and this integral is dominated by its lower limit, most of the variance must come on timescales of a year or more. If one guesses L/L_{E} to be ~ 0.1, the orbital timescale in a thermal disk at the radius where we might expect most 1300 Å photons to be made is ~ 3 d; from this, it is possible to rule out immediately local orbital effects as the source of the variability, although local thermal effects would be possible if $\alpha \sim 0.01$.

Further information can be gleaned by comparing the lightcurves of different bands. Surprisingly, the variations all the way from 5000 to 1300 Å

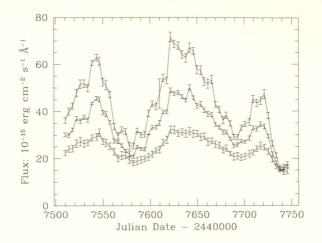

Fig. 7.13   Continuum lightcurves in three UV bands—1337 Å, 1813 Å, and 2670 Å, from top to bottom—for the type 1 Seyfert galaxy NGC 5548 (Clavel et al. 1991). Although continuous lines are drawn to guide the eye, the flux was sampled once every four days, as shown by the error bars. The pattern of variation is very similar in each of the three bands, but the fractional amplitude declines as the wavelength increases.

are essentially simultaneous, to within the resolution (4 d) of the experiment. Because a thermal disk model with L/L_E smaller than a few tenths would predict that the optical photons are made at radii around 0.2–0.4 lt-d, the group velocity of whatever signal coordinates these fluctuations must travel at least as fast as $\sim 0.1c$. The sound speed in the disk is far slower than this, so one is immediately led to the thought that it is photons that carry the signal; that is, a major part of the optical and ultraviolet light may be produced by reprocessing higher frequency photons originally radiated closer to the center. It hardly needs emphasis that if this is true, many of the relations derived earlier connecting emitted flux with local dissipation would need substantial alteration.

In sumary, although there are very strong theoretical reasons why most accretion should take place through a disk, the observational predictions made by the simplest versions of this model agree at best only partially with the AGN spectra we see. While the zeroth-order prediction of strong emission in the UV is vindicated, in general disk atmosphere predictions

194

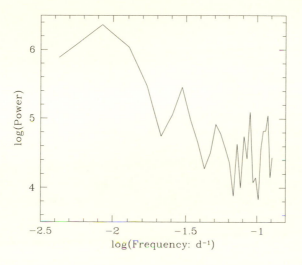

Fig. 7.14   The Fourier power spectrum for the 1337 Å continuum lightcurve shown in fig. 7.13 (Krolik et al. 1991). At frequncies above $\sim (30\ \mathrm{d})^{-1}$, the data are noise-dominated, so the power spectrum is flat.

do not agree with details of the spectra. More importantly, one of the salient characteristics of AGNs is that they emit substantial power over a tremendous span of the electromagnetic spectrum, and no simple accretion disk can come close to doing that. Clearly, the physics of simple accretion disks are an inadequate basis for understanding the continuum emission of AGNs.

8   X-ray and γ-ray Emission

In the previous chapter we showed how the most natural mode of accretion—through a disk—leads naturally to the production of a strong optical/ultraviolet continuum. However, we also know that most AGNs are bright all the way up to 100 keV and sometimes all the way up to 1 GeV or beyond. Clearly, some mechanism other than quasi-thermodynamic equilibrium radiation must be operating in order to produce such high-energy photons. In addition, because the total amount of flux in these high-energy bands can be a significant fraction of the bolometric flux, there must be an efficient mechanism for transferring energy released in the accretion disk to a far hotter plasma radiating the high-energy portion of the spectrum.

At this point we can nominate candidate mechanisms for the radiation processes, given the existence of a hot plasma, but ideas for the energy transfer process remain extremely sketchy. Consequently, most of this chapter will be an enumeration of high-energy radiation processes; for each of these, we must assume that somehow enough energy is channeled its way. In order to create a more specific context for this discussion, we begin by displaying the specific character of the different high-energy spectra encountered.

8.1   Observed Spectra

8.1.1   Radio-quiet AGNs

The high-energy spectra of radio-quiet AGNs are remarkably stereotyped (see fig. 8.1). Three components are present in most examples, and the parameters describing these components vary rather little from case to case (for greater detail, see the review by Mushotzky, Done, and Pounds 1993).

The dominant component is a power law extending from ~ 1 keV up to a cutoff, usually somewhere above ~ 100 keV. This power law has a spectral index in energy units $\alpha_x \simeq 0.9$, with a sample dispersion of only about 0.1, although there are occasional examples (e.g., NGC 4151) with significantly different slopes. The cutoff is rarely well determined, for the number of photons per unit energy drops as $\epsilon^{-\alpha_x-1}$, making it difficult to achieve adequate S/N at very high energies.

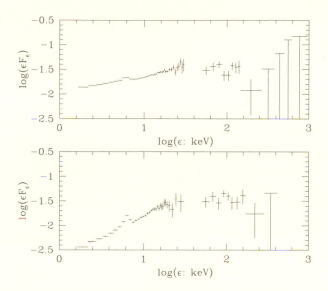

Fig. 8.1   The average X-ray spectra of radio-quiet AGNs, type 1 in the top panel, type 2 in the bottom panel. In terms of ϵF_ϵ, the flux has a broad, flat peak that runs from $\simeq 20$ to 100 keV or beyond. The bump at $\simeq 6$ keV, noticeable even in the compressed scale of these plots, is the Fe Kα line. Note also how sharply the type 2 flux drops below a few keV; this is due to strong absorption on the line-of-sight (§§8.7.1, 8.8.1, Chap. 12). The data are from *Ginga* and *OSSE*, as reported by Zdziarski et al. (1995) and Wozniak et al. (1998).

The contribution of this component to the bolometric luminosity of the AGN has a rather greater dispersion than its shape. Because it is a directly measurable quantity, the amplitude of the X-ray power law is conventionally parameterized by α_{ox}, the spectral index of the power law that would interpolate between the highest frequencies in the optical band (actually 2500 Å in the rest frame) and the soft end of the observed X-ray power law (specifically, 2 keV). AGNs can be found spanning the range $0.8 \leq \alpha_{ox} \leq 2.1$. Most surveys have found the peak of the distribution to be somewhere near 1.4–1.6, but there also appears to be a correlation with luminosity in the sense that α_{ox} slowly increases (*i.e.* the X-ray flux weakens) with rising luminosity. In addition, it should come as no surprise that the shape of the distribution depends rather sensitively on how the sample has been selected (Stocke et al. 1991; Wilkes et al. 1994; Green et al. 1995.)

In terms of the α_{ox} parameter, the ratio between $dF/d\log\epsilon$ at 2 keV and at 2500 Å is $0.05 \times 400^{1.5-\alpha_{ox}}$, so that the energy flux seen by imaging (i.e., soft) X-ray telescopes is typically about a factor of 20 smaller than that seen in the UV. On the other hand, the dynamic range in frequency generally styled "X-rays" is more than a factor of a hundred, and additional components (detailed in the next few paragraphs) often increase ϵF_ϵ at X-ray energies both above and below 2 keV, so the total contribution of the X-ray band to the bolometric luminosity is generally several times greater than ϵF_ϵ at 2 keV.

The next most important component has a spectrum that rises sharply near 10 keV, reaches a broad maximum between 20 and 60 keV and rolls off more gradually at higher energies. In some cases, this component raises the flux at 30 keV by as much as a factor of two above the underlying power law, but there are also examples in which this component is apparently absent. An Fe Kα line whose energy indicates it is excited by fluorescence in Fe less ionized than about Fe^{+16} is generally associated with this component, but it can also appear even without the "bump." The equivalent width of this line is sometimes as large as $\simeq 350$ eV and sometimes essentially undetectable, that is, its equivalent width is sometimes less than 50 eV. Both the "bump" and Fe Kα emission seem to decline in importance as the luminosity rises from the Seyfert galaxy level to that of quasars.

The third component is a *soft excess*, that is, a contribution that rises steeply below 1 keV. There is a great deal of dispersion in the relative amplitude and spectral shape of this component. In fact, even in individual objects its precise spectral shape is very difficult to pin down: spectral resolution is often poor in soft X-ray detectors, calibration can be problematic, and it can be hard to disentangle the intrinsic spectral shape from the effects of interstellar absorption, whether in our Galaxy or in the AGN's host galaxy (see §8.7.1). We do not even know whether this component is better described as continuum or a sum of emission lines. About the most we can say is that in some significant fraction of all radio-quiet AGNs (30%? 50%?), this component dominates below $\simeq 0.5$ keV.

Finally, although there were reports early in the 1980s that several radio-quiet AGNs were detectable at energies above 1 MeV, none of these reports has been confirmed by later experiments.

8.1.2 Radio-loud lobe-dominated AGNs

Radio-loud lobe-dominated AGNs (see §9.1.2 for a discussion of the distinction between "lobe-dominated" and "core-dominated") are qualitatively similar to radio-quiet AGNs in the X-ray band, but with certain clear quantitative differences (fig. 8.2). They share the same three components, but in different proportions. In these sources the power law is also the dominant part of the spectrum, but the spectral index is generally a bit smaller, nearer 0.7 than 0.9. The relative strength of the X-ray power law is also somewhat greater; $\langle \alpha_{ox} \rangle \simeq 1.3$–$1.4$, although the distribution is just as wide as for the radio-quiet AGNs (Zamorani et al. 1981; Brunner et al. 1992).

In the energy range above 50 keV, no individual object is bright enough for a decent S/N spectrum (with the instrumentation available as of 1998). Instead, a sum of a number of individually weaker sources indicates that a simple extrapolation of the power law passes through the observations but is not a particularly good fit, and that any cutoff is at energies greater than $\simeq 150$ keV.

Interestingly, there is no evidence for a "bump" component in any radio-loud AGNs, although the limits are rather weak because radio-loud AGNs are rare enough that even for the brightest we do not yet have a spectrum with excellent S/N.

Evidence exists for soft excesses in radio-loud lobe-dominated AGNs, as in their radio-quiet cousins, but the statistics are even poorer.

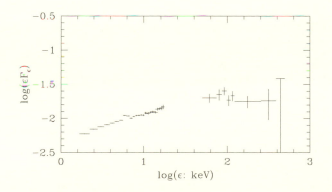

Fig. 8.2   The average X-ray spectrum of radio-loud, lobe-dominated AGNs (Zdziarski et al. 1995; Wozniak et al. (1998). The spectral shape of these AGNs is quite similar to that of type 1 Seyferts but is rather harder in the range 2–20 keV.

8.1.3   Radio-loud core-dominated AGNs

Radio-loud core-dominated AGNs (also sometimes known as *blazars*: §1.3) are very different (as fig. 8.3 makes plain). Their spectra in the normal (i.e., 1–20 keV) X-ray band are also power laws, but their spectral indices span a much greater range, from $\simeq 0.7$ to $\simeq 1.5$ (Sambruna et al. 1994), and there is evidence for neither a 20 keV bump nor a soft excess. In fact, individual objects display substantial spectral variability, with the spectral index changing as much as several tenths, generally in the sense that the spectra are harder when the objects are brighter. The amplitude of the X-ray component relative to the optical is also significantly greater than for radio-quiet AGNs: Brunner et al. (1992) find that for radio-loud core-dominated AGNs, $\langle \alpha_{ox} \rangle \simeq 1.1$, that is, the fluxes per logarithmic frequency interval are approximately the same in the UV and soft X-ray bands.

But the real contrast is at higher energies. Although defining a true time-average picture of their spectra is difficult because of their extreme variability, it is possible to compare the flux in high-energy γ-rays in a single measurement with the available data at other energies at other times. For the overwhelming majority of objects in this category, the *EGRET* experiment discovered that a single sample of the energy flux in photons above 100 MeV is at least comparable to, and possibly an order of magnitude or more greater than, the flux measured in any other band of the spectrum (von Montigny et al. 1995)! Moreover, the shape of the spectrum in the range 100 MeV–1 GeV is typically a power law with index 1 ± 0.5; that is, the flux per logarithmic frequency interval is approximately constant.

As we have already remarked, in a few cases, for example, Markarian 421, the flux at 1 TeV is as great as at 1 GeV. In most other blazars, there are only upper bounds on the TeV flux, and these bounds lie at a level lower than the extrapolation of that object's 100 MeV–1 GeV power law. It is possible that Markarian 421 and the others are simply unusual objects. However, it may also be no coincidence that they are exceptionally close (in the case of Mrk 421, $z = 0.03$). As pointed out by Stecker, de Jager, and Salamon (1992), reasonable estimates of the intergalactic infrared background intensity suggest that it presents a significant opacity to 1 TeV photons due to pair production (see §8.4.1). If this is true, the optical depth over distances of only a few hundredths of a Hubble length may not be enough to diminish the TeV flux substantially, but the optical depth to more distant objects would be great enough to cause most of their photons in that range to be absorbed en route.

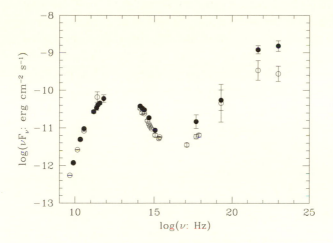

Fig. 8.3 The broad-band spectrum of 3C 279, an optically violently variable quasar (Wehrle et al. 1998). In this case, the data in the different bands are nearly simultaneous, with the open circles representing the spectrum in January 1996, and the filled circles showing the spectrum a month later. At both epochs, the high-energy γ-ray flux was roughly an order of magnitude greater than the flux in any other band.

8.1.4 Observed variability

Although most AGNs vary relatively sedately in the optical and ultra-violet bands, their behavior in the X-ray band is much more active. Significant changes in a few hours are not at all uncommon (a typical lightcurve is shown in fig. 8.4). However, the character of these fluctuations is similar to those seen in lower frequency bands in one respect—their Fourier power spectra are quite smooth and (at least for radio-quiet objects) are generally well described by power laws rather similar in slope to those seen in the ultraviolet. That is, $P(f) \propto f^{-b}$ and $1 \leq b \leq 2$, albeit the range of observed timescales for X-ray fluctuations is typically fractions of an hour to a few days, whereas for UV fluctuations the timescales probed are usually in the range ~ 1–~ 100 d. For a variety of operational reasons, it has been difficult to monitor X-ray fluxes on timescales of days to months, but it is believed that for low-luminosity radio-quiet objects, the variance becomes comparable to the square of the mean flux somewhere in just that range of timescales. In other words, the largest variations take place on timescales of days to months. As usual, the amplitude of variability for BL Lac objects and OVV quasars is much greater than for radio-quiet objects like Seyfert galaxies.

At still higher energies (above several MeV), only BL Lac objects and OVV quasars can be seen at all. True to their nature, they are strongly variable in this band also; contrasts of factors of two in a week or an order or magnitude in a month (e.g., as in fig. 8.3) can often be seen. However, observations in this band are still so sparse that it has not yet been possible to quantify fully the character of these variations.

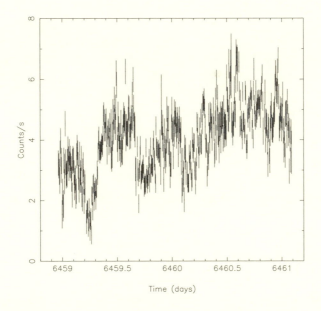

Fig. 8.4 *Exosat* observations of the 1.2–10 keV flux from the type 1 Seyfert galaxy MCG 6-30-15 (Krolik et al. 1993).

8.2   Radiation Physics: Bremsstrahlung and Inverse Compton Scattering

A number of different radiation mechanisms might potentially contribute to the production of the X-ray and γ-ray continuum in AGNs: bremsstrahlung, inverse Compton scattering, synchrotron radiation, and $e^{\pm}$ annihilation are all possible in principle. A basic introduction to the nature of the first three can be found in Rybicki and Lightman (1979). A much more detailed review of these three is given in Blumenthal and Gould

(1970), and an account of the second and fourth from quantum mechanical first principles can be found in Jauch and Rohrlich (1976).

We will postpone a discussion of synchrotron radiation until Chapter 9 because a simple estimate shows that it is unlikely to be important to the high-energy spectra of most AGN. Suppose that the X-rays are made near the innermost rings of an accretion disk. There the energy density in magnetic field might be at most comparable to the total pressure, that is, $B^2/(8\pi) \sim p_{\rm rad} + p_g \sim aT^4/3$, where T is the temperature of the accretion disk. Then we would expect $B \sim 3000T_5^2$ G. The maximum photon energy at which an electron with Lorentz factor γ radiates significantly by the synchrotron mechanism is $\sim \gamma^2 \hbar eB/(m_e c)$ (see §9.2.1); therefore, the electrons responsible for making X-rays of energy ϵ have $\gamma \gtrsim 2 \times 10^4 (\epsilon/10 \text{ keV})^{1/2} T_5^{-1}$. On the other hand, these same electrons also up-scatter the photons made by the disk to energies $\sim \gamma^2 k_{\rm B} T \sim 3$ GeV (see §8.2.2). The ratio of synchrotron luminosity to inverse Compton luminosity is $B^2/(8\pi aT^4)$, which we are supposing is at most ~ 1. Therefore, if synchrotron radiation contributes significantly to X-ray production, there should be a component of high-energy γ-rays in the spectrum of comparable luminosity. As shown in §8.1, this only occurs in the core-dominated radio-loud AGNs; this is why the focus in this chapter will be on the other three processes.

Annihilation radiation is best treated in the context of other pair-related processes. It will therefore be discussed in §8.4.

8.2.1 Bremsstrahlung

Bremsstrahlung radiation is the result of a changing electric charge distribution during a particle-particle collision. Ordinarily, electron-proton (or electron-nucleus) collisions dominate over other sorts of collisions (electron-electron, proton-proton, etc.) because the electron-proton system possesses a significant electric dipole, whereas the others do not. However, when the electron velocities become relativistic, radiation in other moments (magnetic dipole, electric quadrupole, and so on) becomes comparable in importance to electric dipole radiation. In addition, the kinematics of the e–p system change as the momentum of emitted photons becomes significant. Consequently, the standard expressions for e–p bremsstrahlung must be adjusted, and the contributions of several other varieties of collision must be included, when $k_{\rm B} T \sim m_e c^2$.

8.2.1.1 relativistic e–nucleus bremsstrahlung

When electrons of density n_e collide with nuclei of charge Ze whose density is n_Z, they change velocity and radiate. The radiated power produced per unit volume per unit photon energy is $n_e n_Z \Lambda_\epsilon^{eZ}$, where the spectral emissivity is

$$\Lambda_\epsilon^{eZ} = 2\sqrt{\frac{2}{3\pi}} \alpha_{\mathrm{fs}} \sigma_T c Z^2 \Theta^{-1/2} g(\Theta, \varepsilon) e^{-\varepsilon}. \tag{8.1}$$

Here $\sigma_T \simeq 6.65 \times 10^{-25}$ cm^2 is the Thomson cross section, $\Theta \equiv k_B T/(m_e c^2)$, and $g(\Theta, \varepsilon)$ is the Gaunt factor. As this form illustrates, the natural unit for bremsstrahlung emissivity is $\alpha_{\mathrm{fs}} \sigma_T c$; that is, $\Lambda_\epsilon^{eZ} \sim 1$ in these units when $Z \sim \Theta \sim \varepsilon \sim 1$. The reason is that in quantum-mechanical terms, bremsstrahlung is an event involving an exchange of a (virtual) photon and emission of a new photon (fig. 8.6). It therefore resembles Compton scattering with the addition of a photon creation event.

When $10^{-3} < \Theta \ll 1$, the Gaunt factor takes the approximate form

$$g(\Theta, \varepsilon) = \frac{\sqrt{3}}{\pi} \begin{cases} \ln\left[(2/\gamma_E)^{5/2} \Theta^{3/2} (Z\alpha_{\mathrm{fs}})^{-1} (m_e c^2/\epsilon)\right] & \varepsilon \ll Z\alpha_{\mathrm{fs}}^2/(2\Theta) \\ \ln\left[4/(\gamma_E \varepsilon)\right] & \varepsilon \gg Z\alpha_{\mathrm{fs}}^2/(2\Theta), \end{cases} \tag{8.2}$$

where $\gamma_E = 0.5772$, Euler's constant. Thus, $g(\Theta, \varepsilon) \sim 1$ over a large part of parameter space.

However, when Θ is comparable to or greater than 1, the Gaunt factor changes to

$$g_{\mathrm{rel}}(\Theta, \varepsilon) \simeq \left(\frac{3}{2\pi^3}\right)^{1/2} \Theta^{1/2} e^{-1/\Theta} \frac{\ln[4\eta_E(1 + C_1\Theta)/\varepsilon](1 + 2\Theta + 2\Theta^2)}{K_2(1/\Theta)}, \tag{8.3}$$

where $\eta_E = \exp(-\gamma_E) \simeq 0.56$, and $C_1 = \eta_E \exp(5/2)/2 \simeq 3.42$ (Svensson 1984). K_n is the modified Bessel function of order n, and may be approximated by

$$K_2(1/\Theta) \simeq \begin{cases} \sqrt{\pi/2} \Theta^{1/2} e^{-1/\Theta} & \Theta \ll 1 \\ 2\Theta^2 & \Theta \gg 1. \end{cases} \tag{8.4}$$

Although the Gaunt factor is a slowly varying function of order unity when $\Theta \ll 1$, when the typical electron moves with relativistic speed, it becomes a much more important modification to the semi-classical emissivity. In the limit of $\Theta \gg 1$, $g(\Theta, \varepsilon) \propto \Theta^{1/2}$ times a logarithmic function of ε.

By Kirchhoff's law, the emissivity of a population in thermodynamic equilibrium with radiation must be exactly equal to the rate at which it absorbs radiation, that is, the product of its absorptivity times the blackbody intensity at that temperature. The expression presented in equation

7.92 for nonrelativistic electron-ion bremsstrahlung opacity can be (in fact, usually is) derived by this means from equation 8.1.

The total emissivity is, of course, the integral over all ϵ of the spectral emissivity. When we write temperature in units of the electron rest mass, the natural unit for total emissivity is then $\alpha_{\mathrm{fs}}\sigma_{\mathrm{T}}m_ec^3 \simeq 1.2 \times 10^{-22}$ erg cm^3 s^{-1}. Svensson (1982b) found convenient fitting formulae for both relativistic and nonrelativistic temperatures:

$$\Lambda^{eZ} \simeq \alpha_{\mathrm{fs}}\sigma_{\mathrm{T}}m_ec^3Z^2 \begin{cases} (32/\pi^3)^{1/2}\Theta^{1/2}(1+1.781\Theta^{1.34}) & \alpha_{\mathrm{fs}}^2 \ll \Theta \leq 1 \\ [9/(2\pi)]\Theta\left[\ln(2\eta\Theta + 0.42) + 3/2\right] & \Theta \geq 1. \end{cases}$$
$$(8.5)$$

As the temperature enters the relativistic range, the emissivity of e–Z bremsstrahlung increases faster than the $\Theta^{1/2}$ characteristic of the nonrelativistic regime. When $\Theta \gg 1$, the proportionality of the Gaunt factor to $\Theta^{1/2}$ makes the total emissivity rise $\propto \Theta$.

8.2.1.2 e^-–e^- or e^+–e^+ bremsstrahlung

Although generally unimportant at nonrelativistic temperatures for the reasons discussed in the introduction to this subsection, electron-electron (and positron-positron) bremsstrahlung can be as important as electron-ion bremsstrahlung when $\Theta \sim 1$ or more. Svensson (1982b) also found a convenient analytic expression for these rate coefficients (they are identical, of course):

$$\Lambda_\epsilon^{ee} \simeq \frac{2}{\pi}\alpha_{\mathrm{fs}}\sigma_{\mathrm{T}}ce^{-\varepsilon}\ln\left(4\eta_{\mathrm{E}}\frac{C_2 + C_3\Theta^2}{\varepsilon}\right)\frac{(3/5)\sqrt{2}\Theta + 2\Theta^2}{\exp(1/\Theta)K_2(1/\Theta)}, \qquad (8.6)$$

where $C_2 = \exp(29/12) \simeq 11.2$ and $C_3 = \eta_{\mathrm{E}}^2\exp(7/2) \simeq 10.4$. At low temperatures, the electron-electron "Gaunt factor" is $\propto \Theta$, and so falls rapidly relative to electron-ion bremsstrahlung when $\Theta \ll 1$.

The result of integrating over frequency to find the total emissivity for electron-electron bremsstrahlung is well fitted by

$$\Lambda^{ee} = \frac{3}{2\pi}\alpha_{\mathrm{fs}}\sigma_{\mathrm{T}}m_ec^3\Theta$$
$$\times \begin{cases} 5(44 - 3\pi^2)/(9\pi^{1/2})\Theta^{1/2}(1 + 1.1\Theta + \Theta^2 - 1.25\Theta^{5/2}) & \alpha_{\mathrm{fs}}^2 \ll \Theta < 1 \\ 6[\ln(2\eta\Theta) + 5/4] & \Theta \geq 1. \end{cases}$$
$$(8.7)$$

This rate is exactly double the e–p rate for $\Theta > 1$ because both electrons radiate. It therefore scales in the same way with temperature, $\propto \Theta$ when $\Theta \gg 1$.

8.2.1.3 e^--e^+ bremsstrahlung

If one views it in terms of the equivalent single-particle problem, non-relativistic electron-positron bremsstrahlung is identical to nonrelativistic electron-proton scattering but for the fact that the reduced mass is half as large. Consequently, the emissivity of *nonrelativistic* electron-positron bremsstrahlung is exactly $2^{3/2}$ times larger than nonrelativistic electron-proton bremsstrahlung. On the other hand, *relativistic* electron-positron bremsstrahlung is almost identical to relativistic electron-electron bremsstrahlung except for the fact that there are twice as many available final states in electron-positron bremsstrahlung because the final two-particle wave function is not constrained by Fermi-Dirac statistics to be antisymmetric. Therefore, the emissivity of relativistic electron-positron bremsstrahlung is exactly twice as large as the emissivity of relativistic electron-electron bremsstrahlung.

8.2.2 Inverse Compton scattering

In "ordinary" Compton scattering, photons scatter off electrons, generally assumed to be initially stationary. As a result, the photons lose some of their energy in recoil. "Inverse" Compton scattering is exactly the same as ordinary Compton scattering except for the fact that the electrons are not stationary in the interesting reference frame. Consequently, although the kinematics in the electron rest frame are identical to the usual case, in the "lab" frame the photons can gain energy at the expense of the electrons.

8.2.2.1 kinematics: recoils and boosts

As in any collision, the energy transfer is largely determined by kinematics. Suppose that the initial and final electron four-momenta are $p_{i,f}^{\mu}$ and the initial and final photon four-momenta are $k_{i,f}^{\mu}$. Then momentum-energy conservation can be expressed by

$$p_i^{\mu} + k_i^{\mu} = p_f^{\mu} + k_f^{\mu}. \tag{8.8}$$

Isolating p_f on one side of the equation and taking the magnitude squared reduces this four-vector equation to the scalar equation

$$p_i^{\mu} \cdot k_i^{\mu} - p_i^{\mu} \cdot k_f^{\mu} - k_i^{\mu} \cdot k_f^{\mu} = 0. \tag{8.9}$$

For these purposes, we will always take the metric to be that of flat space, so we will adopt the notation $x^{\mu} \cdot y^{\mu} \equiv x^{\mu} y_{\mu} \equiv x^{\mu} \eta_{\mu\nu} y^{\nu}$. Because all the quantities in equation 8.9 are Lorentz scalars, we can choose freely in which frame to evaluate them. The simplest frame (for our first purpose) is the initial electron rest frame. In that frame, equation 8.9 (with c set $= 1$) becomes

$$k_f = \frac{k_i}{1 + (k_i/m_e)(1 - \hat{n}_i \cdot \hat{n}_f)}, \tag{8.10}$$

where $\hat{n}_{i,f}$ are unit vectors in the initial and final photon directions. Equation 8.10 is, of course, the famous Compton recoil formula. From it, we see that, in the initial electron rest frame, photons lose a fraction k_i/m_e of their energy per scatter. Photons of 100 keV or more can drop in energy by a significant amount in a single scattering. The limit $k_i/m_e \ll 1$ (small recoil) is called *Thomson scattering*, and the opposite limit (large recoil) is called *Klein-Nishina scattering*.

Let us now change points of view and examine the same process in a frame (the "lab" frame) in which the electrons move with velocity $\vec{\beta}$. Evaluating equation 8.9 in the lab frame yields

$$k_f = k_i \frac{1 - \vec{\beta} \cdot \hat{n}_i}{1 - \vec{\beta} \cdot \hat{n}_f + (\gamma k_i/m_e)(1 - \hat{n}_i \cdot \hat{n}_f)}. \tag{8.11}$$

With this expression as a base, we can find the outgoing photon energy for fixed γ and k_i averaged over all three directions: $\hat{n}_i$, $\hat{n}_f$, and $\vec{\beta}/\beta$.

First consider the nonrelativistic limit of equation 8.11:

$$k_f/k_i \simeq \left[1 + \vec{\beta} \cdot (\hat{n}_f - \hat{n}_i) \right]. \tag{8.12}$$

That is, the largest immediate effect of scattering a photon with $k_i \ll m_e$ off a nonrelativistic electron is to Doppler shift the photon energy by a fractional amount $\vec{\beta} \cdot (\hat{n}_f - \hat{n}_i)$. Depending on the relative directions, this shift can be either positive or negative. Thus, the largest single effect of Compton scattering in the nonrelativistic limit is to broaden initially narrow spectral features by an amount $\sim \beta$.

In the electron rest frame, the scattering is front-back symmetric, but the transformation to the lab frame introduces a correlation between $\vec{\beta}$ and $\hat{n}_f$ that is $\sim \beta$. Consequently, if the initial directions of the electron and the photon are uncorrelated, the *mean* value of the term first-order in β is $\sim \beta^2$. There is also an additional contribution $\sim \beta^2$ due to higher-order terms in the expansion leading to equation 8.12. As a result, nonrelativistic inverse Compton scattering on average increases photon energies by an amount $\sim \beta^2$. Careful angle-averaging shows that the actual mean fractional increase is $(4/3)\beta^2$.

Next consider the extreme relativistic (but still Thomson) limit. The relativistic beaming is much stronger, so we expect that most outgoing photons will be nearly parallel to $\vec{\beta}$ in the lab frame. Then, because $\beta \simeq 1 - 1/(2\gamma^2)$ for $\gamma \gg 1$, $\langle k_f/k_i \rangle \sim \gamma^2$, with a maximum possible amplification of $4\gamma^2$, achieved when the collision is exactly head-on in the lab frame. This extreme relativistic limit makes it plain that the energy change due to electron scattering can be viewed as the result of two successive Doppler boosts: one from the lab frame into the electron rest frame, and the other back into the lab frame. Again, the result of more careful angle-averaging is that $\langle k_f/k_i \rangle = (4/3)\gamma^2$.

In Klein-Nishina scattering, the factor $\gamma k_i/m_e$ in the denominator of equation 8.10 becomes significant. When it is $\gg 1$, $\langle k_f \rangle \simeq \gamma m_e$. In other words, the electron gives all its energy to the photon in a single interaction.

8.2.2.2 cross section

How many Compton scatters occur depends, of course, on the cross section. It is most simply described in the initial electron rest frame. It is then independent of initial photon energy for $k_i' \ll m_e$ and equal to σ_T. The angular dependence has the usual form for classical electric dipole scattering,

$$\frac{\partial \sigma_T}{\partial \Omega} = \frac{3}{16\pi} \sigma_T \left(1 + \cos^2 \theta\right), \tag{8.13}$$

where $\cos \theta = \hat{n}_i \cdot \hat{n}_f$.

This total cross section combined with the expressions already derived for the mean photon energy amplification per scattering yields the total energy loss rate for electrons scattering in the Thomson limit:

$$\dot{E}_T = \frac{4}{3}\beta^2 \gamma^2 c \sigma_T U_\gamma, \tag{8.14}$$

where U_γ is the total energy density in photons scattering in the Thomson limit, that is,

$$U_\gamma \simeq \int_0^{m_e/\gamma} d\epsilon \, \epsilon n_{\mathrm{ph}}(\epsilon). \tag{8.15}$$

Here $n_{\mathrm{ph}}(\epsilon)$ is the number density of photons with energy ϵ per unit energy.

However, when $k_i' \gtrsim m_e$, the differential cross section for photons of energy k_i to scatter to energy k_f changes to the Klein-Nishina formula

$$\frac{\partial \sigma_{KN}}{\partial k_f} = \frac{3}{8}\sigma_T \frac{m_e}{k_i^2} \left[\frac{k_i}{k_f} + \frac{k_f}{k_i} + \left(\frac{m_e}{k_f} - \frac{m_e}{k_i}\right)^2 - 2m_e \left(\frac{1}{k_f} - \frac{1}{k_i}\right)\right]. \tag{8.16}$$

In the limit of $k_i \gg m_e$, the total cross section becomes

$$\sigma_{\text{KN}}(k_i) = \frac{3}{8}\sigma_{\text{T}}\frac{m_e}{k_i}\left(\log\frac{2k_i}{m_e} + \frac{1}{2}\right). \tag{8.17}$$

When $k_i/m_e \gg 1$ (the extreme Klein-Nishina limit), $\sigma_{\text{KN}} \propto k_i^{-1}$.

As k_i/m_e grows, the angular distribution becomes progressively more concentrated in the forward direction:

$$\frac{\partial\sigma_{\text{KN}}}{\partial\Omega} = \frac{3}{16}\sigma_{\text{T}}(1+\cos^2\theta)\left[1 + (k_i/m_e)(1-\cos\theta)\right]^{-2}$$

$$\times\left\{1 + \frac{(k_i/m_e)^2(1-\cos\theta)^2}{(1+\cos^2\theta)[1+(k_i/m_e)(1-\cos\theta)]}\right\}, \tag{8.18}$$

where θ is the angle between the initial and final photon directions in the initial electron rest frame. When $k_i/m_e \gg 1$, $\partial\sigma/\partial\Omega \propto (1-\cos\theta)^{-1}$.

Combining these two effects (diminishing cross section and growing tendency toward forward scattering) with the Compton recoil we have already discussed, Blumenthal and Gould (1970) found an approximate expression for energy loss in the extreme Klein-Nishina limit, that is, the limit in which $k_i \gg m_e$ in the electron rest frame:

$$\dot{E}_{\text{KN}} = \frac{3}{8}\sigma_{\text{T}}m_e c^3 \int_{m_e/\gamma}^{\infty} \frac{d\epsilon}{\epsilon}n_{\text{ph}}(\epsilon)\left[\ln\left(\frac{4\gamma\epsilon}{m_e}\right) - \frac{11}{6}\right]. \tag{8.19}$$

Thus, the total energy loss rate by inverse Compton scattering can be found by summing the contributions from two classes of photons, those scattering in the Thomson limit and those scattering in the Klein-Nishina limit.

8.3 Output Spectra from Inverse Compton Scattering

Because bremsstrahlung is essentially a single interaction event, prediction of its output spectrum follows immediately from a description of its basic physics. However, because photons can often be scattered multiple times, determining the output spectrum from inverse Compton scattering requires consideration of what happens as a result of numerous successive scatterings. Not surprisingly, the regimes of large and small optical depth produce qualitatively different results.

8.3.1 Optically thin

8.3.1.1 non-relativistic single scatter

We have already computed the angle-averaged energy change for a particular electron velocity. The result of averaging over a thermal distribution of electron velocities, and retaining the first-order correction for small Compton recoil, is the mean amplification

$$A \equiv \langle \frac{k_f}{k_i} \rangle = 1 + 4\Theta \frac{K_3(1/\Theta)}{K_2(1/\Theta)} - \left[3\Theta + \frac{K_1(1/\Theta)}{K_2(1/\Theta)} \right] \frac{k_i}{m_e}. \tag{8.20}$$

For $\Theta \ll 1$, equation 8.20 becomes

$$A = 1 + 4\Theta - \frac{k_i}{m_e}. \tag{8.21}$$

When the intensity-weighted mean energy of the photons is $\Theta m_e c^2/4$, there is zero net energy exchange between photons and electrons. Conversely, if the photon spectrum has a fixed shape, and electrons exchange energy with the photons only by Compton scattering, their temperature evolves toward the value at which $A = 1$, or a state of *Compton equilibrium*. It is for this reason that 1/4 the intensity-weighted mean energy of the photons is called the *Compton temperature*, T_C.

On the other hand, when $T \gg T_C$, the net result of averaging individual scatters is simply to multiply the photons' energies by the factor $1+4\Theta$. A continuous spectrum of seed photons subjected to this sort of scattering retains its shape but is shifted to higher energies by that amount.

8.3.1.2 relativistic single scatter

Because the average photon's energy is increased by factor $(4/3)\gamma^2$ (in the Thomson regime) when scattered by a relativistic electron, the mean amplification due to relativistic electrons is

$$A = \frac{\int d\gamma \, (dn_e/d\gamma)(4/3)\gamma^2}{\int d\gamma \, (dn_e/d\gamma)}. \tag{8.22}$$

Just as in the nonrelativistic case, the photons' energies are shifted by a constant factor. A simple approximation to the power that emerges from a unit volume per unit photon energy is therefore given by

$$j_\epsilon = \sigma_T \int d\gamma \int dk \, \frac{dn_e}{d\gamma} \frac{J_k}{k} 4\pi \delta(4\gamma^2 k/3 - \epsilon)\epsilon \tag{8.23}$$

when the electron velocities are isotropic. In this expression we have retained use of units in which $c = 1$. A more precise description would be

$$j_\epsilon = \sigma_T \int d\gamma \int dk \frac{dn_e}{d\gamma} \frac{J_k}{k} \frac{3\pi}{\gamma^2 k} F_C(\epsilon'), \qquad (8.24)$$

where the Compton scattering spectral distribution is

$$F_C(\epsilon') = 2\epsilon' \ln \epsilon' + \epsilon' + 1 - 2\epsilon'^2, \qquad (8.25)$$

for $\epsilon' = \epsilon/(4\gamma^2 k)$, and $\epsilon' \leq 1$ (Blumenthal and Gould 1970).

The shape of the output spectrum from a single scattering by relativistic electrons clearly depends on both the seed photon spectrum and the electron energy distribution function. Which is of greater influence depends on whether the spread induced by the dynamic range in electron energies $(\gamma_{max}/\gamma_{min})^2$ is greater than or less than the spread induced by the dynamic range in seed photon energy k_{max}/k_{min}. For example, consider the output spectrum produced when both the electron distribution and the initial photon spectrum are power laws over limited ranges. Making use of the δ-function approximation of equation 8.23, one finds

$$j_\epsilon = \frac{2^{\xi-2}}{3^{(\xi-1)/2}} \frac{(1-\xi)}{\gamma_{max}^{1-\xi} - \gamma_{min}^{1-\xi}} \sigma_T n_e U_o \epsilon^{(1-\xi)/2} \int_{\max[k_{min}, 3\epsilon/(4\gamma_{max}^2)]}^{\min[k_{max}, 3\epsilon/(4\gamma_{min}^2)]} dk\, k^{-2-\alpha}, \qquad (8.26)$$

where the total electron density is n_e, the electron distribution function $dn_e/d\gamma \propto \gamma^{-\xi}$ between γ_{min} and γ_{max}, and the seed photon energy density per unit energy is $U_o k^{-\alpha}$. If the initial photon spectrum is very narrow compared to the electron distribution, the output spectrum is a power law with index $(\xi - 1)/2$ extending across a dynamic range $(\gamma_{max}/\gamma_{min})^2$. On the other hand, if the initial range of electron energies is very narrow, the spectral cutoffs are determined by the breadth of the photon distibution on which the (nearly) monoenergetic electrons act; in this case, the output spectrum is best described as the input spectrum translated in energy by a factor $\sim \langle \gamma^2 \rangle$.

8.3.1.3 multiple scatterings in the optically thin limit

Even when the optical depth is small, there may be a spectral range in which multiply scattered photons dominate the spectrum. Such a range may exist if there are no other sources of radiation in this band, and it lies above the upper limit for single scatterings (by electrons with energy $\sim k_B T$ if the distribution is thermal, or by electrons with γ_{max} if the distribution is nonthermal). Using an argument very similar to the one below, it is easy to show that even when the electron distribution function is a power law, at any energy reachable by a single scatter, the single-scatter contribution outweighs that of multiple scatters by lower energy electrons.

To estimate the character of the multiply scattered spectrum, we follow an argument devised by Zel'dovich. First, as we have already done, compute A, the mean amplification of photon energy per scatter. To achieve an energy ϵ, a photon injected with energy k must scatter an average of m times, where

$$A^m \simeq \frac{\epsilon}{k}, \tag{8.27}$$

so that $m \simeq \ln(\epsilon/k)/\ln A$. If the scattering depth $\tau_T \ll 1$, the probability of making m scatters before leaving is $\simeq \tau_T^m$, so the emergent flux has the spectral shape

$$\begin{aligned}
\frac{dF}{d\epsilon} &\propto \int dk \, J_k \frac{\epsilon}{k} \frac{dm}{d\epsilon} \tau_T^m \\
&\propto \int dk \, \frac{J_k}{k \ln A} \tau_T^{\ln(\epsilon/k)/\ln A} \\
&\propto \int dk \, \frac{J_k}{k \ln A} \left(\frac{\epsilon}{k}\right)^{\ln \tau_T / \ln A}.
\end{aligned} \tag{8.28}$$

Thus, a power law output spectrum is generated that is harder for higher temperatures (or flatter nonthermal distributions—the argument is entirely independent of the shape of the electron distribution function) and larger optical depths (provided τ_T remains less than unity). This power law does not, of course, extend to infinite energy. No photon can be scattered to an energy greater than that of the electrons doing the scattering, so the power law cuts off at $\epsilon \sim k_B T$ (for thermal electrons) or $\sim \gamma_{max} m_e c^2$ (for nonthermal relativistic electrons). In addition, when the total range of the Comptonized spectrum, from the seed photon energy to the temperature (or maximum energy) of the electrons, can be spanned with the amplification of a few scatterings, treating m as a continuous variable is no longer a very good approximation. As a consequence, in these conditions the output spectrum is better described by a series of "humps" corresponding to photons scattered once, twice, and so on than by a smooth power law.

8.3.2 Optically thick

8.3.2.1 nonrelativistic (thermal) electrons

When the plasma is optically thick, the evolution of photon density as

212

a function of position can be described by a diffusion equation:

$$\frac{\partial n_{\text{ph}}}{\partial t} = \nabla \cdot \left(\frac{3c}{n_e \sigma_{\text{T}}} \nabla n_{\text{ph}} \right). \tag{8.29}$$

Every time the photons scatter, their energies are changed by Compton scattering. As we have seen, this change can be either up or down, but if $\Theta > \epsilon/(4m_e)$, there is a net drift upwards. If in addition $\Theta \ll 1$, so that the changes, no matter what their signs, are small compared to the photon energy, a Fokker-Planck equation gives a good description of the evolution of the spectrum. Equations of this sort account simultaneously for random walks and net drift when the step sizes are small compared to the gradient scale. The Bose-Einstein character of photons means that the scattering rate to a given final state is enhanced by the factor $1+\mathcal{N}$, where $\mathcal{N}$ is the photon phase space density (occupation number) of the final state. For this reason, the Fokker-Planck equation in this context, known as the Kompaneets equation, is most conveniently written not in terms of the photon number density, but in terms of the photon occupation number:

$$\frac{1}{n_e \sigma_{\text{T}} c} \frac{\partial \mathcal{N}}{\partial t} = \Theta \frac{1}{\varepsilon^2} \frac{\partial}{\partial \varepsilon} \left[\varepsilon^4 \left(\frac{\partial \mathcal{N}}{\partial \varepsilon} + \mathcal{N} + \mathcal{N}^2 \right) \right] + q(\varepsilon), \tag{8.30}$$

where $q(\varepsilon)$ is a source term in units of change in occupation number per Compton scattering time, and $\varepsilon = \epsilon/(k_B T)$, and we have omitted the spatial gradient part. When $\epsilon/m_e < 1$, but not very small compared to unity, the coefficients change slightly as a result of the onset of Klein-Nishina corrections (Ross, Weaver, and McCray 1978).

That the natural unit of time for this equation is the Compton scattering time is obvious. Some of the other terms are not. The overall factor of Θ comes from the fractional energy change per scatter; the ε^4 term comes from a product of the photon density of states and the Compton recoil energy; the term proportional to the gradient in the density of states describes the usual "diffusive smoothing" of a distribution as a result of a random walk process; the term linear in $\mathcal{N}$ describes the secular change in energy; and the term in $\mathcal{N}^2$ accounts for stimulated scattering. In the present context, it is usually negligible (photon intensities at such high energies are usually well below Planckian), but there are contexts in which it can be important (see §9.2.5).

It is obvious both from physical reasoning and from the form of this equation that the importance of Comptonization can be measured by the product of the total number of scatterings and the fractional energy change per scatter. This quantity is called the *Compton y parameter*:

$$y = 4\Theta \max(\tau_{\text{T}}, \tau_{\text{T}}^2). \tag{8.31}$$

If $y \ll 1$, little happens; but if $y \gg 1$, a significant amount of energy is transferred from the electrons to the photons.

We are now equipped to pick up the thread left dangling in §7.5.3, where Comptonization was first introduced. In the inner regions of conventional AGN accretion disks, where radiation pressure is the dominant support, equation 7.117 becomes

$$ y = 0.08 L_{46}^{2/3} \left(\frac{R_R}{\eta R_z} \right)^{2/3} T_5^{10/3} \frac{\varepsilon^2}{(1 - e^{-\varepsilon})^{2/3}}, \tag{8.32} $$

again supposing that the only significant absorptive opacity is free-free, and that the density scale height at the photosphere is equal to the disk thickness. The importance of Comptonization is thus *very* sensitive to the temperature; whether it is trivial or central may depend on the existence of a "chromospheric" region in which the electron temperature rises above the local effective temperature.

Let us, then, consider what happens when photons whose energy is very small compared to the electron temperature are injected into a hot plasma. Although it is possible to obtain analytic solutions to this problem when spatial diffusion is combined with Comptonization (e.g., Sunyaev and Titarchuk 1980), much of the essential physics can be seen with a much simpler approach (as done, e.g., in Shapiro, Lightman, and Eardley 1976). Because the most important result of the spatial diffusion is to define the probability distribution for the number of scatters suffered by emerging photons, and that distribution is peaked fairly narrowly around τ_{T}^2 when the seed photons are injected at the center and $\tau_{\mathrm{T}} > 1$, we can mock up the spectral impact of diffusion by introducing into equation 8.30 a loss-rate term $\propto \mathcal{N}/\tau_{\mathrm{T}}^2$.

With that approximation, it is easy to find the time-steady solution for the Comptonized spectrum in the range of photon energies well above the energy of the injected photons. $\mathcal{N}$ is approximately $\propto \exp(-\varepsilon)$ for $\varepsilon \gg 1$, and is $\propto \varepsilon^w$ for $\varepsilon \ll 1$. The power w is determined from the quadratic equation

$$ w(w + 3) = \frac{4}{y}, \tag{8.33} $$

which has the solution

$$ w = \frac{3}{2} \left[-1 \pm \sqrt{1 + 16/(9y)} \right]. \tag{8.34} $$

This expression can be extended (Zdziarski 1985) to the trans-relativistic regime by changing the $4/y$ term in equation 8.33 to

$$ \frac{\pi^2}{3(\tau_{\mathrm{T}} + 2/3)^2 \Theta K_3(1/\Theta)/K_2(1/\Theta)}. \tag{8.35} $$

For $y \gg 1$, $0 \leq w \lesssim 1/2$, so that $\epsilon n_{\mathrm{ph}}(\epsilon) \propto \epsilon^{\chi} \exp(-\epsilon/\Theta)$ with $3 \leq \chi \lesssim 3.5$. In this limit, the spectrum approaches that found in a Bose-Einstein thermal equilibrium when the number of photons is fixed a priori (if the chemical potential were zero, it would be the Wien limit of the Planck spectrum). In this limit, the mean energy of the photons is comparable to the mean thermal energy of the electrons. By definition, this is the regime in which most photons have been scattered up to energies comparable to the electrons. Therefore, a more precise lower bound on the y required to reach Wien saturation is $y > \ln(\Theta/\epsilon_o)$, where ϵ_o is the mean energy of injected photons.

For $y \ll 1$, the physical root is the one with the negative sign, so that $w \simeq -3/2 - 2/y^{1/2}$, and $\chi \simeq 3/2 - 2/y^{1/2}$. In this limit, the photon spectrum is a power law that extends from the (low) energy of initial photon injection up to the electron temperature, beyond which it is exponentially cut off. When y is not too small, the power law that results can be hard enough that most of its luminosity is carried in the highest energy photons. That is, although a majority of photons are amplified only slightly, a minority can be amplified by such large factors that they dominate the power output.

This subtlety has been the source of considerable confusion in the literature. Many have erroneously argued that because the average photon leaves a Comptonizing plasma increased in energy by a factor e^y when y is not too large, that the output luminosity is simply $e^y L_s$, where L_s is the input luminosity of seed photons. The point made in the previous paragraph (first worked out in detail by Dermer, Liang, and Canfield 1991) demonstrates why this argument is wrong.

For intermediate values of y, the equilibrium spectrum at energies well above the injection energy can be approximated by a sum of power law and Wien contributions:

$$n_{\mathrm{ph}}(\epsilon) \propto \left[\left(\frac{\epsilon}{\Theta} \right)^{\chi} + \frac{n_{\mathrm{W}}}{n_{\mathrm{pl}}} \left(\frac{\epsilon}{\Theta} \right)^3 \right] \frac{e^{-\epsilon/\Theta}}{\epsilon}. \tag{8.36}$$

The ratio between the number of photons in each component is given approximately by

$$\frac{n_{\mathrm{W}}}{n_{\mathrm{pl}}} \simeq \frac{\Gamma(-\chi)}{\Gamma(3 - 2\chi)} P(\tau_{\mathrm{T}}), \tag{8.37}$$

where Γ is the usual gamma function, χ is calculated using the negative square root in the quadratic solution, and

$$P(\tau_{\mathrm{T}}) = 1 - \frac{3}{8\tau_T^3} [2\tau_{\mathrm{T}}^2 - 1 + e^{-2\tau_{\mathrm{T}}}(1 + 2\tau_{\mathrm{T}})] \tag{8.38}$$

is appropriate to the case in which the photon sources are distributed uniformly over a sphere (Zdziarski 1985). If the optical depth is large, the power-law index χ is best found using the solution of the Kompaneets equation; if it is only a few or smaller, an approximation that generalizes equation 8.28 by replacing τ_T with $P(\tau_T)$ does reasonably well.

8.3.2.2 relativistic (nonthermal) electrons

Large optical depth effects are complicated to treat when the electrons are relativistic. This is because the photon energies change by large factors upon scattering, so a "continuum" approach, like the Kompaneets equation, to the evolution of their energy is completely inappropriate. In the limit of small optical depth, the rarity of multiple scatters meant that an approximate analytic estimate could be made of their effect; that is very difficult in this limit, in which most photons are scattered many times. Attempts have been made (e.g., Titarchuk 1994), but they are controversial.

Faced with these difficulties, most calculations in this regime are done by means of Monte Carlo simulations (e.g., Stern et al. 1995a), but genuine (numerical) radiative transfer calculations can also be done when the geometry has sufficient symmetry (Poutanen and Svensson 1996). Although it is harder to gain physical intuition from simulations than from analytic calculations, they have the advantage that their only source of error (if done correctly) is statistical fluctuations. The most efficient algorithm for performing them was invented by Pozd'nyakov, Sobol, and Sunyaev 1977; for sample results, see Zdziarski (1985).

8.4 Electron-Positron Pair Physics

Whenever the electrons have enough energy to make X-rays, they are automatically almost relativistic and therefore have almost enough energy to make $e^\pm$ pairs. It is then necessary to incorporate pair physics (or at least investigate whether pairs might be significant) in almost any picture of X-ray production in AGNs.

8.4.1 Pair production processes and their rates

A number of different mechanisms can produce $e^\pm$ pairs: $\gamma + \gamma \rightarrow e^+ + e^-$; $\gamma + e^\pm \rightarrow e^\pm + e^+ + e^-$; $\gamma + Z \rightarrow Z + e^+ + e^-$; $e^\pm + e^\pm \rightarrow e^\pm + e^\pm + e^+ + e^-$; and $e^\pm + Z \rightarrow e^\pm + Z + e^+ + e^-$. At an order of magnitude level, the cross section for the first mechanism (photon-photon pair production) is $\sim \sigma_T$; the cross sections for the reactions involving a charged particle and a photon are $\sim \alpha_{fs}\sigma_T$; and the cross sections for those processes beginning with two charged particles are $\sim \alpha_{fs}^2\sigma_T$. These relative scales arise from the basic nature of relativistic quantum mechanics. As figure 8.5 illustrates, the Feynman diagrams for photon-photon pair production and Compton scattering are identical up to a rotation; this makes their matrix elements identical for appropriately matched four-momenta. When all electrons involved are relativistic, their density of states is very similar to the density of states for photons, so the cross sections for the two processes must be comparable. Particle-γ reactions (fig. 8.6 top) resemble Compton scattering but involve an extra electron-photon vertex, so their cross section is generically smaller by one power of α_{fs}; particle-particle reactions (fig. 8.6 bottom) add yet another electron-photon vertex, causing their cross sections to be smaller by a second factor of α_{fs}. The last group, therefore, is generally unimportant compared to the first two.

In the following subsections, a brief introduction to the principal pair production mechanisms is given. Further detail can be found in, for example, Svensson (1982b).

8.4.1.1 photon-photon pair production

In the classical picture presented by Maxwell's equations, electromagnetic waves do not interact, and there is no such thing as two waves reacting to create an electron-positron pair. Quantum mechanically, because individual photons can be created and annihilated, the process does exist provided a kinematic threshold is surmounted. That is, the two photons must have at least enough energy to create the rest mass of the pair. Because momentum and energy must also be conserved, this criterion is most easily evaluated in the zero-momentum (or center of mass) frame. In that frame, the total energy is $2\sqrt{p^2 + m_e^2}$, where p is the magnitude of the electron's (or positron's) momentum. Clearly, the least energy is required when the electron and positron are at rest immediately after their creation.

If the total momentum of the two photons is not zero in the lab frame, to find the threshold condition it is necessary to transform from the lab to the zero-momentum frame. The special case in which the two photons are directed head-on in the lab frame is especially easy to analyze. Then the frame in which they have equal and opposite momenta differs from the lab frame by a speed $\beta = (\epsilon_1 - \epsilon_2)/(\epsilon_1 + \epsilon_2)$, where $\epsilon_{1,2}$ are the photon energies in the lab frame. In the zero-momentum frame their total energy is then $2\sqrt{\epsilon_1\epsilon_2}$. For this to exceed the rest-mass threshold, $\sqrt{\epsilon_1\epsilon_2} \geq m_e$. High-energy γ-rays can therefore pair produce when coupled with relatively low-

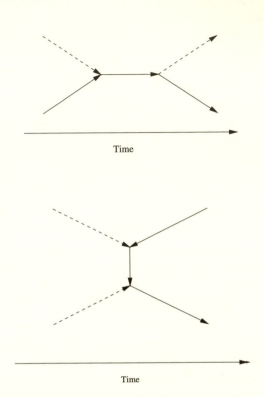

Fig. 8.5   Top panel shows the Feynman diagram for Compton scattering; bottom panel shows the diagram for photon-photon pair production. In both diagrams, electrons are represented by solid lines, photons by dashed lines. The two panels are identical *modulo* a rotation of $\pi/2$ radians.

energy photons; the pairs produced thereby must then be highly relativistic and will tend to travel in the direction of the initial γ-ray.

We note in passing that photon-photon pair production can only proceed if the total angular momentum of the photons in the zero-momentum frame is zero; the electron and positron are then created in a singlet spin state to conserve angular momentum. If the total angular momentum of the photons in that frame were $2\hbar$ (the only other possibility), pair production could not occur because the electron and positron are created in a state of zero orbital angular momentum, and their spins cannot add to anything other than zero or $\hbar$.

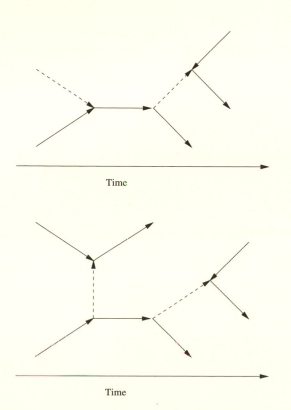

Fig. 8.6   Top panel shows the Feynman diagram for $e^- + \gamma \rightarrow$ $e^- + e^+ + e^+$; bottom panel shows the diagram for $e^- + e^- \rightarrow$ $e^- + e^- + e^+ + e^-$. In both diagrams, electrons are represented by solid lines, photons by dashed lines. Note that, like photon-photon pair production and Compton scattering, the Feynman diagrams for the former process and e–e bremsstrahlung are identical *modulo* a rotation by $\pi/2$, so they, too, share the same matrix element.

The rate of photon-photon pair production is most conveniently described in terms of the opacity presented by one population of photons to another. Assuming photons with zero mean polarization, it is

$$\kappa_{\gamma\gamma}(\epsilon) = \frac{2m_e^2}{\epsilon^2} \int_1^\infty ds\, \sigma_{\gamma\gamma}(s)s \int_{sm_e^2/\epsilon} dY\, m_e \frac{n_{\rm ph}(Y)}{Y^2}, \qquad (8.39)$$

where sm_e^2 is the square of the center of mass energy for photons with energy ϵ and Y. The cross section in the center of mass frame is

$$\sigma_{\gamma\gamma}(s) = \frac{3}{8}\frac{\sigma_T}{s}\left[\left(2+\frac{2}{s}-\frac{1}{s^2}\right)\cosh^{-1} s^{1/2} - \left(1+\frac{1}{s}\right)\left(1-\frac{1}{s}\right)^{1/2}\right].$$
(8.40)

In two interesting limits,

$$\sigma_{\gamma\gamma}(s) \to \frac{3}{8}\sigma_T \begin{cases} \sqrt{s-1} & s-1 \ll 1 \\ \left[\ln(4s)-1\right]/s & s \gg 1. \end{cases}$$
(8.41)

Thus we see that, at least near threshold, the photon-photon pair production cross section is indeed comparable to the Compton cross section.

8.4.1.2 electron-photon pair production

For this reaction, the threshold kinematics are easiest to analyze in the initial rest frame of the electron. The three particles present after the pairs are created cannot remain still, for they must absorb the momentum of the photon. However, it is obvious that the least energy is required when all three move in the direction of the photon, and all three also have the same energy. Then it is easy to show that $\epsilon = 4m_e$ in this minimum energy reaction. The total cross section for the reaction is

$$\sigma_{e\gamma}(\epsilon) = \frac{3}{8\pi}\alpha_{fs}\sigma_T \begin{cases} (\sqrt{\pi}/324)(\epsilon/m_e - 4)^2 & \epsilon/m_e - 4 \ll 1 \\ (28/9)\ln(2\epsilon/m_e) - 218/27 & \epsilon/m_e \gg 4. \end{cases}$$
(8.42)

8.4.1.3 nucleus-photon pair production

Unless ϵ is at least $\sim Am_p \gg m_e$, where A is the atomic mass of the nucleus, the center of mass frame is identical to the rest frame of the nucleus. The least energy is required when the nucleus takes up all of the momentum of the photon, for its ratio of energy to momentum is much smaller than the pairs'. The threshold condition is then trivially $\epsilon \geq 2m_e$.

The cross section for this reaction is (Racah 1934; Maximon 1968)

$$\sigma_{Z\gamma} = \frac{3}{8\pi}\alpha_{fs}\sigma_T Z^2 \begin{cases} (2\pi/3)(\epsilon/m_e - 2)^2 & \epsilon/m_e - 2 \ll 1 \\ (28/9)\ln(2\epsilon/m_e) - 218/27 & \epsilon/m_e \gg 2. \end{cases}$$
(8.43)

It scales $\propto Z^2$ because the photon-proton interactions add coherently.

8.4.2 Pair annihilation

The inverse reaction of photon-photon pair production is, of course, two-photon pair annihilation. As has already been pointed out, this means the pair must be in a spin singlet, and the emitted photons will have oppositely directed spins. The most direct way to obtain the cross section is to apply the principle of detailed balance (Svensson 1983). In the center of mass frame of the pair, both the electron and the positron have energy γ_{cm} and speed β_{cm}. The resulting photons will then also each have energy γ_{cm}. Because the photon-photon cross section depends on the product of the photon energies, we have

$$\sigma_{\text{ann}}(\gamma_{cm}) = \frac{1}{2\beta_{cm}^2} \sigma_{\gamma\gamma}(s = \gamma_{cm}^2). \tag{8.44}$$

Again taking the interesting limits of very slow and very rapid motion, we find

$$\sigma_{\text{ann}} = \frac{3}{16}\sigma_{\text{T}} \begin{cases} (1 + \beta_{cm}^2)/\beta_{cm} & \alpha_{\text{fs}} \ll \beta_{cm} \ll 1 \\ [2\ln(2\gamma_{cm}) - 1]/\gamma_{cm}^2 & \gamma_{cm} \gg 1 \,, \end{cases} \tag{8.45}$$

where the restriction to speeds greater than α_{fs} is necessary to avoid dependence on the details of Coulomb wave functions.

The fact that $\sigma_{\text{ann}} \propto \beta_{cm}^{-1}$ in the nonrelativistic limit means that the annihilation rate coefficient becomes *independent* of pair speeds in that limit:

$$\alpha_{\text{ann}} \simeq \frac{3}{8}\sigma_{\text{T}}c, \tag{8.46}$$

no matter what the electron and positron distribution functions are, so long as they are nonrelativistic (but $\langle\beta_{cm}\rangle \gg \alpha_{\text{fs}}$). On the other hand, if the electrons and the positrons share the same Maxwell-Boltzmann distribution with dimensionless temperature $\Theta \gg 1$, the annihilation rate coefficient is (Svensson 1982a)

$$\alpha_{\text{ann}} \simeq \frac{3}{16}\sigma_{\text{T}}c\Theta^{-2}\ln(2\eta_{\text{E}}\Theta). \tag{8.47}$$

The spectrum of annihilation photons in the center of mass frame is always a (two-photon) line at $\epsilon' = \gamma_{cm}m_e$. In the nonrelativistic limit, the lab frame spectrum is then very close to a delta function at $m_e c^2 = 511$ keV. The nonrelativistic total cooling rate coefficient is, of course, $\Lambda^{ann} = (3/4)\sigma_{\text{T}}m_e c^3$.

As the total energy of the pair increases, the distribution shifts to higher energies. In the opposite extreme, where both members of the pair are extremely relativistic, transforming back to the lab frame results in a spectrum that peaks at $\epsilon \sim (\gamma_+ + \gamma_-)m_e$, where $\gamma_\pm$ are the Lorentz factors of the electron and the positron in the lab frame, but the spectrum has a fairly broad tail to lower energies (Svensson 1982a). If the pairs have a thermal distribution with $\Theta > 4$, the emission coefficient can be fairly approximated by

$$\Lambda_\epsilon^{ann} \simeq \frac{3}{16}\sigma_T c \Theta^{-2}\varepsilon^2 \left[\ln(4\eta\Theta^2\varepsilon) - 1\right] e^{-\varepsilon}, \tag{8.48}$$

that, integrated over energies yields a total cooling rate coefficient

$$\Lambda^{ann} \simeq \frac{3}{4}\sigma_T m_e c^3 \Theta^{-1} \left[\ln(2\eta\Theta) + \frac{1}{4}\right] \tag{8.49}$$

(Svensson 1982a).

8.4.3 Peculiarities of pair dynamics

Before proceeding on to a description of equilibria possibly involving pairs, it is necessary to note several special dynamical properties of pair plasmas.

First, the degree of dynamical coupling between $e^\pm$ pairs and any background normal plasma is highly uncertain. In ordinary astrophysical plasmas, if the electrons were to attempt to move away from the ions, only a very slight displacement would create such a strong electric field that no further relative motion could take place. However, when a positron moves along with each electron, obviously no electric field is generated to restrain their joint motion. Similarly, although large-scale currents due to electron flow past the ions would create strong magnetic fields, which would then pull the ions along, when a positron travels together with each electron, there is no net current and consequently no creation of magnetic field.

Coulomb collisions are a rather inefficient vehicle for dynamical coupling between pair plasmas and normal electron-ion plasmas. The cross section for deflection of an electron with Lorentz factor γ and speed β through an angle of at least $\pi/2$ by an infinitely massive ion of charge Ze is

$$\sigma_{\text{Coul}}(\theta \geq \pi/2) = \frac{3}{8\pi}Z^2 \frac{\sigma_T}{(\beta^2\gamma)^2} \begin{cases} 1 & \beta \ll 1 \\ 1/2 - \ln\sqrt{2} & \gamma \gg 1 \end{cases}. \tag{8.50}$$

Thus, when the electrons are energetic enough to make pairs, the Coulomb deflection cross section is at most comparable to the Thomson cross section and may be rather smaller. As we shall see, the likely Thomson optical depths of these plasmas are rarely more than a few, so Coulomb collisions hardly suffice to bind pairs and ions dynamically.

Consequently, the primary dynamical connection between pairs and adjacent normal plasma may be scattering events in which the electrons and positrons interact with plasma turbulence involving the background normal plasma. Although that can be extremely efficient, the mechanisms are highly speculative, and it is very difficult to be quantitative about it.

The consequences of a lack of coupling can be severe. In the limit of very weak momentum transfer between pairs and normal plasma, the pairs would be impossible to confine gravitationally. As we will show in the next section, the temperatures of pair plamas are almost always relativistic. The pair thermal speeds are therefore always close to c, that is, at least comparable to the escape speed from *any* gravitational potential.

On the other hand, even if the pairs and the background normal plasma are well coupled dynamically, gravitational confinement is still made problematic by the presence of the pairs. For example, if the ratio of positrons to protons $n_+/n_p \equiv z > m_p/(2m_e)$, the sound speed is $\sim \Theta^{1/2}c$, which we expect in these circumstances to be relativistic. Copious mass loss would then result in this situation as well.

The Eddington limit argument (§6.2) is also altered when there are numerous pairs. There we found that the Eddington luminosity is proportional to the mean mass per electron μ_e. Now if the density of unpaired electrons is n_e and the density of positrons is n_+, this becomes

$$\mu_e = \mu_e^* \frac{1 + 2z(m_e/m_p)}{1 + 2z}, \tag{8.51}$$

where $\mu_e^* \simeq 1.2m_p$ is the mass per electron in a normal ionized plasma. Thus, if there are large numbers of pairs, the Eddington luminosity can be reduced by as much as $\mu_e^*/m_e \simeq 2000$. Put another way, large numbers of pairs can greatly enhance the dynamical importance of radiation forces.

8.5   Self-consistent Equilibria

At this point we are equipped to analyze models of the X-ray and γ-ray production regions of AGNs. Given the heating rate, the optical depth, the supply of seed photons, and the geometry, we can solve for the equilibrium temperature and output spectrum. If the heating rate is defined as a function of electron energy, we can predict not just the electron temperature, but also the whole electron distribution function. From this solution we can then find the *net* electron optical depth (i.e., the Compton depth without consideration of electron-positron pairs) such that the *total* optical depth is consistent with pair balance. The last step in model definition is to search for a dynamical context with which such a state might be consistent.

These models naturally divide into two categories, thermal and non-thermal. In the former case, the electron distribution function follows the Maxwell-Boltzmann form, but the photons are not necessarily in thermal equilibrium (nor must the ions be). In the latter case, we will generally take the electron distribution function to be a power law extending to highly relativistic energies. This choice is justified on the physical ground that if the electrons are far from equilibrium, there is no single characteristic energy and on the empirical ground that observed spectra are frequently well described by power laws.

Unfortunately, we do not have a secure physical basis for deciding when the distribution function should be thermal. Electron-electron collisions alone are often insufficient to force relaxation to a thermal distribution. The reason lies once again in the small value of the Coulomb cross section when the electrons are relativistic. On the other hand, as we have already discussed in regard to pair-normal plasma dynamical coupling, the most important relaxation mechanisms are likely to be wave-plasma interactions about which little is known. Synchrotron self-absorption (§9.2.4) can also promote thermalization (Ghisellini, Guilbert, and Svensson 1988).

8.5.1 Thermal electrons

8.5.1.1 temperature balance for fixed electron numbers

The temperature in a static plasma is fixed by the balance between the heating rate and the rate of radiative cooling. Generally speaking, the most we can say about the heating rate is to parameterize it as some bulk rate of energy transfer (see §8.6). Plasma cooling, however, we understand in considerable detail. It takes place through creation of new photons (bremsstrahlung and annihilation) and inverse Compton scattering of both these newly created photons and any additional photons entering the plasma from outside.

Consider first the simplest case, in which a spherical plasma of radius r is given no external photons. Then the conservation of energy equation can be approximated by

$$L_h = (1 - f_W)L_{\text{brems}} + L_W, \qquad (8.52)$$

where L_h is the total heating rate; L_{brems} is the total luminosity in bremsstrahlung, including all four varieties—electron-ion, electron-positron, electron-electron, and positron-positron; f_W is the fraction of photons Compton scattered so many times that they have risen to energies close to the electron temperature; and L_W is the luminosity of these "saturated" photons, whose spectrum must approach the Wien distribution. A more precise description would take into account the fact that the probability of reaching saturation depends on the photon's initial energy, but the approximation embodied in equation 8.52 is satisfactory for an account of the global energy balance.

The bremsstrahlung luminosity is, of course, proportional to the square of the density times the volume. It is more physically revealing, however, to write the density in terms of the Thomson optical depth. When that is done, the conservation of energy equation becomes (in the approximation that all Gaunt factors are unity):

$$\frac{3}{4\pi} l_h = \alpha_{fs} \tau_p^2 \left[1 - f_W + 3\Theta f_W \ln\left(\frac{\Theta}{\epsilon_{min}}\right) \right]$$
$$\times \left[(1+2z)\frac{n_i}{n_e}\phi^{eZ}(\Theta) + (1+2z+2z^2)\phi^{ee}(\Theta) + z(1+z)\phi^{+-}(\Theta) \right].$$
$$(8.53)$$

Several useful notations have been introduced in this equation: $\phi^i = \Lambda^i/(\alpha_{fs}\sigma_T m_e c^3)$ is the dimensionless bremsstrahlung emissivity for variety i (which could be e–Z, e–e, or $+$$-$$-$); $\tau_p = n_p \sigma_T r$ is the Thomson depth due just to the net electrons; and, most important,

$$l_h = \frac{L_h \sigma_T}{r m_e c^3} \qquad (8.54)$$

is the *compactness* (note that it is sometimes defined with an extra factor of 4π in the denominator). The third term in the first square bracket gives the outgoing flux in the Wien component.

Equation 8.53 shows that the compactness is the fundamental scaling quantity for almost all considerations having to do with plasmas whose temperatures are ~ 1. In this case it is clear that (for fixed z) the temperature rises with increasing l_h/τ_p^2, and $\Theta \gtrsim 1$ when $l_h/\tau_p^2 > 1$. We shall see in the following discussion the recurring importance of l_h. For now we simply point out that $l_h \sim 1$ is to be expected in the inner regions of AGNs. In the context of an accretion model, l_h can be estimated by

$$l_h = \frac{L \sigma_T}{r m_e c^3} = \frac{L}{L_E}\frac{\mu_e^*}{m_e}\frac{4\pi}{x}. \qquad (8.55)$$

Because $\mu_e^* \simeq m_p \gg m_e$, we can expect that whenever L/L_E is not too small, there will be an interesting region not far outside the black hole where $l > 1$.

When additional photons are supplied from outside the hot plasma, it is obvious qualitatively that the greater their rate of injection, the more rapidly the electrons will cool, and therefore, the lower their temperature will be. This argument can be formalized in a very schematic way as follows.

Consider a slab of plasma heated at a rate F_h per unit area, and receiving a seed photon flux F_s. If its electrons have a Maxwell-Boltzmann distribution function at temperature Θ and cool predominantly by inverse Compton scattering, then the statement of energy conservation is

$$F_h = \tau_T[A(\Theta) - 1]4\pi J, \qquad (8.56)$$

where J is the mean intensity of the photons inside the plasma. In the limit of extremely small y and τ_T, J would be simply $\simeq F_s/c$ (the exact coefficient depends on the angular distribution of the soft photons). Our discussion is only interesting, of course, when F_s/c is a rather poor *underestimate* of J. The degree to which the mean intensity is raised above this floor depends on both the temperature and the optical depth, so we can write $J = (F_s/c)f_C(\Theta, \tau_T)$. J is a simple multiplicative factor times F_s/c because, as equation 8.20 shows, in the Thomson limit inverse Compton scattering simply multiplies each photon's energy by a given factor that depends on the electron's motion but *not* on the photon's energy. The energy conservation equation then becomes

$$\frac{F_h}{F_s} = \tau_T[A(\Theta) - 1]f_C(\Theta, \tau_T). \qquad (8.57)$$

This form makes it obvious that the equilibrium temperature depends on τ_T and the ratio F_h/F_s, or equivalently, the ratio of the associated compactnesses l_h/l_s but not on F_h by itself. In other words, if some external constraint fixes τ_T and l_h/l_s, the temperature is independent of heating rate! If the heating rate increases, the constraint fixing l_h/l_s ensures that more seed photons are supplied to the hot plasma. Each one is amplified by a fixed amount, but the greater number of photons permits a greater output power.

Figure 8.7 displays the equilibrium temperature as a function of l_h for several values of l_h/l_s, but fixed τ_T. It was computed using the approximations of equations 8.36–8.38. Although these are not exact, they are close enough to be faithful to the qualitative character of the equilibrium. As the figure shows, the temperature is independent of l_h over a wide range of values, confirming the argument leading up to equation 8.57; where it does depend on l_h, it is because Comptonization does not completely dominate the cooling.

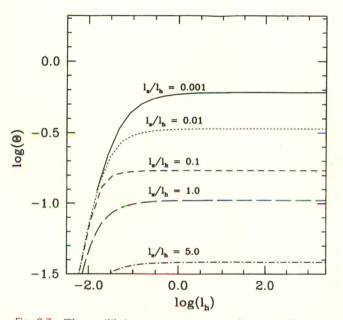

Fig. 8.7 The equilibrium temperature as a function of heating rate for $\tau_p = \tau_T = 1$ (i.e., not including any pairs), and several values of l_s/l_h: 0.001, 0.01, 0.1, 1.0, and 5.0. When l_h is large enough that Compton cooling dominates, the temperature is virtually constant despite many order of magnitude changes in l_h. Greater quantities of seed photons lead to lower temperatures. Courtesy of P. Pietrini.

Moreover, such equilibria can be well described (Pietrini and Krolik 1995) by a simple approximate scaling law

$$\Theta \tau_T \simeq a(l_h/l_s)^{1/4}, \qquad (8.58)$$

which is valid over an extremely wide range of l_h/l_s and l_h, and for τ_T not too great ($\tau_T \lesssim 5$). The coefficient a depends on geometry; it is $\simeq 0.1$ for spheres (with τ_T the optical depth across one radius), $\simeq 0.06$ for slabs (for τ_T the total slab thickness). The only significant deviation from this scaling law comes when $l_h/l_s \sim 1$, where there is a slight secondary dependence on τ_T.

This equilibrium scaling law confirms that the Compton y parameter is effectively regulated to ~ 1 whenever there is any significant source of seed photons. Although the cooling does not genuinely increase $\propto e^y$, it does increase rapidly enough that, over a wide range of parameters, Comptonization is "locked in" to the range where it is always of moderate, but not overwhelming, impact. Put another way, the output spectrum has a strong Comptonized power law, but the Wien contribution, although present, is generally less important.

At the same level of approximation as the thermal equilibrium scaling relation, the spectral index of the output power law depends only on l_h/l_s:

$$\chi = -1.6(l_h/l_s)^{-1/4}. \tag{8.59}$$

In this case even the coefficient is independent of geometry. If some process regulates l_h/l_s, we would then likewise expect the spectral index of the X-ray emission to be fixed. If in addition there is regulation of the optical depth of the plasma, the temperature is fixed as well. In such a case, it would be these two regulatory processes that would control both the slope of the X-ray power law and its high-energy cutoff (for reasons that will become clear in the next subsection, thermal models can never account for high-energy γ-ray emission).

This simple model of thermal Comptonization operating on seed photons produced in fixed ratio to the heating rate is not an idle, abstract suggestion. As first pointed out by Liang (1979), and developed by Haardt and Maraschi (1991), if the hot plasma is located adjacent to a cool surface, such as an accretion disk, approximately half the photons radiated by the hot plasma will be absorbed in the cool surface. Their energy will then be reradiated back into the hot plasma in the form of low-energy photons. In models such as these, if most of the disk's dissipation takes place in the hot plasma rather than in the disk proper, l_h/l_s is regulated entirely by geometry. This ratio is determined by a combination of the solid angle around the hot plasma occluded by cool reprocessing material and the ratio between the typical size of the hot plasma region(s) and its (or their) typical distance to the reprocessor. In the limit of perfect slab geometry, $l_h/l_s \simeq 2$, but other arrangements could push this ratio either up or down.* A large number of small knots near a slablike reprocessor, for example, would raise l_h/l_s, while a smooth corona tracing the edge of a strongly flaring disk would decrease it.

---

*This ratio is not exactly 2 in slab geometry because, as a result of the injection of the seed photons from one side, the Compton upscattered photons are not produced with precise up-down symmetry.

Two further consequences follow from this model. First (see §8.7.4 for a more detailed discussion), although most photons directed at the reprocessing surface are absorbed, a fraction is reflected. Because the albedo is far from "gray," the reflected photons have a distinct spectral signature. Second, the X-ray luminosity is a non-negligible fraction of the bolometric luminosity; half of it is therefore also a significant part of the total. If the reprocessing surface is dense enough to thermalize the energy, the argument resulting in equations 7.3 and 7.4 applies, and the seed photons emerge primarily in the ultraviolet. Thus, reprocessed X-rays may be an important supplement to photons made directly from dissipation inside an accretion disk (cf. the discussion of correlated continuum variability at the end of Chap. 7).

Both the equilibrium temperature and the output spectral index scaling laws apply over a broad, but not infinite, range of parameters. Their domain of validity is defined by the fact that they apply only to that range of l_h and τ_T in which thermal inverse Comptonization dominates the cooling, so that the temperature is independent of heating rate. For example, one might ask whether these scaling relations apply when the optical depth is so small that the equilibrium temperature relation would predict a very high temperature. The answer, as we shall see in the next subsection, is that if, in a plasma with reasonably large l_h, Θ rises much above unity, the pair production rate becomes so high that the optical depth can never be too small.

8.5.1.2 pair balance

As we have seen, the thermal properties of a hot Compton-scattering plasma depend simply on τ_T and l_s/l_h. However, when $\Theta \sim 1$, we can expect significant $e^\pm$ pair production, and therefore τ_T can be increased well above τ_p. To find the time-steady pair density, the pair production mechanisms listed in §8.4.1 must be balanced against pair annihilation (e.g., Svensson 1982b).

Because the photon-photon cross section is larger than either the electron-photon or electron-nucleus pair creation cross sections by a factor α_{fs}^{-1}, yet the number of photons is often comparable to or greater than the number density of electrons, the photon-photon mechanism is often the most important for creating pairs.

The compactness l is then also the key scaling parameter for pair production because it provides a rough measure of both the availability of high enough energy photons to trigger pair production and the efficiency with which they are used. Suppose that photon-photon pair production dominates the pair creation rate, while thermal annihilation dominates the rate of destruction. The equilibrium condition is

$$n_+(n_+ + n_p)\alpha_{\mathrm{ann}}(\Theta) = \int d\epsilon \int_{m_e/\epsilon} d\epsilon'\, c\sigma_{\gamma\gamma}(\epsilon\epsilon')n_\gamma(\epsilon)n_\gamma(\epsilon'). \qquad (8.60)$$

If we approximate the spectrum by a pair of mono-energetic peaks at ϵ_s and ϵ_h, this equation can be written in nondimensional form as

$$z(1+z)\Phi_{\mathrm{ann}}(\Theta) = l_h l_s f(\tau_{\mathrm{T}}) \frac{m_e^2}{\epsilon_s \epsilon_h} \left(\frac{\sigma_{\gamma\gamma}(\epsilon_h \epsilon_s)}{\sigma_{\mathrm{T}}} \right), \qquad (8.61)$$

where $\alpha_{\mathrm{ann}} = \Phi_{\mathrm{ann}}\sigma_{\mathrm{T}}c$. Provided that $\epsilon_h \epsilon_s$ is not too much bigger than m_e^2, the pair production cross section (eq. 8.40) is not much smaller than σ_{T}. Here $f(\tau_{\mathrm{T}})$ is a factor that accounts for optical depth effects that alter the soft photon density relative to $L_s/(4\pi r^2 c)$ (photons with energy greater than m_e both scatter very weakly and lose much of their energy to recoil if they do scatter). In this context, l_s plays the role of an estimator of the optical depth to pair production created by soft photons, while l_h describes the rate of energy injection in the form of hard photons.

The actual value of the compactness at which pairs become important depends on (net lepton) optical depth and seed photon supply. Greater optical depth and seed photon supply both lead to lower temperatures at fixed l_h and therefore postpone the onset of significant pair production, sometimes to $l_h \gg 1$. These principles are illustrated in figure 8.8, whose data were computed using the same set of approximations used for figure 8.7. When $\Theta < 1$, these approximations may slightly underestimate the pair production (Stern et al. 1995b), but the trends are correct.

An important property of thermal balance in the presence of pairs can be seen in figure 8.9. This property is most dramatically seen in the limit of $l_s/l_h \ll 1$. In that limit, when l_h is comparatively small, the paucity of photons available for up-scattering means that in order to carry off the required luminosity, all photons must be scattered up to Compton saturation, that is, they form a Wien spectrum. Therefore, the cooling rate is simply proportional to temperature and an increase in the heating rate (i.e., an increase in l_h) is directly translated into an increase in temperature.

However, when l_h is great enough that significant numbers of pairs are created, the additional electrons can create more photons through thermal bremsstrahlung (at these temperatures, both the e–p and e–e versions contribute comparably). The cooling rate can then increase *without* an

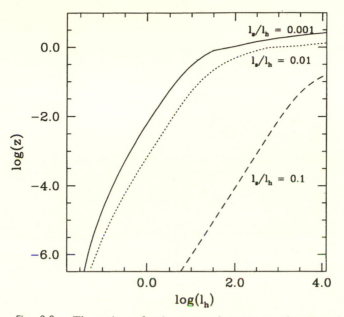

Fig. 8.8 The ratio z of pairs to net electrons as a function of l_h for fixed $\tau_p = 1$, but several values of l_s/l_h: 0.001, 0.01, and 0.1. The lower electron temperatures found at larger values of l_s/l_h lead to weaker pair production for fixed l_h. When τ_p is as large as this, it is difficult to make z terribly large and even to make $z \sim 1$ with $l_s/l_h \sim 1$ requires an extremely large value of l_h. Courtesy of P. Pietrini.

increase in temperature. In fact, when the pair production rate is high enough, the equilibrium temperature actually *decreases* with increasing l_h, as is shown at the high l_h end of figure 8.9.

As a corollary, for any particular rate of external seed photon supply (including zero) and for any particular τ_p, there is a maximum equilibrium temperature. When there are actually no externally supplied photons, only locally created bremsstrahlung photons are available for up-scattering, and the maximum temperature is $\simeq 1.5$ MeV if $\tau_p = 1$. The larger the number of external seed photons, the lower the maximum temperature; when there are enough photons that unsaturated inverse Compton scattering suffices for the radiation, the maximum as a function of l_h is very broad and flat, as we have already discussed. Similarly, as τ_p decreases, the maximum temperature slowly increases, reaching $\simeq 25$ MeV for $\tau_p = 10^{-4}$.

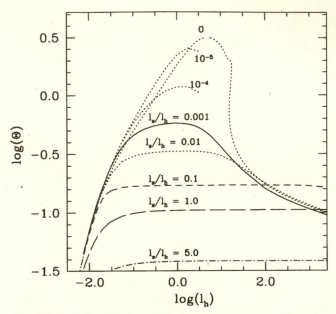

Fig. 8.9 As in fig. 8.7, the equilibrium temperature as a function of l_h for fixed $\tau_p = 1$, and $l_s/l_h = 0$, 10^{-5}, 10^{-4}, 0.001, 0.01, 0.1., 1, and 5, but including pair balance. Even when there are no externally supplied soft photons, $\Theta \leq 3$. Note also that l_s/l_h must be very small in order for locally generated photons to be more important than those supplied from outside. The small l_s/l_h curves cross the larger l_s/l_h curves at large l_h because they become pair-dominated at much smaller l_h. As a result, their optical depths are substantially greater, and they cool more efficiently. Figure courtesy of P. Pietrini.

8.5.2 Nonthermal electrons

8.5.2.1 feedback on the distribution function

The character of the back-reaction on the electron distribution function due to Compton scattering losses is a bit different in the case of a nonthermal distribution function. Thermal distribution functions have a fixed shape scaled by a single parameter T; nonthermal distributions can change their shape. So long as the scattering takes place in the Thomson limit, we can describe the evolution of the electron distribution function in terms of a smooth flow of electrons in energy space because the change each scattering makes in the electron energy is small compared to its initial energy. This follows from the fact that $\Delta\epsilon/\epsilon \sim (\gamma + 1)k/m_e$. We have defined the Thomson limit by $k'/m_e \sim \gamma k/m_e \ll 1$, so the small $\Delta\epsilon/\epsilon$ criterion is always met whenever the Thomson limit criterion is. Given this approximation, we can write down a continuity equation for the electron density in energy space

$$\frac{\partial n_e(\gamma)}{\partial t} + \frac{\partial (\dot{\gamma} n_e)}{\partial \gamma} = Q(\gamma) \tag{8.62}$$

where $Q(\gamma)$ is the rate at which electrons are injected (or destroyed) with energy γ.

As we have already shown in §8.3.1.2, $\dot{\gamma} = (4/3)\gamma^2 \sigma_T U_{ph}/(m_e c)$. Using this relation, we can define the cooling time $t_{cool} \equiv \gamma/\dot{\gamma} \propto \gamma^{-1}$, while we call the time for the distribution function to change for any other reason (injection, adiabatic expansion, electron escape, etc.) t_{evol}. The divergence in energy space term in equation 8.62 dominates the simple rate of change term whenever $t_{evol} > t_{cool}$. The characteristic γ^2 dependence of the inverse Compton energy loss rate means that generically the cooling time will be shorter than other timescales at high energies and longer at low energies.

With this distinction in mind, first consider the electron distribution function evolution equation in the limit $t_{evol} < t_{cool}$, that is, at low energies. It has the trivial solution $n_e = \int dt\, Q(\gamma)$. On the other hand, in the limit $t_{cool} < t_{evol}$ (the high-energy limit), we have

$$\frac{\partial (\dot{\gamma} n_e)}{\partial \gamma} = Q(\gamma) \tag{8.63}$$

so that

$$\begin{aligned} n_e(\gamma) &= \frac{1}{\dot{\gamma}} \int_\gamma^\infty d\gamma'\, Q(\gamma') \\ &= \frac{3 c m_e}{4\gamma^2 \sigma_T U_{ph}} \int_\gamma^\infty d\gamma'\, Q(\gamma'). \end{aligned} \tag{8.64}$$

This last equation has a very simple interpretation: the number of electrons at energy γ is just the integrated production rate of all electrons at higher energies, divided by the rate at which they leave energy γ. Because the cooling rate increases monotonically to higher energies, if there has been enough time for electrons to cool from 2γ to γ, there has also been enough time for them to cool from 100γ to γ, or from any higher energy.

Consider, for example, the interesting special case in which the injection function is a power law $Q_o \gamma^{-\psi}$. Then

$$n_e(\gamma) = \frac{3 c m_e}{4 \sigma_T U_{ph}} \frac{Q_o}{\psi - 1} \gamma^{-\psi - 1}. \tag{8.65}$$

That is, cooling has the effect of steepening the electron distribution function by one unit in log-log space.

We can combine this result with our previous derivation of the shape of the inverse Compton spectrum produced by a power law electron distribution acting on a relatively narrow seed photon spectrum (eq. 8.26). Because $F_\epsilon \propto \epsilon^{(1-\xi)/2}$ when the logarithmic slope of the electron distribution function is ξ, the spectrum radiated by electrons injected with slope ψ that have been in the source long enough to cool has the form $F_\epsilon \propto \epsilon^{-\psi/2}$. In other words, compared to what it would have been without cooling, Compton losses steepen the output spectrum by half a unit.

8.5.2.2 pair cascades

There is also the possibility that a small number of very energetic particles injected into the high-density environment of an AGN central engine could create a very large number of more modest energy $e^\pm$ pairs. Although there are differences in detail, it does not in practice matter greatly whether the initial particles are electrons, protons, or photons: the end results tend to be much the same.

Consider the photon-initiated mechanism first. Suppose that somewhere in the system there is a source of γ-rays with energy $\epsilon_h \gg m_e$. If these high-energy γ-rays encounter lower energy photons, they can create $e^\pm$ pairs by photon-photon pair production. This will be done efficiently whenever the compactness l_s in photons at or above the pair production threshold is large, that is, when

$$l_s = \int_{m_e/\epsilon_h} d\epsilon_s \, \frac{dL_s}{d\epsilon} \, \frac{\sigma_{\rm T}}{r m_e c^3 \epsilon_s} \, \frac{\sigma_{\gamma\gamma}(\epsilon_s, \epsilon_h)}{\sigma_{\rm T}} > 4\pi. \qquad (8.66)$$

For the purposes of this estimate we have approximated the correct cross section by $\sigma_{\rm T}$ (cf. equation 8.41). As a corollary, of course, when $l_s > 4\pi$, γ-rays cannot escape their source region without significant spectral modification.

Once the pairs are produced, the electron and the positron each have lab frame energies $\simeq (\epsilon_h + \epsilon_s)/2 \simeq \epsilon_h/2$. Two channels are available to them for losing energy: radiative losses (e.g., synchrotron radiation and inverse Compton scattering) and Coulomb scattering with other charged particles in the plasma. Which one dominates depends on circumstances. Annihilation occurs only once the pairs have thermalized because the cross section rises steeply as the energy falls (eq. 8.45). To the extent that inverse Compton scattering is important, it can result in the promotion of soft photons to energies high enough that they can initiate further pair production. At the same time, radiation losses of all sorts can increase

l_s well above what it might have been in the absence of high-energy photon injection. If the energy of the original particle is high enough, many generations of pair production can ensue.

This nonlinearity can make calculations of the ultimate equilibrium quite involved. Nonetheless, several fairly robust general conclusions can be made. For example, we can estimate the equilibrium pair density by matching the pair creation and annihilation rates:

$$\left(1 - e^{-\tau_{\gamma\gamma}}\right) \frac{L_h}{\epsilon_h} = \frac{4\pi}{3} r^3 n_+ (n_+ + n_p) \Phi(\Theta) \sigma_T c, \tag{8.67}$$

where $\Phi(\Theta) \simeq 3/8$ over a fairly broad range of temperatures. This leads to an estimate of the Compton optical depth due to pairs:

$$\tau_T (\tau_T - \tau_p) \simeq \frac{8}{3} \left(1 - e^{-\tau_{\gamma\gamma}}\right) \frac{l_h m_e}{\epsilon_h}. \tag{8.68}$$

From this estimate we can conclude that when $l_s \gg 1$ and pairs are common,

$$\tau_T \simeq \left(\frac{8 l_h m_e}{3 \epsilon_h}\right)^{1/2}. \tag{8.69}$$

In fact, as we have already remarked, $l_h > 1$ is often enough to make $l_s > 1$, so in general high compactness automatically leads to a significant Compton depth, even if none would exist without high energy photon injection. This optical depth then scales $\propto l_h^{1/2}$.

The compactness has yet another interesting interpretation in this context. Consider the ratio between the time for an electron to lose energy by Compton scattering and the time it takes to cross the region:

$$\frac{t_{\text{cool}}}{t_{\text{dyn}}} \sim \frac{m_e c^2 4\pi c r^2}{\gamma \sigma_T c L} \frac{c}{r} \sim \frac{4\pi m_e c^3 r}{\gamma L \sigma_T} = \frac{4\pi}{\gamma l}. \tag{8.70}$$

That is, when the compactness is large, the Compton loss time is always short compared to the free-flight time across the region; for electrons with substantial Lorentz factors, the Compton loss time at large compactness is much shorter than the free-flight time. Because the electron escape time must be at least as long as the crossing time (a variety of scattering events could slow down its escape), when $l > 1$, the electron distribution function is primarily determined by events inside the region, rather than by particle escape.

Several qualitative inferences follow from these estimates. First, when the compactness is large, Comptonization of low-energy photons must certainly be important. Second, in this regime downscattering of photons with

235

energy of 100 keV or more will also be important. In fact, because of severe Compton recoil losses, even in the high-compactness regime where the pair density is very high annihilation features are unlikely to be prominent in the emergent spectrum.

It also follows that the X-ray spectrum is surprisingly insensitive to parameters. Over a wide range of conditions, nonthermal pair plasmas all produce X-ray spectra of the form $F_\epsilon \propto \epsilon^{-\alpha_x}$ with $0.5 \leq \alpha_x \leq 1$ (Lightman and Zdziarski 1987). The lower limit applies in the limit of "unsaturated" pair cascades, that is, those with smaller l_h in which there is only a single generation of pair creation; the upper limit applies in the opposite limit, the "saturated" cascades in which l_h is large and many generations are created.

To understand this result, consider the lower limit on α_x first. The hardest possible electron injection spectrum is one that is essentially monoenergetic: $Q(\gamma) \simeq Q_o \delta[\gamma - \epsilon_h/(2m_e)]$. As we have already shown, the Compton cooling time is short compared to a crossing time, so the equilibrium electron distribution function is

$$
\begin{aligned}
n_e(\gamma) &= \frac{3cm_e}{4\gamma^2 \sigma_{\mathrm{T}} U_{\mathrm{ph}}} \int_\gamma d\gamma' Q(\gamma') \\
&= \frac{3cm_e}{4\gamma^2 \sigma_{\mathrm{T}} U_{\mathrm{ph}}} Q_o
\end{aligned}
\tag{8.71}
$$

for $\gamma m_e < \epsilon_h/2$. From the results we have already obtained, when the electron distribution function has an energy power-law index of ξ, we know that the photon output spectrum has a power law with index $\alpha_x = (\xi - 1)/2$. Here that translates to $\alpha_x = 1/2$. This is the hardest possible X-ray spectrum produced by single Compton scattering.

The physical origin of the upper limit can be understood in a looser way (Bonometto and Rees 1971, Blandford 1990). In the saturated limit, there are many electrons and positrons created at all energies below the energy of the first generation. Therefore, although the *number* of electrons is not conserved as they lose energy, their total energy is. Even though electron energy is continually being injected and removed by pair creation and annihilation at all energies below the maximum, the *net* flow of energy downwards from the energy of the first generation is conserved. We can then write the electron distribution evolution equation as

$$
\frac{\partial \dot{\gamma} n_e}{\partial \gamma} = \gamma Q(\gamma),
\tag{8.72}
$$

so that

$$n_e \propto \gamma^{-3} \int_\gamma d\gamma' \, \gamma' Q(\gamma'). \tag{8.73}$$

At energies well below where the initial electrons are created, $n_e \propto \gamma^{-3}$, so that the Compton-scattered spectrum is $\propto \epsilon^{-1}$.

8.6 Sources of Energetic Electrons

Having discussed the various ways hot electrons may produce a high-energy photon spectrum, it is now time to consider why such hot electrons should be present. If AGNs generated their energy solely by quiescent accretion of matter in thermodynamic equilibrium, the highest temperatures present would be only $\sim 10^5$ K, and we would expect negligible X-ray emission. Somehow there must be a way to bleed off some of the accretion energy into a plasma component that is far from thermal equilibrium with the ambient radiation.

8.6.1 Hot thermal electrons

If the Sun had no corona, we might find it shocking that so much very hot plasma exists near an accretion disk. That the Sun, whose surface temperature is only 5500 K, is surrounded by gas at 2×10^6 K, is, however, a source of great reassurance that mechanisms exist that can maintain such an arrangement. At the same time, however, the fact that we are far from understanding the heating of the Solar corona is a caution against too ready acceptance of models for heating accretion disk coronae.

One way or another, the most likely source of energy for coronal heating is the dissipation of magnetic fields. These fields could be created deep inside the accretion disk by, for example, the magneto-rotational instabilities described in §7.2.3. A net upwards Poynting flux might result from magnetic field "buoyancy", so that energy released by accretion in the center of the disk ultimately finds its way into magnetic structures on the surface of the disk. Mechanisms of this variety are, however, extremely speculative.

Whatever magnetic energy is supplied to the disk edge may be dissipated into heat if the field gradients (and plasma resistivity) are great enough (see Appendix B). Strong field gradients are automatically associated with current sheets, within which resistivity can act. After all,

$$\vec{J} = \frac{c}{4\pi} \nabla \times \vec{B} \tag{8.74}$$

when $\partial\vec{E}/\partial t$ is negligible, so there must be a large current density wherever the direction of the field changes sharply. Numerous pathways may lead to these regions (see, e.g., the review papers in Ulmschneider, Priest, and Rosner 1991). Nonlinear mode-mode coupling in MHD turbulence may pour energy into short wavelength modes; alternatively, turbulent fluid motions dragging the footpoints of magnetic loops may create magnetic X-points or tangential discontinuities. The essential requirement for strong local heating is that magnetic field lines of different orientation must be forcibly brought close together.

8.6.2 Relativistic electrons

When an AGN displays substantial flux in γ-rays, there can be little doubt that relativistic particles of some sort are present. As we have just seen, X-ray emission by itself may suggest relativistic electrons, but it does not necessarily require energies terribly much greater than $m_e c^2$. In fact, it is extremely difficult to drive thermal plasmas to temperatures much above $m_e c^2$, and so they are very unpromising sources for emission of truly energetic photons.

Because only a fraction of AGNs do produce large quantities of genuine high-energy γ-rays, we are in the position of trying to imagine what mechanisms might operate in some AGNs, but not necessarily all. Here, too, analogy with Galactic sources has provided some candidate mechanisms. Radio pulsars seem to be very efficient at generating extremely relativistic electrons, most likely by direct electromagnetic acceleration. On the other hand, Galactic cosmic-ray electrons may have been accelerated by repeated crossings of shock fronts.

8.6.2.1 direct electromagnetic acceleration

When considering black hole physics, we estimated the scale of the electric field that could be created by a rotating magnetized black hole: $10^{19} M_8 (B/10^4 \text{ G})$ V (§5.2.3). Although all sorts of plasma processes might be expected to short out such an enormous potential drop, even a small fraction of such a strong field could still give charged particles very large energies. Unfortunately, beyond this vague qualitative statement, the complexities of general relativistic plasma physics make it very difficult to be terribly more quantitative.

8.6.2.2 shock acceleration

We know little enough about the hydrodynamics of material very near an accreting black hole, but highly supersonic speeds certainly seem plausible. As we have seen in our discussion of accretion disks, orbital speeds $\sim O(0.1c)$ are quite likely, carrying material whose sound speed may be as low as ~ 30 km s^{-1}. If any of these flows converge, shocks will be created.

Because shock acceleration is widely believed to be the origin of Galactic cosmic rays, there is a tremendous literature on it. An excellent review is that by Blandford and Eichler (1987).

Elementary thermodynamics teaches us that fluids in thermodynamic equilibrium heat up when they are compressed. This is also true of nonthermal fluids, for reasons that, although not identical, are nonetheless closely related. Shocks are a particularly clean realization of this effect because they can often be thought of as infinitesimally thin with respect to the collision lengths of energetic superthermal particles, even while their thicknesses are (by definition) a few collision lengths for the thermal particles comprising the bulk of the fluid.

Imagine, therefore, a shock in a fluid containing a few particles whose mean free paths are much longer than the shock thickness (a short summary of the principal properties of shock waves can be found in Appendix C). For this argument to be self-consistent, the pressure due to these particles must be small compared to the postshock pressure. If the mean free path of these special particles is short compared to the total length of fluid on either side of the shock, once such a particle crosses the front it will scatter and will ultimately be reflected back across the front. In the frame of the particle, the upstream and downstream fluids look like mirrors that are moving toward it. Consequently, each time it is reflected its momentum grows by a factor $\sim 1 + u/v$, where u is the "mirror"'s speed of approach, and v is the velocity of the particle. Clearly, the net result is a steady growth in energy of those particles with long mean free paths, and this energy comes from the reservoir of energy driving the shock.

Let us now make this argument (originally due to Fermi) more quantitative. Suppose that we are in a frame for which the shock is stationary and located at $x_s = 0$. We can describe the spatial evolution of the particle distribution function f by the diffusion equation

$$\frac{\partial f}{\partial t} + u\frac{\partial f}{\partial x_s} = \frac{\partial}{\partial x_s}\left(D\frac{\partial f}{\partial x_s}\right), \qquad (8.75)$$

where $D = v\lambda_{\mathrm{mfp}}/3$ is the particles' spatial diffusion coefficient. Here $u = u_s$ upstream of the shock ($x_s < 0$) and $u = u/R$ downstream ($x_s > 0$), where R is the compression ratio of the shock. On the upstream side, there is a simple time-steady solution

$$f(x_s) = f_u + [f(0) - f_u] \exp\left[-\int_{x_s}^0 dx'_s \frac{u(x'_s)}{D(x'_s)} \right], \qquad (8.76)$$

where we have assumed that the gradient in f goes to zero as $f \to f_u$ at $x_s \to -\infty$. On the other hand, downstream of the shock, both diffusion and advection tend to take particles toward $x_s = +\infty$, so the only possible solution is $f(x_s) = f_d$, where f_d is the downstream limit of f.

These two solutions must now be joined across the shock front. So long as $u \ll v$, the particle density must be continuous across the shock because the net flux is $\sim O(uf)$, whereas the "one-way" flux is $\sim O(vf)$. In addition, particle number flux must be continuous because no particles can be created at the shock front. This statement translates to

$$\left(-D\frac{\partial f}{\partial x_s} - \frac{u}{3}p\frac{\partial f}{\partial p} \right)_{upstream} = \left(-D\frac{\partial f}{\partial x_s} - \frac{u}{3}p\frac{\partial f}{\partial p} \right)_{downstream}, \qquad (8.77)$$

where the flux of particles with momentum p in the shock frame is computed to lowest order in u by the diffusive flux at momentum p in the fluid frame supplemented by another term giving the correction due to the Lorentz transformation between the fluid frame and the shock frame. This correction term arises from the combination of two facts. On the one hand, the particle distribution is more nearly isotropic in the fluid frame than in the shock frame, so it is a better approximation to take the distribution function as depending only on the magnitude of the momentum in the fluid frame; on the other hand, particles having momentum p in the shock frame have a range of momenta $\sim (u/c)p$ in the fluid frame. Consequently, the usual "nv" expression for the flux is replaced by an expression that amounts to an integral of the distribution function over this range of momenta. Evaluating the flux continuity condition at $x_s = 0$ gives a differential equation for $f(p)$ at $x_s = 0$:

$$\frac{\partial f}{\partial \ln p} = \frac{3r}{r-1}[f_u(p) - f(p)], \qquad (8.78)$$

whose solution is

$$f(p) = qp^{-q} \int_0^p dp'\, f_u(p')p'^{q-1}, \qquad (8.79)$$

where

$$q = \frac{3R}{R-1}. \qquad (8.80)$$

Thus, particles with a δ-function distribution of energies brought into the shock acquire a new a power-law distribution with index $q = 3R/(R-1)$ upon passing through it. A power-law distribution function is required because the frame-jumping introduces a scale-free dependence on the slope of the distribution function.

A strong adiabatic nonrelativistic shock has a compression ratio $R = 4$, so the power-law index in momentum for shock-accelerated particles is $q = 4$; the equivalent index in energy space is $q_E = 2$. Thus, we expect that in the simplest instance the injection distribution function is $\propto E^{-2}$.

It is important to notice that the argument we have just given is completely independent of the nature of the particle or of the scattering agents. Electrons, protons, or photons (cf. Neufeld and McKee 1988) will do, and any sort of scattering that changes the particles' directions but does not cost (an appreciable amount of) energy will satisfy the criteria. The most important conditions are simply that the particles' mean free paths should be long compared to the shock thickness (so they perceive it as a sharp discontinuity), and the particles' pressure should not be necessary for maintenance of the shock. Obviously, one cannot draw energy from the postshock pressure to accelerate particles and then recycle it into those same particles.

In addition to the shape of the accelerated particle distribution, we would also like to know its amplitude and any cutoffs. To find the amplitude of the accelerated particle distribution function, we need to learn about how seed particles are injected and also how particles are lost or lose energy. Unfortunately, our state of knowledge on these subjects is extremely primitive. It is at best possible only to make a few crude estimates.

For example, a particularly plausible mechanism for scattering relativistic charged particles is resonant pitch-angle scattering on Alfvén waves. The associated diffusion coefficient is $\sim cr_L(\delta B/B)^{-2}$, where r_L is the particle Larmor radius and δB is the rms magnitude of magnetic field perturbations with wavenumber $k \sim r_L^{-1}$. In the rest frame of the particle, waves of this form have stationary fields, and the scattering is very efficient.

If this sort of scattering is the dominant process, and the scale of the shock region X_s is as small as $r_L(\delta B/B)^{-2}$, particles are likely to escape from the shock before scattering enough times to affect their energy appreciably. In the context of the inner regions of AGNs, we have already estimated that field strengths $\sim 10^4$ G might be expected. Therefore, electrons with Lorentz factors $\gamma > 10^{14}(X_s/r_g)(\delta B/B)^2 M_8(B/10^4 \text{ G})$ will not be efficiently accelerated by this mechanism.

A variant of shock acceleration may also contribute to relativistic electron production (Begelman 1991). If protons and other nuclei are accelerated more efficiently than electrons, it is possible for them to create very energetic electrons via reactions of the form $p + \gamma \to p + e^+ + e^-$ and $p + \gamma \to p + \pi^+ + \pi^-$. The latter reaction is followed by the successive decays $\pi^+ \to \mu^+ + \bar{\nu}_\mu$ and $\mu^+ \to e^+ + \bar{\nu}_\mu + \nu_e$ and an analogous chain for the π^-.

8.7   X-ray and γ-ray Propagation

So far we have studied the spectrum as it is radiated. However, this spectrum can be altered by processes occurring en route between the source and us. Although the interactions between hard photons and matter involve rather fewer complicated atomic physics effects than for lower energy photons, the interactions are strong enough to require serious consideration.

8.7.1   The nature of soft X-ray opacity

In contrast to valence shell transitions in neutral atoms, whose energies are generally in the ultraviolet, inner shell transitions characteristically have energies in the X-ray band when $Z \sim 10$, as for the most common heavy elements. That this should be so is a simple consequence of fundamental atomic physics: the "Rydberg" for a hydrogen-like atom with $Z > 1$ is $(1/2)(Z\alpha_{\rm fs})^2 m_e c^2$. Thus, the ionization potential for H-like C is $36 \times 13.6 = 490$ eV, for H-like O it is $64 \times 13.6 = 870$ eV, and for H-like Fe it is $676 \times 13.6 = 9.19$ keV (the last number requires a correction $\sim (Z\alpha_{\rm fs})^2 \sim 0.036$ for relativistic effects). Because outer shell electrons do not greatly diminish the effective nuclear charge seen by inner shell electrons, the threshold for ionization from the K-shell even in neutral C, O, or Fe is not much smaller than the numbers just given (e.g., for neutral O it is $\simeq 500$ eV).

Inner shell ionization can be accomplished by any of a number of mechanisms. The most common in astrophysical environments are collision with an electrically charged particle, absorption of a photon, and Compton scattering of a photon of such high energy that the Compton recoil exceeds the ionization potential. When an atom with at least one outer shell electron has been ionized by removal of an inner-shell electron through any of these mechanisms, it finds itself in a highly excited state, for there are electrons in higher energy states than the vacancy in the K-shell. One of these electrons then quickly drops to fill the vacancy.

The energy made available by that electron's de-excitation is generally greater than the valence shell ionization potential because the newly created ion is in a multiply excited configuration. Indeed, for this reason, these excited states are called "auto-ionizing" states. Two different mechanisms compete to carry away this excess energy: an outer shell electron can leave the atom (this is called Auger ionization), or a photon may be radiated. Their relative importance changes with Z. The rate of Auger ionization is proportional to the square of the two-electron matrix element connecting valence shell electrons to the continuum and to the K-shell, times the volume of phase space available to the newly ionized electron. The former quantity hardly depends on Z, but the latter rises $\propto Z^3$ because the energy available is $\propto Z^2$. On the other hand, the rate of photon emission (fluorescence) scales $\propto Z^6$ because its matrix element (the single-electron transition from valence shell to K-shell) is likewise nearly independent of Z, whereas the photon phase space increases $\propto E^3 \propto Z^6$. Consequently, the ratio of fluorescence rate to Auger rate is $\propto Z^3$, so that Auger ionization dominates for the lower Z elements, and fluorescence dominates for higher Z elements. The transition point comes at atomic numbers slightly higher than the Fe group: for neutral Fe the fluorescence yield is 0.34.

Although the presence of other electrons does affect the K-shell photo-ionization cross section at the level of factors of a few, the hydrogenic cross section (eq. 7.95) for an atom with atomic number Z is not a bad first approximation. That is, for photon energies above the K-shell ionization threshold ϵ_Z,

$$\sigma_K \sim 6 \times 10^{-18} Z^{-2} (\epsilon/\epsilon_Z)^{-a} \text{ cm}^2, \tag{8.81}$$

where the multiple electron effects make the energy dependence a bit more gradual than hydrogenic, so that $-3 \lesssim a \lesssim -2$. In a rough manner of speaking, the photo-ionization cross section at threshold is $\sim f \alpha_{\text{fs}} a_*^2$, where the dimensionless correction factor $f \sim 10\text{--}100$, and a_* is the characteristic scale of the electron orbit. Because σ_K declines sharply with increasing energy above threshold, any given species presents significant opacity over at most a factor of several in energy.

To find the total opacity of a gas with a variety of elements, we sum the individual opacities with weights equal to their relative abundances. These, too, tend to fall with increasing Z, though there are of course striking blips on that curve corresponding to elements one of whose isotopes is a multiple α-particle nucleus (e.g., C, N, O) or a doubly magic nucleus (e.g.,

Fe). Assembling all the data for the different heavy elements, the opacity of Solar abundance neutral material has a net trend that is roughly $\propto \epsilon^{-3}$ all the way from 13.6 eV to about 10 keV (fig. 8.10). Superimposed on this trend there are sharp edges where the opacity can jump by as much as a factor of two (the edge with the greatest contrast is that due to O). As a crude rule of thumb, a column density of neutral atoms with Solar chemical composition that has 10^{21} H atoms cm^{-2} will absorb all photons from 13.6 eV to almost 1 keV.

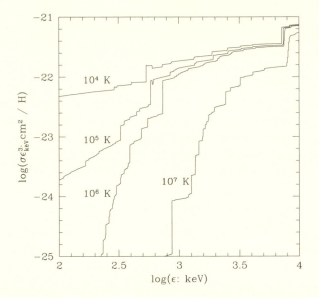

Fig. 8.10 The opacity per hydrogen atom times $(\epsilon/\mathrm{keV})^3$ as a function of photon energy for various temperatures, assuming collisional ionization equilibrium (Krolik and Kallman 1984). For this purpose, $T = 10^4$ K is essentially equivalent to $T = 0$.

Frequently the medium-Z elements that contribute the most to X-ray opacity are absent from interstellar gas, having been largely bound into dust grains. If the dust grains were large enough to be optically thick to X-rays, these atoms would then contribute less than their normal share to the mean opacity because of the grains' self-shielding. However, as demonstrated by Fireman (1974), most interstellar grains are probably too small for this to be a significant effect.

If the gas is ionized, whether through high-temperature collisional ionization or photo-ionization, the soft X-ray opacity does diminish, but this diminution proceeds in steps, as different elements are stripped of their electrons. Neither the K-shell photo-ionization cross section nor its threshold changes very greatly until all the outer electrons have been removed. Thus, little happens to the opacity in the several hundred eV range until C and N are either He-like, H-like, or completely ionized; similarly, the few keV range is affected only after O has been fully stripped and Si, S, Ne, and Ar are down to their K-shells. When the elements with edges in a given band are fully stripped, photoelectric opacity in that portion of the spectrum disappears. Because Z maps to ionization potential, as the temperature rises, the lowest Z atoms tend to be stripped first, and the opacity to lower energy photons is removed first (see fig. 8.10).

The Solar abundance of Fe creates an interesting numerical coincidence: the photoelectric opacity for cosmic gas just above the Fe edge (7–9 keV, depending on the ionization stage) is always within a factor of two the same as the Compton opacity. In fact, the Compton opacity at these energies is independent of the ionization state of the gas because the recoil loss when a 10 keV photon scatters a free electron is $\simeq$ 200 eV, which is much larger than the typical scale of valence shell ionization potentials. Therefore, even bound electrons (except inner shell electrons in high-Z atoms) behave as if they were free when scattering X-rays with energy greater than roughly 3–4 keV.

Soft X-ray opacity of the sort just described affects AGNs in several different ways. First, it absorbs photons from any AGN whose line-of-sight happens to lie in the local Galactic plane. Typical vertical column densities are several times 10^{20} cm^{-2} and increase as csc b, where b is the Galactic latitude. Consequently, even for AGNs well above the Galactic plane, it is often difficult to see the intrinsic spectrum at energies much below 500 eV, and local opacity makes the spectral range from 13.6 to roughly 100 eV essentially black on *all* lines of sight. If there is any additional opacity in the interstellar medium of the host galaxy (as there might well be), the situation is made even worse. Sadly, we may *never* have a direct indication of the true spectra of AGNs in this energy range (in our reference frame); high-redshift AGNs are our only hope for learning about EUV emission from these sources.

8.7.2   Fe Kα emission

Many fewer prominent lines can be seen in the X-ray band than in the ultraviolet or optical bands because most of the high-Z elements whose lines fall in that energy range are comparatively rare. Because detection of a small equivalent width line requires an instrument with high spectral resolution, the technical difficulties of X-ray spectroscopy further discourage the study of X-ray lines. However, there is one great exception to this rule—the Fe Kα line. Fe is sufficiently abundant (in the Sun, its fractional abundance relative to H is $\simeq 3 \times 10^{-5}$, only one order of magnitude less than C, N, and O) that its line emission can be quite strong.

When Fe is in any ionization state with at least four electrons, K-shell photo-ionization can be followed by either Auger ionization or fluorescence. Although the branching ratio for fluorescence increases slowly with increasing ionization, the dependence is weak: the fluorescence yield $Y = 0.34$ for neutral Fe and rises only to $\simeq 0.5$ in the higher ionization stages.

The physics of line emission is different in the highest ionization stages but can be treated in a very closely analogous way. In Fe^{+23} the ground state configuration is 1s^{2}2s; when a 1s electron is removed, the direct decay route would be 2s$\rightarrow$1s, but that is strongly forbidden. However, configuration mixing still ultimately permits the decay, with $Y \simeq 0.1$. In Fe^{+24} and Fe^{+25}, the process is better described as recombination. Still, because the rate of emission is proportional to the rate of K-shell photo-ionization, it can be described in terms of an effective yield, but in this case one that, depending on temperature (and, in the case of Fe^{+24}, some uncertain atomic physics), lies in the range $\simeq$ 0.5–0.7. Similarly, the line energy, which is 6.4 keV for neutral Fe, rises only slightly with increasing ionization until the atom is mostly stripped. Even then, the line energy is not tremendously higher: 6.7 keV in He-like Fe and 6.9 keV in H-like Fe.

The total fluorescence rate depends, of course, on the covering factor C of material around the X-ray continuum source and its optical depth at the Fe K-edge τ_K. Written as an equivalent width, for $\tau_K < 1$ the strength of the Fe Kα line relative to the continuum is

$$W(K\alpha) \simeq \frac{C\langle\tau_K\rangle}{\alpha} Y \left(\frac{\epsilon_{K\alpha}}{\epsilon_K}\right)^\alpha \epsilon_{K\alpha}, \qquad (8.82)$$

where ϵ_K is the energy of the edge, $\epsilon_{K\alpha}$ is the energy of the line, the averaging is with respect to solid angle, and the continuum spectrum has $F_\epsilon \propto \epsilon^{-\alpha}$. Because the absorptive opacity presented to Kα photons is not greatly less than the opacity facing K-edge photons, and because Thomson scattering can deflect Fe K-continuum photons if there are any holes in the irradiated material, in practice it is hard to arrange conditions in which

$W(K\alpha)$ can exceed 250 eV unless the exciting continuum is hidden from our view or the Fe abundance exceeds Solar.

8.7.3 Consequences of Comptonization and pair production for γ-rays

We have previously discussed the effects on soft photons of Compton scattering by comparatively energetic electrons. Now it is time to turn the question around and ask about energetic photons scattering off cool electrons. Compton recoil is the most important effect in this case, and for photons of energy greater than a modest fraction of $m_e c^2$, it can have an appreciable impact.

Photons more energetic than $\gtrsim 100$ keV can lose a significant part of their energy in a single scattering event. Consequently, our usual analytic tools, which use differential equations to model smooth changes, are irrelevant. It is generally necessary to resort to Monte Carlo calculations in order to see what happens in any given instance. The only clear qualitative statement that can be made is that a modest optical depth can wipe out the spectrum in this range of energies.

At somewhat lower energies a continuum description works better. Energy loss by recoil is a statistical process, so the evolution of the photon energy loss cannot be described deterministically. However, in this limit the expectation value for the photon's energy obeys the simple equation

$$\frac{d\epsilon}{dt} = -n_e \sigma_T \frac{\epsilon^2}{m_e c^2}, \tag{8.83}$$

which has the solution

$$\epsilon(t) = \frac{\epsilon_o}{1 + n_e \sigma_T t \epsilon_o / (m_e c^2)}. \tag{8.84}$$

This has a very clear physical interpretation: the Compton loss time is the Compton scattering time multiplied by $m_e c^2/\epsilon_o$; after that time, the photon energy declines $\propto t^{-1}$. Equivalently, we might write this solution as

$$\epsilon(N_{\text{scatt}}) = \frac{\epsilon_o}{1 + (\epsilon_o / m_e c^2) N_{\text{scatt}}}. \tag{8.85}$$

That is, once the photon energy has been reduced by a factor of order unity, its energy after subsequent scatters decreases as N_{scatt}^{-1}. Thus, if the photons are trying to make their way through a medium of depth $\tau_T > 1$, few photons will emerge with more energy than $m_e c^2 / \tau_T^2$.

A second hazard exists for more energetic photons: $\gamma - \gamma$ pair production (§8.4.1.1). Although it is not too hard to estimate the mean intensity of γ-rays deep inside a compact (i.e., $l > 1$) pair plasma, it is much harder to solve the problem of how much γ-ray flux escapes from such a plasma. The difficulty is in coupling the radiation transfer with a local calculation of the pair balance, which in turn requires determining the local thermal properties of the plasma. Thus, prediction of the emergent γ-ray spectrum requires a complex calculation, tuned to the specifics of the case in point.

8.7.4 X-ray reflection

These propagation issues join in an application of considerable relevance to AGNs: the study of how a "cool" optically thick slab reflects X-rays incident upon it (Lightman and White 1988). The plausibility of locating the X-ray source in AGNs near the putative accretion disk makes this a problem worth considering (George and Fabian 1991). The motivation for this idea is further bolstered by the fact that type 1 Seyfert galaxy X-ray spectra exhibit a spectral "hardening" near 10 keV that may have its origin in this effect (Pounds et al. 1990; §8.1.1). Almost the same physics applies if AGNs are surrounded by geometrically and optically thick "obscuring tori" (Krolik, Madau, and Życki 1994; Ghisellini, Haardt, and Matt 1994; see Chap. 12). Here we will briefly discuss the physical issues involved.

To pose the problem cleanly, imagine a plane-parallel semi-infinite slab exposed to X-rays on its open face. If it were completely ionized, then all the photons would be reflected, but those high-energy photons that scatter more than $m_e c^2 / \epsilon_o$ times would be reduced to an energy $m_e c^2 / N_{\text{scatt}}$. The probability distribution for the number of scatters before reflection is $\propto N_{\text{scatt}}^{-3/2}$ for large N_{scatt} (Lightman and Rybicki 1979), so the "filter function" is approximately $1 - (\epsilon / m_e c^2)^{1/2}$. In practice, incident photons of less than 30 keV are reflected with essentially unit albedo. The reflected spectrum would then look very much like the incident below 30 keV, but would roll-off at higher energies and would include very few photons above 100 keV.

On the other hand, if the slab were completely neutral, low-energy photons would suffer very severe attenuation. The average reflected photon scatters a few times before reflecting, so the average pathlength traversed by a reflecting photon while in the slab is several times 10^{24} cm^{-2}. As figure 8.10 demonstrates, the absorption optical depth associated with such a large column density is greater than unity for all photons below ~ 10 keV. As a result, the "mirror" is highly absorptive below that energy; even at

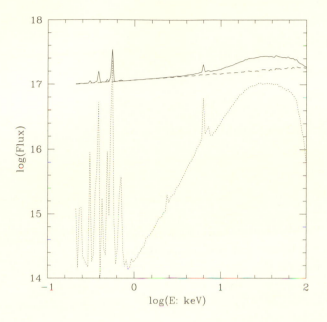

Fig. 8.11   The effects of X-ray reflection (Życki et al. 1994). The dashed line represents the incident continuum, chosen to have $F_\epsilon \propto \epsilon^{-0.9}$. The solid line above it is the spectrum as observed, displaying the reflection "bump" peaking around 20–50 keV and several emission lines. The dotted curve shows the pure reflection spectrum. Below 1 keV, emission lines are a very important part of the reflection spectrum, but just which lines are present depends on the exact state of the reflecting gas. Here, in order to mimic AGN accretion disk conditions, the reflecting material was bathed in 10^5 K blackbody radiation.

20–30 keV, the albedo is still only $\simeq 0.5$. The combination of photoelectric opacity at the low-energy end of the spectrum and Compton recoil at the high end thus eliminates both extremes of the spectrum, leaving only a "bump" between 20 and 50 keV (fig. 8.11).

A major part of the absorptive opacity from 7–20 keV is, of course, Fe K-edge opacity. Substantial Fe Kα fluorescence is therefore a corollary of X-ray reflection.

In real accretion disks, it is possible for the X-ray flux incident on the disk to be so intense that the upper layers are highly ionized. When this is the case, the soft X-ray opacity can be greatly reduced in the layers of the disk lying at such small Compton depth from the surface that the intensity of X-rays is not sensibly reduced by reflection. Deeper into the disk, the X-ray intensity diminishes, and the material begins to recombine. Only those soft X-rays penetrating that deep are then absorbed. In situations such as these, the mean albedo is rather greater than when the material is cold, and there is also substantial reemission of absorbed energy in the form of X-ray emission lines and recombination continua (Życki et al. 1994).

8.8   Comparison with Observations

8.8.1   X-rays

As we have already discussed, radio-quiet AGNs seem to have very stereotyped high-energy spectra: the spectra are generally fairly well described by a sum of a cut off power law $F_\epsilon \propto \epsilon^{-\alpha_x} \exp(-\epsilon/\epsilon_c)$, with $\alpha_x \simeq 0.9$ and ϵ_c ill-determined, but perhaps ~ 100 keV or more, and sometimes a bump component that peaks in the neighborhood of 20–50 keV. In addition, they frequently—but not always—have Fe Kα emission lines with equivalent widths up to a few hundred eV.

Because they generically produce cutoff power laws, thermal Comptonization models easily reproduce such a continuum shape. If thermal balance is required in the model, two sorts of solutions are possible: either $l_s/l_h \sim 0.1$ and the optical depth may be almost anything between a few tenths and several, or $l_s/l_h = 0.5$ and $\tau_T \simeq 0.1$. These two alternatives can be distinguished by the cutoff energy, for in thermal equilibrium $\tau_T \propto \Theta^{-1}$. For example, if $l_s/l_h = 0.5$ and $\tau_T \simeq 0.1$, $\Theta m_e c^2 \simeq 400$ keV. Smaller ϵ_c would then favor models in which $l_s/l_h \sim 0.1$.

The ratio l_s/l_h is controlled by three separate factors: the ratio of intrinsic dissipation in the regions producing the seed photons and the X-rays, the fraction of the X-rays reprocessed into low-energy seed photons, and the fraction of the seed photons that end up entering the X-ray production region. The latter two factors are purely geometrical. For example, if the X-ray "corona" were a slab sitting on top of the accretion disk, roughly half its emitted energy would end up being reprocessed, and all of those reprocessed photons would reenter the "corona." To achieve l_s/l_h as small as 0.1, there must be some combination of a smaller covering fraction for the reprocessor around the hot plasma, and a smaller fraction of the soft photons entering the "corona." Perhaps the "corona" occupies a hole in the disk or is broken up into small regions that do not completely surround the disk.

The bump component and Fe Kα emission are easily reproduced by Compton reflection, with contributions possible from the accretion disk proper (Pounds et al. 1990), the flow inside the marginally stable orbit (Reynolds and Begelman 1997, Życki, Done, and Smith 1997), or matter farther out (Krolik et al. 1994; Ghisellini et al. 1994). On the one hand, many Fe Kα lines are observed to be several tens of thousands of km s^{-1} broad and significantly redshifted (Mushotzky et al. 1995; Tanaka et al. 1995: fig. 8.12)–orbital motion deep in a relativistic gravitational potential (cf. §5.4) seems to be the only intepretation for line profiles of this sort. Indeed, in the long run lines like this may be the best evidence for the existence of massive black holes in AGNs. Moreover, analysis of their profiles (and of the corresponding shifts and smearing of the Compton reflection features) may provide detailed diagnostics of dynamics at the very center of relativistic accretion disks. On the other hand, some lines are (or have components) narrower than 1000 km s^{-1} across and vary in flux far more slowly than the continuum, suggesting that they are made far from the black hole (Weaver et al. 1996).

Several different possibilities may account for the "soft excess" that is sometimes seen. One simple expedient is to attribute it to Comptonization of disk photons by passage through a warm layer. If the slope of this soft excess is $d \ln F_\epsilon / d \ln \epsilon \simeq -2$, equation 8.33 can be solved to indicate a Compton $y \simeq 0.4$. Because the electrons in this layer must have enough energy to boost photons up to ~ 1 keV, their temperature must be at least $\sim 10^7$ K. The optical depth of such a layer would then be $\lesssim 10$ in order to keep $y \simeq 0.4$. In most cases, disks are optically thick enough that this layer would be only a small fraction of the total column density (see the equilibria worked out in §7.3.4). Presumably some fraction of the (unknown) heat source that supplies the source region for the hard X-rays could be diverted to keep this Comptonizing layer warm. It is also possible that the soft excess is merely a merging of large numbers of X-ray lines, as shown in figure 8.11.

Many type 1 Seyfert galaxies show evidence for significant interven-ing absorption, often in gas that appears to be ionized (Reynolds 1997). These cases have been dubbed *warm absorbers*. The typical signatures are sharp photoelectric edges with optical depths ~ 0.1–1 in species such as O VII, O VIII, and ionized Fe. The location and source of the gas responsible for these features are controversial, but the presence of such highly ionized species indicates that the gas is both warm ($\sim 10^5$–10^6 K) and strongly pho-toionized. The column densities of H atoms are $\sim 10^{22}$–10^{23} cm^{-2}. Where and how these features are formed will be considered in Chapters 11 and 12.

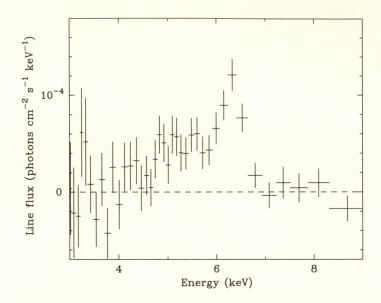

Fig. 8.12 Residuals after a power law plus reflection bump model has been subtracted from the data given by *ASCA* observations of the type 1 Seyfert galaxy MCG -6-30-15 (data from the *ASCA* archive, reduced and plotted by K. Weaver). The emission line peaks near 6.4 keV but extends upwards to $\simeq 6.7$ keV and has a very broad wing extending down to almost 4 keV. Such a profile strongly suggests an origin in material orbiting deep inside a relativistic potential.

Type 2 Seyfert galaxies generally exhibit strong soft X-ray absorption. The associated H column density is usually at least 3×10^{22} cm^{-2} (with little evidence for ionization), and it may be that in the typical case the absorbing matter is optically thick to Compton scattering. When the column density of obscuration is that great, the entire X-ray continuum is suppressed. We will discuss possible locations for this absorption in Chapter 12.

Because few quasars are as bright as the brighter Seyfert galaxies, the available spectra for them are generally of poorer quality and less can be said. However, there are indications that intrinsic absorption is rarer in radio-quiet quasars than in Seyfert galaxies (Laor et al. 1994).

Due to their rarity, considerably less is known about the X-ray properties of radio-loud AGNs. Nonetheless, it does appear that the physics of their X-ray emission (at least in the lobe-dominated subclass) is qualitatively similar to what we infer occurs in their radio-quiet cousins. However, the systematically harder spectra they display, if interpreted in the vein of thermal Comptonization models, suggest that l_s/l_h in these AGNs is perhaps a factor of three smaller than in the radio-quiet variety. Their products of $\Theta\tau_T$ would then be a corresponding factor of 1.3 larger. Too little is known about details of their spectra to say whether "bumps" and "warm absorbers" are as strong and common as in their radio-quiet brethren, although there are individual examples (e.g., Mathur et al. 1994).

In the case of radio-loud core-dominated AGNs, the observed diversity of spectral slope and the variations in spectral slope seen in individual objects indicate the need for significant physical changes from the radio-quiet paradigm. If thermal Comptonization is the dominant radiation mechanism in these AGNs (and the evidence is considerably weaker than for the radio-quiet AGNs because high-energy cutoffs have not been seen), then l_s/l_h must vary over a much greater range. However, it is, if anything, more likely that X-rays in these AGNs are due to nonthermal electron distributions, so inferences drawn from thermal models may be irrelevant.

8.8.2 γ-rays

Although radio-quiet AGNs are very faint above several hundred keV, those radio-loud AGNs with strong core emission (see Chap. 9) are in general quite strong above 1 MeV (§8.1.3). Their power-law spectra extending out to GeV energies or higher clearly point to an origin in inverse Compton scattering by a power-law distribution of relativistic electrons. A mechanism of this sort would naturally produce a power-law spectrum, and, in any event, the electrons responsible must certainly be highly relativistic.

Relativistic beaming of the γ-rays is also very likely. If the source were stationary, the observed variability timescales (sometimes as short as 1 d) would place approximate upper bounds on the size of the source region through causality arguments like the one presented in §4.3.1. Given an observed flux, we could then estimate the compactness of the source:

$$l \sim \frac{F_\gamma 4\pi D^2 \sigma_T}{m_e c^4 \Delta t}. \tag{8.86}$$

In many (perhaps most) cases, l estimated in this way is $\gg 1$. If this were true, it would be hard to understand how the γ-rays could escape without suffering catastrophic losses to pair production. Although new γ-rays would be created by pair annihilation, because the annihilating pairs tend to have the lowest energies, the output spectrum would no longer be a power law, nor would it extend to energies $\sim 10^3 m_e c^2$.

These inferences are dramatically altered by relativistic beaming. To understand its effect, we begin by estimating the optical depth to pair production. Optical depth is a Lorentz invariant, so we can calculate it in any frame we choose. Because it enables us to avoid worrying about angle-dependent energy thresholds, the most convenient frame is the one in which the photons whose energies are near threshold for pair production with the γ-rays (these photons are likely to be in the X-ray band) are nearly isotropic. We call the boost from that frame to ours δ_x and label noninvariant quantities measured in that frame by primes. It is important to recognize that the γ-ray source frame may be different from the frame in which the X-rays are isotropic. The opacity to pair production is determined by the photon density at the pair production threshold

$$n'_{\rm ph}(\nu'_{\rm th}) = \frac{1}{hc} \int d\Omega' \, I'_{\nu'_{\rm th}}. \tag{8.87}$$

We estimate the intensity in the moving frame from the intensity we measure in our frame via the relativistic invariant I_ν/ν^3 (§5.3), finding that

$$n'_{\rm ph}(\nu'_{\rm th}) = \frac{1}{hc} \int d\Omega' \, I_{\nu_{\rm th}} \delta_x^{-3}, \tag{8.88}$$

where $\delta_x = \nu/\nu'$. Because $I'_{\nu'}$ is isotropic, the integral can be done immediately. The threshold energy in the moving frame is defined by $h\nu'_{\rm th} = 1/\epsilon'_\gamma = \delta_x/\epsilon_\gamma$, making $\nu_{\rm th} = \delta_x^2/\epsilon_\gamma$. Thus, if the observed spectrum is a power law $\propto \nu^{-\alpha_\gamma}$, the photon density at threshold in the moving frame ends up scaling very strongly with the boost factor:

$$n_{\rm ph}(\nu'_{\rm th}) \propto \delta_x^{-2\alpha-3} \epsilon_\gamma^\alpha. \tag{8.89}$$

In addition, the density of photons at threshold increases with increasing γ-ray energy because the energy of the threshold decreases, and there are generally more photons at lower energies.

If the size of the pair production region is determined by causality arguments based on variability timescales, its proper size is $\sim \delta_x c\Delta t$, so we arrive at the conclusion that $\tau_{\gamma\gamma} = \delta_x^{-2\alpha-4} c\Delta t \tau_o$, where τ_o is the optical depth estimated ignoring relativistic effects.

In many sources, $\tau_o \sim 10^4$, and $\alpha \simeq 1$, so boosts with $\delta \simeq 5$ from the frame in which the X-rays are isotropic to the lab frame would allow the γ-rays to escape. Because this is also roughly the order of magnitude of the boosts estimated from superluminal motion (§9.3.3.2), relativistic kinematics is a strong candidate for explaining the emergence of strongly variable γ-ray emission. Indeed, although the first arguments for relativistic motion in AGNs were made on the basis of radio phenomenology, even if we knew nothing about their radio emission, the escape of high-energy γ-rays would by itself be a compelling argument for its existence. Moreover, this argument can also be turned around: for a given relativistic boost, the γ-rays we see cannot emerge from a size smaller than some minimum "gamma-sphere" (Blandford and Levinson 1995).

Beaming may also explain why not every radio-loud core-dominated AGNs can always be seen in high-energy γ-rays. Relativistic beaming tends to enhance variability, so the undetected ones may simply have been in temporary low states when observed. Alternatively, the Lorentz factor of the region responsible for the γ-ray emission may be somewhat larger than the Lorentz factor of the region responsible for the core radio emission. In that case, there may be objects for which we are within the favorable cone for radio emission, but not for γ-ray emission (see §9.3.4).

Further details, however, remain obscure and controversial. Many different candidates have been nominated as the source of the seed photons—synchrotron photons generated inside the relativistic plasma, ultraviolet photons from the accretion disk, emission line photons—while the origin of relativistic jets and the nature of particle acceleration inside them are subjects about which equally little is known. In fact, the issues regarding the origin and nature of jets form a major part of the next chapter (§9.3). We will postpone discussion of more complete models for high-energy γ-ray emission in jets until that section.

9   Radio Emission and Jets

9.1   Phenomenology

The phenomenology of radio emission in AGNs is a rich and complex subject. The fraction of the bolometric luminosity that emerges in the radio, its spectral shape, and the morphology of the radiating region(s) all vary substantially from one object to another.

9.1.1   Relative strength of radio emission

Because AGNs emit their power over such a broad range of frequencies, even defining the fraction of the bolometric luminosity emerging in the radio band introduces several subtle problems. For a start, the range of frequencies we define as "radio" is broad enough, and there are enough variations in radio spectral shape, to make the choice of radio frequency at which $L_{\rm r}$ is measured a significant one. Next, the difficulty of truly measuring the bolometric luminosity is great enough that this goal is often abandoned, and the much more restricted goal of comparing the radio luminosity to the optical luminosity is substituted in the hope that $L_{\rm opt} \propto L_{\rm bol}$. Even this restricted approach can go wrong—there are numerous AGNs (see Chap. 12) whose optical (and also ultraviolet and X-ray) emission is largely transformed into infrared by dust; if we tabulate $L_{\rm r}/L_{\rm opt}$ for these, the result will be a very bad indicator of $L_{\rm r}/L_{\rm bol}$. Another subtlety (as we will discuss at greater length in §9.3.4.3 and again in Chap. 12) is that, particularly in radio-loud objects, the different bands can be radiated strongly anisotropically. To make a fair estimate of the fraction of the luminosity radiated in the radio band would then require observations from different directions, which is generally a rather difficult feat for us. Finally, AGNs often vary in the radio, as well as in the other bands, and the variations are often not synchronized. Consequently, the ratio of radio to total flux changes from time to time.

Nonetheless, we may temporarily put aside these qualms and look at some data. The conventional measure of "radio loudness" is the quantity $R \equiv F_{\nu_r}/F_\nu(4400\text{Å})$, where ν_r is the observing frequency in the radio band. In terms of monochromatic luminosity, $\nu_r L_{\nu_r}/L_{\rm opt} = 1.47 \times 10^{-6}(\nu_r/1$ GHz$)R$. Figure 9.1 shows the distribution of R for the optical color-selected

BQS (also known as PG) sample and $\nu_r = 5$ GHz. Although the best value of R to choose as the division between radio-loud and radio-quiet is not perfectly determined by this distribution, the distinction is clearly a useful one. If the divide is placed at $R = 30$, $\simeq 15\%$ of the sample is radio-loud, that is, has $L_r/L_{\rm opt} > 2 \times 10^{-4}$. Thus, even those designated "radio-loud" still produce only a small fraction of their luminosity in the radio band. It is also noteworthy that the distribution of R does not appear to depend strongly on $L_{\rm opt}$, although there may be a tendency for the radio-loud fraction to diminish when $L_{\rm opt}$ is as small as a typical galaxy's stellar luminosity (Peacock, Miller, and Longair 1986). Whether the radio-loud fraction changes with redshift is uncertain (McMahon 1991; Hooper et al. 1996).

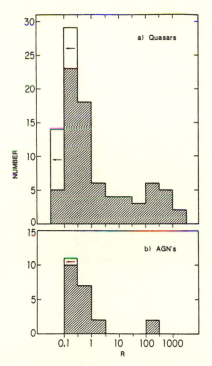

Fig. 9.1   The distribution of the ratio of radio to optical flux density in the optical color-selected PG sample (Kellermann et al. 1989). (Top) the distribution for quasars ($M_B < -26$); (bottom) that for lower luminosity AGNs. The two distributions are statistically indistinguishable.

How to interpret the distribution of $L_r/L_{\rm opt}$ is a controversial matter.

If we (warily) allow the optical luminosity to serve as a stand-in for the bolometric luminosity, one possibility is that the radio luminosity scales with the bolometric luminosity, but the ratio between the two has a distribution described by $f(R)$. In that case the radio luminosity function could be written as

$$\frac{dn}{dL_{\rm r}} = \int dL_{\rm opt} \int dR \, \frac{dn}{dL_{\rm opt}} f(R) \delta \left(L_{\rm r} - R L_{\rm opt} \right). \tag{9.1}$$

Another possibility is that the probability of having a given radio luminosity is independent of the probability of having a given bolometric luminosity, so that the joint luminosity function is separable into a product of two functions, one depending on $L_{\rm r}$ alone and the other on $L_{\rm opt}$ alone. With the freedom to choose $f(R)$ arbitrarily, the two are equivalent; however, to the degree that $f(R)$ has one or a few narrow peaks, these two pictures become distinct from one another. In the limit that $f(R)$ is quite narrow, the distribution of objects in the $L_{\rm r}$–$L_{\rm opt}$ plane is a thin band; in the limit that $f(R)$ is very broad, the distribution fills the entire plane.

9.1.2 Morphology

Unlike (almost) any other band in AGN spectra, the radio emission can be resolved with current technology. The image varies considerably, however, from object to object and as a function of frequency. Among the radio-loud AGNs, especially at low frequencies (i.e., a few hundred MHz), in most instances the radio emission comes from a pair of extremely large lobes on opposite sides of the host galaxy. These lobes are often several hundred kiloparsecs in length and can be separated from the galaxy by a similar distance (see plate 2). Strikingly, the axes of the two lobes and the center of the galaxy generally lie along a common line. However, there are other radio-loud objects in which, especially at higher frequencies (i.e., several GHz), the region responsible for the bulk of the emission is essentially unresolved on VLA (i.e., $1''$ scales), so that the source must be smaller than $\sim 20 (D_A H_o/c)(h/0.75)^{-1}$ kpc (e.g., fig. 9.2). Intermediate cases also exist, in which the core and lobe fluxes are more nearly comparable, at least at some frequencies.

Even within the lobe-dominated class, there is another subdivision in morphology. In some cases, the lobes are edge-brightened and have very bright, very small hot spots near the outer surfaces of the lobes (plate 3). Often spots of infrared or even optical continuum emission can be seen at the same locations as the radio hot spots. In other cases, the lobes are brightest near the galaxy and grow dimmer as one looks farther away

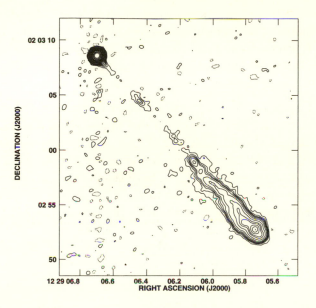

Fig. 9.2   The core-dominated radio-loud quasar 3C 273 at 15 GHz. The contour levels are quasi-logarithmic. As this figure vividly shows, we receive far more flux from the unresolved core than from the extended jet. Unpublished data courtesy of R. Perley, H.-J. Roeser, and K. Meisenheimer.

(plate 4). Objects in this class often have lobes that are bent. Interestingly, this division in morphology corresponds very closely to a division in radio luminosity that is dependent upon the *optical* luminosity of the host: the edge-brightened ones almost all have luminosities at 1.4 GHz (defined as νL_ν) greater than $6 \times 10^{40} L_{R,44}^2$ erg s^{-1}; the edge-darkened ones almost all have smaller radio luminosities, where L_R is the R band luminosity of the host (see fig. 9.3, taken from Ledlow and Owen 1996). Enshrining this division, Fanaroff and Riley (1974) designated as class 1 those radio galaxies for which the ratio of the distance between the two brightest spots on either side of the nucleus to the overall source size was less than 0.5 and as class 2 those for which the ratio was greater than 0.5. Occasionally there are objects that fall into one class according to morphology while having a luminosity more typical of the other, but by and large the division is a fairly clean one.

Both for lobe-dominated radio-loud AGNs and for radio-quiet AGNs, there is often interesting structure on the $1''$ scale (the classic example is shown in fig. 9.4). This generally, but *not* always, takes the form of elongated low-luminosity emission regions, often very well aligned with the outer lobe structure. Note that in radio-quiet objects there generally are no outer lobes. When the overall emission is core-dominated, if one looks with great enough dynamic range, elongated structure is also usually present. These elongated structures are often (tendentiously) called *jets*.

In lobe-dominated objects, the morphology on these smaller (i.e., kpc) scales displays another correlation with the large-scale morphological distinction that defines the FR1 and FR2 classes: in FR1 objects, the brightening toward the center seen on much larger scales extends in to the kiloparsec scale, and a "jet" can be seen on both sides of the nucleus; in FR2's, the "jet" is relatively weak compared to the lobes and is virtually always only on one side. In both FR1's and FR2's the jet is usually well aligned with the outer structure, though in FR1's that means connecting smoothly with the outer structure, because bends are often present. Sometimes the "jets" are smooth, but sometimes they can be resolved into discrete lumps, usually called "knots."

In radio-quiet AGNs, the small-scale morphology is much more various. Jets, lumps, multiple structures, and more complicated pictures can all be seen. Because the physical lengthscale associated with the angular resolution of any particular telescope depends on distance, the presence or absence of resolvable structure is strongly correlated with the redshift of the object.

Interestingly, kpc-scale jets can sometimes be seen in visible light and X-rays, as well as in the radio band. Indeed, even detailed features are often closely aligned in all the different bands (see fig. 9.5).

At radio frequencies ultra-high resolution is available for bright enough sources in the form of Very Long Baseline Interferometry (VLBI). In this technique, stations separated by intercontinental distances (or the even greater distance from Earth to an orbiting satellite) simultaneously record the signal from a source. Phase differences are then searched for computationally in order to achieve a partial reconstruction of the wavefronts at Earth. Given the size of the Earth, resolutions of milliarcsec or less can be achieved in the several GHz band.

Because the minimum S/N required for VLBI to be successful corresponds to a fairly high brightness temperature in the source (at least $\sim 10^{10}$ K), this technique has only been applied to objects with bright cores. The very fact that VLBI can be done teaches us an important

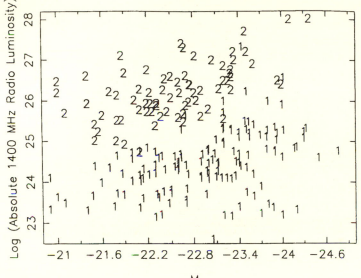

Fig. 9.3   The distribution of FR1 and FR2 galaxies in the radio luminosity—host galaxy luminosity plane (Ledlow and Owen 1996). Here radio luminosity (in W Hz^{-1}) is measured at 1.4 GHz and optical luminosity is defined as the absolute magnitude in the R band integrated down to a surface brightness of 24.5 mag per square arcsec in the rest frame. Radio galaxies designated as FR1 or FR2 on the basis of lobe morphology are very well separated; there is at most a band of perhaps a factor of 10 in radio luminosity within which the two classes are mixed. The radio luminosity dividing the two classes increases roughly $\propto L_{\mathrm{opt}}^2$.

qualitative fact: sources this bright must radiate by a highly nonthermal mechanism.

Since the invention of the VLBI technique in the late 1960s, numerous sources have been observed in this way. The results have been spectacular. When the surface brightness on the milliarcsec scale (i.e., $20(\theta/1 \text{ milliarcsec})(D_A H_o/c)(h/0.75)^{-1}$ pc) is great enough for the source to be detected, the structure is generally rather blobby and often has an overall linear organization. In core-dominated sources, the emission seen on this scale often accounts for the majority of the flux. Most surprisingly, the separations between the blobs change with time, and in almost every case the apparent transverse velocity is several to ten times c! A typical example (for which the transverse speeds of individual features are indeed of this magnitude) is shown in figure 9.6.

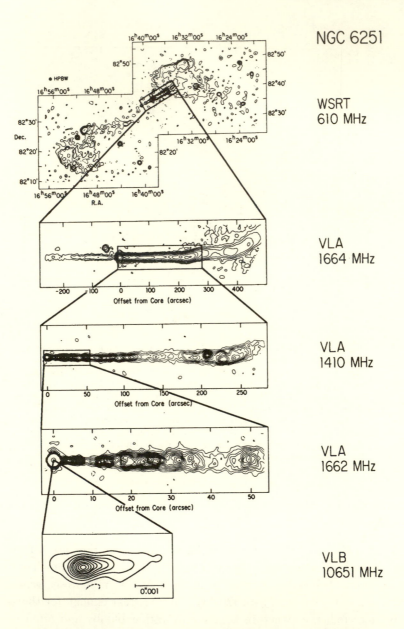

Fig. 9.4   The FR1 radio galaxy NGC 6251 at a succession of resolutions and frequencies (courtesy A. Bridle). Although small displacements occur from scale to scale, overall the jet retains a remarkably constant alignment over a dynamic range in lengthscale of 10^6!

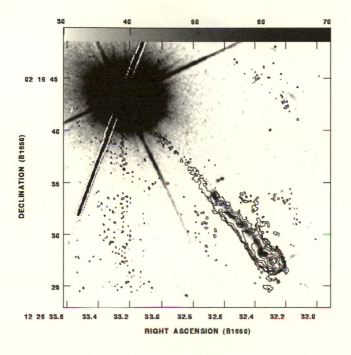

Fig. 9.5 The radio and optical jet in the quasar 3C 273 (Bahcall et al. 1995). The gray scale shows the optical surface brightness, contours (spaced by factors of two) show the radio (compare fig. 9.2, which shows the radio emission alone). The overexposed dark circle with the diffraction spikes and CCD charge leakage bar is the quasar proper.

9.1.3 Spectrum

The division between lobe- and core-dominated also correlates with spectrum. When the flux is dominated by extended lobes, the spectrum is usually fairly well described by a power law with $F_\nu \propto \nu^{-\alpha}$ and $0.5 \lesssim \alpha \lesssim 1$. Interestingly, the steeper sources tend to be at higher redshifts. On the other hand, most core-dominated sources have spectra that are more nearly flat in F_ν, though bumps and wiggles can sometimes be seen.

It should come as no surprise that the relative numbers of "flat" and "steep" spectrum sources depend strongly on the selection procedure: low-frequency selection finds more objects with steep spectra, high-frequency

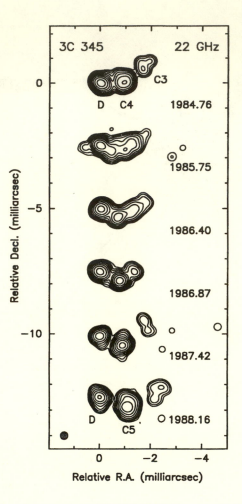

Fig. 9.6   A succession of 22 GHz VLBI images of the core of the quasar 3C 345 (Zensus et al. 1995). Contours represent factors of two in surface brightness. The small shaded circle in the lower left is the beam size.

selection finds those with flatter spectra (fig. 9.7). In addition, there is a tendency for the spectrum to flatten as the band over which its shape is measured moves to higher frequencies.

Within the subset having flat spectra, there are significant correlations between high-frequency spectral slope and optical polarization and variability. Greater optical polarization and variability (as found particularly in BL Lac objects, highly polarized quasars, and OVV quasars) correlate with

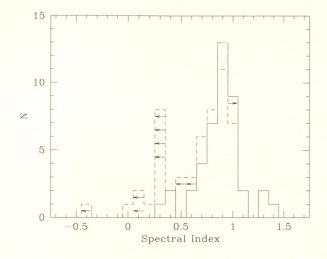

Fig. 9.7   The 408–4850 MHz spectral index distribution from the portion of the 5C survey analyzed by Maslowski et al. (1984). The solid line shows how many sources were found with flux at 408 MHz ≥ 120 mJy; the dashed line shows the sources with flux at 4850 MHz ≥ 17 mJy. Arrows are used for those sources found at one frequency but not at the other. Significantly more flat-spectrum sources are found when the selection is made at 4850 MHz than when it is made at 408 MHz.

a flat radio spectrum extending to higher and higher frequency, often well beyond 100 GHz (Tornikoski et al. 1993). In these highly polarized objects, the high-frequency radio spectrum blends smoothly into the infrared and optical continuum, suggesting that one physical mechanism accounts for all of them (Landau et al. 1986; §9.3.4). The spectral index α increases so gradually with increasing frequency that while there is a peak in νF_ν, it is very broad and, depending on the object, can be found anywhere from the mm-band to soft X-rays. It is also exactly this set of objects in which there is generally a second, often much higher, peak in νF_ν at energies $\sim$ 100 MeV (§8.1.3).

The correlation between morphology and spectrum is not quite perfect, however. In some radio galaxies (the so-called "Gigahertz Peaked Sources," or GPSs—see O'Dea, Baum, and Stanghellini 1991, and "Compact Steep Spectrum Sources," or CSSs—see Fanti et al. 1990), the radio emission is confined to a rather small region (10–1000 pc in the GPSs, 1–10 kpc in the CSSs), yet there is no flat section to the spectrum. Instead, there is a sharp transition from a steeply rising spectrum at low frequencies to a steeply falling power law at higher frequencies. About 25% of all FR2 galaxies found in low-frequency catalogs fall into the category of "compact steep-spectrum sources" (Dallacasa et al. 1995).

9.1.4 Polarization

In all instances where the radio structure (either core or lobes) can be resolved, the radiation is linearly polarized at levels of a few to a few tens of percent (fig. 9.8). As the example in the figure vividly shows, the fractional polarization can change radically from place to place within a single object (contrast the top and bottom panels). The direction of polarization may also depend on position (as it certainly does in this example).

On the other hand, it is a bit tricky to go directly from these images to clear statements about the intrinsic polarization in the source. For a start, the fact that the polarization direction changes from place to place immediately means that observations of different angular resolution can report different fractional polarization. Crude angular resolution can mix the light of different regions, each having high polarization, but with different orientations; the resultant total polarization might be much smaller than in each region separately.

In addition, Faraday rotation can introduce a frequency-dependent offset between the observed polarization direction and the intrinsic one (see also §7.5.6). In radio observations, the amount of Faraday rotation is conventionally parameterized in terms of the rotation measure

$$RM = 8.1 \times 10^5 \int d\vec{l} \cdot \vec{B}(n_e - n_+) \text{ rad cm}^{-2}, \qquad (9.2)$$

where $|\vec{B}|$ is measured in G, distance in pc, and electron density in cm^{-3}. Positrons (density n_+) counteract the effect of ordinary electrons. With this definition, the actual rotation angle is $\Delta\phi = RM\lambda_{cm}^2$. Both the intrinsic angle ϕ_o and RM can be determined by fitting the expression $\phi(\lambda) = \phi_o + RM\lambda^2$ to the polarization angle measured at a number of different wavelengths.

9.1.5 Variability

Not surprisingly, the extended radio lobes do not vary on any humanly observable timescale. However, the more compact regions can, and do, vary quite strongly. As discussed in §1.3, there is a subset of radio-loud AGNs—those with high polarization—that are extremely variable in all bands of the spectrum.

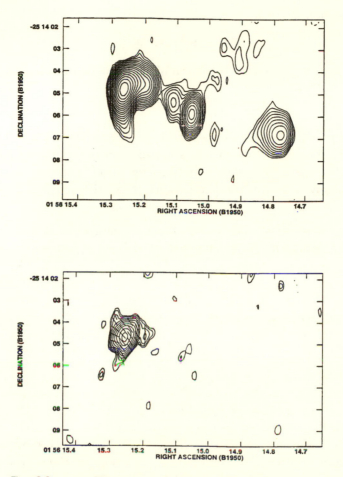

Fig. 9.8   4.7 GHz images of the $z = 2.09$ AGN 0156-252, from
Carilli et al. (1997). (Top) The total intensity; (bottom) the linearly
polarized intensity. In both cases the contour levels are logarithmic,
rising by $\sqrt{2}$ at each step, and the beam size is $0.''44 \times 0.''90$. The line
segments in the polarized intensity image show the local direction of
the $\vec{E}$-vector.

It is customary in the radio astronomy community to describe variations in the flux F not in terms of their Fourier power spectra, but in terms
of their "structure function"

$$S(\tau) = \langle [F(t) - F(t + \tau)]^2 \rangle, \tag{9.3}$$

where the average is over all measurements at times t that can be compared
at lag τ. The structure function is closely related to the more familiar

autocorrelation function

$$S(\tau) = 2\sigma^2 \left[1 - A(\tau)\right],\tag{9.4}$$

which in turn is the Fourier transform of the power spectrum

$$A(\tau) = \sigma^{-2} \int df\, e^{2\pi i f \tau} |\hat{F}(f)|^2.\tag{9.5}$$

Here σ is the *rms* fluctuation measured for F, and the Fourier transform of F is $\hat{F}$. The preference for the structure function is rooted partly in the fact that the autocorrelation and Fourier power spectrum are, strictly speaking, only defined if the fluctuations are statistically stationary and partly in custom.

Over the observed range of timescales (from a week to 10 years), the structure functions for both flat-spectrum radio-loud quasars and BL Lac objects are roughly similar (Hughes, Aller, and Aller 1992: observations at 5, 8, and 15 GHz; Lainela and Valtaoja 1993: observations at 22 and 37 GHz). Although there are significant exceptions, $S(\tau) \propto \tau$ from the shortest timescales up to a few years and levels off at longer lags. The corresponding power spectrum is roughly f^{-1} at high frequencies, rolling off to a flat (i.e., white noise) spectrum for $f < (\text{few yr})^{-1}$.

Given the size of the structures seen in VLBI images (~ 10 pc commonly), it is not surprising to see variations on several year timescales. The strength of shorter timescale variations is, however, somewhat unexpected. Indeed, their existence is one of the (now numerous) arguments for relativistic motion in these sources (§9.3.4.3). What is still more surprising is the existence of modest, but clearly real, variations on timescales as short as ~ 1 d (Quirrenbach et al. 1992). In the most common case, where the intraday variations are only $\sim 1\%$, it is quite plausible to attribute them to interstellar scintillation. This is especially likely when the majority of the several GHz flux comes from within an area on the sky smaller than a milliarcsec across. However, in about 25% of all flat-spectrum radio-loud AGNs the fluctuations are large enough (several tens of percent) that they are more likely to be intrinsic.

As we have already mentioned, in the highly polarized objects the spectrum between the GHz band and the optical/ultraviolet band is so smooth that we are tempted to think that it is created by a single mechanism. This thought is made even more attractive by the facts that in these objects the optical/ultraviolet light is also highly polarized and strongly variable. It is not uncommon for these objects to change in optical flux by factors of order unity within a few hours; in one spectacular case, the optical flux grew by a factor of 100 over a few months! Quiescent, quasi-thermal emission (created perhaps in conditions like those explored in Chap. 7) does not seem anywhere near appropriate for these objects.

9.2 Synchroton Theory

Unlike either the optical/ultraviolet or X-ray bands, the basic mechanism for radio emission in AGNs is very well established. Synchrotron radiation was first suggested by Alfvén and Herlofson (1950), and independently by Shklovsky (1953), to be the radiation mechanism responsible for the then newly discovered cosmic radio sources. The list of radio sources in the early 1950s included local supernova remnants as well as AGNs; today we are confident that synchrotron radiation is responsible for the radio emission in all of them. As an historical note, it is remarkable to see how quickly synchrotron radiation passed from prediction as a physical phenomenon (Schwinger 1949) to astrophysical application.

For a modern account of the basic physical theory of synchrotron radiation, the best place to look is Jackson (1975). To find many astrophysically useful formulae worked out, see Rybicki and Lightman (1979) or Pacholczyk (1970).

Strong linear polarization was the clue that first pointed astrophysicists toward considering synchrotron radiation, and it remains the single strongest argument in its favor. Certainly no thermal mechanism could produce strong polarization, and the means by which nonthermal electrons couple efficiently to radiation are relatively few, namely, synchrotron radiation and Compton scattering. Synchrotron radiation passes the polarization test because the acceleration of the charge is always perpendicular to the field, thus establishing a preferred direction for the wave $\vec{E}$-field (this fact also allows determination of the field direction from polarization observations). Compton scattering, although it can produce net polarization if either the electrons or the seed photons have a preferred direction, does not necessarily result in significant polarization. In addition, in the case of the giant double radio sources, it is easy to see that Compton scattering would be a very inefficient way to generate the radio power. The reason is that the local photon density is dominated by the microwave background, itself already predominantly in photons of energy *greater* than the radio photons we are seeking to explain. Consequently, even if there are much lower energy photons available for scattering up to radio frequencies, the electrons responsible will lose much more energy to Compton scattering the microwave background.

9.2.1 Optically thin emission

In fact, viewing a magnetic field as a collection of virtual photons of energy $\epsilon_B \equiv \hbar eB/(m_e c)$ allows one to see that synchrotron radiation and inverse Compton scattering are actually very similar processes (Lieu and Axford 1995). Relativistic boost effects transform the electron Larmor frequency $eB/(\gamma m_e c)$ into a characteristic radiated photon frequency ν_c that is larger by a factor γ^3, so the typical energy of the photons produced by synchrotron radiation is $\epsilon_c = \gamma^2 \epsilon_B$. Thus, the energy of synchrotron photons bears the same relation to their "seeds" as does the energy of inverse Compton scattered photons to theirs. Moreover, the rate at which an electron loses energy by synchrotron radiation (averaging over an isotropic pitch angle distribution) is $(4/3)\gamma^2 \beta^2 \sigma_T c (B^2/8\pi)$, closely analogous to the inverse Compton energy loss rate $(4/3)\gamma^2 \beta^2 \sigma_T c U_{\rm ph}$. Indeed, an important corollary follows immediately from this parallel behavior: the ratio between the energy loss rate by synchrotron radiation to the energy loss rate by inverse Compton scattering is exactly the ratio of the magnetic field energy density to the radiation energy density (assuming that Compton scattering is not limited by Klein-Nishina effects).

Because $\epsilon_c \propto \gamma^2$ for both processes, the synchrotron spectrum radiated by relativistic electrons with a given distribution function has the same shape as the inverse Compton spectrum produced by the same electrons scattering on effectively monochromatic seeds: if $dn/d\gamma \propto \gamma^{-\xi}$ between $\gamma_{\rm min}$ and $\gamma_{\rm max}$, $F_\nu \propto \nu^{(1-\xi)/2}$. More completely, if $dn/d\gamma = K\gamma^{-\xi}$, where

$$K = n_e \frac{1-\xi}{\gamma_{\rm max}^{1-\xi} - \gamma_{\rm min}^{1-\xi}}, \tag{9.6}$$

the synchrotron emissivity can be estimated by a procedure in close analogy with the δ-function approximation used in §8.3.1.2 to derive the inverse Compton spectrum:

$$j_\nu \simeq \int d\gamma\, K\gamma^{-\xi} \int d\epsilon_o\, h\nu \sigma_T c \frac{B^2}{8\pi\epsilon_o} \delta\left(\epsilon_o - \frac{\hbar eB}{m_e c}\right) \delta\left(\frac{4}{3}\gamma^2 \epsilon_o/h - \nu\right). \tag{9.7}$$

Here the fictitious seed photon spectrum is simply a δ-function at $\hbar eB/(m_e c)$. Evaluating the integral leads to the expression

$$j_\nu \simeq \frac{2^{\xi-1}\pi}{3^{(1+\xi)/2}} K \frac{e^3 B}{m_e c^2} \left(\frac{2\pi m_e c\nu}{eB}\right)^{(1-\xi)/2}. \tag{9.8}$$

Note that this is the power integrated over all solid angle per unit frequency per unit volume.

However, the analogy cannot be stretched forever. Two effects distinguish the output spectra of synchrotron radiation and inverse Compton scattering: synchrotron radiation depends on the pitch angle θ of the electron with respect to the magnetic field (electrons traveling exactly parallel to the field do not radiate at all), and the shapes of the radiated spectra with respect to the characteristic energy differ. Allowing for these distinctions, the synchrotron spectrum of a single electron can be written in a form closely resembling that of inverse Compton scattering:

$$P_\nu = \frac{4}{3} \sin \theta \gamma^2 \frac{\sigma_T c}{\nu_c} \frac{B^2}{8\pi} F_s \left(\frac{\nu}{\nu_c} \right),$$
(9.9)

where

$$\nu_c = \frac{3}{2} \gamma^2 \sin \theta \frac{eB}{2\pi m_e c}.$$
(9.10)

The dimensionless function $F_s(\nu/\nu_c)$ (contrast eq. 8.25) for synchrotron radiation takes the form

$$F_s \left(\frac{\nu}{\nu_c} \right) = \frac{3^{5/2}}{8\pi} \frac{\nu}{\nu_c} \int_{\nu/\nu_c}^{\infty} dy \, K_{5/3}(y)$$

$$\simeq \begin{cases} \frac{9}{2\Gamma(1/3)} \left(\frac{\nu}{2\nu_c} \right)^{1/3} & \nu \ll \nu_c \\ \sqrt{\frac{243}{128\pi}} \left(\frac{\nu}{\nu_c} \right)^{1/2} \exp\left(-\nu/\nu_c \right) & \nu \gg \nu_c . \end{cases}$$
(9.11)

Taking into account this spectral shape, we rewrite the previous expression for the emissivity (eq. 9.8) as

$$j_\nu = E(\xi) \frac{K e^3 B}{m_e c^2} \left(\frac{2\pi m_e c \nu}{eB} \right)^{(1-\xi)/2},$$
(9.12)

where

$$E(\xi) = \frac{3^{\xi/2} \pi^{1/2}}{2(\xi+1)} \frac{\Gamma\left(\frac{3\xi+19}{12} \right) \Gamma\left(\frac{3\xi-1}{12} \right) \Gamma\left(\frac{\xi+5}{4} \right)}{\Gamma\left(\frac{\xi+7}{4} \right)},$$
(9.13)

assuming an isotropic distribution of electron pitch angles (Blumenthal and Gould 1970). The function $E(\xi)$ is generally $\sim O(1)$.

Because electrons radiating by the synchrotron mechanism lose energy at a rate proportional to γ^2, just as do electrons cooling by inverse Compton radiation, the effect on their distribution function is exactly the same as for inverse Compton radiation (see §8.5.2.1). The cooling time is again $\propto \gamma^{-1}$, and the distribution function at energies above the cooling break is one unit steeper than below. This result leads, of course, to the same expectation for the radiated spectrum: it should steepen by half a unit above the frequency radiated by electrons that have had just enough time to lose half their initial energy.

There is also one special property of synchrotron radiation that it shares with inverse Compton scattering only when the seed photon spectrum is very narrow: for each output photon energy, there is a narrow range of electron energies primarily responsible for creating those photons. We call this energy

$$\gamma_s(\nu) \simeq \left(\frac{2\pi m_e c\nu}{eB}\right)^{1/2}.$$
(9.14)

If we now compare the predicted synchrotron spectrum with actual spectra emitted by extended radio emission regions of AGNs, we see that they match well provided only that we are able to adjust the single free parameter ξ. For spectral slopes α between 0.5 and 1, the slope of the electron distribution function $\xi \simeq 2$–3. Substituted in equation 9.13, we can use this inference to verify our remark that $E(\xi) \sim 1$; when $\xi = 2.5$, $E(\xi) = 1.42$.

9.2.2 Polarization

We justified the identification of the radio emission mechanism with synchrotron radiation on the basis that it naturally produced significant linear polarization. It is now time to make that statement quantitative.

A single electron of fixed pitch angle with respect to the magnetic field produces an elliptically polarized wave, but averaging over an ensemble of electrons with different pitch angles reduces the polarization to linear, oriented such that the $\vec{E}$-field of the wave is perpendicular to the magnetic field direction. The magnitude of this polarization (for an ensemble with different pitch angles but the same energy) is

$$P(\nu/\nu_c) = \frac{G_s(\nu/\nu_c)}{F_s(\nu/\nu_c)},$$
(9.15)

with

$$G_s\left(\frac{\nu}{\nu_c}\right) = \frac{3^{5/2}}{8\pi}\frac{\nu}{\nu_c}K_{2/3}\left(\frac{\nu}{\nu_c}\right) \simeq \begin{cases} \frac{9}{4\Gamma(1/3)}\left(\frac{\nu}{\nu_c}\right)^{1/3} & \nu \ll \nu_c \\ \sqrt{\frac{243}{128\pi}}\left(\frac{\nu}{\nu_c}\right)^{1/2}\exp\left(-\nu/\nu_c\right) & \nu \gg \nu_c . \end{cases}$$
(9.16)

Thus, when $\nu/\nu_c \ll 1$, $P \simeq 1/2$, while in the limit of $\nu \gg \nu_c$, $P \simeq 1$. This is very substantial polarization indeed.

Because the polarization at a given frequency depends on electron energy, the net polarization after averaging over the electron energy distribution depends on the shape of that distribution. However, the narrowness of the electron energy range dominating the emission at any given frequency means that the net polarization is still substantial for almost any electron distribution function.

The main reason why the observed polarization is often closer to a few to 10% than $\simeq 1/2$ is the aperture-averaging effect already mentioned in §9.1.4. The more tightly tangled the field lines are relative to the image resolution, the smaller the net polarization we measure.

9.2.3 Estimating the magnetic field strength: the minimum energy argument

There is an important element of the story that we have ignored up to this point: the magnetic field. How can we estimate its strength? Unfortunately, in isolated optically thin sources, it is very difficult to do this. Given the extent and (comparatively) low surface brightness of the extended radio sources, we expect (and these expectations can be vindicated post hoc: §9.2.4) that extended sources are all optically thin and therefore all fall into this category.

However, as Burbidge (1958) pointed out, we can at least estimate the value of the magnetic field that minimizes the total energy in field plus electrons:

$$\mathcal{E} = \frac{B^2}{8\pi} + \int d\gamma \, \frac{dn_e}{d\gamma} \gamma m_e c^2. \tag{9.17}$$

Because typically $2 < \xi < 3$, the integral over the electron energy distribution is usually dominated by electrons with lower energies than the ones responsible for the frequencies we observe. Consequently, if we set the bounds of the integral to the range of electrons whose radio emission we actually see, we obtain only a *lower* bound on the total electron energy.

Despite the likelihood that the total energy may be considerably greater, it is conventional when applying this estimation method to be conservative and restrict the range of integration in the electron energy integral in exactly that way. The integral's upper and lower limits are simply $\gamma_s(\nu_u)$ and $\gamma_s(\nu_l)$, where we observe across the range of frequencies $\nu_l \leq \nu \leq \nu_u$. We can also write the normalization of the electron power law in terms of the observed emissivity per unit volume at frequency ν_l, $j_\nu(\nu_l)$. Because the energy integral is dominated by the low-energy end, we find that

$$U_e = \frac{6j_\nu(\nu_l)}{\sigma_T} \left(\frac{em_e c\nu_l}{2\pi}\right)^{1/2} B^{-3/2}.\qquad(9.18)$$

Here we have taken the δ-function approximation to the individual electron synchrotron spectrum. Adding U_e to the magnetic energy density $B^2/(8\pi)$ gives the total energy as a function of B, parameterized by the observed quantities j_ν and ν_l. Minimizing the total energy density with respect to B gives the field strength at which the total energy is least:

$$B_m = \left[\frac{36\pi j_\nu(\nu_l)}{\sigma_T} \left(\frac{em_e c\nu_l}{2\pi}\right)^{1/2}\right]^{2/7}.\qquad(9.19)$$

Because we have already obtained the electron energy density in terms of B, we can now find the minimum total energy density:

$$\mathcal{E}_m = \left[\frac{j_\nu(\nu_l)}{\sigma_T} \left(\frac{em_e c\nu_l}{2\pi}\right)^{1/2}\right]^{4/7} \left[\left(\frac{9}{16\pi}\right)^{4/7} + \pi^{-4/7}\right].\qquad(9.20)$$

The left-hand term in the sum comes from the magnetic field energy density and the right-hand term from the electron energy density. Clearly, to within factors of order unity, they are equal when the total energy is minimal. For this reason, B_m is often called the "equipartition" field.

In the absence of any better evidence, this minimum energy argument is often used to estimate the field. However, one should always bear in mind that this is merely a *minimum* energy; the field is not necessarily the one that allows the least total energy to be used. It is also important to bear in mind that the lowest frequency we observe is not necessarily the lowest frequency radiated. Because the total energy scales $\propto [j_\nu(\nu_l)\nu_l^{1/2}]^{4/7} \propto \nu_l^{(4/7)(0.5-\alpha)}$ and $\alpha > 0.5$, the equipartition energy estimate from the *observed* flux is generally less than the equipartition energy estimate from a hypothetical completely observed radio spectrum.

To obtain a sense of scale, we evaluate B_m for parameters typical of FR2 radio galaxies:

$$B_m \simeq 2.4 \times 10^{-6} \left(\frac{L_{43}}{r_{100}^3}\right)^{2/7} \nu_{l9}^{(1-2\alpha)/7} \text{ G},\qquad(9.21)$$

where L_{43} is the monochromatic luminosity at 1 GHz in units of 10^{43} erg s^{-1}, $\nu_{l9} = \nu_l/1$ GHz, and r_{100} is the lengthscale of the source in units of 100 kpc. Amusingly, the associated energy density is very similar to the energy density of the cosmic microwave background. In other words, the rate at which relativistic electrons lose energy by inverse Compton scattering the background photons is comparable to the rate at which they lose energy by synchrotron radiation. This effect is likely to be particularly important at high redshift because the energy density of the cosmological background scales $\propto (1+z)^4$.

Another way to put this energy density into perspective is to note that the minimum total energy stored in the radio emission region is

$$E_m \sim 2 \times 10^{57} L_{43}^{4/7} r_{100}^{9/7} \nu_{l9}^{(2-4\alpha)/7} \text{ erg.} \tag{9.22}$$

In other words, the minimum energy content of the radio emission region is equivalent to $\sim 3 \times 10^6$ yr of a typical galaxy's stellar luminosity.

If the field strength is a few μG (as suggested by the equipartition argument), the Lorentz factors of the electrons responsible for 1 GHz photons are $\sim 10^4$. The associated synchrotron cooling time is

$$t_{\text{cool}} \sim 2.4 \times 10^9 \gamma_4^{-1} B_{\mu\text{G}}^{-2} \text{ yr.} \tag{9.23}$$

In §9.3.3.3 we will see how to infer the lifetime of a source. We find there that typical ages are rather less than 10^8 yr, so that the electrons producing radiation at or below 1 GHz still have most of the energy they were injected with. Consequently, the mean *total* power injected into the source is substantially (at least an order of magnitude) greater than the observed radio luminosity. If the magnetic field differs from B_m, the total power put into the radio source must be even larger.

In addition, there must be reinjection and/or reacceleration of electrons at the energies responsible for frequencies not too much greater or there would no longer be radiation at those frequencies. We would also expect the spectrum to steepen by half a unit above the frequency at which the source age equals the electron cooling time (§9.2.1). In fact, just this sort of spectral steepening can sometimes be seen in the lobes of FR2 radio galaxies: The spectrum is flattest in the hot spots and gradually steepens with distance away from them, just as one might expect if the electrons are accelerated in the hot spots and then stream away, slowly losing their energy.

At the beginning of this section, we argued that inverse Compton scattering of cosmic microwave background photons could not explain the radio emission, for the photons produced would have energies that were too great. Given favorable observational circumstances, this fact can be turned around to provide an actual *measurement* of the magnetic field strength. We have just estimated the energy of the electrons responsible for making the observed radio emission ($\gamma \sim 10^4$); these electrons scatter microwave background photons to ~ 100 keV. The ratio between the X-ray luminosity produced by inverse Compton scattering and the radio luminosity produced by synchrotron radiation is, of course, simply the ratio of the microwave background energy density to the energy density in magnetic field. Combining equations 8.23 and 9.12, we predict that this ratio should be

$$\frac{L_{\epsilon_x}}{L_{\epsilon_r}} \sim \left(\frac{aT_{CMB}^4}{B^2}\right)\left(\frac{\hbar eB/m_ec}{k_B T_{CMB}}\right)^{(3-\xi)/2}\left(\frac{\epsilon_x}{\epsilon_r}\right)^{-(\xi-1)/2}. \tag{9.24}$$

If the X-ray flux is bright enough to detect, measurement of this ratio then permits inference of the field strength. In the lobes of the radio galaxy Fornax A, the inferred magnetic field is, in fact, close to B_m (Feigelson et al. 1995).

9.2.4 Optically thick emission

In compact sources, there is additional information we can bring to bear that allows us a more detailed view of the conditions inside synchrotron emitters. When synchrotron sources are sufficiently compact, they are also likely to be opaque. By analyzing the impact on the emergent spectrum of opacity as well as emission, we gain an additional constraint.

The most direct way of finding the synchrotron opacity is by making use of the well known relationship between the emission and absorption coefficients

$$\frac{g_l}{g_u}\sigma_{lu}c\mathcal{D}_\nu = A_{ul}, \tag{9.25}$$

where l, u refer to the lower and upper states, $g_{l,u}$ are the statistical weights of the two states, σ_{lu} is the cross section for the lower state to absorb a photon and arrive in the upper state, $\mathcal{D}_\nu$ is the photon density of states, and A_{ul} is the Einstein spontaneous emission coefficient. This relationship can be viewed equally well as a consequence of Kirchhoff's law and thermodynamics, or as a consequence of the time symmetry of quantum mechanical transition matrices. The net absorption coefficient requires a further adjustment to allow for stimulated emission:

$$\sigma_{\text{net}} = \sigma_{lu}\left(1 - \frac{n_u/g_u}{n_l/g_l}\right). \tag{9.26}$$

Translating these expressions into the language of synchrotron radiation yields a net absorption rate per unit length

$$\rho\kappa_\nu = \frac{c^2}{8\pi h\nu^3}\frac{1}{m_e c^2}\int d\gamma\, P_\nu(\gamma)\gamma^2 h\nu\frac{d}{d\gamma}\left(\gamma^{-2}\frac{dn_e}{d\gamma}\right). \qquad (9.27)$$

In the extreme relativistic limit, the electron density of states is $\propto \gamma^2$, so the quantity $\gamma^{-2}dn_e/d\gamma$ is proportional to the electron occupation number per state. The difference in occupation number between the upper and lower states for the radiation event is the product of the energy of the radiated photon and the gradient with respect to energy in the electron occupation number.

If $dn_e/d\gamma = K\gamma^{-\xi}$, the integral in equation 9.27 can be directly computed (Blumenthal and Gould 1970). Again assuming an isotropic pitch angle distribution, it is

$$\rho\kappa_\nu = A(\xi)K\frac{e}{B}\left(\frac{2\pi\nu}{eB/m_e c}\right)^{-(\xi+4)/2}, \qquad (9.28)$$

where, as shown by Blumenthal and Gould (1970),

$$A(\xi) = \frac{\pi^{3/2}}{12}3^{(\xi+3)/2}\frac{\Gamma\left(\frac{3\xi+2}{12}\right)\Gamma\left(\frac{3\xi+22}{12}\right)\Gamma\left(\frac{\xi+6}{4}\right)}{\Gamma\left(\frac{\xi+8}{4}\right)}. \qquad (9.29)$$

The quantity $A(\xi) \sim 10$; for example, $A(2.5) \simeq 10.4$. Because $\kappa_\nu \propto \nu^{-(\xi+4)/2}$ and $\xi \simeq 2$–3, synchrotron self-absorption declines very rapidly with increasing frequency.

Further insight into the nature of synchrotron self-absorption can be gleaned by estimating the cross section per electron, whose most natural definition is

$$\sigma_e = \frac{(d/d\gamma)(\rho\kappa_\nu)}{dn_e/d\gamma}. \qquad (9.30)$$

Differentiating the right-hand side of equation 9.27 with respect to γ then gives

$$\sigma_e \simeq \frac{4\pi^2}{9}(\xi+2)\frac{e}{B}\gamma^{-5}\left(\frac{\nu}{\nu_c}\right)^{-2}F\left(\frac{\nu}{\nu_c}\right). \qquad (9.31)$$

One interpretation for this cross section is that it is $\sim \gamma^{-6}r_L r_e$, that is, γ^{-6} times the product of the electron Larmor radius r_L and the classical electron radius r_e. Clearly, more energetic electrons rapidly become less effective at synchrotron self-absorption. A dependence on the shape of the electron distribution function persists in this estimate of the cross section for an individual electron. Although at first glance this might seem strange, it stems from the importance of the stimulated emission correction, which depends, of course, on the shape of the electron distribution function.

Applying the estimated numbers for the extended portions of FR2 radio galaxies, we find that the cross section in the frequency range of interest is coincidentally $\sim \sigma_T$, that is, $\sim 10^{-24}$ cm^2. If the equipartition assumption applies, the optical depth across a region r across is

$$\tau_{\text{eq}} \sim \frac{\pi^2}{36}(\xi+2)\frac{r}{c}\left(\frac{eB_m}{2\pi m_e c}\right)^4 \nu_l^{-3}\left(\frac{\nu}{\nu_l}\right)^{(4+\xi)/2}. \tag{9.32}$$

Incorporating our earlier numerical estimate of B_m (and setting $\xi = 2.5$), we find

$$\tau_{\text{eq}} \sim 5 \times 10^{-11} L_{43}^{8/7} r_{100}^{-17/7} \nu_{l9}^{-25/7}\left(\frac{\nu}{\nu_l}\right)^{-(4+\xi)/2}. \tag{9.33}$$

We thus confirm our original guess that the giant extended lobes are quite optically thin.

With such strong scaling in r the situation changes dramatically in compact cores. Although these regions can have luminosities nearly as large as the extended sources, their sizes (as measured by VLBI) are 10^4 times smaller. As a result, their equipartition field strengths are larger by $\sim 10^{3.4}$, and the energy of the "responsible" electrons decreases by $10^{1.7}$. Using the scaling just derived, we find $\tau_{\text{eq}}(\nu_l) \sim 0.3 L_{43}^{8/7} r_{100}^{-17/7} \nu_{l9}^{-25/7}$. If equipartition has anything to do with the true physics, synchrotron self-absorption should definitely be apparent in the spectra of these sources, particularly if $\nu_l < 1$ GHz or $L > 10^{43}$ erg s^{-1}.

When a source is optically thick, its intensity I_ν deep inside is equal to the source function $j_\nu/(4\pi\rho\kappa_\nu)$. Applying that relation to self-absorbed synchrotron sources gives

$$I_\nu = \frac{1}{4\pi}\frac{E(\xi)}{A(\xi)}\frac{e^2 B^2}{m_e c^2}\left(\frac{2\pi m_e c\nu}{eB}\right)^{5/2}. \tag{9.34}$$

Because self-absorption leads to a saturation in the intensity, the surface intensity is not much less than the interior intensity. Thus, we expect that a self-absorbed synchrotron source should have a surface brightness $\propto B^{-1/2}\nu^{5/2}$. We can also rewrite the expression for the intensity found in equation 9.34 in a physically more transparent form:

$$I_\nu = \pi\frac{E(\xi)}{A(\xi)}\frac{\gamma_s(\nu)m_e c^2}{\lambda^2}. \tag{9.35}$$

That is, the brightness temperature at wavelength $\lambda = c/\nu$ is of order the energy of the electrons responsible for frequency ν. We began this derivation by calling upon an identity that can be derived from thermodynamic equilibrium; this last result should not therefore be too surprising.

At frequencies low enough for the source to be optically thick, the spectrum rises $\propto \nu^{5/2}$; on the other hand, at frequencies high enough for the source to be optically thin, it falls $\propto \nu^{-\alpha}$. In between, the spectrum should have a well defined peak near the frequency

$$\nu_t = \frac{eB}{2\pi m_e c} \left[\frac{(\xi - 1)A(\xi)e\tau'_T}{B\sigma_T} \right]^{2/(\xi+4)}$$
$$= 2.4 \times 10^{11} B^{0.692} \tau'^{0.308}_T \text{ Hz}, \tag{9.36}$$

for which the optical depth is unity. Here τ'_T is the electron scattering optical depth across a radius of the source if the electron distribution function continues with constant slope down to $\gamma = 1$. In the numerical evaluation of equation 9.36, as well as the rest of the equations of this section, we set $\xi = 2.5$. We can estimate the brightness temperature at ν_t by supposing that the surface brightness is equal to the source function (because the optical depth is only unity at ν_t, this is a bit of an overestimate):

$$T_{bt} = \frac{\pi}{2} \frac{E(\xi)}{A(\xi)} \frac{m_e c^2}{k_B} \left[\frac{(\xi - 1)A(\xi)e\tau'_T}{B\sigma_T} \right]^{1/(\xi+4)}$$
$$= 3.8 \times 10^{11} \left(\frac{\tau'_T}{B} \right)^{0.154} \text{ K}. \tag{9.37}$$

Thus, the brightness temperature at the peak of the spectrum scales in proportion to a very small power of τ'_T/B. If the dynamic range of τ'_T/B is not too great, we can expect most sources to have very similar peak brightness temperatures.

Spectra of compact radio cores are generally much flatter than those of extended sources, but none has ever shown a spectrum rising as sharply as $\nu^{5/2}$. $F_\nu \simeq constant$ is more typical. Nonetheless, they have generally been interpreted as self-absorbed synchrotron sources, with the spectral discrepancy attributed to inhomogeneity (see §9.3.4.1). This interpretation is a fairly natural one because the spectrum of each component peaks so sharply at ν_t, and ν_t rises with increasing τ'_T and B. One might then imagine that regions of progressively greater optical depth and magnetic field produce higher frequency portions of the spectrum. This is illustrated in figure 9.9, in which three separate optically thick regions combine to produce a spectrum approximately flat over an order of magnitude in frequency.

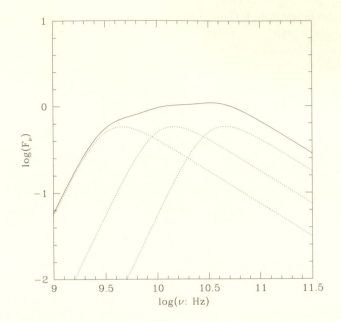

Fig. 9.9   A "flat" spectrum made by superposition. Each of the dotted curves represents the emisison from a single region; the solid curve is their sum. Each region has the same $F_\nu(\nu_t)$, but the values of ν_t are 3.16, 10, and 31.6 GHz.

With this interpretation, it is possible, using equations 9.36 and 9.37, to work backwards from the observations to find the physical conditions inside the source. Suppose, then, that we wish to find the optical depth and magnetic field in the region responsible for the flux at the frequency we designate ν_t. Suppose, too, that VLBI measurements give us the brightness temperature at that frequency, T_{bt}. Then equations 9.36 and 9.37 can be solved for the magnetic field and optical depth of the emitting plasma:

$$B = \frac{2\pi m_e c}{\nu_t}\left[\frac{\pi m_e c^2}{2k_B T_{bt}}\frac{E(\xi)}{A(\xi)}\right]^2$$
$$= 5.8 \times 10^{-4}\nu_{tGHZ}T_{bt12}^{-2}\ \text{G},$$

(9.38)

and

$$\tau_T' = \frac{16\pi^2}{3(\xi-1)A(\xi)}\frac{e^2\nu_t}{m_e c^3}\left[\frac{2k_B T_{bt}}{\pi m_e c^2}\frac{A(\xi)}{E(\xi)}\right]^{\xi+2}$$
$$= 0.34 T_{bt12}^{4.5}\nu_{tGHz}.$$

(9.39)

Once again, the numerical values pertain to $\xi = 2.5$.

280

There are some subtleties in carrying out this program. First, if the observations at hand do not properly resolve the source, the measured T_{bt} is, of course, only a lower bound. Second, the resolution of interferometric mapping depends on frequency, so there can be objects that are resolved at high frequencies but not at low, leaving the determination of ν_t somewhat ambiguous. On the other hand, if the overall flat spectrum is due to a superposition of regions, we might expect the angular scale of the source to grow smaller at higher frequencies. Finally, τ'_T depends very strongly on T_{bt}, so any uncertainty in measurement of the brightness temperature at the turnover frequency leads to a much greater uncertainty in τ'_T. This fact is simply a corollary to the insensitivity of the predicted T_{bt} to τ'_T (eq. 9.37).

Despite these qualms, it is still possible to at least make estimates. Core-dominated sources have spectra that are relatively flat (i.e., $\alpha < 0.5$) over a range of frequencies that may extend from 1–100 GHz, so that if we adopt the hypothesis of inhomogeneity, there are regions with ν_t anywhere in that range. Because $T_b \propto F_\nu / \nu^2$, the greatest brightness temperature in a flat spectrum is at the lowest frequencies. Modulo the cautions expressed in the previous paragraph, the great majority of peak brightness temperatures fall between 10^{11} and 3×10^{12} K (fig. 9.10), very nicely in line with the estimate from theory. One then finds optical depths τ'_T in the range 0.01–10 and magnetic fields $\sim 10^{-4}$–10^{-3} G.

On the basis of these rough inferences, we can estimate the energy requirements for self-absorbed synchrotron sources. The electrons we know about directly have $\gamma \sim 600(\nu_{\rm GHz}/B_{-3})^{1/2}$; therefore, to find the total energy we must guess about how the electron distribution function extrapolates outside the observed range. If we suppose that it has the same logarithmic slope down to a low-energy cutoff at $\gamma_{\min}$, then the total energy density is

$$U_e = 0.40 \frac{\xi - 1}{\xi - 2} \tau'_T r_{pc}^{-1} \gamma_{\min}^{2-\xi} \text{ erg cm}^{-3}. \tag{9.40}$$

The ratio of the magnetic energy density U_m to U_e depends exceedingly strongly on the measured peak brightness temperature:

$$\frac{U_m}{U_e} = 3.3 \times 10^{-8} T_{bt12}^{-8.5} \nu_{t\rm GHz} r_{pc} \gamma_{\min}^{0.5}, \tag{9.41}$$

where the numerical values assume $\xi = 2.5$. Despite the very small coefficient on the right-hand side of equation 9.41, a modest decrease (a mere factor of 8) in T_{bt} could bring the magnetic energy density close to equipartition. In fact, as we shall see later in this chapter (§9.3.4.3), just such a

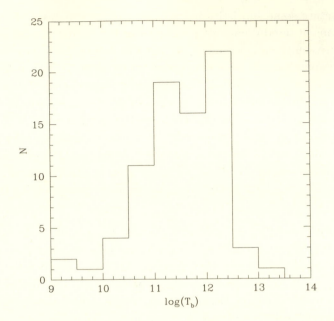

Fig. 9.10 The distribution of observed peak brightness temperatures in a large but heterogeneous sample of core-dominated radio-loud AGNs compiled by Ghisellini et al. (1993).

reduction may occur when we reconsider the energetics of these sources *in their own rest frames*.

9.2.5 Spontaneous and stimulated Compton scattering in optically thick synchrotron sources

Relatively weak magnetic field in a region of high photon density suggests that inverse Compton scattering may be important. The ratio of inverse Compton luminosity to synchrotron luminosity is simply the ratio of photon energy density to magnetic energy density. Phrased in terms of observables, this ratio is

$$\frac{U_{\rm ph}}{U_m} = \frac{256}{\pi^4} \frac{e^2 \nu_t}{m_e c^3} \left[\frac{A(\xi)}{E(\xi)}\right]^4 \left(\frac{k_{\rm B} T_{bt}}{m_e c^2}\right)^5 \left\{\frac{2}{7} + \frac{2}{3-\xi}\left[\left(\frac{\nu_{\max}}{\nu_t}\right)^{(3-\xi)/2} - 1\right]\right\},$$
$$(9.42)$$

where $\nu_{\max}$ is the high-frequency cutoff of the optically thin synchrotron spectrum. When $\xi < 3$, so that $\alpha < 1$, the *high-frequency end* of the synchrotron spectrum dominates the total emitted power. No matter what the shape of the synchrotron power law is, the ratio of inverse Compton to

282

synchrotron luminosity is a very steep function of the brightness temperature at the turnover. Substituting the actual coefficients indicates that the inverse Compton luminosity surpasses the synchrotron luminosity when $T_{bt} \gtrsim 10^{12}$ K. It was on this basis that Kellermann and Pauliny-Toth (1969) suggested an explanation (often called the *inverse Compton catastrophe*) for the narrow distribution of observed brightness temperatures: that if a source were made with such high electron energy density and magnetic field that its brightness temperature exceeded this limit, the energy would be rapidly drained from the electrons by inverse Compton scattering. The radio brightness temperature would then fall back to the commonly observed level.

This explanation is, however, somewhat unsatisfying. We do not know anything about the particle acceleration processes in these sources, so it is hard to say what they could or could not support. In principle there is no reason why we should not see high brightness temperature compact radio sources that are associated with much higher luminosity X-ray sources—and, in fact, we do. As we shall discuss shortly (§9.3.4.2), AGNs with bright compact synchrotron cores very often emit far more power in X-rays and γ-rays than at radio frequencies, and it is likely that the high-energy photons are created as a result of inverse Compton scattering by the very same relativistic electrons generating radio photons by the synchrotron mechanism.

Several alternative explanations may be more relevant to the brightness temperature limit. One is simply that T_{bt} depends so weakly on the physical parameters that even a wide range of intrinsic τ'_T/B may not be enough to create many sources with greater brightness temperature (see equation 9.37). Another (Readhead 1994) is that these sources are moving toward us at relativistic speeds (as is likely for numerous other reasons: §§8.8.2, 9.3.3.2, 9.3.4.3), so that relativistic boosting enhances the observed brightness temperature by the order of magnitude or so required for the magnetic field to be in equipartition. In that case, it is whatever physics regulates the field strength to be in equipartition, in combination with the regulation of the relativistic bulk dynamics, that controls the distribution of observed brightness temperature.

There is also another possibility, the effects of stimulated Compton scattering. We already know that the photon occupation number $\mathcal{N}$ inside compact radio sources is very high, for $\mathcal{N} = k_B T_b/(h\nu) \sim 10^{13} T_{b12}/\nu_{GHz}$. Consequently, the Bose-Einstein scattering rate enhancement factor $1 + \mathcal{N}$ is a very large number indeed. Thus, one might expect this process to be important anywhere free electrons are exposed to the radiation from self-absorbed synchrotron sources.

Its effect, however, is subtler than the crude enhancement factor indicates. Suppose that photon scattering (e.g., by the Compton mechanism) were exactly elastic in the electron rest frame (and for radio frequency photons, that's a pretty good approximation). Then the net rate at which photons of energy ϵ leave direction $\hat{n}_1$ to go to direction $\hat{n}_2$ is

$$
\begin{aligned}
\frac{d\mathcal{N}_1}{dt} &= -n_e \frac{\partial \sigma_{\mathrm{T}}}{\partial \Omega} \left(\hat{n}_1 \cdot \hat{n}_2 \right) c \left[\mathcal{N}_1 \left(1 + \mathcal{N}_2 \right) - \mathcal{N}_2 \left(1 + \mathcal{N}_1 \right) \right] \\
&= -n_e \frac{\partial \sigma_{\mathrm{T}}}{\partial \Omega} \left(\hat{n}_1 \cdot \hat{n}_2 \right) c \left(\mathcal{N}_1 - \mathcal{N}_2 \right) ;
\end{aligned}
\tag{9.43}
$$

that is, the stimulated part of the scattering identically cancels.

This is not the end of the story, however, for Compton scattering is not exactly elastic. When the electrons are initially at rest, the photons lose energy on average at the rate of $\epsilon^2/(m_e c^2)$ per scatter. Rewriting the previous equation to take into account the recoil losses, and focusing attention on photons with energy ϵ traveling in any direction, we find

$$
\begin{aligned}
\frac{d\mathcal{N}(\epsilon)}{dt} =& n_e \sigma_{\mathrm{T}} c \int d\epsilon' \; - \mathcal{N}(\epsilon) \left\{ 1 + \mathcal{N} \left[\epsilon \left(1 - \frac{\epsilon}{m_e c^2} \right) \right] \right\} \delta \left[\epsilon' - \epsilon_f(\epsilon) \right] \\
& + \frac{\mathcal{D}\left[\epsilon \left(1 + \frac{\epsilon}{m_e c^2} \right) \right]}{\mathcal{D}(\epsilon)} \mathcal{N} \left[\epsilon \left(1 + \frac{\epsilon}{m_e c^2} \right) \right] \left[1 + \mathcal{N}(\epsilon) \right] \delta \left[\epsilon - \epsilon_f(\epsilon') \right] ,
\end{aligned}
\tag{9.44}
$$

where $\mathcal{D}(\epsilon)$ is the density of photon states at energy ϵ and $\epsilon_f(\epsilon)$ is the mean final energy after Compton scattering for a photon with initial energy ϵ. There is no need for a density of states ratio in the term describing photons scattering out of energy ϵ because the rate of scattering for them is simply proportional to the number of photons per state at energy ϵ times the usual Bose-Einstein factor. On the other hand, there is a correction factor in the rate of photons scattering *into* energy ϵ because for equal $\mathcal{N}$ the total number of photons starting out at energy $\epsilon(1 + \epsilon/m_e c^2)$ is greater than the number at energy ϵ by the ratio of their densities of states. For $\epsilon \ll m_e c^2$, we can expand both $\mathcal{D}$ and $\mathcal{N}$ to first order, so that

$$
\frac{d\mathcal{N}(\epsilon)}{dt} = 2 n_e \sigma_{\mathrm{T}} c \mathcal{N}^2(\epsilon) \frac{\epsilon}{m_e c^2} \left(2 + \frac{d\ln \mathcal{N}}{d\ln \epsilon} \right) .
\tag{9.45}
$$

Now it is apparent that unless the spectrum has a special form ($\mathcal{N} \propto \epsilon^{-2}$, i.e., $F_\nu \propto \nu$), the effects of stimulated scattering do not identically cancel (the $\mathcal{N}^2$ term in the Kompaneets equation, eq. 8.30, can be derived from just this sort of calculation).

Relative to spontaneous scattering, stimulated scattering occurs at a rate $\sim \mathcal{N}\epsilon/(m_e c^2) \sim k_B T_b(\epsilon)/(m_e c^2)$. Thus, it becomes important when the brightness temperature $T_b > m_e c^2/k_B \simeq 5 \times 10^9$ K. But the conventional analysis of optically thick synchrotron sources predicts that in *any* such source, the brightness temperature reaches as high as $\sim \gamma_s(\nu_t) m_e c^2/k_B$, so stimulated scattering *always* has a greater impact on the spectrum near the synchrotron peak than spontaneous scattering does. In the models we have just worked out, stimulated scattering is more important by a factor ~ 100 (Wilson and Rees 1982). Put another way, the very high brightness temperature of self-absorbed synchrotron sources makes it very easy for them to "light up" even small amounts of surrounding gas, for its effective scattering opacity is increased by that same factor ~ 100 (Coppi, Blandford, and Rees 1993).

Before analyzing the impact of stimulated scattering *inside* the synchrotron source region, a further subtlety must be exposed: the equation for the rate of change of the photon occupation number was derived in the electron rest frame, yet inside the source, the electrons we know about are highly relativistic. The Lorentz transformations in and out of the electron rest frames introduce a number of correction factors that result in a diminution of the strength of the effect for highly relativistic electrons (Sincell and Krolik 1994). When the occupation number spectrum is sharply peaked, the most important of these is due to the fact that Compton scattering is very nearly elastic in the rest frame of the electron. That means when a photon whose energy in the rest frame places it near the peak of the occupation number spectrum encounters an electron of energy γ, it is only strongly induced to scatter by those photons traveling within a solid angle $\sim \gamma^{-2}$ of the photon's initial direction. Roughly speaking, then, the efficiency per electron of stimulated scattering is reduced by $\sim \gamma^{-2}$.

Because the consequences of stimulated scattering all have their origin in electron recoil, its net effect is always to drive photons to lower energies (and conversely to push electrons to higher energies). For this reason, in some respects it can be thought of as an absorption mechanism. When the brightness temperature is high, it can, in fact, be very powerful. Defining its effective opacity by $\sigma_T \gamma_{\min}^{-2} k_B T_b/(m_e c^2)$, one finds that the ratio of stimulated scattering opacity to synchrotron self-absorption opacity is $\sim 4 \times 10^4 \gamma_{\min}^{-(\xi+1)} \nu_9^{\xi/2+2} B_{-4}^{-1+\xi/2} T_{b12}$. The extra factor of $\gamma_{\min}^{1-\xi}$ comes from the fact that synchrotron opacity is produced by the electrons we see directly at γ_s, whereas stimulated scattering opacity is dominated by the more numerous lowest energy electrons. If the classic self-absorbed synchrotron spectrum obtains—and if there are enough low-energy electrons—stimulated scattering completely dominates synchrotron self-absorption near the turn-over peak but falls rapidly in importance toward lower frequencies.

Overall, the effect of stimulated scattering is to shave down the height of the turnover peak by driving photons to lower frequencies, where they are ultimately absorbed by synchrotron opacity. The equilibrium level in the turnover region is then set by a balance between stimulated scattering opacity and synchrotron emission:

$$\kappa_{stim,eff} I_\nu = j_\nu. \tag{9.46}$$

The only frequency dependence of the stimulated scattering effective opacity is through $T_b \propto I_\nu/\nu^2$, so if $I_\nu \propto \nu^{-\alpha'}$ when $j_\nu \propto \nu^{-\alpha}$, we have $\alpha' = (\alpha - 2)/2$. For $0.5 \leq \alpha \leq 1$, stimulated scattering will flatten the spectrum to an index between -0.75 and -0.5, that is, F_ν *rises* slowly with frequency. This equilibrium continues up to frequencies high enough that the brightness temperature (which still falls with increasing frequency) is small enough to make the source effectively optically thin. This last criterion means, of course, that $\tau_T \gamma_{min}^{-2} k_B T_b/(m_e c^2)$ must be < 1. If γ_{min} is not too large, the flat peak in the spectrum extends to frequencies high enough that the brightness temperature has fallen by one to two orders of magnitude from its peak. More precise radiation transfer calculations (Sincell and Krolik 1994; fig. 9.11) confirm this rough estimate. It is possible, then, that stimulated scattering explains the observed upper limit on the brightness temperature of these sources and also makes it easier to understand their flat spectra.

9.3   Jets

So far we have deliberately avoided the questions of what shapes the radio emission regions, how the electrons in those regions get their energy, or, for that matter, what determines how much energy goes into radio emission in the first place. Now it is time to ask, and attempt to answer, those questions.

Many years ago, when giant double radio sources were first discovered, it was thought that the lobes were a pair of plasma blobs that had been expelled from the central galaxy in a single explosion. However, that thought fell into disfavor when it was realized how difficult it would be to supply so much energy in a single event. When Blandford and Rees (1974) constructed an example of a mechanism that might channel plasma out to the

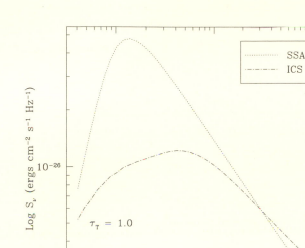

Fig. 9.11 A comparison (taken from Sincell and Krolik 1994) of the predictions of simple self-absorbed synchrotron theory (dotted curve) and the full theory including stimulated Compton scattering (dot-dashed curve). The parameters chosen are fairly typical for those inferred for compact radio cores in AGNs, but with the important assumption that $\gamma_{min} = 2$; for these numbers, stimulated scattering reduces the peak flux by a factor of four, pushes it to a frequency higher by a factor of three, and flattens the spectral peak so that it stretches across a decade in frequency.

lobes in a continuous fashion, and later observations with the Cambridge 5 km telescope and the VLA discovered linear structures on small distance scales, a firm consensus formed around a picture in which energy is continuously transported from the nucleus out to the radio emisison regions.

However, the actual mechanics of this energy supply still remain quite murky. That jets are visible in so many sources clearly favors some sort of continuous feed, but its nature is not at all well known. Such basic issues as the content of the stream (Is it normal plasma or a pair plasma? What is the ratio of magnetic flux to particles?), the mass density, and the velocity all remain unresolved. Even less is known about their source of power and the forces that collimate them. At the moment, the most popular schemes generally involve MHD winds driven by rapidly rotating black holes, but these are still quite speculative. As usual, the most fundamental issues are the least understood: why do jets exist at all? And why, comparing different AGNs, are there such vast disparities in jet power relative to the total luminosity of the central engine?

9.3.1 Content

It is difficult to say whether the jet material is a normal or a pair plasma. If we normalize to the observed radio emission in the jet itself, the least energy is required by a jet composed of only electrons and positrons. In fact, for the very least energy, there would be no electrons in the jet with energies less than the minimum needed to explain the observed radio emission. Neutralizing the electrons' electric charge with protons increases the total energy by a factor $[(2 - \xi)/(1 - \xi)]m_p/(\gamma_{\min}m_e)$. On the other hand, if we normalize by the total momentum and energy fluxes required to sustain the outer structure, the total energy is fixed independently of the positron/proton ratio. One might think that we could test for positron content by searching for annihilation radiation, but the predicted fluxes are well below current upper limits.

Another open question is the origin of the relativistic electrons that create the radiation we see. Just as in the case of the sources of X-rays and γ-rays discussed in Chapter 8, little is known about how such large quantities of very energetic electrons are produced; instead, once again shocks and the other mechanisms of §8.6 are invoked.

Somewhat stronger diagnostics exist for the magnetic field. We have already shown how to estimate the magnitude of the magnetic field in regions that are optically thick to synchrotron self-absorption, and how to find a minimum energy estimate for the magnitude of the field in optically thin regions. One might expect the field geometry to be controlled by the jet motion itself. Even if the field in the inner part of the jet were tangled, magnetic flux conservation would make $B_\parallel \propto (r\theta)^{-2}$, where r is distance along the jet and θ is its opening angle, while $B_\perp$ would be $\propto r^{-1}$. Consequently, at distances large compared to the launch radius of the jet, one might expect the field to be predominantly perpendicular to the jet axis. Polarization angle measurements are consistent with this sort of scaling in lower luminosity jets. However, at higher luminosities, the field direction appears to stay largely longitudinal far away from the nucleus. Many mechanisms could potentially complicate the simple argument just presented (and possibly account for its failure to correctly predict the observed field directions). For example, if the initial field configuration contains some transverse loops, and there is a velocity shear across the jet, axial field would be created as the loops are stretched.

A second diagnostic arises from Faraday rotation of the synchrotron radiation's plane of polarization (§9.1.4). Unfortunately, this provides only a *lower* limit to the field strength (if the electron density is known independently) because the field can double back and forth along the line-of-sight. In addition, if charge neutralization is accomplished by positrons rather than protons, the Faraday rotation is canceled. And, of course, the plasma screen could be located somewhere along the line-of-sight and not inside the synchrotron source at all.

9.3.2 Acceleration and collimation

In principle one could imagine that accretion onto a black hole occurs without any outflow at any point. Put another way, we do not know why jets exist. However, granted that we see them, it is also possible to imagine that part of the accretion energy or some of the rotational energy of the black hole is tapped to expel matter and Poynting flux outward. We might then ask, "Under what circumstances does this occur? What controls the rate of mass loss? What determines its shape and speed?" None of these questions can be answered with any confidence at present. However, it is possible to identify some of the underlying issues. The key points are that any candidate mechanism must, at least potentially, be able to accelerate material to bulk Lorentz factors ~ 10 and collimate the flow to angles ~ 0.1 radians. If those criteria aren't met, there is no way the proposed mechanism could account for images such as figures 9.4–9.6.

The first force one might look to in order to drive mass loss is ordinary gas pressure gradients. Anywhere gas is heated to a temperature high enough that the mean energy of particles approaches their gravitational binding energy, substantial mass loss will occur. There are several plausible ways this might occur in AGNs. The continuum spectrum created in the central regions has a large enough contribution from high-energy photons that any gas in Compton equilibrium with it (see §§8.3.1 and 10.5.1) will have a temperature near or above the escape temperature at radii outside $\sim 10^3 r_g$ (Begelman, McKee, and Shields 1983). Alternatively, as we have already seen (§8.4.3), if the X-ray-emitting region contains a pair-dominated plasma, and if the pairs are not coupled by wave-particle scattering to normal plasma, they cannot be gravitationally bound.

Collimation by gas pressure forces is also possible, though perhaps not likely. The original scheme suggested by Blandford and Rees (1974) used external gas pressure to create a "de Laval nozzle" through which hot gas might be channeled outward into a supersonic flow. However, subsequent work (e.g., VLBI images) showed that jets are collimated on very small scales. Consequently, the collimating gas must have very high pressure and therefore would cool rapidly. Even if it is reheated at a sufficient rate, the luminosity associated with this cooling would be visible in the observed spectrum, and its absence rules out the model.

One might next try radiation forces. As is also discussed in §8.4.3, even if the pairs are well coupled to normal plasma, the Eddington limit can be substantially diminished when there are large numbers of pairs, making the creation of a radiation-driven wind much easier. This, too, however, seems not to be the answer. The most fundamental problem is that once the flow reaches mildly relativistic speeds, a medley of effects make further acceleration by a directed component of radiation extremely inefficient, while any isotropic radiation acts as a source of drag (Phinney 1987).

The simplest way to see these effects is to compute the mean change in energy of a photon scattered by a relativistic electron (cf. §8.2.2.1):

$$\Delta k = k_i \left[1 - \gamma^2 (1 - \vec{\beta} \cdot \hat{n}_i)(1 + \vec{\beta} \cdot \hat{n}'_f) \right], \tag{9.47}$$

where $k_{i,f}$ are the initial and final photon energies, $\hat{n}_{i,f}$ the initial and final photon directions, and primed quantities refer to the electron's initial rest frame, while unprimed quantities are measured in the lab frame. In the Thomson limit, the angular distribution of scattered photons is forward-backward symmetric in the electron rest frame, $\langle \vec{\beta} \cdot \hat{n}'_f \rangle = 0$. The rate coefficient for scattering is $(1 - \vec{\beta} \cdot \hat{n}_i)\sigma_T c$, so the electron's energy changes according to

$$
\begin{aligned}
m_e c^2 \frac{d\gamma}{dt} &= \int d\nu \int d\Omega I_\nu(\nu, \hat{n}_i) \sigma_T (1 - \vec{\beta} \cdot \hat{n}_i) \left[1 - \gamma^2 (1 - \vec{\beta} \cdot \hat{n}_i) \right] \\
&= \int d\nu \int d\Omega I_\nu(\nu, \hat{n}_i) \sigma_T \beta \gamma^2 \left(1 - \vec{\beta} \cdot \hat{n}_i \right) \left(\hat{n}_i \cdot \hat{\beta} - \beta \right),
\end{aligned}
\tag{9.48}
$$

where $\hat{\beta}$ is a unit vector in the direction of the velocity. The second form of the equation illustrates the fact that unless the photons are directed sufficiently parallel to the velocity, they *decelerate* the flow rather than *accelerate* it. The problem is that unless they push directly parallel to the motion, the momentum the photons gain as a result of scattering exceeds what they deliver.

To see this point another way, suppose that the frequency-integrated angular distribution of the radiation can be decomposed into a directed component and an isotropic component:

$$I(\hat{n}_i) = \frac{F}{2\pi}\delta(\hat{n}_i - \hat{z}) + J_*. \tag{9.49}$$

For this angular distribution, the mean intensity $J = J_* + F/2$. Evaluating the angular integral in equation 9.48 yields

$$m_e c^2 \frac{d\gamma}{dt} = \sigma_{\mathrm{T}} \left[\frac{\beta(1-\beta)}{1+\beta} F - \frac{16\pi}{3} \beta^2 \gamma^2 J_* \right]. \tag{9.50}$$

In the limit that $\beta \to 1$, this becomes

$$m_e c^2 \frac{d\gamma}{dt} = \sigma_{\mathrm{T}} \left(\frac{1}{4\gamma^2} F - \frac{16\pi}{3} \gamma^2 J_* \right). \tag{9.51}$$

Thus, an equilibrium is reached at

$$\gamma = \left(\frac{3}{64\pi} \frac{F}{J_*} \right)^{1/4}. \tag{9.52}$$

As this form makes clear, the ratio F/J_* would have to be $\sim 10^4$ (i.e., the radiation would have to be extremely well collimated) in order for radiation to drive a flow with $\gamma \sim 10$. If some nuclei are mixed in with the electron-positron pairs, equilibrium is achieved at the same γ but is approached more slowly.

We do not know enough about the details of the radiation distribution deep inside AGNs to specify F/J_* as a function of position, but we might well make the plausible guess that it is ~ 1 near the black hole, increases as one moves outward, but then decreases at larger radii where significant fractions of the intrinsic luminosity are reprocessed (e.g., in the broad emission line region: Chap. 10; and in an obscuring dust torus: Chap. 12). Achieving a ratio $\sim 10^4$ anywhere seems quite unlikely; in fact, radiation is more likely to *decelerate* a jet accelerated by some other means than it is to *accelerate* the jets we see.

A third possibility is magnetic forces, that is, the relativistic generalization of the sort of MHD wind we considered briefly in §7.2.2 as a torque mechanism for accretion disks. Instead of focusing on the angular momentum loss, we now concentrate on the wind itself. MHD processes have a number of advantages for creating relativistic flows, starting with the fact that if the energy flux is dominated by Poynting flux rather than kinetic energy associated with matter, the speed is automatically $\sim c$. In addition, if the magnetic field energy density at the base of the wind (presumably near the center of the accretion disk) is in equipartition with the total pressure there, it automatically has the energy density of the radiation, and so would carry a luminosity comparable to the radiative luminosity if it moves at c.

Collimation also seems to be a generic feature of MHD winds, although exactly how much is difficult to specify. The basic reason for this tendency is that one expects the footpoints of magnetic field lines to be fixed in the matter of the accretion disk. The field lines will then rotate with the orbital frequencies of their footpoints, creating toroidal field even if there was none to begin with. Anywhere the field is predominantly toroidal, magnetic "hoop stress" exerts an inward force. Consider a simple model in which $\vec{B} = B(r,z)\hat{\phi}$, where r is now cylindrical radius. Equation B.26 can then be written in the form

$$\rho\left[\frac{\partial \vec{v}}{\partial t} + \vec{v}\cdot\nabla\vec{v}\right] = -\frac{B^2}{4\pi r}\left[\left(\frac{1}{2} + \frac{\partial \ln B}{\partial \ln r}\right)\hat{r} + \frac{r}{z}\frac{\partial \ln B}{\partial \ln z}\hat{z}\right] - \nabla p + \vec{F}_{\text{ext}}.$$

$$(9.53)$$

Even in the absence of field gradients, the net magnetic force has a component pushing toward the axis because plasma motion inward, carrying the field lines along, allows the total field energy to decrease.

Toroidal field with a consistent sense of winding implies (through Ampère's law) that within a cylindrical radius r of its center the jet must carry a net current $I(r) = (cr/2)B_\phi(r)$. A return current must then exist somewhere, but it is very hard to pin down whether that current flows just outside the jet or very far away.

Because the return current flows back to the central engine via different field lines than the outflowing current, the current must eventually cross field lines, that is, there must be enough dissipation somewhere to allow the necessary breakdown in the flux-freezing condition (there is further discussion of this point in Appendix B). Similarly, there must also be a dissipative region at the base of the flow where the return current is turned around and joins the outward-going current. If this latter dissipative region is associated with the event horizon of the black hole, MHD winds of this character can be a realization of the schematic "black hole battery" picture developed in §5.2.3 (Lovelace 1976; Blandford and Znajek 1977). In principle, rotating black holes *without any accretion* may be able to produce jets and radio emission. That dissipative regions play an essential role in determining the luminosity of the jet also means that no pure MHD model can explain even the most basic properties of these jets.

9.3.3 Propagation and morphology

In this section we set out the basic principles for understanding the motion—and therefore the appearance—of fluid jets, starting from the smallest observable scales and working our way outwards.

The most basic fact about jet propagation is that, as a jet works its way out through the host galaxy and then the intergalactic medium, it must push aside any gas it finds in its way. As a result of the momentum the jet loses to the interstellar gas, its tip cannot move outward as fast as the jet material itself moves. Two shocks are then formed, one moving out into the interstellar gas, and the other moving backwards into the jet, slowing it down. To find how fast the outermost shock (the "working surface") moves, we must balance mass flux and momentum flux across each of three boundaries: the outer shock, the reverse shock, and the contact discontinuity separating shocked external gas from shocked jet gas. Shocks may also be the device through which the energy carried in jets is transformed into random motions of energetic electrons (§8.6.2.2), and thence into photons we can observe. Consequently, both for their role in defining the basic structure of jets and for their role in making jets shine, shock dynamics are central to an understanding of jet propagation and appearance (the basic properties of nonrelativistic shocks are summarized in Appendix C).

9.3.3.1 relativistic fluid dynamics and shocks

Very close to the nucleus, jets often appear to be traveling at relativistic speeds. We must therefore begin by developing the machinery of relativistic hydrodynamics. A fuller account of this subject than is presented here can be found in Landau and Lifshitz (1985).

Relativistic hydrodynamics is most simply described in terms of four-vectors and tensors, so that the forms of the equations are independent of the reference frame. In this language, the mass continuity equation becomes

$$\frac{\partial}{\partial x^\mu}\left(nu^\mu\right) = 0, \tag{9.54}$$

where u^μ is the four-velocity $(\gamma, \gamma\vec{\beta})$ and n is the proper particle density, that is, the density in the fluid rest frame. Because particle-antiparticle pairs can be created when the temperature is relativistic, the appropriate density to use here is one that is truly conserved, for example, lepton number density rather than the total electron density. Note that if the metric is not constant (i.e., gravity is significant), the partial derivative should be replaced by a covariant derivative.

Momentum and energy conservation are expressed by requiring the energy-momentum tensor $T^{\mu\nu} = (\mathcal{E} + p)u^\mu u^\nu - pg^{\mu\nu}$ to have zero (covariant) divergence. Here $\mathcal{E}$ and p are the proper energy density and pressure, with $\mathcal{E}$ including the rest-mass energy. An equation of state can give p as an explicit function of $\mathcal{E}$; in the limit of strongly relativistic random motions in the fluid rest frame, $p = \mathcal{E}/3$. Stresses due to source-free electromagnetic fields can be very easily accounted for by simply adding the four-dimensional Maxwell stress tensor

$$
\mathcal{M}^{\mu\nu} = \begin{pmatrix}
u & \vdots & & \vec{g} & \\
\cdots & \cdots & \cdots & \cdots & \cdots \\
& \vdots & & & \\
\vec{g} & \vdots & & M_{ij} & \\
& \vdots & & &
\end{pmatrix}
\tag{9.55}
$$

to the energy-momentum tensor. Here $u = (E^2 + B^2)/(8\pi)$, the usual energy density, $\vec{g} = \vec{E} \times \vec{B}/(4\pi)$, the Poynting vector, and M_{ij} is the nonrelativistic Maxwell tensor.

In addition to the reasons given above, shocks are of special interest in the context of relativistic jets because of the strong variations often seen in the radiation they produce. These surely imply the existence of large-scale disturbances, likely in the form of shocks.

In direct analogy with nonrelativistic shocks, the jump conditions for relativistic shocks are given by the conservation of particle number, energy, and momentum:

$$
\begin{aligned}
[nu_z] &= [\gamma\beta n] = 0, \\
[T_{0z}] &= [\gamma^2\beta(\mathcal{E} + p)] = 0, \\
[T_{zz}] &+ [\gamma^2\beta^2(\mathcal{E} + p) + p] = 0,
\end{aligned}
\tag{9.56}
$$

where the z-axis is defined by the shock normal, $\mathcal{E}$, p, and n are proper values, and the speeds are evaluated in the shock front's rest frame. The notation $[X]$ means the difference between X immediately before and immediately after the shock. If we designate the preshock gas by subscript 1 and the postshock gas by subscript 2, the last two equations can be solved for β_1 and β_2:

$$
\begin{aligned}
\beta_1 &= \left[\frac{(p_2 - p_1)(\mathcal{E}_2 + p_1)}{(\mathcal{E}_2 - \mathcal{E}_1)(\mathcal{E}_1 + p_2)}\right]^{1/2} \\
\beta_2 &= \left[\frac{(p_2 - p_1)(\mathcal{E}_1 + p_2)}{(\mathcal{E}_2 - \mathcal{E}_1)(\mathcal{E}_2 + p_1)}\right]^{1/2}.
\end{aligned}
\tag{9.57}
$$

These relations can be combined to give the constraints

$$\beta_1 \beta_2 = \frac{p_2 - p_1}{\mathcal{E}_2 - \mathcal{E}_1} \tag{9.58}$$

and

$$\frac{\beta_1}{\beta_2} = \frac{\mathcal{E}_2 + p_1}{\mathcal{E}_1 + p_2}. \tag{9.59}$$

The compression ratio follows from the particle conservation condition:

$$\frac{n_2}{n_1} = 2 \left(\frac{\mathcal{E}_2}{\mathcal{E}_1}\right)^{1/2} \left(\frac{1 + p_1/\mathcal{E}_2}{1 + 3\mathcal{E}_1/\mathcal{E}_2}\right)^{1/2}$$

$$\times \left[\frac{1 - (3/2)(\mathcal{E}_1/\mathcal{E}_2) + (3/2)(p_1/\mathcal{E}_2) + (9/8)p_1\mathcal{E}_1/\mathcal{E}_2^2}{1 + p_1/\mathcal{E}_1 - (3/2)(\mathcal{E}_1/\mathcal{E}_2) + (3/2)p_1^2/(\mathcal{E}_1\mathcal{E}_2)}\right]^{1/2}, \tag{9.60}$$

where it has been assumed that the postshock gas is extremely relativistic, so that $p_2 = \mathcal{E}_2/3$. Written in this form, it is apparent that in this limit $n_2/n_1 \sim (\mathcal{E}_2/\mathcal{E}_1)^{1/2}$.

It is worthwhile examining two special cases. First consider the limit in which the preshock gas is "cold," that is, $p_1 = 0$ and $\mathcal{E}_1 = mn_1$, but the postshock state is extremely relativistic. Here m is the mean rest-mass energy per conserved particle. Then

$$\beta_1 = [(1 + 3mn_1/\mathcal{E}_2)(1 - mn_1/\mathcal{E}_2)]^{-1/2},$$

$$\beta_2 = \frac{1}{3}\left(\frac{1 + mn_1/\mathcal{E}_2}{1 - mn_1/\mathcal{E}_2}\right)^{1/2}, \tag{9.61}$$

$$n_2/n_1 = 2\left(\frac{\mathcal{E}_2}{mn_1}\right)^{1/2}\left(\frac{1 - (3/2)mn_1/\mathcal{E}_2}{1 + mn_1/\mathcal{E}_2}\right)^{1/2}.$$

Such shocks must engulf new material at speeds close to c, but the flow out of the front (as seen in the front's frame) is always $\simeq c/3$. The compression ratio is unlimited, scaling $\propto [\mathcal{E}_2/(mn_1)]^{1/2}$.

The other case of special interest is the limit in which the gas is extremely relativistic before the shock as well as after. One might imagine that this is the limit of greatest relevance to shocks running through relativistic jets. In this limit,

$$\beta_1 = \left[\frac{1 + \mathcal{E}_1/(3\mathcal{E}_2)}{1 + 3\mathcal{E}_1/\mathcal{E}_2}\right]^{1/2},$$

$$\beta_2 = \frac{1}{3}\left[\frac{1 + 3\mathcal{E}_1/\mathcal{E}_2}{1 + \mathcal{E}_1/(3\mathcal{E}_2)}\right]^{1/2}, \tag{9.62}$$

$$n_2/n_1 = \left(\frac{3\mathcal{E}_2}{\mathcal{E}_1}\right)^{1/2}\left[\frac{1 + \mathcal{E}_1/(3\mathcal{E}_2)}{1 + 3\mathcal{E}_1/\mathcal{E}_2}\right]^{1/2}.$$

Once again, while pre-shock material is swept into the shock at nearly c, the postshock flow travels at $c/3$ in the front frame. Similarly, $n_2/n_1 \propto (\mathcal{E}_2/\mathcal{E}_1)^{1/2}$.

9.3.3.2 superluminal motion

The appearance on the sky of relativistically moving fluid can be very strange. Different segments of the jet can appear to separate at speeds greater than c, as was actually predicted (Rees 1966) five years *in advance* of the discovery of superluminal motion (Whitney et al. 1971; Cohen et al. 1971).

Consider a radio source that consists of two blobs, one stationary in our frame, the other moving with velocity $\vec{\beta}$ at angle θ with respect to our line-of-sight (fig. 9.12). Suppose that at $t = 0$ in our frame, the moving blob is at the same location as the stationary blob, a distance D away from us, and sends us a signal. That signal arrives at the time D/c. After time Δt in our frame, the moving blob has moved a distance $\beta \Delta t \sin \theta$ in the transverse direction and sends us a second signal. This one arrives at $t = \Delta t + D/c - \beta \Delta t \cos \theta$. Thus, its apparent transverse velocity is

$$\beta_{\text{app}} = \frac{\beta \sin \theta}{(1 - \beta \cos \theta)}. \tag{9.63}$$

When $\theta \ll 1$ and $\gamma \gg 1$, we can expand both the trigonometric functions and β:

$$\beta_{\text{app}} \simeq \frac{(1 - \frac{1}{2\gamma^2})\theta}{1 - (1 - \frac{1}{2\gamma^2})(1 - \theta^2/2)} \simeq \frac{2\theta}{\gamma^{-2} + \theta^2}. \tag{9.64}$$

When $\theta \ll 1/\gamma$, $\beta_{\text{app}} \simeq 2\gamma^2\theta < 2\gamma$; when $1/\gamma < \theta \ll 1$, $\beta_{\text{app}} \simeq 2/\theta$. Until $\theta \gtrsim 1$, the apparent speed is larger than unity. For typical observed numbers, β_{app} is in the range of a few to twenty, suggesting Lorentz factors of roughly the same magnitude.

Although this example demonstrates how relativistic bulk motion can produce apparent superluminal velocities, it is not the only way relativistic kinematics can produce this result. In fact, for the purpose of the preceding example, the blob need not be a physical cloud of plasma; it could just as well be a pattern of excitation, as created, for example, by a relativistic shock front. Moreover, although one-dimensional relativistic shock fronts must travel slower than c, in two dimensions it is easy to construct examples of front patterns that, through projection, travel faster than c.

The degree of Doppler boosting and relativistic beaming of the radiation (§9.3.4.3) are controlled by the velocity of the postshock fluid, which

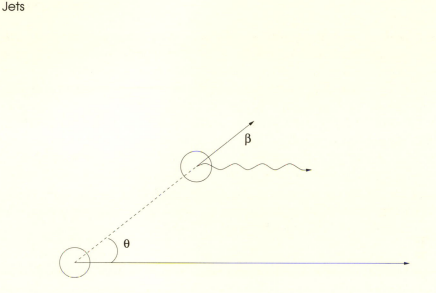

Fig. 9.12   Geometry of apparent superluminal motion.

must always be less than c, and can be quite different from the front velocity. For example, in the simple case of a one-dimensional relativistic shock
moving through material that is already relativistic, the postshock fluid
speed behind a front moving at β_s in the lab frame is

$$\beta_r = \frac{\beta_s - \beta_2}{1 - \beta_s \beta_2} \simeq \frac{\beta_s - 1/3}{1 - \beta_s/3}. \tag{9.65}$$

The Lorentz factor of the radiating fluid is

$$\gamma_r = \frac{3}{2\sqrt{2}}(1 - \beta_s/3)\gamma_s, \tag{9.66}$$

which is $\gamma_s/\sqrt{2}$ in the limit that $\beta_s \to 1$. In this example the pattern speed
is the shock speed β_s, but the speed appropriate to the Lorentz transformations involving radiation intensities is β_r. Other propagation speeds may
also be relevant (as explained in a delightful review of relativistic jets: see
Phinney 1985).

9.3.3.3   nonrelativistic jet morphology

At greater distances from the nucleus, it is quite possible that even
jets that started out relativistic are slowed down (e.g., by entrainment of

adjacent material) to subrelativistic speeds. Their appearance, too, can be quite complicated, but fortunately many properties of interest can be described in terms of lumped parameters independent of the detailed internal nature of the jet. The absolute scale of the jet depends on its luminosity L_j and thrust (i.e., momentum flux) T_j. In terms of these quantities, a nonrelativistic jet moves with speed $v_j = 2L_j/T_j$. The jet's shape depends on four dimensionless numbers: the ratio of jet inertia density to the density in the external medium ρ_j/ρ_e; the Mach number of the jet relative to its internal sound speed $\mathcal{M}_j = v_j/c_s$; the ratio between the jet pressure (in its rest frame) and the pressure in the external medium p_j/p_e; and the plasma $\beta_p = 8\pi p_j/B^2 = 2c_s^2/(\gamma_{ad}v_A^2)$, where γ_{ad} is the ratio of specific heats in the jet material and v_A is the Alfvén speed in the jet.

For example, we can find the speed v_s at which the working surface of the jet advances by enforcing conservation of mass and momentum at both the forward and reverse shocks and balancing the pressure of the shocked interstellar and jet fluids. In the limit that all the shocks are strong and nonrelativistic (so that $\rho_2 = 4\rho_1$),

$$v_s \simeq \frac{v_j}{1 + \sqrt{\rho_e/\rho_j}}. \tag{9.67}$$

A simple estimate of the jet speed can be found from the equipartition argument (Longair, Ryle, and Scheuer 1973). Equipartition gives the minimum pressure in the "hot spots" where the jet crashes into the intergalactic medium; the luminosity of the jet must then be at least $\sim (v_j/2)p_{min}A$. For FR2 radio galaxies, typical values of $p_{min}A$ are $\sim 10^{35}$ dyne, so that L_j must be at least $\sim 3 \times 10^{45}(v_j/c)$ erg s^{-1}. Presumably $L_j \geq L_r$, so $v_j \gtrsim 0.01c$ if $L_r \gtrsim 10^{43}$ erg s^{-1}. However, if the true pressure is greater than the minimum, v_j could be slower. On the other hand, if $v_j > L_j/(p_{min}A)$, there would be a large amount of excess energy delivered to the radio lobes that does not result in radiation.

A lower bound can similarly be placed on v_j by comparing the associated mass flux to what we might deem "reasonable." Because $\dot{M} \geq p_{min}A/v_j$, we find

$$\dot{M} \geq 5 \left(\frac{p_{min}A}{10^{35}\ \text{dyne}} \right) \left(\frac{v_j/c}{0.01} \right)^{-1} M_\odot\ \text{yr}^{-1}. \tag{9.68}$$

Jets that are too slow must carry away an uncomfortably large amount of mass.

The age of the source can be estimated from v_s and the linear extent of the lobes:

$$t \sim 3 \times 10^7 \left(\frac{v_s/c}{0.01} \right)^{-1} \left(\frac{l}{100 \text{ kpc}} \right) \text{ yr.} \qquad (9.69)$$

A rough description of the jet's shape at large distances from the nucleus follows from a very simple argument: If it has been tightly collimated close in, the fastest it can expand in the transverse direction is $\sim c_s$. The opening angle at large distances would then be $\simeq \tan^{-1}(c_s/v_j)$, which is $\sim \mathcal{M}_j^{-1}$ for $\mathcal{M}_j \gg 1$. This effect is sometimes called *inertial confinement*.

Numerical simulations give us the ability to study the shapes of jets in greater detail and to see how the underlying physical parameters influence their structure (see, e.g., the review by Burns, Norman, and Clarke 1991). Much of the morphology of radio jets, including the bends seen in FR1 radio galaxies, can be reproduced in simulations if appropriate parameter choices are made.

The most consistent and dramatic result of these simulations is that when $\rho_e \gg \rho_j$, $\mathcal{M}_j \gg 1$, and $\beta_p \gg 1$, a large "cocoon" develops, composed of shocked jet material that is deflected backward and soon surrounds the comparatively narrow cone occupied by the jet proper (fig. 9.13). This material expands rapidly sideways because the forward shock has raised its pressure far above the ambient pressure. Roughly speaking, its sideways expansion speed is $\sim v_s$, which is likely to be very supersonic relative to the external medium. As the cocoon gas flows backward along the outside of the jet, its elevated pressure helps to confine the jet, but the shear between it and the stationary external gas feeds Kelvin-Helmholtz instabilities. The transverse motions associated with these instabilities ruffle the outer surface of the cocoon and also push the jet back and forth (this last effect is only visible in three-dimensional simulations that do not force the jet axis to coincide with the symmetry axis).

On the other hand, if $\beta_p \ll 1$, so that magnetic forces are strong, the picture changes substantially (fig. 9.14). Now the shocked jet gas, instead of expanding sideways and being left behind, is magnetically confined to a "nose cone" that is pushed ahead of the terminal Mach disk. A cocoon does form around the sides of the jet, but it is made of jet material that has expanded sideways.

A major gap separates these simulations from direct comparison with radio images: We do not know how to connect jet density, the quantity followed in the simulation, with synchrotron surface brightness, the observed

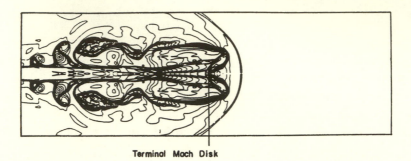

Terminal Mach Disk

Fig. 9.13   A two-dimensional fluid simulation of a jet pushing through surrounding interstellar gas (Clarke et al. 1986). For this simulation, $\rho_j/\rho_e = 0.1$, $p_j/p_e = 1$, and $\mathcal{M}_j = 6$. Many of the qualitative features seen here—the terminal shock, the cocoon of shocked interstellar gas—are very reminiscent of the features seen in FR2 radio jets.

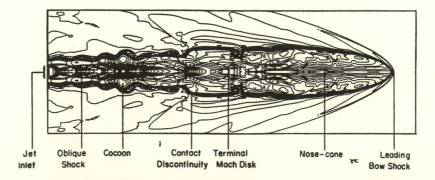

| Jet Inlet | Oblique Shock | Cocoon | Contact Discontinuity | Terminal Mach Disk | Nose-cone | Leading Bow Shock |

Fig. 9.14   A two-dimensional MHD simulation of a jet pushing through surrounding interstellar gas (Clarke et al. 1986). With the exception of the plasma β (here 0.2 at the cylindrical edge of the jet, there ∞), the parameters of this simulation are identical to those of the simulation shown in fig. 9.13, but the shape of this jet is very different. The position of the terminal Mach disk does, however, advance at exactly the same rate in both simulations.

quantity. As already discussed in §8.6.2.3, shocks may be efficient accelerators of nonthermal particles, but we have no good way of quantitatively connecting the shocks in simulations with relativistic electron production rates. Nor (in purely fluid simulations) is there any better method than equipartition to estimate magnetic field strengths. In the next section, therefore, the radiative properties of jets are posed in terms of more or less plausible guesses for how the emissivity may vary from place to place.

9.3.4   Jet radiation

9.3.4.1   gradients in synchrotron emissivity

We earlier remarked that one way to achieve a flat integrated spectrum despite the sharply peaked character of the homogeneous self-absorbed synchrotron spectrum (without allowance for stimulated scattering) is to superpose a number of optically thick regions with different synchrotron turnover frequencies. A jet model provides a very natural context in which this might happen.

Suppose, then, that relativistic electrons flow out a conical jet. If the number of electrons is conserved (i.e., no pair processes), their density falls $\propto r^{-2}$. To allow for pair creation and annihilation, we will generalize that to $n_e \propto r^{-a}$. The magnetic field is also likely to decline outwards; we write its dependence as $B \propto r^{-b}$. If simple flux-freezing controls the field strength, $\vec{B}$ becomes dominated by its perpendicular component and $b = 1$. The same scaling results if the ratio of magnetic energy density to electron energy density is held constant, and electron number is conserved. Despite their seemingly quantitative character, it is important to remember that these guessed scalings contain no physics; they merely parameterize our ignorance.

The simplest case is one in which the jet speed is subrelativistic, or more precisely, the maximum relative speed between one part of the jet and another is $\ll c$. Suppose that the solid angle of the jet varies with radius as $\Delta\Omega(r)$. To describe the synchrotron spectrum, we will make the approximation that at frequencies above ν_t the optically thin emissivity applies (eq. 9.12), while the emissivity is zero at frequencies below ν_t:

$$L_\nu \propto \int_{r_{\min}} dr\, \Delta\Omega(r) n_e(r) B^{(1+\xi)/2}(r) \nu^{(1-\xi)/2}, \qquad (9.70)$$

where $r_{\min}$ is defined by $\nu = \nu_t(r_{\min})$. This integral is schematically illustrated in figure 9.9. It is a good approximation because the frequencies nearest ν_t dominate the flux radiated in the optically thick portion of the spectrum, whereas the highest frequencies often dominate the integrated flux of the optically thin part of the spectrum. We see that

$$\nu_t(r) \propto B^{(\xi+2)/(\xi+4)}(n_e r)^{2/(\xi+4)},$$ (9.71)

so

$$r_{\min} \propto \nu^{1/[2(b-a+1)/(\xi+4)-b]}.$$ (9.72)

Because we expect $\xi \simeq 2\text{–}3$ and $b \sim 1$, $r_{\min}$ generally decreases as ν increases, as expected. If $\Delta\Omega$ is independent of r (i.e., the jet is conical), the integrated spectrum has the shape

$$\frac{d\ln L_\nu}{d\ln \nu} = s = \frac{1-\xi}{2} - \frac{3-a-b(1+\xi)/2}{b-2(b-a+1)/(\xi+4)}.$$ (9.73)

The favored case we have already identified ($a = 2$, $b = 1$, and a conical jet) has several special properties (Blandford and Königl 1979). The minimum radius at which any given frequency is radiated efficiently is $\propto \nu^{-1}$, so that the size of the radiating region scales in exactly the same way with frequency as the resolution of an interferometric image. Defining the smallest dimension of the jet segment at r as r_t, the peak brightness temperature is independent of radius because $T_{bt} \propto (n_e r_t/B)^{1/(\xi+4)}$ and $r_t \propto r$ in a conical jet. Finally, because T_{bt} is constant, while $\nu_t \propto r^{-1}$ and the radiating surface area $\propto r^2$,

$$s = 0,$$ (9.74)

independent of ξ. Thus, this very simple model provides a quite plausible way of explaining flat radio spectra; it is not by any means, however, a unique solution.

Note, also, that the synchrotron continuum can extend up to very high frequencies, possibly all the way into the X-ray band. As can be seen in figure 9.15, the lower frequency half of the broad-band continuum distributions of BL Lac objects and flat-spectrum radio quasars can be neatly described as a single synchrotron component. Most often, the peak in νL_ν can be ascribed to the cooling break described in §9.2.1, but the location of this peak varies widely from object to object. Examples are known in which the peak frequency is as low as $\lesssim 10^{13}$ Hz and as high as $\sim 10^{17}$ Hz (Sambruna, Maraschi, and Urry 1996). Not surprisingly, those with low-frequency peaks tend to be found in radio surveys, while those with higher frequency peaks are more common in optical or X-ray surveys. Ascribing this portion of the continuum to synchrotron radiation may also explain two other striking features of this class of objects: the strong (and variable) optical polarization they display and the close correspondence sometimes seen (e.g., fig. 9.5) between radio and optical features in extended jets.

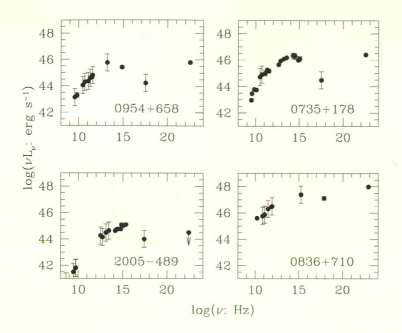

Fig. 9.15   Four examples of the possible broad-band continuum shapes shown by BL Lac objects and flat-spectrum radio quasars (data courtesy R. Sambruna; originally published in Sambruna et al. 1996). From radio frequencies up to the optical band, and sometimes all the way to soft X-rays, there appears to be a well defined and smooth distribution of power output. At higher frequencies a different component appears to emerge, possibly due to inverse Compton scattering.

9.3.4.2   synchro-Compton radiation

In §9.2.5 we discussed the effect of Compton scattering inside the source on the emergent radio spectrum, remarking in passing that considerable X-ray luminosity could result from inverse Compton scattering of synchrotron photons by the relativistic electrons. It is now time to consider that up-scattering more carefully.

Again taking the limit of a subrelativistic jet, the photon energy density at a given location in the jet can be approximated in a way closely analogous to the way we estimated the integrated synchrotron spectrum:

$$U_\nu(r) \propto \int_{r_{\min}(\nu)}^{r_{\max}(\nu)} dr' \, \min[1, (r'/r)^2] n_e(r') \gamma_{\min}^{\xi-1} B^{(1+\xi)/2}(r') \nu^{(1-\xi)/2},$$

$$(9.75)$$

where $r_{\max}(\nu)$ is defined implicitly by $\nu = 4\gamma_{\max}^2 eB(r)/(2\pi m_e c)$. If $a + b(1+\xi)/2 > 2$ (as is generally likely), the energy density at ν is dominated by those photons produced near $r_{\min}$, not by locally produced photons. We can then write the emissivity for inverse Compton photons as

$$J_\epsilon(r) \propto \sigma_T \int d\gamma \, n_e(r) \gamma_{\min}^{\xi-1} \gamma^{-\xi} \int d\nu \, \frac{U_\nu(r)}{h\nu} \epsilon\delta\left(\frac{4}{3}\gamma^2 h\nu - \epsilon\right). \qquad (9.76)$$

Suppose that $U_\nu(r) = [U_o(r)/\nu_o](\nu/\nu_o)^{-s}$. Then evaluating the integral over ν with the delta function yields

$$J_\epsilon(r) = (\xi - 1)\gamma_{\min}^{\xi-1} n_e \sigma_T c \frac{U_o(r)}{h\nu_o} \left(\frac{\epsilon}{h\nu_o}\right)^{-s} \int_{\gamma_{\min}}^{\gamma_{\max}} d\gamma \, \gamma^{2s-\xi}. \qquad (9.77)$$

This expression is valid for

$$\gamma_{\min}^2 h\nu_{\min} \leq \epsilon \leq \min[\gamma_{\max} m_e c^2, \gamma_{\max}^4 \hbar eB_{\max}/(m_e c)]. \qquad (9.78)$$

We have retained the dependence on $\gamma_{\min}$ and $\gamma_{\max}$ in this expression in order to emphasize that the amplitude of the inverse Compton spectrum, as well as its range of energy, depend on these (unknown) cutoffs to the electron distribution function. For usual parameter values, the dominant dependence is on $\gamma_{\min}$. The up-scattered spectrum has exactly the same shape as the original synchrotron spectrum because it is the same electrons that are responsible for both, and inverse Compton scattering shares the property of synchrotron radiation of multiplying the "seed" photon energy by γ^2.

In §9.2.4 we estimated that the "observed" electrons in optically thick synchrotron sources had Lorentz factors $\sim 600(\nu_{\mathrm{GHz}}/B_{-3})^{1/2}$; the inverse Compton photons that result will then range from ~ 1 to $\sim 100B_{-3}^{-1}$ eV if the spectrum is flat from ~ 1 to ~ 100 GHz. The luminosity of the inverse Compton component can easily be comparable to or greater than the synchrotron luminosity (see eq. 9.42). Because the Compton optical depth of optically thick synchrotron sources is generally $\lesssim 1$ (eq. 9.39), the ratio of up-scattered photon energy density to sychrotron photon energy density is comparable to their luminosity ratio. Thus, if the first-generation inverse Compton luminosity exceeds the radio luminosity, we can expect the second-generation Compton-scattered luminosity to exceed the first generation. The multiplication of energy only ceases when the Klein-Nishina limit is reached.

In fact, it seems quite likely that effects of just this sort occur. Radio-loud AGNs are somewhat stronger X-ray emitters than radio-quiet AGNs, and the core-dominated radio-loud AGNs are extremely powerful γ-ray sources.

Detailed consideration of the spectra of BL Lac objects and OVV quasars also supports models of this sort, at least in general terms. The spectra shown in figure 9.15 are reasonably well described by this sort of model, for example. Numerous sources may contribute the seed photons for inverse Compton scattering. Synchrotron radiation from the jet itself, accretion disk photons diffused by an external scattering medium, line photons from photoionized gas, or infrared photons produced by dust warmed by higher frequency photons are all plausible candidates. The high-energy γ-rays could well be the product of second-order inverse Compton scattering, in which relativistic electrons amplify for a second time the energy of photons boosted once already from the radio or infrared bands into the ultraviolet or X-ray regimes. Thus, in these cases, the optical, ultraviolet, and even X-ray emission may be dominated not by the sorts of thermal and quasi-thermal processes discussed in Chapters 7 and 8, but by the highly nonthermal mechanisms of synchrotron radiation and relativistic inverse Compton scattering.

9.3.4.3   relativistic beaming and boosting

Relativistic beaming and Doppler boosting can powerfully alter the angular distribution, spectrum, and time behavior of the radiation produced by matter in relativistic motion. It is now time to be quantitative about these effects. We begin with the time-steady properties. With jets in mind, imagine a source in which relativistic material streams down a fixed channel at a fixed rate, and each fluid element in this jet emits a fixed amount of energy (as measured in the fluid rest frame) by the time it passes some marker in the channel. Suppose, too, that the radiation is isotropic in the fluid rest frame and has a power-law spectrum with spectral index α.

The frequency-specific fluences from each fluid element received by lab frame observers at different directions are given by

$$S_\nu = \frac{1}{D^2} \frac{dE_\nu}{d\Omega} = \frac{1}{D^2} \frac{dE'_{\nu'}}{d\Omega'} \frac{d\Omega'}{d\Omega} \frac{dE_\nu}{dE'_{\nu'}}, \tag{9.79}$$

where primed quantities are measured in the fluid frame, unprimed quantities are measured in the observer's frame, and D is the distance between the source and the observer in question. The effect of relativistic beaming is

$$\frac{d\Omega'}{d\Omega} = \frac{d\phi'}{d\phi} \frac{d\mu'}{d\mu}, \tag{9.80}$$

where the polar angle θ, for which $\mu = \cos\theta$, is defined with respect to the jet velocity. Because the azimuthal angle ϕ refers to transverse dimensions, it is an invariant: $d\phi' = d\phi$. The transformation for μ is

$$\mu' = \frac{\mu - \beta}{1 - \mu\beta}, \tag{9.81}$$

so

$$\frac{d\mu'}{d\mu} = \left[\frac{1}{\gamma(1 - \mu\beta)}\right]^2 \equiv \delta^2. \tag{9.82}$$

The quantity δ is the Doppler boost:

$$\nu' = \delta\nu = \gamma(1 - \beta\mu)\nu. \tag{9.83}$$

When $\gamma \gg 1$ and $\theta \ll 1$,

$$1 - \beta\mu \simeq \frac{1}{2\gamma^2} + \frac{\theta^2}{2}, \tag{9.84}$$

so that

$$\delta \sim \begin{cases} \gamma & \theta < 1/\gamma \\ \gamma\theta^2 & 1/\gamma < \theta \ll 1 \\ \gamma^{-1} & 1 \lesssim \theta . \end{cases} \tag{9.85}$$

Thus, frequencies are boosted when viewed close to the forward direction, and de-boosted when viewed at large angles. An immediate corollary is that approximating θ by the angle between our line-of-sight and the polar axis of the flow is valid only when the opening angle of the flow is narrower than $\sim 1/\gamma$. Otherwise, it is necessary to integrate over the range of angles between our direction and the velocities of the radiating matter.

E_ν is (almost) an invariant because energy and frequency are both boosted by exactly the same amount. If we were asking about the energy per unit frequency received at whatever frequency the photons of interest were shifted to, it would be exactly invariant. However, because our observing frequency is fixed in our frame, the photons we receive had a variety of frequencies in the fluid frame: $\nu' = \nu/\delta$. For power-law spectra, this adjustment produces another factor of δ^α. Finally, the total luminosity an observer receives is the fluence times the rate at which fluid elements pass through the jet. Because that rate is measured in our frame, there is no need to transform it; if our picture were instead one in which a moving blob radiates with a given luminosity in its rest frame, the transformation would require an additional Doppler boost. Combining all these transformations, we find that the flux per unit frequency F_ν observed at a fixed frequency ν from a steady jet scales $\propto \delta^{2+\alpha}$.

The pace of time variations follows the same transformation as for frequency, so variability is similarly made more rapid in the forward direction and slowed down at large angles from the direction of motion. In this way, observed variations on few week timescales can be interpreted as variations in the source frame on timescales of a year or so, granted $\gamma \sim 10$ and a view from the favored direction. The remarkable variations seen in BL Lac objects and OVV quasars at all frequencies from radio through γ-rays may well be in part due to this "Doppler acceleration" of variability, as relativistic boosting both accelerates and amplifies the magnitude of changes in the radiated flux due to shock waves and other disturbances propagating down the jet.

The great enhancement of the intensity in the forward direction has several important consequences. First, it provides a promising explanation for why jets seen in both core-dominated radio galaxies and lobe-dominated FR2 radio galaxies are generally only found on one side of the nucleus. A counterjet probably exists but is so strongly deboosted relative to the one moving toward us that we cannot see it. Second, "Doppler favoritism" strongly biases flux-limited samples to those objects in which a relativistic jet is pointed toward us. The impact of these and other effects of relativistic kinematics on the statistics of radio-loud AGNs will be discussed more fully in Chapter 12. Third, it can explain how it is that nonthermal continuum emission can mask any signature of thermal emission associated with accretion (and sometimes also emission lines) in BL Lac objects and OVV quasars.

The brightness temperature is proportional to the occupation number times the frequency, so it transforms in the same fashion as frequency: $T_b \propto \delta T_b'$. This fact allows us to use the ratio of inverse Compton luminosity to radio luminosity as an indicator of the size of the boost. As demonstrated in equation 9.42, the ratio $L_x/L_r \propto T_{bt}^5$. A measurement of L_x/L_r can therefore be turned into a (model-dependent) inference of T_{bt} *in the source frame*, under the assumptions that the radio and X-ray photons are boosted by the same amount, the photon intensity at radio frequencies is not substantially altered by stimulated scattering losses, and that γ_{min} (and γ_{max} if necessary) can be estimated (see eq. 9.75). Studies of this sort often infer boost factors ~ 10.

The distortions produced by relativistic effects are even greater when we do not have a direct measurement of the source size. Normally, $I_\nu =$

$F_{\nu}D^2/r_{\perp}^2$, but if high-resolution measurements do not exist, we must use some other means to estimate $r_{\perp}$. One way to guess $r_{\perp}$ is through a causality argument. If the source changed flux by order unity in a time Δt, then presumably whatever signal coordinated this change could not have traveled faster than c, so the source cannot be larger than $\sim c\Delta t$. However, if the source is moving relativistically, something that takes $\Delta t'$ in its rest frame occupies a time $\Delta t = \Delta t'/\delta$ in an external observer's frame. This effect means that we may overestimate the brightness temperature in the source's rest frame by two more powers of δ if $r_{\perp}$ is estimated from the variability timescale. Note that other signal geometries may result in different powers of δ, but the sense of the effect is always as described.

Perhaps the greatest challenge to the power of relativistic beaming and boosting is posed by intraday variability in radio galaxies (Quirrenbach et al. 1992). If we estimate a brightness temperature on the basis of a causality limit for the source region without allowance for relativistic effects, we find that

$$T_b \sim 1 \times 10^{18} \left(\frac{F_{\nu}}{1 \text{ Jy}} \right) \lambda_{cm}^2 \left(\frac{D_A H_o}{c} \right)^2 \left(\frac{h}{0.75} \right)^{-2} \left(\frac{\Delta t}{1 \text{ d}} \right)^{-2} \text{ K!} \quad (9.86)$$

If T_b' is held to $< 10^{12}$ K (by stimulated or inverse Compton scattering, or by the constraint of equipartition), the Lorentz factor necessary to explain these objects is ~ 100.

10   Emission Lines

The great majority of all AGNs display strong emission lines in their optical, UV, and even X-ray spectra. Only the relatively rare BL Lac objects exhibit no lines at all (or, more precisely, have lines whose equivalent widths are at most very small).

AGN emission lines have consistently attracted a great deal of attention both because they are strong and because we understand so much about atomic physics that it is possible to use them with comparative confidence as diagnostics of physical conditions. The latter reason is a bit reminiscent of why the drunk looked for his lost keys under the street lamp: it wasn't where he had lost them, but at least there was light. This statement is not a pejorative one: as this and subsequent chapters will show, it has been possible to reflect some of the "light" from emission line studies so that it shines on other, perhaps more intrinsically important areas of the subject, which regrettably have no direct illumination of their own.

10.1   Observables

10.1.1   Line lists and relative strengths

Within several broadly defined categories, the lines that are observed, and their relative strengths, are remarkably stereotyped. In fact, it is this stereotypical character that has permitted the construction of a rather well defined AGN classification system based on simple properties of the emission lines.

10.1.1.1   broad lines

When any emission lines in the spectrum have "broad" profiles, that is, FHWM of at least 1000–2000 km s^{-1}, essentially every member of the following list is present, and has a broad component in its profile: Lyα, Hα and Hβ, Paα and Paβ, Brα; He II 1640, He II 4686, He I 5876, and He I 10830; C III] 1909 and C IV 1548,1551; N V 1239,1243; O VI 1032,1038, O IV] 1400, O I 1305, and O I 8446; Si IV 1394,1403; Mg II 2796,2804; and many multiplets of Fe II, which are found mostly (but not entirely) in four clusters: 2200–2600 Å, 3000–3400 Å, 4500–4600 Å, and 5250–5350 Å. The only stipulation is that the data should have the wavelength coverage and signal/noise to see them. If typical wavelength coverage were broader and the signal/noise ratio greater, this list would undoubtedly extend to include more H I lines, as well as many weaker lines from other elements. In contrast to the strongest lines in typical galactic emission line nebulae such as H II regions and supernova remnants, it is notable that forbidden lines are entirely absent from this list.

Several doublets were listed in the previous paragraph. Because the lines are generally broadened by more than the intradoublet separation, they are often treated as a single line. When this is done, the C IV doublet is usually described as falling at 1549 Å, N V at 1240 Å, O VI at 1034 Å, Si IV at 1400 Å, and Mg II at 2800 Å. We will adopt this convention here.

Apart from the Fe II lines, variations from object to object in the relative strengths of these lines are typically only at the factor of two level (see table 10.1). Lyα is generally the strongest, with C IV 1549 no more than a factor of two behind, and sometimes as strong, or even stronger than, Lyα. Hα is typically almost as strong as C IV 1549. C III] 1909 and Mg II 2800 are in the next rung, a factor of two smaller than C IV 1549. N V 1240, the Si IV/O IV] 1400 blend, He II 1640, and Hβ are somewhat weaker than the last set, with O VI 1034 (often the dominant partner in a blend with Lyβ) perhaps another factor of two still fainter. The rest of the list is weaker yet.

We excluded the Fe II multiplets from the generalization of the previous paragraph because their fluxes are so difficult to measure that we really don't know in any quantitative sense how much they vary from one object to the next. There are certainly objects in which they appear relatively stronger or weaker, but it is very difficult to attach numbers to those contrasts. For example, there seems to be a trend for the Fe II lines to be weaker in radio-loud objects, especially those which are lobe-dominated (Phillips 1977; Boroson and Green 1992), but this correlation is difficult to quantify.

The reason for this difficulty is the sheer number of the Fe II lines. When broadened by a few thousand km s^{-1}, the lines in each cluster merge to form a quasi continuum. The lines near 3000 Å are further blended with the Balmer continuum, forming a spectral feature called the *little blue bump*. Attempts to solve this problem by developing templates that predict the shapes of the blends for specific physical conditions (Wills et al. 1985) have not been able to settle these questions because their application introduces substantial model dependence, and no satisfactory goodness-of-fit criterion has been developed to indicate when the templates are appropriate. This unsatisfactory situation is made even more frustrating because it is within the range of possibilities (e.g., Netzer 1985) that the total flux in FeII lines may by itself be greater than the total flux of all the other broad lines.

Table 10.1: Mean Relative Broad Line Fluxes

Line	Relative Flux
Lyα + Nv 1240 + Ov 1218	100.
Hβ	22.
HeII 1640	18.
CIII] 1909	29.
CII] 2326	6.0
CIV 1549	63.
OI 1305	3.5
OVI 1034	9.3
MgII 2800	34.
SiIV 1400 + OIV] 1400	19.

Notes: Relative line fluxes from the composite quasar spectrum of Francis et al. (1991) based on the LBQS survey. Note that the single strongest emisison line feature is a blend, Lyα + Nv 1240 +Ov 1218. Certain other strong lines, such as Hα, are omitted from this list because of a lack of wavelength coverage in the LBQS sample.

Like the line ratio distribution, the equivalent width distribution (for spectra in which the nuclear continuum clearly dominates the total continuum) has a reasonably well defined center, but factor of two fluctuations are common. Figure 10.1 shows the distribution for one well defined sample, the PG quasars; although the distributions for other varieties of broad-line AGNs may differ in detail, they are not qualitatively different. That this distribution spans only a factor of a few, while the luminosity range (in this subset of the PG sample) is about three orders of magnitude, immediately demonstrates that there is a strong (and nearly linear) correlation between emission line and continuum luminosity. If the PG sample can be taken as a guide to what is typical for all broad emission line AGNs, about 2% of the optical light is in Hβ alone. Because the total broad line flux (excluding FeII) is $\simeq 14\times$ the Hβ flux (see table 10.1), and the integrated optical/ultraviolet flux is often $\simeq 5 \times \nu F_\nu$ measured near the frequency of Hβ, roughly 6% of the total optical/ultraviolet flux coming out in emission lines (again, excluding FeII!) is a reasonable fiducial number.

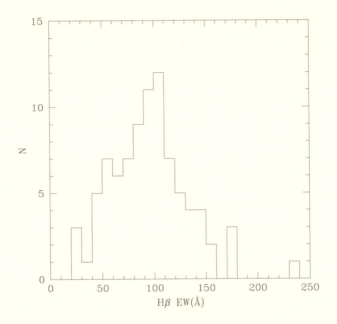

Fig. 10.1   The distribution of rest-frame EW for Hβ in the low-redshift half of the PG sample. Data are taken from Boroson and Green (1992).

Correlations between the relative line strengths and the absolute luminosity exist (crudely in the sense that higher ionization lines can be somewhat stronger at lower luminosity) but are fairly weak (Mushotzky and Ferland 1984; Baldwin, Wampler, and Gaskell 1989; Boroson and Green 1992). On the other hand, there are tantalizing hints of a stronger relation between the *equivalent width* of the broad CIV 1549 line and the luminosity (fig. 10.2). Baldwin (1977) first pointed out that the equivalent width of CIV 1549 appears to decline with increasing luminosity for high-luminosity quasars. Studies of well defined samples have since shown that indeed $W(CIV\ 1549) \propto L_{opt}^{-0.6}$ for flat-spectrum radio-loud quasars, but the correlation is much weaker both for other broad lines, and for other categories of AGNs (Baldwin et al. 1989).

If these correlations were better established, they would be of great importance because the easily measured equivalent width of the emission lines would serve (at least statistically) as an indicator of the absolute luminosity. Because quasars are so common at high redshifts, this relationship could then allow a direct measurement of the cosmological parameters Ω_M and Λ through the strong dependence of the luminosity distance on them when $z > 1$ (Appendix F). Unfortunately, the unclear significance of the *Baldwin effect* has stymied such efforts.

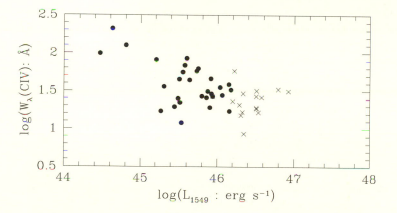

Fig. 10.2 The equivalent width of broad C IV 1549 versus luminosity in the adjacent continuum. The crosses are those PG quasars with $M_B < -24$; the filled circles are an optical flux-limited subset of flat-spectrum radio quasars from the PKS±4° survey. These data are from Baldwin et al. (1989). Note that the continuum luminosities are computed assuming $h = 0.5$ and $q_o = 1$; the redshift range for these objects is great enough that changing to a different q_o could significantly alter the plot.

10.1.1.2 narrow lines

The repertory of "narrow" lines is similarly stereotyped but is closer to the usual interstellar list (see table 10.2). Forbidden lines such as [O III] 4959,5007 and [O II] 3727 are particularly prominent, along with the [N II] 6548,6583 doublet, [O I] 6300, and the [S II] 6716,6731 doublet. The H I Lyman and Balmer series, as well as C IV 1549, also appear with narrow components, but, unlike H$_{\text{II}}$ regions, [O$_{\text{III}}$] 5007 is usually the strongest line overall except for Lyα. Unlike the broad lines, however, in both radio-quiet and radio-loud objects the narrow line equivalent widths (when they can be measured relative to the nuclear continuum) appear to diminish with increasing continuum luminosity, so that they are virtually undetectable in high-luminosity quasars. Moreover, the rest-frame ultraviolet narrow lines seem to diminish even faster with increasing luminosity than do the rest-frame optical narrow lines; because high-luminosity quasars are almost all found at $z > 1$, the lines most commonly observed in them are almost entirely free of narrow components (Wills et al. 1993).

For this reason, the equivalent width distribution of the narrow lines is a much less useful quantity than it is for the broad lines. It is only in low-luminosity nuclei that the narrow lines are prominent, but those are exactly the objects in which contributions from the host galaxy can be important, both to the continuum and to the lines.

Stellar processes in the host also complicate the task of defining the relative line strengths for the narrow lines. In many nearby AGNs it is possible to map the narrow emission line region, and it is not always easy to separate those regions primarily associated with the nucleus from those regions dominated by adjacent stars. It is sometimes also necessary to correct for interstellar reddening by dust in the host galaxy.

In objects (and lines) possessing both broad and narrow components, there is the further complication of separating the flux in the two components. Because the broad component, when it exists, generally dominates the total flux in the line, the uncertainty in its flux due to possible contamination from the narrow component is usually $\sim 10\%$ or less; the corollary of this fact, of course, is that the fractional uncertainty in the flux of the *narrow* component due to blending with the broad part can easily be order unity.

Finally, although it is controversial whether dust reddening has any significant effect on the broad emission lines, the evidence that it is important for narrow lines is much stronger. As always, the impact on ultraviolet lines is greater than on optical, but unfortunately it is difficult to deredden in a way one can prove to be reliable. One possible method is to determine the reddening by comparing the fluxes of two recombination lines (such as He$_{\text{II}}$ 1640 and He$_{\text{II}}$ 4686) that are well separated in wavelength and whose ratio one might believe to be relatively impervious to extraneous effects (in this example because they share the same upper state, and both lower states are very unlikely to be populated enough to create much optical depth). However, as we shall see later, the evidence for inhomogeneity in the narrow line region is so strong that even if this procedure gives the correct extinction on the line-of-sight to the material responsible for most HeII emission, it may not be correct for all the matter.

For all these reasons, the entries in the following table listing typical relative line strengths for narrow lines are considerably less well defined than the entries in the previous table for broad lines. In fact, in order to eliminate the problem of blending with broad components, the study from which these data were taken ignored all broad line objects. Anecdotally it appears that the relative strengths of the narrow components in broad line objects are not greatly different from those found in narrow line objects, but this statement is difficult to quantify. One should also be aware that, other than the ratio [OIII] 5007/[OIII] 4959, whose value is set by atomic physics, the line ratios vary by factors of several from object to object.

10.1.2 Line profiles

10.1.2.1 broad lines

Even within the category of broad lines, there is a fair range of variation in profile shape. Early observations indicated that these lines could all be well fitted by an analytic form $F_\nu = -A_l \ln(|\nu - \nu_o|/\Delta\nu)$, and that the width parameter $\Delta\nu/\nu_o$ was the same for all lines in a given object, but better data have shown that to be too facile a conclusion.

Different lines within a single object do tend to have profiles more similar than examples of the same line in different objects, but the correspondence is far from perfect. Whereas the FWHMs for the CIV 1549 and CIII] 1909 lines in the same object are generally the same to within $\simeq 25\%$, the FWHM of the MgII 2800 line can easily be a factor of two different, usually in the sense that MgII 2800 is narrower (Brotherton et al. 1994). The range of broad line profile widths is illustrated in figure 10.3.

One concise way of characterizing the profile shapes is to describe them as a linear combination of two Gaussians whose widths vary relatively little from case to case (Brotherton et al. 1994). The relative proportions of these two Gaussian components do vary considerably, both from line to line, and, even more so, from object to object. The narrower component has a typical FWHM $\simeq 2000$ km s^{-1}, while the broader has a typical FWHM $\simeq 7000$ km s^{-1} and is blueshifted by ≥ 1000 km s^{-1}. The lower ionization lines appear to have relatively greater proportions of the narrower component, and the high-ionization lines relatively more of the broader one. As a

Table 10.2: Mean Relative Narrow Line Fluxes

Line	Relative Flux
Lyα	55.
Hα	3.2
Hβ	1.0
Ciii] 1909	5.5
Civ 1549	12.
[Oii] 3727	3.2
[Oiii] 4959	1.75
[Oiii] 5007	5.0
[Oi] 6300	0.50
[Nii] 6583	3.0
[Neiii]3869	1.4
[Nev] 3426	1.2
Mgii 2800	1.8
[Sii] 6724	1.6

Notes: A composite of two sets of mean line ratios: those with wavelengths from Hβ to the red are the mean of those from Seyfert galaxies in the CfA sample without broad components (data from Osterbrock and Martel 1993); the rest are from the heterogeneous sample of Ferland and Osterbrock (1986).

result, the higher ionization lines are characteristically both broader than the lower ionization lines, and blue-shifted relative to them (Gaskell 1982, Wilkes 1986). All the broad lines appear to be fairly red-blue symmetric, but asymmetries do appear at the 10% level.

10.1.2.2 narrow lines

The so-called "narrow lines" are narrow only by comparison with the truly broad lines. Nelson and Whittle (1995) showed that the FWHM of the [Oiii] 5007 line is generally quite similar to the FWHM of the host galaxy's stellar velocity distribution, although the [Oiii] line width distribution does have a high-velocity tail not present in the stellar velocity distribution. This fact leads one to think of the narrow line emission (or at least much of it) as

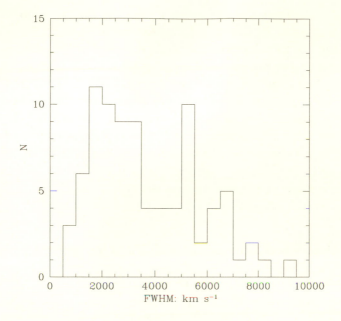

Fig. 10.3   The distribution of FWHM for the Hβ line in the low-redshift half of the PG quasar sample. Data are taken from Boroson and Green (1992).

due to ordinary interstellar clouds in the host galaxy that happen to be "lit up" by the nucleus. The exceptions help support this interpretation—they are all AGNs with bright radio jets. It is a plausible guess that the higher speeds seen in these AGNs are due to forces exerted by the jets.

However, a closer examination of the profiles shows that such a simple story may not account for all aspects of narrow line kinematics. The [OIII]5007 narrow line is often rather strongly asymmetric and favors the blue side of the systemic redshift more often than the red (Whittle 1985; Vrtilek and Carleton 1985).

10.1.3   Classification on the basis of emission line properties

As we remarked above, emission line properties can provide a very clean way to divide AGNs into groups. The presence of lines with broad profiles is the principal defining property for quasars (both radio-quiet and radio-loud), type 1 Seyfert galaxies, and broad line radio galaxies. AGNs with only narrow lines are called type 2 Seyfert galaxies, narrow line radio galaxies, or LINERs. Although its physical significance is unclear, a finer division can also be created (Osterbrock 1989): as the proportion of narrow line flux in the Balmer lines grows, one can define Seyferts of types 1, 1.2, 1.5, 1.8, 1.9, and 2. In the pure type 1, no narrow core is visible; in type 1.2 it is present, but small; in type 1.5 the difference in F_ν between the peak of the broad component and the continuum is comparable to the difference in F_ν between the peak of the broad component and the absolute peak in the line; in type 1.8 weak broad components can be seen in both Hα and Hβ; in type 1.9, the broad component is so weak it is visible only in Hα; and in type 2.0 there is no broad component apparent at all. Given the time variability that is often seen in broad line flux, not to mention the dependence of one's ability to detect a feature on the S/N of the data, this finer classification scheme must be treated as a bit "fuzzy."

A second problem arises when considering galaxies with only narrow emission lines: whether they are truly AGNs or not. For this purpose, one can often seek guidance in a set of line ratio diagrams first introduced by Baldwin, Phillips, and Terlevich (1981). They suggested that AGNs should generically have greater [OIII] 5007/Hβ than galaxies whose emission lines are due to stellar processes because the creation of [OIII] by photo-ionization demands photons with energies above 35 eV—these photons are relatively rare in stellar spectra, while common in AGN continua. To further enhance the contrast between AGNs and normal galaxies, other line ratios (preferably at nearly identical wavelengths so as to eliminate reddening effects; see Veilleux and Osterbrock 1987) can also be selected. For example, [NII]6583/Hα tends to be stronger in AGNs than in normal galaxies. These trends are illustrated in figure 10.4.

A different sort of narrow line AGN can be distinguished at the low ionization end of the scale. LINERs (Low Ionization Nuclear Emission line Regions) are defined by satisfying either of two sets of criteria:

$$[\text{OII}]\ 3727/[\text{OIII}]\ 5007 \geq 1,$$
$$[\text{OI}]\ 6300/[\text{OIII}]\ 5007 \geq 1/3$$

or

$$[\text{OIII}]\ 5007/\text{H}\beta < 3,$$
$$[\text{OI}]\ 6300/\text{H}\alpha > 0.05,$$
$$[\text{SII}]\ 6716,6731\ /\text{H}\alpha > 0.4,$$
$$[\text{NII}]\ 6583/\text{H}\alpha > 0.5.$$

The second definition is preferable because it eliminates problematic de-reddening.

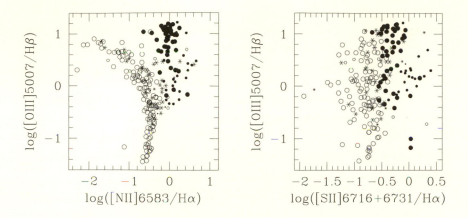

Fig. 10.4 Two diagnostic line ratio diagrams. (Left) [NII] 6583/Hα is plotted against [OIII] 5007/Hβ; (right) [SII] 6716,6731/Hα also versus [OIII] 5007/Hβ. Open circles represent "HII galaxies" and "starburst" galaxies, whose emission line spectra are expected to be dominated by HII regions; filled circles represent narrow line AGNs, both Seyfert galaxies and LINERs; stars are "Narrow Emission Line Galaxies." This last category is often used for galaxies whose type is difficult to determine. Both the data and the classifications in these figures are courtesy of S. Veilleux.

As can be seen in figure 10.4, these galaxies are grouped with the genuine AGNs in terms of the low-ionization line ratios, but have much weaker [OIII] 5007 relative to Hβ. This ambiguity makes their relationship to other galaxies uncertain. Indeed, the degree to which this set of line ratios defines a homogeneous class of objects is unclear.

10.2 Physical Overview

Within a few years of the discovery of AGNs, the primary physical mechanism for the production of these emission lines was identified: photo-ionization by the AGN's own continuum. The original arguments that led to this conclusion have only been bolstered over the years:

1. Excluding BL Lac objects, the equivalent widths of the lines show a relatively small dispersion from object to object across a very wide range of luminosity. This suggests that somehow the production of lines is proportional to the overall energy release, which in turn suggests that one causes the other.

2. A wide range of ionization stages is seen, including some species that require overcoming ionization potentials far higher in energy than the temperatures inferred from the line ratios. Collisional ionization is therefore ruled out. Photo-ionization, on the other hand, seems quite reasonable, as the observed continuum supplies many photons across the entire broad range of ionization potentials required. If photo-ionization explains the ionization equilibrium, why not let it also provide the energy?

Since these arguments were developed, several new ones have further strengthened the conclusion that photo-ionization is the most important mechanism producing AGN emission lines:

1. Detailed models based on photo-ionization are able to reproduce approximately the relative line strengths with only a few free parameters.

2. Fluctuations in the strength of most broad lines are proportional to fluctuations in the continuum flux, if a delay is allowed in the comparison. Such a delay would be a natural consequence of the light travel time between the continuum source and the line production region, and the delays that are found are roughly consistent with the delays predicted by time-averaged photo-ionization models (§10.5.3.1.2).

3. In many nearby Seyfert galaxies, the narrow line region is seen to occupy a region with a (sometimes truncated) triangular projection on the sky. This would be a natural corollary of photo-ionization excitation if some structure close to the nucleus collimates the ionizing radiation. As will be discussed in Chapter 12, there is in fact good reason to think that just such a structure exists.

4. In some radio galaxies, the narrow line region is seen to extend *beyond* the end of the kiloparsec-scale radio jets, eliminating jet-driven shocks as an excitation mechanism.

These arguments are strong enough that we now believe we understand the line production mechanisms fairly well (albeit with certain technical problems). The photo-ionization picture also provides us with a clear and simple analogy for understanding the line emission process: think of the emission line region as a fluorescent lampshade. A small amount of mass (the actual quantity can be inferred on the basis of photo-ionization models) occupies a modest fraction of the solid angle around a central broad-band light source (the central continuum engine). Using its photo-ionization opacity, this thin screen filters the light that hits it and reradiates the energy in light of different colors.

In fact, the covering factor of the line-emitting gas can be derived by means of a very simple argument: If photo-ionization provides the energy for the lines, and the line luminosity is several percent of the continuum luminosity, then the gas must absorb several percent of the continuum. If the individual bits of gas are optically thick across most of the ionizing band, then the geometrical covering factor is that same several percent. In the simplest model—a distribution that is, at least in a statistical sense, spherically symmetric—we would then expect that the average Lyman continuum break should likewise be several percent. Depending on whether the individual bits of gas were individually optically thin and covered a large part of solid angle, or were individually optically thick and covered a small solid angle, this mean break would be realized either through a weak but consistent Lyman edge, or a strong but rare edge.

Despite the many successes of the photo-ionization model, important problems still remain. Most importantly, we do not understand the life cycle of the emission line gas or its dynamics. In addition, despite the strength of the arguments for photo-ionization as the "prime mover," one must recognize certain cautions about them. The narrow lines (almost) never vary on humanly observable timescales, so the argument based on variability analysis does not apply to them. That models are successful does not mean they are *uniquely* so. There may be both subsets of lines that are not generated by photo-ionization, and subsets of AGNs whose lines are also exceptions to the general conclusion. For example, there are cases in which "knots" in the radio structure correspond closely to enhancements in the narrow line surface brightness, suggesting that at least locally shocks can power line emission (see §10.6). In LINERs as a class the case for photo-ionization is most weakly supported.

10.3   Line Production by Photo-ionization

We began this chapter with the remark that one of the reasons AGN emission lines have received so much attention is the same reason why a drunk looks for his lost keys under the nearest lamppost. In this section, by laying out the physical processes relevant to the production of emission lines by photo-ionization, and summarizing the quantitative techniques necessary to account for them in a predictive calculation, we will demonstrate the depth of our understanding of this subject, and therefore why we can justifiably hope to find so much illumination on this subject. By its nature, this is a topic of voluminous detail, so the material presented here will necessarily be only an introduction; fortunately there is also a very large literature on this subject. Reviews can be found in Davidson and Netzer (1979), Osterbrock (1989), and Netzer (1990).

Much of the physics relevant to photo-ionization excitation of line emission also pertains to shock excitation. Indeed, shocks are particularly efficient at generating line emission when the immediate postshock gas radiates a continuous spectrum much of whose energy is in the EUV, so that these photons can be absorbed nearby and their energy transmuted into lines. Therefore, a large part of the apparatus, and even some of the results, of photo-ionization modeling can be carried over to the prediction of line emission from shock waves (§10.6). As just discussed, there may be some AGN contexts in which shock waves really are the prime source of energy for line emission.

Those wishing to explore the results of detailed photo-ionization calculations have a choice of several publically available codes, including Ferland's CLOUDY and Kallman's XSTAR. Comparison of the results of a number of codes given the same input parameters shows that they agree at the qualitative level, but close examination often reveals quantitative differences (Ferland 1995). These discrepancies illustrate the uncertainty level in the modeling and are variously due to different physical assumptions, different numerical algorithms, and different atomic data. Manuals for the public codes generally include pointers to tables of the particular atomic data they use; the contrasting choices made by different code authors point up some of the gaps in our knowledge of atomic properties. For tables containing frequently used atomic data, one can consult Allen (1973) or Osterbrock (1989). The most complete (and visually enlightening) collection of atomic levels and transitions is in Bashkin and Stoner (1975).

Line production depends on the density (or pressure) of the gas responsible; its ionization state, temperature, and distribution of atomic state populations; the background photon intensity; and the ability of the emitted photons to escape. Thus, the state of affairs in each of these respects must be specified in order to make predictions about the emergent line spectrum. Indeed, the very existence of the emission lines is intimately bound up with these processes. Thermal balance is achieved in large part by the emission of line photons; ionization balance entails a recombination casacade. Consequently, the prediction of line strengths is part-and-parcel of a calculation of cooling, ionization state, and so on. Most often one assumes that each process is in a time-steady state, but this is not always an appropriate assumption. In the following subsections we will examine how a time-steady state is achieved in each of these respects. At the end of this discussion, we will examine the conditions under which stationarity might actually be reached.

10.3.1   Geometry and internal dynamics

The geometrical disposition and pressure of the gas is ultimately determined by the history of forces acting upon it. We know little for sure about this history (see §10.7). Instead, to do calculations one generally assumes the simplest possible case, a plane-parallel slab. This might be a very good approximation if the true geometry of the gas is sheetlike clouds, or thin layers on the surface of an accretion disk, but the true configuration might well be something quite different.

The run of pressure inside this slab must be determined on the basis of external considerations. For example, if it floats in free space, one might expect its pressure to be constant if the time for sound waves to cross it is short compared to the continuum variability time, and the pressure scale length in whatever gas surrounds it is long compared to the slab thickness. On the other hand, if the sound crossing time is long compared to the variability time, differential heating effects might well drive continuing pressure fluctuations. Alternatively, if the slab rests on a massive object (it might be the atmosphere of an accretion disk or a star), the gravitational pressure scale length might be short compared to the cloud thickness, and the pressure might vary according to the usual hydrostatic law. Further complications come from radiation pressure (see also §10.7). The ram pressure of the incident continuum can be comparable to the gas pressure, so that any absorption by the gas introduces a substantial pressure difference from front to back. Unless balanced by a countervailing gas pressure gradient, this causes, of course, a net acceleration, whose consequences will be further discussed in §10.7. In addition, the energy density of trapped line photons can be large enough to contribute significantly to the internal pressure. Faced with these varying possibilities, some modelers throw up their hands and assume the gas maintains a constant density; although no identifiable physical process would produce this state, it (slightly) simplifies the calculation.

Whether or not the pressure varies internally, the temperature, the intensity of the incident continuum, and various other important properties can all vary by large amounts. To resolve these changes, one simply divides up the slab into as many subunits are necessary, and finds the emission line production from each one separately (fig. 10.5).

10.3.2   Ionization equilibrium

Many different processes affect the ionization state of atoms in the gas phase. To balance them all in the general case, and find a time-steady

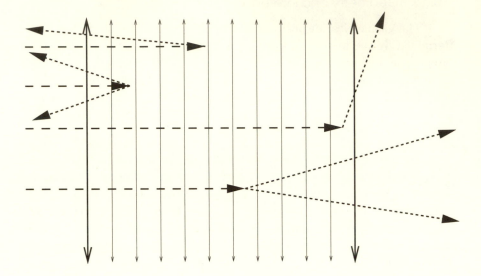

Fig. 10.5   A schematic view of a photoionized cloud. Ionizing continuum photons (long-dashed lines) strike it from one side; when they are absorbed, their energy is reemitted in other sorts of photons (short-dashed lines). Physical conditions such as temperature, ionization state, etc., must be calculated separately in each of the zones (separated by thin lines) inside the cloud.

solution for $n_{Z,j}$, the density of the jth ionization stage of the element with atomic number Z, one must solve an equation of the form

$$
n_{Z,j-1}\left[\int_{I(Z,j-1)} d\nu\, 4\pi \frac{J_\nu}{h\nu}\sigma_{\mathrm{ph},Z,j-1}(\nu) + \int_{I_K(Z,j-1)} d\nu\, 4\pi \frac{J_\nu}{h\nu}Y_{\mathrm{fl}}(Z,j-1)\sigma_{K,Z,j-1}(\nu)\right]
$$

$$
+ n_{Z,j-2}\int_{I_K(Z,j-2)} d\nu\, 4\pi \frac{J_\nu}{h\nu}Y_A(Z,j-2)\sigma_{K,Z,j-2}(\nu) + n_{Z,j+1}n_e\alpha_{\mathrm{rec},Z,j}(T)
$$

$$
+ n_{Z,j-1}n_eC_{Z,j-1}(T) + n_e^2 n_{Z,j+1}\alpha_{3,Z,j} = n_{Z,j}n_{Z',l}R_{\mathrm{ce}} - n_{Z,j+1}n_{Z',l-1}R'_{\mathrm{ce}}
$$

$$
+ n_{Z,j}\left[\int_{I(Z,j)} d\nu\, 4\pi \frac{J_\nu}{h\nu}\sigma_{\mathrm{ph},Z,j}(\nu) + \int_{I_K(Z,j)} d\nu\, 4\pi \frac{J_\nu}{h\nu}\sigma_{K,Z,j}(\nu)\right.
$$

$$
\left. + n_e\alpha_{\mathrm{rec},Z,j-1}(T) + n_eC_{Z,j}(T) + n_e^2\alpha_{3,Z,j-1}\right].
$$

$$(10.1)$$

Here n_e is the electron density, $\alpha_{\mathrm{rec},Z,j}$ is the recombination coefficient (radiative plus dielectronic) from stage $j+1$ to stage j, $\sigma_{\mathrm{ph},Z,j}$ is the valence shell photo-ionization cross section to go from stage j to stage $j+1$, $I(Z,j)$ is the ionization potential of stage j, $\sigma_{K,Z,j}$ is the K-shell photo-ionization cross section, $I_K(Z,j)$ is the K-shell ionization threshold, $Y_{fl}(K,j)$ is the fluorescence yield after K-shell ionization (see §8.7.1), $Y_A(K,j)$ is the Auger ionization yield, $C_{Z,j}$ is the collisional ionization rate coefficient for electrons to ionize stage j, $\alpha_{3,Z,j}$ is the rate coefficient for its inverse process, three-body recombination to stage j, and R_{ce} is the rate coefficient for charge exchange between species (Z,j) and (Z',l). The mean intensity J_ν usually contains only contributions from the incident continuum but can sometimes contain significant contributions from locally generated recombination and bremsstrahlung continua, or lines. Note that not all these rates and cross sections exist for every species; for example, Auger processes are irrelevant unless the ionization stage has at least one electron in a shell higher than the K-shell. Note also that in the conditions relevant to AGN emission line radiation, it is often, *but not always* a good approximation to count only ionizations from the ground state.

The ionization balance equations for individual species are supplemented by a single global charge balance equation:

$$n_e = \sum_{Z,j} (j-1)n_{Z,j}. \tag{10.2}$$

It is immediately apparent from the form of equation 10.1 that bookkeeping is a major issue in performing these calculations. It is also immediately apparent that the results depend on the atomic data, that is, the expressions one uses for the various rate coefficients and cross sections. The reliability of these numbers varies widely. It is a never-ending source of surprise how difficult it is to either calculate or measure quantities that are seemingly quite simple. Once an atom or ion has more than a few electrons, the complexity of possible atomic states makes calculations extremely computationally intensive; on the other hand, most of these ions exist for significant times only in the extremely low densities of astrophysical environments, making experimental work in terrestrial laboratories very difficult.

However, the job of collecting reliable atomic data is made easier by the existence of some self-consistency requirements, known as *Milne relations*, that can be imposed upon these coefficients. In a state of thermodynamic equilibrium, all processes and their inverses must be in detailed balance. Because photo-ionization and radiative recombination are mutual inverses, the radiative recombination cross section can be derived from the photo-ionization cross section:

$$\sigma_{rec,Z,j}(v) = \frac{g_{Z,j}}{g_{Z,j+1}} \frac{\epsilon^2}{m_e^2 c^2 v^2} \sigma_{\mathrm{ph},Z,j}(\epsilon), \tag{10.3}$$

where $\sigma_{rec,Z,j}(v)$ is the cross section for an electron of speed v to radiatively recombine with the ion $(Z, j + 1)$. Reciprocity between the ionized and recombined states is assured by setting the photon energy $\epsilon = I(Z,j) + (1/2)m_e v^2$. To estimate the valence shell photo-ionization cross section $\sigma_{\mathrm{ph},Z,j}$, we can use the hydrogenic photo-ionization cross section (eq. 7.95) with Z set to $Z - j + 1$. Then the recombination coefficient integrated over a Maxwellian distribution scales as

$$\alpha_{\mathrm{rec},Z,j} \sim \frac{I(Z,j)^2 \sigma_{\mathrm{ph},Z,j}[I(Z,j)]}{m_e c^2 (m_e k_B T)^{1/2}} \sim Z_{\mathrm{eff}}^2 \alpha_{\mathrm{fs}}^4 \left[\frac{I(1,1)}{k_B T} \right]^{1/2} c a_o^2. \tag{10.4}$$

In practice, the dimensionless coefficient of the final expression is several tens, so that

$$\alpha_{\mathrm{rec},Z,j} \simeq 2 \times 10^{-13} \left(Z - j + 1 \right)^2 T_4^{-1/2} \ \mathrm{cm}^3 \ \mathrm{s}^{-1}. \tag{10.5}$$

Using a rate coefficient found in this way assures that if the density is high enough to bring this transition into detailed balance, the rates predicted will obey the rules of thermodynamics even if there are errors in the rate coefficients.

In the simplest case, the ionization of H, the ionization balance equation simplifies greatly, and reveals a general principle that applies to all the elements:

$$n_e \alpha_{\mathrm{rec},1,1}(T) n_{1,2} = n_{1,1} \int_{I(1,1)} d\nu 4\pi \frac{J_\nu}{h\nu} \sigma_{\mathrm{ph},1,1}. \tag{10.6}$$

Because H is by far the most abundant element in most astrophysical environments, the charge balance equation can be approximated by

$$n_e \simeq n_{1,2} \tag{10.7}$$

when H is mostly ionized. With this approximation, the H ionization balance equation can be rewritten

$$\begin{aligned}
\frac{n_e}{n_{1,1}} &= \frac{\int_{I(1,1)} d\nu 4\pi \frac{J_\nu}{h\nu} \sigma_{\mathrm{ph},1,1}}{\alpha_{\mathrm{rec},1,1}(T)} \\
&= \frac{4\pi J_{\mathrm{ion}}}{n_e} \frac{\int_{I(1,1)} d\nu \frac{J_\nu}{h\nu} \sigma_{\mathrm{ph},1,1}}{\alpha_{\mathrm{rec},1,1}(T)},
\end{aligned} \tag{10.8}$$

where $\mathcal{J}_\nu$ is the spectral shape normalized to unity when integrated from $I(1,1)$ to ∞. This relation demonstrates that the neutral fraction of H is inversely proportional to the ratio J_{ion}/n_e with a proportionality constant that is mostly determined by atomic constants but does depend (weakly) on the spectral shape.

This ratio, or quantities closely related to it, is called the *ionization parameter* and plays a fundamental role in determining the properties of any photoionized gas. Although the H ionization balance depends on it in a very direct and simple way—assuming a typical temperature $\sim 10^4$ K (see §10.3.3 for *why* this is a typical temperature),

$$\frac{n_{1,1}}{n_e} \sim 10^{-4} \left[\frac{4\pi J_{\mathrm{ion}}}{cI(1,1)n_e} \right]^{-1}, \tag{10.9}$$

it plays an equally strong role in the ionization balance of the other elements, as can be seen by dividing through the general ionization balance equation by n_e. Provided the spectrum extends reasonably smoothly to high enough energies that all interesting ionization potentials are covered, the ionization balance of all the elements is well predicted by the ionization parameter, with only minor adjustments due to details of the spectral shape. The reason why it is this integral quantity that matters, rather than the intensity at any particular ionization edge, is that to ionize up to state Z^{+n} it is necessary first to ionize Z, Z^{+1}, Z^{+2}, and so on up to $Z^{+(n-1)}$. Each of these is most efficiently ionized by photons just above its own ionization potential. Consequently, the best measure of photoionizing effectiveness is an integral over a broad range of frequencies.

Several different definitions for the ionization parameter exist in the literature. They are all conceptually the same, but each has its own technical advantages and disadvantages, as well its own promoters and detractors. The first to be used was $L/(n_{\mathrm{H}}r^2) \equiv \xi$. Note that this is formally a ratio of radiation flux to density, rather than intensity to density. When the source of radiation is a distant point source and there is little scattering along the way (as many imagine the situation to be in AGN emission line regions), the flux is identical to the intensity, but this is not necessarily always so. The next historically was $\int_{I(1,1)} d\nu\, J_\nu/(ch\nu n_{\mathrm{H}})$, originally called Γ, but more recently called U. This ionization parameter is the ratio of the number density of ionizing photons to hydrogen nuclei. For purposes of calculating the ionization equilibrium, this is probably the most useful form because photo-ionization rates depend on the *numbers* of photons rather than their energies, and this quantity is an integral over the entire ionizing range. Some have preferred, however, to use the quantity

$J_\nu(\nu_H)/(hn_H) \equiv U_1$ because the H ionization balance is governed primarily by photons right at the edge, and because this quantity is closer to being an observable than the integral over the whole ionization band. Finally, there is the form $J_{ion}/(2.3n_H k_B T c) \equiv \Xi$. This form gives the ratio of ionizing radiation pressure to gas pressure (the factor 2.3 is appropriate when the gas is fully ionized and nearly all the free electrons come from H and He in cosmological proportions). It is most useful when regions of different temperature are juxtaposed. Hydrodynamics may then impose pressure balance, and a form of the ionization parameter that is the same in the two adjacent equal-pressure systems is useful.

Several ionization-related processes can be important in AGN emission line regions that are often negligible in other astrophysical contexts. These are: dielectronic recombination, three-body recombination, and charge exchange. In contrast to radiative recombination, in dielectronic recombination no immediate photon is emitted, so the initial energy of the recombined atom is above its ionization potential:

$$e^- + Z^{+n} \rightarrow Z^{+(n-1)^*}. \tag{10.10}$$

The excitation energy is divided over two electrons, so that both can be in bound states. Later, the two electrons de-excite by separate radiative transitions.

Three-body recombination is the inverse process to electron collisional ionization:

$$e^- + Z^{+n} \rightleftharpoons Z^{+(n+1)} + 2e^-. \tag{10.11}$$

Because it requires three particles (two electrons and an ion) to come together simultaneously, it is only important at high densities, but these can occur in AGN emission line regions. Just as for ordinary radiative recombination, we can find the recombination coefficient in this instance by using the principle of detailed balance

$$\alpha_{3,Z,j} = \frac{n_{Z,j}(n_e, T)}{n_e n_{Z,j+1}(n_e, T)} C_{Z,j}(T), \tag{10.12}$$

where the density ratio is evaluated using the Saha equation at temperature T and electron density n_e.

Charge exchange reactions occur efficiently when there are fortuitous energy matches between states of the system Z^{+j}, Y^{+l} and $Z^{+j\pm1}, Y^{+l\pm-1}$. The cross section can be quite large when the resonance is nearly exact; in addition, when $j = 1$ and $l = 0$ (or the other way round), there is an attractive polarization that enhances the cross section. Because the energies are nearly the same, these reactions can go either way even at modest temperatures and, absent other ionization or recombination processes, quickly reach an equilibrium determined only by the relative statistical weights.

Perhaps the single most important charge-exchange reaction is

$$O + H^+ \rightleftharpoons O^+ + H. \tag{10.13}$$

This reaction is particularly strong because the energy difference is only 0.19 eV; under a wide range of conditions it enforces an ionization ratio

$$\frac{n_{8,1}}{n_{8,2}} = \frac{9}{8} \frac{n_{1,1}}{n_{1,2}} \tag{10.14}$$

because the ground state term of O is ^3P and the ground state term of O$^+$ is ^4S, whereas the ground state term of H is ^2S.

10.3.3 Thermal equilibrium

10.3.3.1 heating

Photoionization delivers heat by ionizing atoms with photons having energy greater than the ionization threshold. The heating rate per unit volume can therefore be written as

$$H = \int d\epsilon \, 4\pi \frac{J_\epsilon}{\epsilon} \left\{ \sum_{Z,j} n_{Z,j} \sigma_{\mathrm{ph},Z,j}(\epsilon) \left[\epsilon - I(Z,j)\right] + \right.$$

$$\left. \sum_{Z,j} \sigma_{K,Z,j}(\epsilon) \left[\epsilon - I_K(Z,j) - Y_{\mathrm{fl}}\epsilon_{\mathrm{fl}}(Z,j) - Y_A(Z,j)I(Z,j+1)\right] \right\}. \tag{10.15}$$

Note that in this form of the energy tally the energy absorbed from the continuum that goes into satisfying ionization potentials has been subtracted off. The electrons freed by photo-ionization carry off the difference in energy between the photons that created them and the relevant ionization potential, and, provided $n_e/n_{1,1} \gtrsim 0.1$, quickly share it by Coulomb collisions with the other electrons (this is almost as good an assumption for Auger electrons as for electrons ionized from the valence shell).

Other heating processes can also operate, though they tend to be less important in line-emitting regions. Photons can also provide heat through Compton scattering (§§6.3.3, 8.3.1.1), though this is usually only important compared to photo-ionization if the gas is highly ionized because the cross section is so small. At the other end of the spectrum, free-free absorption can sometimes contribute a significant amount of energy. When the temperature is comparatively low, but there is still a significant ionized fraction, there may also be enough H$^-$ to create significant opacity in the red and infrared continuum. It is likewise possible to imagine situations in which the gas is compressed adiabatically, or is warmed by conduction from a nearby hotter gas, or there is some sort of wave damping, or ... When appropriate, extra processes such as these can always be added to H.

10.3.3.2   collisionally excited line emission

In most cases, the strongest cooling mechanisms are electron collisional excitation of atomic and ionic lines and several forms of continuum radiation. Consider the collisionally excited lines first. Most lie in the ultraviolet. The reason is that the characteristic energy of valence shell electronic transitions is $\sim (Z_{\rm eff}\alpha_{\rm fs})^2 m_e c^2$, where $Z_{\rm eff}$ is the effective nuclear charge felt in the valence shell. Typically $Z_{\rm eff}$ is only a few, so that the characteristic energy is several to several tens of eV.

To find the rate of excitation, we must first find the cross section for external electrons to promote the atom from state l to state u. This is most simply done by means of the Born approximation, which, to be proper, is usually restricted to cases in which the energy E of the external electron is much larger than the excitation threshold.* In that limit,

$$\sigma_{lu} = \frac{a_o^2(\mu/m_e)^2}{4\pi^2} \frac{|\vec{k}'|}{|\vec{k}|} \int d\Omega \left| \int d^3x\, U_{lu}(\vec{x}) \exp(-ia_o\vec{q}\cdot\vec{x}) \right|^2 . \quad (10.16)$$

Here we have retained a factor μ/m_e (the ratio of the reduced mass of the scattering system to the electron mass) so that this expression can also be applied to collisions in which the scattering particle is not an electron. The integration over solid angle refers to the direction of motion of the incident particle after scattering; the volume integral (in units of the Bohr radius a_o^3) is taken over the extent of the scattering potential U (in units of e^2/a_o). $\vec{q} = \vec{k}' - \vec{k}$ is the change in wavenumber of the incident electron, and the perturbation potential

$$U_{lu}(\vec{x}) = \int d^3x'\, \psi_u^*(\vec{x}')V(\vec{x},\vec{x}')\psi_l(\vec{x}') \quad (10.17)$$

---

*Technically, the Born approximation is justified by one of two conditions: When the de Broglie wavelength of the scattering particle is much longer than the extent of the scattering potential a, the Born approximation requires that the scattering potential $|V| \ll \hbar^2/(\mu a^2)$. Alternatively, (and more commonly) when the de Broglie wavelength is short compared to a, the Born approximation requires $|V| \ll \hbar v/a$, where v is the speed of approach.

with $V(\vec{x}, \vec{x}')$ the potential at $\vec{x}$ due to the atomic nucleus and an electron at $\vec{x}'$. The functions $\psi_{l,u}$ are the wave functions for the lower and upper states. From the form of the volume integral in equation 10.16, it is clear that the cross section can be written in the simpler form

$$\sigma_{lu} = \pi a_o^2 \left(\frac{\mu}{m_e}\right)^2 \frac{\Omega_{lu}(k)}{g_l(ka_o)^2}$$
$$= 1.2 \times 10^{-15} \left(\frac{\Omega_{lu}}{g_l}\right) \left(\frac{\mu}{m_e}\right) E_{eV}^{-1} \text{ cm}^2, \tag{10.18}$$

where the dimensionless quantity $\Omega_{lu}(k)$, called the *collision strength*, is a slowly varying function of k that is generally ~ 1. The statistical weight of the lower state g_l has been factored out so that the upward and downward rates are both $\propto \Omega_{lu}$, without any apparent dependence on statistical weights. Thus, the basic scale of atomic collision cross sections is $\sim \pi a_o^2$. Because $k^2 = 2 m_e E / \hbar^2$, $\sigma_{lu} \propto E^{-1}$ for electrons well above threshold.

Taking the electron distribution to be Maxwell-Boltzmann at temperature T, the rate coefficient (i.e., the product of the cross section and electron speed averaged over the electron distribution function) for collisional excitation is

$$R_{\text{ex},lu} = \frac{\hbar^2}{g_l m_e^{3/2}} \left(\frac{2\pi}{k_B T}\right)^{1/2} \int_{\epsilon/k_B T} dy \, e^{-y} \Omega_{lu}(y)$$
$$= 8.6 \times 10^{-8} g_l^{-1} T_4^{-1/2} \int_{\epsilon/k_B T} dx \, e^{-x} \Omega_{lu}(x) \text{ cm}^3 \text{ s}^{-1}, \tag{10.19}$$

where $y = E/k_B T$. It is apparent that

$$R_{\text{ex},lu} \propto \langle \Omega_{lu} \rangle T^{-1/2} \exp\left(-\frac{\epsilon_{lu}}{k_B T}\right). \tag{10.20}$$

As we shall see in the next paragraph, the argument of the Boltzmann exponential is often ~ 10, so that typical excitation rate coefficients are $\sim 10^{-9}$ cm^3 s^{-1}. Because collision strengths are almost always order unity (unlike photon transitions, there are no "selection rules" that forbid collisional transitions between states with certain quantum numbers), the rates depend almost entirely on the Boltzmann factor, which gives the fraction of electrons with enough energy to be above threshold. Forbidden and semiforbidden lines can be excited just as easily as dipole permitted lines, provided only that their energies are not too high.

The exponential dependence of the rates on temperature leads to a very robust qualitative result: the temperature of photoionized gases in radiative balance almost always lies within the range $\simeq 8000$ K to $\simeq 30,000$ K. To see this, suppose that the cooling is dominated by a single collisionally excited line. Then the equation balancing heating and cooling is

$$C = n_e n_l \epsilon_{lu} \frac{\langle \Omega \rangle \hbar^2}{g_l m_e^{3/2}} \left(\frac{2\pi}{k_B T} \right)^{1/2} \exp \left(-\frac{\epsilon_{lu}}{k_B T} \right) = H. \qquad (10.21)$$

Although H is generally rather insensitive to T, the left-hand side of this equation depends on it very strongly:

$$\frac{\partial \ln C}{\partial \ln T} = -\frac{1}{2} + \frac{\epsilon_{lu}}{k_B T}. \qquad (10.22)$$

The dependence of the cooling rate on temperature is so strong because $\epsilon_{lu}/k_B T \sim 10$ for ultraviolet lines radiated by a gas with $T \sim 10^4$ K. Consequently, even large changes in the heating rate result in relatively small changes in T when T is in this favored range.

In the preceding discussion it has been assumed that any collisional excitation ultimately leads to a photon that escapes from the gas. This is not always so. If an electron collisionally *de-excites* the atom while it is in the excited state, the heat returns to the gas. In the limit of zero optical depth (see §10.3.5.1 for a more general discussion), the probability of this happening is the ratio of the collisional rate to the decay rate:

$$P_{deex} = \frac{n_e R_{\text{deex},ul}}{A_{ul}}, \qquad (10.23)$$

where A_{ul} is the Einstein A coefficient (a convenient representation of A_{ul} in "natural" atomic units is given in eq. 10.49), and

$$R_{\text{deex},ul} = R_{\text{ex},lu} \exp \left(\frac{\epsilon_{lu}}{k_B T} \right) \frac{g_l}{g_u} \qquad (10.24)$$

because the de-excitation threshold energy is zero. Thus, there is a critical density $n_{\text{crit}} \equiv A_{ul}/R_{\text{deex},ul}$ above which de-excitation by collision is more likely than de-excitation by photon emission. As we have already seen, typical atomic collision cross sections are comparable to atomic sizes, so at $T \sim 10^4$ K, $R_{\text{deex},ul} \sim 10^{-9}$ cm^3 s^{-1}. Because transitions permitted by the electric dipole (E1) selection rules have $A_{ul} \sim 10^8$ s^{-1}, in the optically thin limit $n_{\text{crit}} \sim 10^{17}$ cm^{-3} for them. As we will see in §10.3.5.1, optical depth effects can reduce n_{crit} far below this characteristic value. Semiforbidden and forbidden transitions can have critical densities many orders of magnitude smaller due to their much smaller radiative decay rates. The

[OIII] 5007 line, a strongly forbidden line, has a critical density $\sim 10^6$ cm^{-3}, for example.

One other tacit assumption must also be discussed: that the electron distribution function follows the Maxwell-Boltzmann form. Although this is essentially always true for the majority of the free electrons, in conditions of small ionized fraction (less than about 0.1), most of the line excitation is due to the minority of electrons found in a high-energy nonthermal tail. Electrons created by photo-ionization are born "fast" because their characteristic energies are of order the ionization potential of their parent atom, and these are generally one to two orders of magnitude larger than $k_B T$. When the electrons are made by K-shell photo-ionization of heavy elements, the contrast with $k_B T$ is even greater. If there are numerous other free electrons with which to share energy, Coulomb scattering rapidly degrades them to quasi-thermal energies. However, when there are relatively more atoms than free electrons, the free electrons are more likely to lose energy by Coulomb scattering of *bound* electrons, whether to the continuum (i.e., collisional ionization) or to excited bound states. The secondary electrons created by collisional ionization typically have energies of order their ionization potentials, so, relative to the typical temperature of photoionized gas ~ 1 eV, they, too, are "fast." Only the remainder of their energy below the threshhold for exciting the more common bound-bound transitions is delivered to the thermal electron pool.

10.3.3.3 continua

Radiative recombination is usually the most important continuum coolant in AGN emission line regions. Recombination lines (see §10.3.5.2), though important to compute, do not contribute to cooling if the ionization energy is subtracted out of the heating rate. On the other hand, the portion of the recombination energy that exceeds the ionization potential of the state to which the recombination occurs does count as cooling. On average, this is $\simeq (0.8\text{--}0.9)k_B T$ per recombination (it is somewhat less than $(3/2)k_B T$, the mean energy per electron, because the radiative recombination coefficient for hydrogenic atoms is $\propto E^{-1}$). This energy is carried away either by a directly generated recombination continuum photon if the recombination is radiative, or by line photons created after the atom relaxes if the recombination is dielectronic. Although the Lyman continuum is often the most important of the numerous possible recombination continua, the emergent flux in the Balmer continuum can also sometimes be important.

As already discussed in §7.5.1, free-free emission is very closely related

333

to radiative recombination. Which is a more important coolant depends on the ratio of the temperature to the H ionization potential. Because the photon radiated has energy $\lesssim k_B T$ in the former case, but $\simeq I(i,j) + k_B T$ in the latter, the phase space for the latter becomes rather greater than for the former when $k_B T$ is smaller than the typical ionization potential. As we have just seen from the argument about collisional line excitation, this is generally the case for gases in photo-ionization equilibrium. Consequently, even though the cooling attributed to the recombination radiation only includes the $k_B T$ piece, it still generally dominates free-free radiation as a cooling mechanism.

10.3.3.4 cooling functions

The sum of all the rate coefficients for cooling processes is called the *cooling function*. Because most cooling processes scale in proportion to the electron density times the ion density (e.g., collisional excitation of lines or bremsstrahlung), the total cooling rate per unit volume can be written as $C = n_H^2 \Lambda(T, \Xi)$. In this form, Λ contains all the information about the ionization balance and the atomic data, as well as the temperature dependence, but is only weakly dependent on density.

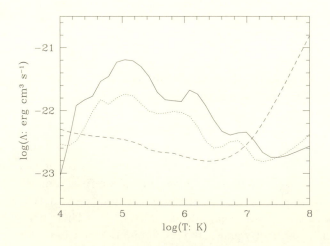

Fig. 10.6   The cooling function as a function of temperature for several values of Ξ. The solid curve is for $\Xi = 10^{-3}$, the dotted curve $\Xi = 0.1$, and the dashed curve $\Xi = 10$. In this plot temperature is in units of K, and the cooling function Λ is in erg cm^3 s^{-1}. The sharp rise at high temperature seen in the $\Xi = 10$ curve is due to Compton cooling.

334

When the ionization state of the gas provides large numbers of ions and atoms with line transitions to excite, the cooling function rises to a sharp peak at $T \sim 10^5$ K, as predicted by the arguments of §10.3.3.2 (fig. 10.6). The cooling function diminishes at higher temperatures as the increase in temperature leads to larger collisional ionization rates, and a diminution in the abundance of species with excitable transitions. As the temperature rises past 10^7 K, all that is left is continuum processes—bremsstrahlung (which by itself would produce a $\Lambda \propto T^{1/2}$) and inverse Compton cooling ($\propto T^2 \Xi$ in these units). However, as the ionization parameter increases, the abundance of species with internal transitions to excite diminishes even when the temperature is as low as 10^5 K. As a result, the height of the peak in the cooling function falls. The consequence this has for the nature of photo-ionization equilibria is discussed further in §10.5.1.

10.3.4   Radiation transfer

Everything we have done so far assumed a purely local treatment. However, the scattering lengths for many of the photons we follow can be very small compared to the size of the emission region. Consequently, the radiation intensity can vary substantially within the emission region, and we must solve the radiation transfer problem for these photons.

10.3.4.1   continuum transfer

When many of the heavy element atoms are chemically bound into grains, dust opacity can affect a very broad range of continuum frequencies. For the present purposes it suffices to summarize a few global properties of dust opacity. More details can be found in reviews such as Mathis (1990) or Whittet (1992).

Small solid grains can both reflect and absorb light incident upon them; both scattering and absorption can be polarization dependent if the grains are not spherical. When the grains have a size and composition distribution similar to those in our Galactic interstellar medium, their total opacity scales roughly $\propto \lambda^{-1}$ in the optical and near-ultraviolet, but there can be significant deviation from this simple law farther into the ultraviolet, as well as systematic changes in the wavelength-dependence when the *total* extinction increases (fig. 10.7 and Cardelli, Clayton, and Mathis 1989). At higher energies, the grains gradually become more transparent; in the soft X-ray band, the opacity of grains is essentially just the summed photoelectric opacity of their constituent atoms. In the V band (5000 Å), the total opacity in low-density regions of our local interstellar medium is $\simeq 5 \times 10^{-22}$ cm^2 H^{-1} $\simeq 200$ cm^2 gm^{-1}.

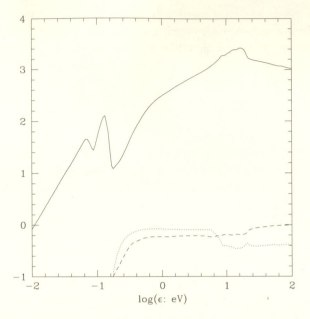

Fig. 10.7   Dust extinction as a function of wavelength. The solid curve is the logarithm of the total extinction in units of the Thomson opacity; the dotted curve is the logarithm of the albedo; the dashed curve the logarithm of the mean cosine of the scattering angle. These curves are based on the "astronomical silicate" model of Laor and Draine (1993) normalized to the local extinction at 5000 Å. The grain-size distribution is $dn/da \propto a^{-3.5}$ from $a = 0.005$ to $1~\mu$.

How much of this extinction is absorption, and how much scattering, is somewhat model dependent, but it is likely that the two contributions are roughly comparable over a wide range of frequencies (see fig. 10.7 for the predictions of a particular model). There is a general trend for the scattering component to become more and more forward-throwing as the photon energy increases.

In fact, in the conditions prevailing in the gas radiating the broad emission lines, there probably is essentially no dust. Although the characteristic temperature is one to two orders of magnitude below the temperature at which ion collisions efficiently destroy grains by sputtering (Tielens et al. 1994), the density can still be high enough that grains would be destroyed rapidly. In addition, the radiation intensity is likely to be high enough that the equilibrium temperature of dust grains would be higher than the sublimation temperature ($\sim$ 1500–2000 K) for even the most refractory grains.

By contrast, in the gas radiating the narrow emission lines, dust is far more likely to survive. Densities are smaller by factors of 10^6 or more, while the dynamical time is greater by factors more like 10^4 or so. The radiation intensity is also smaller by factors of at least 10^6, so the equilibrium temperature (which scales like intensity to the $1/6$ power rather than $1/4$ because the grains are too small to radiate efficiently) is reduced to several hundred K. Dust in these clouds can therefore be expected to have a significant impact on the emergent spectrum. Its effect is greatest when the ionization parameter is relatively large because it can then be the dominant opacity at energies above 1 Rydberg as well as below: for the standard dust/gas ratio, dust dominates the opacity even in the ionizing continuum when $n(1,1)/n_H < 5 \times 10^{-4}$, which comes when $\Gamma = U > 0.02$. Depletion of heavy elements from the gas phase by condensation onto grains can also affect the emission from photoionized regions by altering the abundances of the different atoms available for excitation (Binette et al. 1993).

If there is no dust, at energies below 1 Rydberg there is generally only weak continuum opacity. In certain conditions, particularly those of large optical depth and density, there can be significant opacity in the red and infrared from H^-, or in the near-ultraviolet due to H excited state photo-ionization, but the opacity above 1 Rydberg due to ground state H photo-ionization is always far greater. At slightly higher energies, He photo-ionization can also be significant. Above a few hundred keV, K-shell photo-ionization becomes important, but (as we have already seen when discussing X-ray radiation: §8.7.1) the cross section per H atom is much smaller.

The simplest model for continuum transfer, then, is to approximate the opacity as being solely H ground state photo-ionization. As Stromgren showed for HII regions, because the Lyman continuum intensity regulates its own opacity, there can be very sharp transitions in the intensity of this band. In starlight-ionized HII regions, this transition can be extremely sharp; in AGN-ionized regions, the transition region is somewhat broader because of the wider energy range of ionizing photons.

To define the width of the ionization transition zone, all that is necessary is to solve the (very simple) transfer equation for the ionizing continuum:

$$\frac{\partial I_\nu}{\partial \tau_L} = -\left(\frac{\nu_H}{\nu}\right)^3 I_\nu, \tag{10.25}$$

where the incident radiation has been assumed to enter the slab parallel to its normal, and τ_L is the optical depth at the Lyman edge. The frequency dependence of the opacity simply mirrors the frequency dependence of the

Lyman photo-ionization cross section. The intensity of the incident beam as a function of optical depth is then

$$I_\nu(\tau_L) = I_\nu(0) \exp\left[-\tau_L \left(\frac{\nu}{\nu_H}\right)^{-3}\right]. \tag{10.26}$$

So long as the neutral fraction is small, the optical depth grows with distance z into the cloud as

$$\frac{d\tau_L}{dz} = n_{1,1}\sigma_{\text{ph},1,1}(\nu_H) = \frac{n_e^2 \alpha_{\text{rec},1,1}}{\int_{\nu_H} d\nu\,\sigma_{\text{ph},1,1}(\nu)I_\nu/(h\nu)}. \tag{10.27}$$

The photo-ionization rate integral is simply evaluated when the incident spectrum is a power law:

$$\int_{\nu_H} d\nu \left(\frac{\nu}{\nu_H}\right)^{-3} \frac{I_\nu}{h\nu} = h^{-1}\int_1 dx\, x^{-4-\alpha} I_\nu(0,\nu_H)\exp\left(-\frac{\tau_L}{x^3}\right)$$
$$= \frac{I_\nu(0,\nu_H)}{h(1+\alpha/3)}\tau_L^{-1-\alpha/3}\gamma(1+\alpha/3,\tau_L), \tag{10.28}$$

where α is the power-law index (in flux units), and $\gamma(a,t)$ is the incomplete gamma function. When $a > 0$ and $t > 1$, $\gamma(a,t)$ is slowly varying, so the dominant dependence is the power law in τ_L. Thus, unlike the case of a stellar (exponential) spectrum, whose photo-ionization rate drops exponentially with Lyman edge optical depth, the photo-ionization rate due to a power-law spectrum declines only as a power law in τ_L.

However, this also means that for as long as the gas is mostly ionized,

$$\frac{d\tau_L}{dz} \propto \tau_L^{1+\alpha/3}, \tag{10.29}$$

so

$$\tau_L \propto (z_o - z)^{-3/\alpha}. \tag{10.30}$$

That is, from the point at which the Lyman edge first becomes optically thick to the point z_o at which even very much harder photons are absorbed and the neutral fraction becomes appreciable, the optical depth rises very rapidly. In other words, although the ionization transition is comparatively gradual on the optical depth scale, it is still fairly sharp in terms of physical distance.

The distance at which the ionization transition occurs can also be estimated by a different simple argument: It is the depth at which the integrated recombination rate equals the rate at which ionizing photons arrive

$$\mathcal{N}_{\rm ion} = \int_0^{z_o} dz\, n_e^2 \alpha_{\rm rec,1,1},$$
(10.31)

where $\mathcal{N}_{\rm ion}$ is the number flux of ionizing photons. This can be rewritten as

$$N_e \simeq \frac{\mathcal{N}_{\rm ion}}{n_e \alpha_{\rm rec,1,1}} = \frac{\Gamma c}{\alpha_{\rm rec,1,1}} \simeq 1.5 \times 10^{23} \Gamma \ {\rm cm}^{-2},$$
(10.32)

where N_e is the column density of ionized electrons and n_e is the electron density at the exposed edge. Thus, the column density that can be kept ionized is proportional to the incident ionization parameter.

Both of the preceding arguments can be improved quantitatively by a small correction known as the "on-the-spot" approximation. When the total Lyman edge optical depth of the gas is small, most recombination photons escape. However, when $\tau_{\rm L,tot} > 1$, the photons produced by most recombinations to the ground state are reabsorbed inside the cloud. This effect can be neatly taken into account in an approximate way by defining an effective recombination rate coefficient:

$$\alpha_{\rm rec-eff,1,1} = \alpha_{\rm rec,1,1}(n \ge 2) +$$
$$\begin{cases} 0.5\alpha_{\rm rec,1,1}(n=1) & \tau_{\rm L}(z) < 1 \text{ and } \tau_{\rm L,tot} > 1 \\ 0 & \tau_{\rm L}(z) > 1 \text{ and } \tau_{\rm L,tot} - \tau_{\rm L}(z) > 1 \\ 0.5\alpha_{\rm rec,1,1}(n=1) & \tau_{\rm L}(z) > 1 \text{ and } \tau_{\rm L,tot} - \tau_{\rm L}(z) < 1. \end{cases}$$
(10.33)

The physical meaning of this approximation is that the effective total recombination rate coefficient is reduced *locally* by about a factor of two to account for the absorption of ground-state recombination photons that might actually occur elsewhere. The ground state recombination rate is cut in half near the illuminated surface of the cloud because diffuse Lyman continuum photons created there and directed outward can escape, while those directed inwards can not.

10.3.4.2   line transfer

It is very easy for resonance lines to be extremely optically thick. The integrated oscillator strength for photo-ionization is generally comparable to the oscillator strength for an individual resonance line, but their cross sections differ by the ratio of their bandwidths:

$$\frac{\sigma_{\rm ph}}{\sigma_l} \sim \frac{c}{\Delta v_{\rm th}} \sim 10^4 T_4^{-1/2},$$
(10.34)

where $\Delta v_{\rm th}$ is magnitude of the atomic thermal speed. Consequently, if the Lyman edge is optically thick, the Lyman lines will be very optically thick. Because we frequently encounter neutral fractions $\sim 10^{-4}$, so that the abundance of neutral H is comparable to the abundance of C, N, or O, the resonance lines in the most abundant ionization stages of these ions are also likely to be very optically thick.

Because transitions between different bound states can only occur for specific photon frequencies, the cross section for a photon to interact with an atom through such a transtion depends very strongly on frequency. The opacity profile can be found by convolving the Lorentzian profile of an individual atom (in its own rest frame) with the Maxwell-Boltzmann velocity distribution for the gas. For frequencies within a few thermal Doppler widths of line center, the result is a Gaussian:

$$
\sigma_l(\nu) = 4\pi^{3/2}a_o^2\frac{f}{q}\left(\frac{I_H}{k_BT}\right)^{1/2}\left(\frac{Am_p}{m_e}\right)^{1/2}\exp\left[-\left(\frac{\nu-\nu_l}{\Delta\nu_D}\right)^2\right]
$$

$$
= 1.06\times10^{-13}\frac{f}{q}T_4^{-1/2}A^{1/2}e^{-x^2}\,\mathrm{cm}^2,
$$
(10.35)

where f is the transition oscillator strength, q is its energy in Rydbergs, A is the atomic mass in proton mass units, and x is the deviation of the photon frequency ν from line center ν_l in units of the *thermal Doppler width*

$$
\Delta\nu_D = \left(\frac{2k_BT}{Am_pc^2}\right)^{1/2}\nu_l.
$$
(10.36)

Thus, the opacity drops dramatically when the fractional frequency offset from line center is greater than $\sim \Delta\nu_D$. Farther from line center, the opacity is dominated by the damping wings; taking this broadening into account changes the line profile from the pure Gaussian $\exp(-x^2)$ to approximately

$$
\exp\left(-x^2\right) + \frac{a}{\sqrt{\pi}x^2},
$$
(10.37)

where

$$
a = \frac{\Gamma_{lu}}{4\pi\Delta\nu_D}.
$$
(10.38)

Here Γ_{lu} is the total width of the line, whether due to radiative decay of the upper state or collisions. The damping wings dominate when $x \gtrsim \sqrt{\ln(a/x^2)}$, which is typically $\simeq$ 3–5.

The same random velocities ($\vec{\beta}_{\mathrm{th}}c$, where $|\vec{\beta}_{\mathrm{th}}| \sim (k_BT/Am_pc^2)^{1/2}$) that set the breadth of the resonant portion of the line also create photon frequency changes as a result of scattering. Energy is conserved in scattering, so a photon's outgoing frequency *in the atom frame* is identical (up to the very small loss due to recoil) to its incoming frequency. That is, to use the jargon phrase, the process is *Coherent in the Frame of the Atom*. However, *relative to the lab frame*, a fractional frequency change of order β_{th} results because of the change in direction of the photon. To compute this frequency change, one must relate the results of two Lorentz transformations between the atom frame and the lab frame, one of the incoming photon's energy, and one of the outgoing photon's energy. The result is

$$\frac{\Delta\nu}{\nu} = \vec{\beta}_{th} \cdot (\hat{n}_f - \hat{n}_i) \frac{1}{\gamma(1 - \vec{\beta}_{th} \cdot \hat{n}_i)(1 - \vec{\beta}_{th} \cdot \hat{n}_f)}, \tag{10.39}$$

where $\hat{n}_{i,f}$ are the initial and final photon directions and γ is the usual Lorentz factor.

As long as the photon remains within the thermal core of the line, the redistribution probability is reasonably well approximated mathematically by a physically very different picture: that the outgoing frequency is entirely uncorrelated with the incoming frequency and instead is simply drawn freely from the Gaussian emission profile produced by the Maxwell-Boltzmann velocity distribution (the jargon term for this is *Complete Redistribution in the Frame of the Fluid* or CRFF). When this is the case, the probability for emission farther from line center than x is $\simeq \text{Erfc}(x) \sim \exp(-x^2)$.

Ordinarily one might expect that the number of scatterings to escape would be the usual $N_{\text{scatt}} \sim \tau^2$ that comes from random walk theory. Resonance line photons are different, however, precisely because of the frequency shifts that occur at each scattering.

To see this contrast, we estimate how far from line center a photon must be in order to see the region as transparent. In thermal Doppler units, x must be $\geq x_c$, where

$$x_c \simeq \begin{cases} \sqrt{\ln \tau_{lc}} & \text{thermal core} \\ (a\tau_{lc})^{1/2}/\pi^{1/4} & \text{damping wings,} \end{cases} \tag{10.40}$$

and τ_{lc} is the line center optical depth. Achieving such a large offset has a probability per scatter $\sim \tau_{lc}^{-1}$ when the CRFF approximation applies, and so requires on average $\sim \tau_{lc}$ scatterings before it happens. For large τ_{lc} this is many fewer than the τ_{lc}^2 scatterings required for escape by spatial diffusion. Therefore, frequency wandering is more likely to drive resonance line escape than conventional spatial wandering. This "diffusion" in frequency space (it is really a Fokker-Planck process, not a diffusion process) thus creates a second channel for escape. Instead of wandering in space until it crosses the edge of the gas, a photon is more likely to wander in frequency until it sees the gas as optically thin.

The exact *escape probability* per scattering ϵ_{esc} depends on the geometry of the system. For a slab whose optical depth is not great enough for the damping wings to be important and in which continuous opacity is also negligible, it is

$$\epsilon_{esc} = \left[1 + \tau_{lc}\sqrt{2\pi \ln(2.13 + \tau_{lc}^2)}\right]^{-1} \qquad (10.41)$$

(Hollenbach and McKee 1978).

A corollary of the fact that the escape probability scales roughly $\propto \tau_{lc}^{-1}$ is that the total pathlength traversed by a resonance photon before it leaves a slab of thickness l is not greatly enhanced over the free flight distance. Each flight between scatters has length $\sim l/\tau_{lc}$ (when the frequency wandering does not take the photon out of the line core), and there are $\epsilon_{esc}^{-1} \sim \tau_{lc}\sqrt{\ln \tau_{lc}}$ of them, so the total path is $\sim l\sqrt{\ln \tau_{lc}}$. A more careful computation would allow for the specific geometry of the region, as well as a more careful averaging over the frequency-dependent mean free path, but would not alter this result to order of magnitude.

When the optical depth is great enough that the damping wings become optically thick, the conceptual basis of the escape probability method weakens. In that limit, complete redistribution in the fluid frame is no longer a good approximation. Instead, a better description is that of a Fokker-Planck process in frequency space with an *rms* frequency change per scatter of about one thermal width. Because photons can stay in the damping wings a long time, the mean free path can be fairly long, and spatial diffusion can contribute significantly to escape. If reabsorption takes place at a significant rate, this means heat can be transferred from one region to another, which is a major complication to the modeling effort.

An important consequence of resonance line trapping is the associated buildup of pressure. In most cases, the largest contribution to this pressure comes from Lyα; it is also particularly easy to estimate the magnitude of the pressure due to this line. Photon pressure is, of course, 1/3 the energy density when the angular distribution of the photons is isotropic, while the energy density is simply the emissivity times the escape time. Consequently, the radiation pressure due to Lyα at a point z from the illuminated edge of a cloud is

$$p_\alpha(z) \simeq \frac{1}{3}\epsilon_{Ly\alpha}n_e^2\alpha_{rec,1,1}(n \geq 2)\frac{z}{c}\sqrt{\ln \tau_{lc}}. \qquad (10.42)$$

Using equation 10.31, this can be transformed to:

$$p_\alpha(z) \simeq \frac{1}{3}\epsilon_{\mathrm{Ly}\alpha}n_e\Gamma\frac{z}{z_o}\sqrt{\ln\tau_{\mathrm{lc}}}.$$ (10.43)

In ratio to the gas pressure, we find

$$\frac{p_\alpha(z)}{p_g} \simeq \frac{1}{3}\frac{\epsilon_{\mathrm{Ly}\alpha}}{\langle\epsilon_{\mathrm{ion}}\rangle}\Xi\frac{z}{z_o}\sqrt{\ln\tau_{\mathrm{lc}}},$$ (10.44)

where $\langle\epsilon_{\mathrm{ion}}\rangle$ is the number-weighted average energy of the ionizing photons. Detailed calculations show that this estimate exaggerates the quantity of trapped Lyα by about a factor of 10 because it ignores both collisional de-excitation and escape out the back side (§10.5.2.1), however the scaling is correct. Consequently, the trapped radiation pressure can be significant if $\Xi \sim 10$ and the Lyman edge optical depth is great enough for most of the ionizing photons to be absorbed.

10.3.4.3 line fluorescence

Line transfer is further complicated by occasional wavelength coincidences between different transitions. When that occurs, a photon emitted by one sort of atom can be absorbed by another of a quite different sort and possibly transformed into several new photons if the upper level of the resonant transition in the new atom de-excites to the ground state through a series of steps. In the astronomical literature, this process is called the *Bowen mechanism*.

A good case in point is the resonance between Lyβ and O$\textsc{i}$ (fig. 10.8). There can be circumstances in AGN emission line regions in which a combination of both Lyman and Balmer series opacity creates a substantial intensity in Lyβ over a spread of several thermal widths around line center. To excite O$\textsc{i}$ atoms from the $J = 2$ level of the $2s^2 2p^4\,^3P$ ground term to a $2s^2 2p^3 3d\,^3D$ excited term using Lyβ photons requires a redshift of the Lyβ photons of only 15 km s^{-1}. Therefore, the probability that a photon emitted as Lyβ is absorbed by an O$\textsc{i}$ atom can be significant. Because the excited state in O$\textsc{i}$ has multiple pathways down to the ground state, in some cases the result is the substitution of three new photons for the original Lyβ photon, at 11287 Å, 8446 Å, and 1302, 1305, or 1306 Å.

10.3.5 Excited state population equilibrium, recombination cascade, and the approach to thermodynamic equilibrium

So far we have discussed only the simplest sort of line emission: that in which the upper state of the level is populated by collisional excitation. In

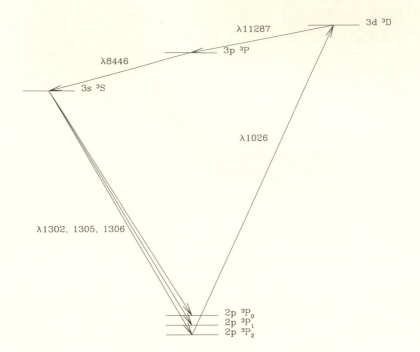

Fig. 10.8   OI terms relevant to the Lyβ fluorescence mechanism. Interterm energies are drawn to scale, but fine-structure splitting is exaggerated. The numbers are the wavelengths in Å of the corresponding transitions.

fact, we have only treated part of that problem: the rate of excitation *per lower state atom.* Moreover, we have tacitly assumed that any interesting lower state was the atomic ground state, so that its density was very nearly the same as the density of the parent ion. Although this is often true, it is not always so. There are a number of important transitions whose lower states are excited. In addition, there are other ways of populating upper states, notably recombination. To find the radiation rate in either case, collisional excitation from excited states or recombination, we must calculate excited state populations for at least some atoms and ions. Given the very large number of states and transitions that can exist, this can be a very complicated exercise. Here we set out the ground rules, with a particular focus on H, for which these effects are especially important.

10.3.5.1 general principles

A single theme unifies the discussion of this section: the approach to thermodynamic equilibrium. In the extreme limit of high density and large optical depth (e.g., inside a star), every state is occupied with its thermodynamic equilibrium population at the local temperature. In the opposite extreme of "nebular equilibrium," virtually every atom can be found in its ground state because the radiative transitions couple to a background that is not far from the cosmological background temperature. AGN emission line regions are an intermediate case in which groups of states are in thermodynamic equilibrium with each other, but not all the temperatures are the same, and not all states belong to these equilibrium groupings.

These different "temperatures" are concisely described by a quantity called the *excitation temperature*. It is simply the temperature that would obtain if the actual population ratio of two states were the thermal equilibrium value; that is,

$$k_B T_{\text{ex},lu} = \frac{\epsilon_{lu}}{\ln[n_l g_u/(n_u g_l)]}. \tag{10.45}$$

A simple example of its utility is provided by a process we have already discussed: collisional de-excitation of an atom in an excited state before it can radiatively de-excite. In our earlier discussion, this process was important because it limited the cooling rate; from the perspective of thinking about atomic excited state populations, this process can be viewed as the first step toward thermal equilibrium. The upward transition rate is dominated by collisions with the thermal electrons; when the (effective) downward rate is also dominated by collisions with those same electrons, by the principle of detailed balance $T_{\text{ex},lu}$ automatically approaches the electron temperature T. If, in addition, the line is optically thick, the radiation brightness temperature approaches the excitation temperature. Collisional de-excitation of a collisionally excited line can therefore be called "thermalization" because it brings both the excitation temperature and the radiation brightness temperature into equilibrium with the electron temperature.

To compute the emergent flux when a collisionally excited line is partially thermalized, one doesn't need to calculate explicitly the population of the upper state. However, there are other circumstances, most notably the emission of recombination lines in dense and optically thick gases, in which such a calculation is necessary. In consonance with the theme of progressive approach to thermal equilibrium, this calculation is most transparently presented in terms of quantities called the *departure coefficients*, defined by

$$b_m = \frac{n_m}{n_m^*(T, n_e)}. \qquad (10.46)$$

That is, b_m is the ratio between the actual population of the state m and the population n_m^* it would have in thermal equilibrium with the free electrons. The distance from thermal equilibrium is then immediately displayed by the deviations of the b_m from unity.

Written in the language of departure coefficients, the excited state population balance equations are sums of terms, each of which goes into detailed balance when the two states it couples have the same departure coefficients. For example, if we are concerned with the population balance in state m, the net rates of all processes involving electron collisions have the form

$$n_e \langle R(m \to m') \rangle (T) (b_{m'} - b_m), \qquad (10.47)$$

where R is the rate coefficient. If m' in this process is the continuum (e.g., as in collisional ionization), $b_{m'} \equiv 1$.

Similarly, the net rates of radiative transitions between state m and higher energy states m' all take the form

$$\mathcal{R} f_{mm'} q_{mm'}^2 \left\{ \exp\left[-q_{mm'} \frac{I(1,1)}{k_B T} \right] b_{m'} (1 + \mathcal{N}_{mm'}) - b_m \mathcal{N}_{mm'} \right\}, \qquad (10.48)$$

where $\mathcal{R} = (1/2)\alpha_{\text{fs}}^5 m_e c^2 / \hbar = 8.03 \times 10^9 \text{ s}^{-1}$ is the natural rate for atomic radiative transitions, $f_{mm'}$ is the absorption oscillator strength for the transition, $q_{mm'} = \epsilon_{mm'}$ in units of $I(1,1)$ (one Rydberg), and the photon occupation number $\mathcal{N}$ has been averaged over solid angle. In this notation

$$A_{m'm} = \mathcal{R} \frac{g_m}{g'_m} f_{mm'} q_{mm'}^2. \qquad (10.49)$$

Such a transition finds itself in detailed balance when

$$\begin{aligned}
\mathcal{N}_{mm'} &= \frac{1}{\exp[q_{mm'} I(1,1)/k_B T] b_m / b_{m'} - 1} \\
&= \frac{1}{\exp[q_{mm'} I(1,1)/k_B T_{\text{ex},mm'}] - 1}.
\end{aligned} \qquad (10.50)$$

That is, detailed balance in a radiative transition occurs when the photon brightness temperature matches the excitation temperature. The form of equation 10.50 also displays the fact that $T_{\text{ex},mm'} = T$ when $b_m = b'_m$, whether or not they are both unity. In the absence of any external illumination at the relevant frequency, the local photon density is determined by a balance between creation and escape whose form is almost identical to the expression shown in equation 10.48 except that $b_{m'}$ is replaced by $b_{m'}(1 - \epsilon_{esc})$. Thus, we see that (in the absence of external photon sources) optically thick transitions reach detailed balance as soon as the photon intensity becomes time steady. In fact, when equation 10.50 is substituted into the rate terms, it reduces them (in the limit of no external irradiation) to

$$\mathcal{R} f_{mm'} q_{mm'}^2 \exp\left(-q_{mm'} \frac{I(1,1)}{k_B T}\right) b_{m'} \epsilon_{esc}. \qquad (10.51)$$

That is, the *effective* radiative decay rate of a given state can be reduced by many orders of magnitude when the brightness temperature of the radiation matches the excitation temperature of the transition.

Because the excitation temperature due to radiative recombination cascade can be substantially different from the electron temperature, the eventual approach to thermodynamic equilibrium for individual transitions can be seen as a competition between radiative and collisional transition rates. In §10.3.3.2 we found that a collisionally excited transition might not produce line photons if the collisional de-excitation rate exceeds the radiative decay rate. When the escape probability is small, the *effective* radiative decay rate (as given by eq. 10.51) is reduced considerably, making it much easier for collisional rates to win this competition. Because ϵ_{esc} is, very roughly, $\propto 1/\tau_{lc}$, thermalization sets in for optically thick transitions when the product $n_e \tau_{lc} \sim 10^{17} f q^2$ cm^{-3} or more.

When n_e and/or τ_{lc} become large enough, true thermodynamic equilibrium between radiation and matter must be reached eventually. However, it does not come in any uniform fashion. Rather, those narrow strips of frequency that are the most opaque are the first to come into equilibrium with the temperature of the corresponding matter transitions. Gradually, as the density and optical depth increase, more and more of these transitions come into equilibrium with the temperature of the free electrons. Because both the Einstein A coefficient and the scattering cross section are proportional to the oscillator strength of the transition, the thermalization criterion of the previous paragraph translates directly into a lower bound for the abundance of the lower state necessary to thermalize a transition. This bound depends on atomic data only through the line energy:

$$X_l > 10^{-5} n_{e,12}^{-1} N_{H,24}^{-1} A^{-1/2} q^3. \qquad (10.52)$$

Many elements have abundances greater than 10^{-5}, including, for Solar abundances, H, He, C, N, O, Ne, Mg, Si, S, and Fe. In fact, at densities and column densities that are within the range of possibilities for the broad line region, transitions arising from *excited* states can be thermalized.

Higher temperatures postpone thermalization because they are associated with the emission of more energetic lines. Equation 10.52 viewed in a different way shows that the product of electron density and atomic column density required for thermalization rises $\propto q^3$.

A corollary of the fact that the outgoing flux of any line saturates at the Planck value is that the contribution to cooling due to a thermalized line depends almost purely on its energy relative to $k_B T$. The only way optical depth enters the picture is through the bandwidth that is optically thick. This width follows the usual *curve-of-growth*. That is, $\Delta\nu_{\text{thick}} = 2\Delta\nu_D x_c$, where x_c is defined in equation 10.40. Over a wide range of optical depths, x_c grows extremely slowly with increasing optical depth; it is for this reason that the output is nearly independent of τ_{lc}. Because the number of transitions whose lower states have abundances greater than some minimum rises *very* rapidly as that minimum falls, accurate portrayals of line emission (and thermal balance) in regions with high density and column density can require *extremely* large atomic databases.

Ultimately, genuine thermodynamic equilibrium is approached when free-free and bound-free opacity are great enough to make much of the thermal continuum optically thick. It is at this stage of the process that emission in the continuum comes to dominate emission in lines. Rescaling the free-free opacity defined in equation 7.92 to conditions more characteristic of line emission regions than disks, we find that the gas is optically thick for all frequencies up to

$$\varepsilon = 0.91 \left(n_{e,12} N_{e,24}\right)^{1/2} T_4^{-7/4}, \tag{10.53}$$

where ε is the frequency normalized to $k_B T/\hbar$, and the expression given is valid for $\varepsilon < 1$. The continuum surface brightness is then

$$S_c \simeq 0.03\varepsilon^3 \sigma T^4. \tag{10.54}$$

If there are N_l optically thick lines, and their energies are $\sim k_B T$, their total surface brightness is

$$S_l \simeq 2x_c N_l \left(\frac{\Delta\nu_D}{\nu}\right) \sigma T^4. \tag{10.55}$$

Taking the ratio between these two surface brightnesses, we find that

$$\frac{S_c}{S_l} \simeq 5 n_{e,12}^{3/2} N_{e,24}^{1/2} N_{\text{H24}} T_4^{-23/4} \left(\frac{N_l}{100}\right)^{-1} \frac{A^{1/2}}{x_c}, \tag{10.56}$$

where A is the typical atomic mass of the line-emitting species. N_l is, of course, a very rapidly growing function of n_e and N_H. Thus, in the temperature range of optical line-emitting gases, thermodynamic equilibrium—and an output spectrum dominated by continuum, rather than line, emission—are achieved when $n_{e,12} N_{e,24} \gtrsim 1$.

10.3.5.2   recombination cascade in H

Recombination cascade in the H atom provides a clear illustration of these principles. It is also an important example to consider because a large part of the energy absorbed by photo-ionization goes into overcoming the ionization potential of H; as a corollary, a large part of the absorbed energy ultimately leaves the gas as recombination line emission. Two very general overviews of the numerous mechanisms that might affect H recombination are presented in Krolik and McKee (1978) and Hummer and Storey (1987).

Consider first a very low density gas that is optically thin in all transitions. When an H atom recombines radiatively (dielectronic recombination is not an option unless the recombining ion has at least one electron), there is a probability distribution for the state in which it initially finds itself. This probability distribution can be derived from quantum mechanical first principles, for it depends only on the matrix elements between various free electron states in the Coulomb field of the nucleus and corresponding bound states, and on the phase space (i.e., degeneracy) associated with these states. Typically, the highest l state at each n is favored relative to lower l states.

After the H atom forms, it de-excites by a succession of radiative transitions. These follow the E1 selection rules because the transitions they single out are, of course, by far the quickest. Many different paths are possible down to the $n = 1$, $l = 0$ ground state, but it is possible to work out the probability distribution for them. Because the initial recombinations tend to be in fairly high l states, and the E1 selection rule is $|\Delta l| = 1$, the stream of de-exciting atoms tends to move down the path of maximal l at each n. As a result, there is a fairly strong concentration of line production in the α transition of each series, most notably Lyα. In the conventional system of normalizing all other lines to the flux in Hβ, in this low-density, low-optical depth limit Hα/H$\beta \simeq 2.8$, slowly declining with increasing temperature, and Lyα/H$\beta \simeq 28$, almost independent of temperature.

As the density increases, the situation gradually changes. The first step toward thermal equilibrium is taken by the angular momentum substates within energy levels whose principal quantum number $n \gg 1$. "l-mixing collisions with positive ions in the plasma can cause transitions with $|\Delta l| = 1$ and $\Delta n = 0$. When the rates of these transitions are faster than any other rates affecting these levels, their relative populations equilibrate so that the ratios of population in the substates are the same as the ratios of their statistical weights. This occurs when

$$n_e > 2.8 \times 10^{13} n^{-7} T_4^{1/2} \text{ cm}^{-3}. \qquad (10.57)$$

Thus, this process can be important for the upper states even at densities for which these processes are completely irrelevant for lower states. These collisions with positive ions have very large rates for a combination of two reasons. First, because the energy threshold (in a hydrogenic atom) is nil, all charged particles can contribute. Second, again because the energy threshold is nil, the cross section is very well described by the Born approximation. As we have already seen (eq. 10.18), the Born cross section for a particle of mass M scattering on an atomic-sized potential is $\propto (M/k)^2 \propto M/E$, so that the ion cross section is larger than the electron cross section by their mass ratio.

These collisions alter the radiated spectrum because recombining atoms no longer follow the path of maximum l. With the atoms spread more evenly over the different orbital angular momentum states, it is possible for some to decay by transitions directly to low-n levels. This diminishes the strength of the upper series α transitions, increasing the relative strength of Lyα to Lyα/H$\beta = 33$. Hα/Hβ, on the other hand, is hardly changed at all.

The next step toward thermal equilibrium also occurs in the high n states. These couple very strongly to the free electrons via two pairs of mechanisms, stimulated radiative recombination and photo-ionization, and collisional ionization and three-body recombination. The rates for both increase strongly with n. On the one hand, if the spectrum of the incident continuum $F_\nu \propto \nu^{-\alpha}$, the occupation number for photons at the ionization threshold for level n increases $\propto n^{2(\alpha+3)}$; on the other, the cross section for collisions increases $\propto n^4$. The result is that the $b_{nl} \to 1$ as n becomes $\gg 1$.

Optical depth in the recombination lines is another mechanism pushing the excited state populations, and therefore the radiation, closer to thermal equilibrium. As the column density of HI grows, the first transitions to become optically thick are the Lyman series. When these lines are thick, photons in them must scatter before leaving the region. Viewing a scattering event as an absorption followed by reemission, there is a significant probability per scatter that the atom de-excites not by reemission of a single Lyman series photon, but by emission of a combination of photons inluding a lower Lyman series photon. As a result, photons in Lyβ and higher Lyman series lines are efficiently destroyed, ultimately to reappear as Lyα plus one or more Balmer and other lines. Absolute transparency in the Lyman series is known in the trade as *Case A*; column density great enough to make the Lyman lines optically thick is called *Case B*.

Substantial population in $n = 2$ also follows as a result of optical depth in the Lyman series. For example, taking the simplified model that $b_{2s} = b_{2p} = b_2$, and assuming that b_2 is governed by a balance between recombination and Lyα emission, one finds that

$$b_2 = 6.5 \times 10^{-3} \frac{T_4^{0.7}}{\epsilon_{esc}(\tau_{Ly\alpha})} \exp\left[3.95\left(1 - 1/T_4\right)\right], \tag{10.58}$$

where the factor $T_4^{0.7}$ comes from a power-law fit to the scaling of the recombination rate coefficient with temperature. Escape probabilities for Lyα as small as $\sim 10^{-3}$ are not at all uncommon, so it is easy to elevate b_2 to unity or greater.

Large optical depth in the Lyman continuum is sometimes called *Case C*. In this condition, every recombination to the ground state produces a photon that is reabsorbed elsewhere, nullifying the event (and also reducing the effective cooling rate due to Lyman recombination). Thus, the only way a recombination really creates a net new neutral atom is if it occurs to a level with $n \geq 2$. In effect, the recombination rate is decreased by several tens of percent, and the efficiency of Lyα production is increased because there are no longer any recombinations that automatically avoid Lyα production. This case prevails in conditions such that the "on-the-spot" approximation (eq. 10.33) is useful.

Where the Lyman continuum is optically thick, the intensity in the Lyman continuum plummets. It is not difficult for it to fall far below the Planckian value; the result, of course, is that the excitation temperature between the 1s state and the continuum $T_{ex,1c}$ becomes $\ll T$, that is, b_1 becomes $\gg 1$. If the density is high enough that collisional de-excitation of Lyα dominates the effective decay rate, while the electron temperature is high enough to maintain at least some collisional excitation from $n = 1$ to 2, $T_{ex,12}$ becomes locked to T and $b_2 = b_1 \gg 1$. When b_2 is large enough, collisional excitation out of $n = 2$ can supplement recombination as a source of Balmer and other photons. It is even possible for the opacity in the Balmer (and other upper series) continua to be significant.

10.4   Conditions for Achieving Equilibrium

Now that we have surveyed the different physical processes occurring in AGN line emission regions, it is time to check under what circumstances the equilibrium assumptions conventionally made are justified.

First, consider radiation diffusion equilibrium. In those bands that are optically thin, the time for any change to be communicated to the entire region is simply a light crossing time, which is generally the shortest response timescale for anything. However, some bands can be optically thick; for them, the photon density front propagates at a rate limited by the opacity.

The propagation of disturbances in the ionizing flux is of special interest because of the central role these photons play in the whole system. Imagine, then, a slab of gas initially neutral suddenly struck by a large flux of ionizing photons. Once the photons arrive, they ionize the gas on the exposed edge of the slab, which sharply decreases its opacity and permits later photons to travel deeper. This ionization front moves at a speed controlled by the rate at which new ionizing photons arrive, that is, the flux of newly ionized matter crossing the front must match the flux of ionizing photons arriving: $n_e v_f = \mathcal{N}_{\text{ion}}$. If the ionizing photons are all directed normal to the edge of the gas, $v_f = \Gamma c$ (up to a maximum speed of c, of course). Thus, the ionization parameter has yet another interpretation: it fixes the speed at which ionization fronts move, and consequently the speed with which an emission line region responds to any change in its illumination.

Emission line photons must also diffuse through the slab, but, as we have discussed, the mechanisms governing their diffusion are quite different. So long as frequency wandering dominates escape, the time for equilibration of the line photon density is greater than the light crossing time by a factor $\sim \sqrt{\ln \tau_{\text{lc}}}$, which, of course, is at most order unity.

Thus, continuum propagation is usually slower than equilibration of the line photon density (for $\Gamma < 1$), and takes a time

$$t_{\text{cont}} \sim \Gamma^{-1} \begin{cases} 30(l/10^{12} \text{ cm}) \text{ s} & \text{broad line gas} \\ 3 \times 10^6 (l/10^{17} \text{ cm}) \text{ s} & \text{narrow line gas.} \end{cases} \qquad (10.59)$$

The fiducial lengthscales for the two regions are justified in §10.5.3.

Approach to H ionization equilibrium is generally governed by the recombination timescale:

$$t_{\text{rc}} \sim (n_e \alpha_{\text{rec},1,1})^{-1} \sim \begin{cases} 10^3 (n_e/10^{10} \text{ cm}^{-3})^{-1} \text{ s} & \text{broad line gas} \\ 10^9 (n_e/10^4 \text{ cm}^{-3})^{-1} \text{ s} & \text{narrow line gas .} \end{cases} \qquad (10.60)$$

Again the fiducial numbers are taken from §10.5.3. Some care must be taken in interpreting these timescales, however. While they *are* a good estimate of the time it takes a fully ionized gas to recombine to neutrality, they are *not* necessarily a good estimate of the time required to change the *neutral* fraction by a factor of order unity. If we can abbreviate the time-dependent equation for the abundance of $n_{Z,j}$ by

$$\frac{dn_{Z,j}}{dt} = n_e n_{Z,j+1}\alpha_{\text{rec},Z,j} + \cdots - \cdots, \qquad (10.61)$$

it is apparent that

$$t_{\text{rc},Z,j} \sim \left(\frac{d\ln n_{Z,j}}{dt}\right)^{-1} \sim \frac{n_{Z,j+1}}{n_{Z,j}}\frac{1}{n_e\alpha_{\text{rec},Z,j}}. \qquad (10.62)$$

That is, the timescale to make a major change in the density of $n_{Z,j}$ by recombination is the "simple" recombination time *multiplied* by the abundance ratio of the species once-more-ionized to the species in question. When $n_{Z,j}$ is a minority species and more highly ionized stages are more abundant, this ratio can often be as large as several orders of magnitude.

Relaxation to thermal balance takes place on the cooling timescale,

$$t_{\text{cool}} \sim \frac{3k_B T}{n_e\Lambda} \sim \begin{cases} 50(n_c/10^{10}\ \text{cm}^{-3})^{-1}\ \text{s} & \text{broad line gas} \\ 5\times 10^7(n_e/10^4\ \text{cm}^{-3})^{-1}\ \text{s} & \text{narrow line gas,} \end{cases} \qquad (10.63)$$

where we have used the fact that when the temperature is around 10^4 K, the cooling function is usually $\sim 10^{-23}$ erg cm^3 s^{-1} . Cooling, then, can be somewhat faster than recombination. This comparison changes dramatically when the gas is so ionized that line cooling becomes unimportant. In regimes of that sort, recombination is far faster than cooling.

Finally, excited state population balance proceeds on a combination of a collisional timescale and a radiative decay timescale. Both can vary dramatically with conditions, as collisional excitation depends very sensitively on temperature, and radiative decay rates can vary drastically, depending on how they conform to selection rules, and how great the line's optical depth is. In other circumstances, the rate-controlling transition connects to the continuum (e.g., recombination). Consequently, examining the quality of the equilibrium assumption for excited state populations must be done on a case-by-case basis.

Overall, we see that, relative to the timescales on which the intrinsic continuum changes (weeks to years), the broad line gas is able to reequilibrate quite effectively, but the narrow line gas is often out of step.

10.5   Results of Photo-ionization Calculations

10.5.1   The T–Ξ diagram

The most fundamental result of photo-ionization calculations is that the temperature and ionization state of a photoionized gas depend primarily on the ionization parameter, and only weakly on the shape of the ionizing spectrum (within certain limits) or the pressure. This dependence is illustrated in figures 10.9 and 10.10. Moreover, over a wide range in ionization parameters, the temperature is nearly always $\sim 10^4$ K.

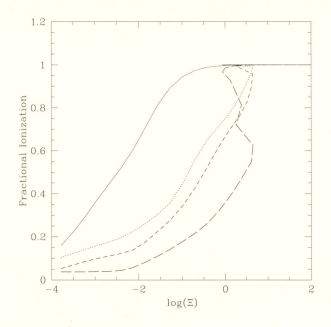

Fig. 10.9   The ionization state as a function of Ξ for a power-law continuum with $\alpha = 1$ when the gas is also in thermal balance. Each curve shows the ratio between the mean charge state of a given element and its fully stripped charge. The solid line is for He, the dotted line for C, the short-dashed line for O, and the long-dashed line for Fe. Places where the curves double-back correspond to the places where the temperature is a multiple-valued function of Ξ: see fig. 10.10. All converge to unity at large Ξ.

Provided the spectrum is fairly broad (i.e., has no sharp cutoffs between ~ 10 and $\sim 10^4$ eV), the ionization state depends on the *frequency-integrated* ionization parameter more than on any other parameter because the most important pathway to higher ionization stages is by valence-shell ionizations of each of the lower stages. Similarly, the rate of recombination to a given stage depends on the abundance of the next most ionized stage, which in turn depends on the ionization rate of this and higher stages. Therefore, the abundance of a species such as O^{+5} does *not* depend primarily on the mean intensity at 114 eV, the ionization potential of O^{+4}, but instead on the intensity over the entire range from 13.6 eV (the ionization potential of O) up past 138 eV, the ionization potential of O^{+5}.

There are several reasons why the temperature is so strongly stabilized near 10^4 K. One has already been explained in §10.3.3: Because the strongest atomic and ionic lines have energies ~ 10 eV, which is $\sim 10k_B T$ when $T \sim 10^4$ K, the inverse exponential dependence on the temperature in the Boltzmann factor makes the excitation rates exquisitely sensitive to T. This is the origin of the dramatic hump in the cooling curve near 10^5 K when Ξ is not too large (fig. 10.6).

Other effects strengthen this "thermostat." For example, when most H is ionized, the heating rate due to photo-ionization of H is almost independent of Ξ because it is $\propto J_{\rm ion} n_{1,1} \propto p\Xi n_{1,1}$, but $n_{1,1} \propto \Xi^{-1}$. The only reason the heating rate increases at all with Ξ (or Γ or ξ or any other form of the ionization parameter) in this regime is that heating by photo-ionization of He does increase with increasing ionization parameter so long as most He is either HeI or HeII. Even when the He ionization balance shifts to mostly HeIII, so that its photo-ionization heating rate saturates, the temperature continues to rise slowly because the ionization balance of the heavier elements shifts to higher stages of ionization whose resonance line energies tend to be somewhat higher. For all these reasons, the temperature stays within a factor of three (from perhaps 8000 to 25,000 K) over four orders of magnitude increase in Ξ (fig. 10.10).

A dramatic change occurs when Ξ approaches $\simeq 10$. At this level, even at relatively low temperature the ionization balance of C, N, and O becomes dominated by H- and He-like species (the Li-like ions are a few percent or less of the total). The ground-state transitions for H- and He-like ions are much higher in energy than the ground-state transitions in ions with L-shell electrons—~ 100 eV rather than ~ 10 eV—so that the cooling rate is sharply suppressed until the temperature rises close to 10^6 K (fig. 10.6). In addition, the line energies are now no longer much smaller than the ionization potentials. Once T becomes high enough to allow line cooling, it is also high enough to drive significant collisional ionization. The ionization balance then shifts to yet more highly ionized species, and eventually C, N, and O are completely stripped. When that happens, the only abundant element offering *any* opportunities for collisional excitation

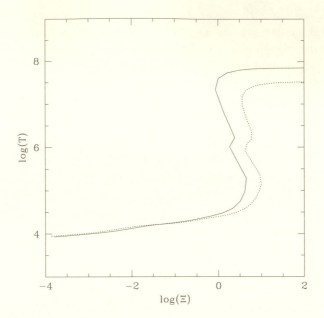

Fig. 10.10   The temperature as a function of Ξ for two different continuum spectra: a power law with $\alpha = 1.0$ from 0.01 eV to
500 keV (solid curve), and the composite radio-quiet quasar spectrum of Elvis et al. (1994) cut off at 1 eV on the low-frequency end
(see fig. 1.3), and extrapolated to 500 keV on the high-frequency end
(dotted curve). Despite the substantial difference in spectral shape,
the low-temperature portions of the curves are nearly identical.

of emission lines is Fe, and these are too few and too weak to prevent a
temperature runaway.

Depending on the shape of the spectrum, it is sometimes possible for
the temperature balance curve to actually double back, producing a narrow
band of Ξ with *three* (or, for very narrow ranges of Ξ, five or even seven)
equilibria. Spectra with relatively high Compton temperature extend the
width of the multiple solution range; when T_C is too small, the multiple
solution region disappears altogether. The intermediate temperature solutions, when they exist, are due to a balance between bremsstrahlung (at
high temperatures) or Fe line emission (at lower temperatures) and Compton heating. Wherever the slope of the curve is negative, the equilibria
are unstable to isobaric perturbations of exactly the same variety that we
discussed in §6.3.3.

When $\Xi \gg 10$, the only equilibrium possible is Compton equilibrium;
for typical AGN spectra, $T_C \sim 10^7$–10^8 K. However, the exact value of T_C

in the broad emission line region cannot be unambiguously derived from observations of the total continuum spectrum because there may be large quantities of infrared photons produced at much larger radii (Chap. 12). That is why in our examples we use the Elvis et al. spectrum cut off at 1 eV.

A corollary of the fact that the curve in figure 10.10 defines where radiative heating and cooling are in balance is the fact that they are in *imbalance* everywhere else in the diagram. Any gas finding itself to the left and above the curve has net radiative cooling; any gas in the region to the right and below has net radiative heating. Whenever the timescale for nonradiative heating or cooling (adiabatic expansion or contraction) is competitive with the radiative timescale, it is possible for gas to wander off the equilibrium curve.

10.5.2   Generic models

The results of the previous subsection pertain specifically to regions sufficiently optically thin that passage through them does not significantly alter the continuum spectrum. It is of greater interest, of course, to examine what happens inside a region of greater optical depth. Although specifics vary with the model parameters, the generic features of optically thick regions photoionized by an AGN continuum can all be illustrated by reference to a few generic cases, one for broad line gas and one for narrow line conditions.

10.5.2.1   broad lines

Consider then, a slab of photoionized gas with properties designed to be more or less representative (see §10.5.3) of broad emission line gas. To define its properties, several parameters must be specified, of which the most important are the ionization parameter (in any of its forms), the pressure or density, and the thickness of the region, either in distance or column density. The shape of the incident continuum spectrum and the elemental abundances must also be specified, but they have a weaker impact on the outcome. For this example, we choose $\Xi = 1$, $p = 0.01$ dyne cm^{-2}, column density $N_{\rm H} = 3 \times 10^{22}$ cm^{-2}, a continuum whose shape is the mean radio-quiet quasar continuum found by Elvis et al. (1994) modified as described in the caption to figure 10.10, and Solar abundances.

At the illuminated edge, the temperature is $\simeq 2.5 \times 10^4$ K (fig. 10.11). In this region, H is so thoroughly ionized that most of the heating is due to photo-ionization of HeII (fig. 10.12), even though most He is HeIII (which

357

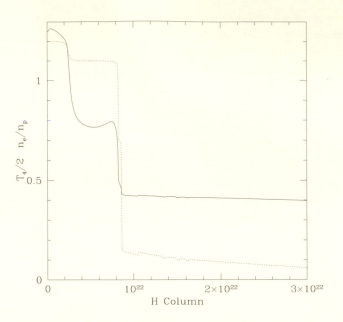

Fig. 10.11 Temperature in units of 2×10^4 K (solid line) and ratio of electron density to H nucleus density (dotted line) as functions of H column density in nuclei per cm^2. The small "glitches" between 10^{22} and 2×10^{22} cm^{-2} are numerical artifacts due to zoning within the slab that is not quite fine enough.

is why the ratio of electrons to H nuclei is 1.2). The most important coolant is Ovi 1034, but Nv 1240 also contributes (fig. 10.14). C is so highly ionized that Cv is the most abundant C ionization stage, and there is more Cvi than Civ (fig. 10.13). As one moves deeper into the cloud, gradually the Heii edge becomes optically thick, and species that require ionization energies more than 55 eV, such as Ovi and Nv, disappear. Civ then grows in abundance, and Civ 1549 becomes the main coolant. Once there are no more Heii continuum photons, Hi and Hei photo-ionization dominate the heating, and Ciii] 1909 rises in importance in the thermal balance as the abundance of Ciii climbs past that of Civ. The temperature in this zone is held on a "shelf" at 1.5–1.7×10^4 K, growing slightly with increasing depth because the mean energy of the photons being absorbed rises slowly. Near the local temperature maximum, the trapped radiation pressure reaches a maximum at $\simeq 7\%$ of the gas pressure. Because the maximum radiation pressure is roughly proportional to Ξ, it reaches a level where it can seriously interfere with the maintenance of dynamical equilibrium at about the critical Ξ for runaway heating.

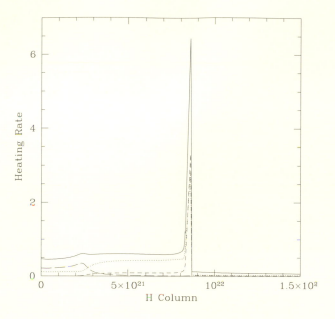

Fig. 10.12   The heating rate (in units of 10^{-4} erg cm^{-3} s^{-1}) as a function of column density for the front half of the generic broad line cloud. The solid curve is the total heating rate, the long-dashed curve is the part due to HeII photo-ionization, the dotted curve HI photo-ionization, and the short-dashed curve the HeI photo-ionization contribution.

As we have already discussed in §10.3.4.1, the transition layer between the zone where H is mostly ionized and the zone where H is mostly neutral is thicker than in HII regions, but still quite sharp in geometric terms. However, because this is the place where most of the continuum energy carried in photons between 13.6 and 55 eV is absorbed (see the spike in the total heating rate in fig. 10.12), this is also the location of a large part of the line emission. As energy conservation requires, where radiation is absorbed, it must also be emitted. Consequently, this geometrically thin layer is responsible for most of the production of Lyα, the Balmer lines, some CIII] 1909, and much of MgII 2800.

Once the ionization transition has been passed, the entire range of continuum energies from 1 Rydberg up to several hundred eV rapidly becomes essentially black. The only source of heating and ionization in the zone where H is mostly neutral is K-shell photo-ionization by soft X-rays, but this is rather weak because the cross sections are so much smaller than for valence shell photo-ionization. Very large optical depths in the Ly-

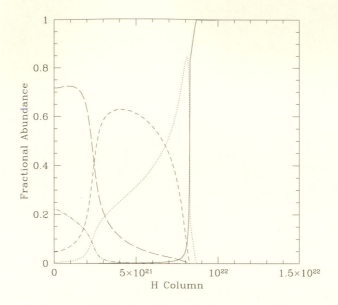

Fig. 10.13 Carbon ionization stages as a function of column density for the front half of the generic broad line cloud. At the exposed edge Cv (He-like C), represented by the long-dashed curve, is the most abundant species, although there is also some Cvi (dot-dashed curve). As the optical depth increases, C passes from Civ (short-dashed curve) to Ciii (dotted curve), and finally, where H is mostly neutral, C becomes almost entirely Cii (solid curve).

man series accrue due to the high density of Hi. These depths are so large that Lyα is very nearly thermalized (fig. 10.15), raising the population in H($n = 2$) to the point that it is comparable to the population in some of the more abundant heavy elements:

$$\frac{n(n=2)}{n_{1,1}} = 4 \exp\left(-\frac{11.84}{T_{\mathrm{ex},4}}\right) = 2.9 \times 10^{-5} \exp\left[11.84\left(1 - T_{\mathrm{ex},4}\right)\right], \quad (10.64)$$

where the excitation temperature is for Lyα. In this region, collisional excitation of the Balmer and other excited H line series can be important, as well as a large number of Feii multiplets. The strength of these lines relative to, for example, Lyα is proportional to the relative strength of the X-ray flux compared to the Lyman continuum.

Looking back at this short discussion of the principal subregions of a typical broad line emission region, it is clear that there are *no* lines radiated with constant relative strength from all parts of an optically thick

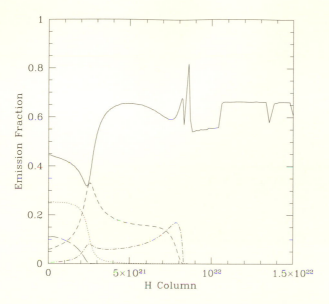

Fig. 10.14 The fraction of total radiated power due to various species as a function of column density in the front half of the generic broad line cloud. The solid curve is H; the dotted curve is HeII, the short-dashed curve CIV, the long-dashed curve OVI, and the dot-dashed curve CIII. Note that because recombination radiation contributes to the emission, but not to the cooling per se, neither HI nor HeII should be considered the dominant coolant. The "spiky" features in the H emission curve around $N_H \simeq 8 \times 10^{21}$ and 1.5×10^{22} cm^{-2} are numerical artifacts.

cloud. Rather, certain regions specialize in emitting certain lines. Similarly, different continuum bands are primarily absorbed in different places, so that each subregion is particularly dependent for its heat on a specific band of energies in the continuum. These regional contrasts have an important corollary: individual lines can be (at least in a rough sense) identified with definite locations and can be regarded as drawing their strength from particular continuum bands. Recombination lines can be associated with the portion of the continuum just above the ionization potential of their parent species; collisionally excited lines are more closely associated with the continuum segment from which their heat is drawn. For example, most of the Lyman continuum is absorbed in the ionization transition layer, and this is, of course, where most of the Lyα and other recombination lines are produced. On the other hand, the HeII continuum is mostly absorbed in the highly ionized region at the front edge of the cloud, where the cooling

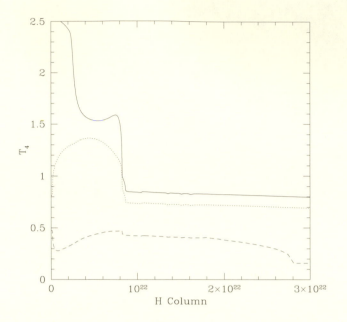

Fig. 10.15   The approach to thermal equilibrium: the excitation temperatures of Lyα (dotted line) and Hα (dashed line) compared to the free electron temperature (sold line) as functions of H column density. All three temperatures are in units of 10^4 K. From the ionization transition on, $T_{\mathrm{ex}}(\mathrm{Ly}\alpha)$ is very nearly equal to T; $T_{\mathrm{ex}}(\mathrm{H}\alpha)$ saturates quickly at $\simeq 3000$ K.

is dominated by Ovi 1034 and Nv 1240. These identifications of lines with continuum bands mean that the relative strengths of different lines can depend on the shape of the EUV continuum as well as on the mix of ionization parameters, densities, and so on in the emission regions.

The relative line strengths, as computed in this generic broad line model, for all those lines radiating at least 0.3% of the total are given in table 10.3. Comparison with table 10.1 shows that such a model provides at least a starting point toward identifying the physical conditions in which these lines are made. As we will discuss in §10.5.3.1, however, *no* single set of physical conditions is able to entirely reproduce all the line ratios.

10.5.2.2 narrow lines

Next consider a generic narrow line cloud. The parameters for this model are $\Xi = 0.03$, $p = 10^{-8}$ dyne cm^{-2}, and $N_{\mathrm{H}} = 10^{21}$ cm^{-2}, and it also uses the modified Elvis et al. mean radio-quiet quasar spectrum and Solar abundances.

Table 10.3: Relative Fluxes from a Generic Broad Line Cloud

Line	Relative Flux
Lyα + Ov 1218	100.
Hα	11.
Hβ	4.0
Hei 10830	13.
Heii 304	42.
Heii 1640	3.8
Ciii] 1176	3.0
Ciii 2297	8.6
Ciii 977	3.5
Ciii] 1909	23.
Civ 1549	90.
Niv] 1486	5.4
Nv 1240	6.1
Oiii]1663	8.1
Ovi 1034	16.
Mgii 2800	9.1
Siiv 1400	2.3

Notes: Relative line fluxes from a cloud with $\Xi = 1.0$, $p = 0.01$ dyne cm^{-2}, and $N_H = 3 \times 10^{22}$ cm^{-2}.

In gross terms, the internal structure of this generic narrow line cloud resembles that of the broad line cloud: it, too, is highly ionized on its illuminated edge and has an ionization transition roughly 10% of the way into the cloud, beyond which H is mostly neutral and the heating rate is predominantly due to X-ray photo-ionization of K-shell electrons from heavy elements (Figs. 10.16 and 10.17). However, there are also a number of notable contrasts. First, because the typical ionization parameter is rather smaller, and because forbidden lines such as [Oiii] 5007 can help cool the gas, the temperature at the illuminated face is considerably lower: $\simeq$ 11,000 K. On the other hand, unlike the broad line cloud, the temperature

363

in the mostly neutral zone is only slightly lower—$\simeq$ 9000 K—which is, in fact, a bit *higher* than the temperature in the corresponding zone of a broad line cloud. The reason is that the heating rates per atom are essentially the same in the neutral zones of the broad and narrow line clouds (in both cases the mechanism is X-ray photo-ionization), but the narrow line cloud is unable to use collisional excitation of Balmer lines for cooling because its much lower density is unable to support much population in $n = 2$.

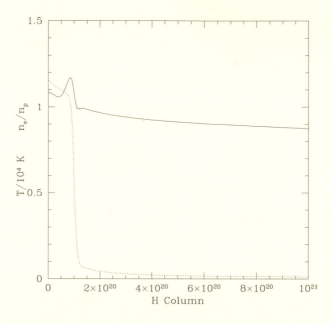

Fig. 10.16 Temperature (in units of 10^4 K: solid line) and electron density relative to H nucleus density (dotted line) as a function of H column density into the cloud for a generic narrow line cloud. The temperature varies rather little with depth, but there is a sharp ionization transition.

The heating rate as a function of depth in a narrow line cloud behaves much as it does in a broad line cloud: HeII dominates at the illuminated edge but is overtaken by HI photo-ionization part way through the ionized zone. At the ionization transition the heating rate per unit volume jumps upward as the HI density rises sharply, but after all the Lyman continuum has been absorbed, the heating rate falls to the much lower level given by the relatively small X-ray opacity.

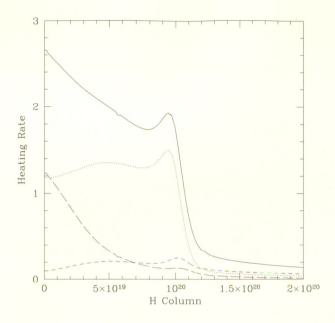

Fig. 10.17   The heating rate (in units of 10^{-16} erg cm^{-3} s^{-1}) as a function of H column density for the front 20% of a generic narrow line cloud. The total heating rate is shown by a solid line; HI photo-ionization by a dotted line, HeII photo-ionization by a long-dashed line, and HeI photo-ionization by a short-dashed line. Beyond the ionization transition at $\simeq 1 \times 10^{20}$ cm^{-2}, heavy element photo-ionization by X-rays is the single largest contributor to the heating.

In the ionized zone, most of the emission is in recombination lines of HI and HeII, and most of the energy removed from the thermal electron pool is consumed exciting the [OIII] 5007 transition (fig. 10.18). In the neutral zone, HI recombination lines continue to be generated, though at a rather slower rate, of course, while genuine cooling is due to such transitions as [OI] 6300 and [SII] 6724. Unlike broad line clouds, nowhere in typical narrow line clouds does the excitation temperature of any permitted lines approach anywhere near the electron kinetic temperature.

Summing over the entire cloud, and including line photons emitted through either face, the relative strengths of all lines predicted to carry at least 0.3% of the total are listed in table 10.4.

10.5.3   Model construction

So far we have begged the question of how to determine just which conditions actually obtain in AGN emission line regions. In principle one

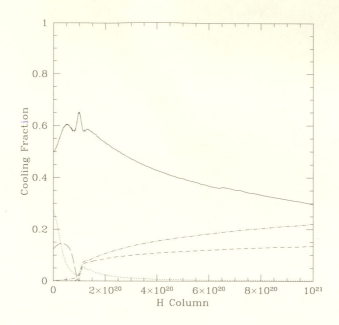

Fig. 10.18   The fraction of the radiative output due to various
species as a function of column density. HI is the solid curve; HeII
is the dotted curve, OIII the long-dashed curve, OI the short-dashed
curve, and SII the dot-dashed curve.

might expect that regions comprising a wide range of physical conditions
would all contribute to radiating emission lines. From the host galaxy
to the edge of the black hole spans 8–10 orders of magnitude in distance
scale, and there is no a priori reason why all could not participate. Even
in more specific locales there might well be significant inhomogeneities.
Nonetheless, the fact that the profiles seem to divide so cleanly into broad
and narrow components suggests that there are at least two special regions
(broadly defined) that are particularly prolific sources of line emission. We
will discuss models for them separately, as they present somewhat different
problems.

10.5.3.1   broad line region

10.5.3.1.1   single-zone models

We begin by attempting to go as far as possible with the simplest
possible picture for the broad line region: that the great majority of the
lines are created within a narrow range of physical conditions. To find those

Table 10.4: Relative Fluxes from a Generic Narrow Line Cloud

Line	Relative Flux
Lyα	100.
Hα	6.0
Hβ	2.2
He$\textsc{i}$ 10830	9.6
He$\textsc{ii}$ 304	35.
He$\textsc{ii}$ 1640	3.9
C$\textsc{ii}$] 2326	2.8
C$\textsc{iii}$] 1909	4.9
[N$\textsc{ii}$] 6550,6585	12.
[O$\textsc{i}$] 6300	4.4
[O$\textsc{ii}$] 3726,3729	21.
[O$\textsc{iii}$] 4959	12.
[O$\textsc{iii}$] 5007	34.
Mg$\textsc{ii}$ 2800	9.2
[S$\textsc{ii}$] 6718,6733	9.5
Fe$\textsc{ii}$ 2600	4.2

Notes: Relative line strengths predicted for a cloud with $\Xi = 0.03$, $p = 10^{-8}$ dyne cm^{-2} and $N_\mathrm{H} = 10^{21}$ cm^{-2}. Fe$\textsc{ii}$ 2600 refers to a blend of multiplets in the vicinity of 2600 Å.

conditions that do the best at reproducing the relative line strengths, one computes a grid of models, varying the free parameters Ξ (or ξ or Γ or U or equivalent); the pressure p (or almost equivalently the density n_H); and the column density N_H. Although the elemental abundances and the spectral shape of the continuum can vary considerably without having a dramatic impact on the results, they, too, must be specified, and the predictions do vary somewhat depending on how they are chosen.

We postpone discussing the degree to which line strengths are sensitive to the abundances to the end of this subsection, but it is appropriate to discuss the major problems involved in choosing a continuum shape here.

First, without the aid of substantial cosmological redshifting, we have no way to observe the actual shape of the ionizing continuum, and even in the most favorable circumstances ($z \simeq 4$ and spectra taken from space), the highest UV energy observable is still only $\simeq 50$ eV. Without UV observations in our own reference frame, life would be even harder. Lyα is shifted into the visible when $2 \leq z \leq 5$, and the Lyman edge appears when $3 \leq z \leq 7$; although there is little intervening material at smaller redshifts, the density of optically thick H$\textsc{i}$ clouds increases so rapidly with redshift that great chunks of flux are removed from the spectra of more distant quasars observed from the ground. Whether the spectral shape that emerges from the AGN plus host galaxy system is identical to the shape that impinges on the emission line regions is another difficult question. Finally, although the cleanest procedure logically would be to attempt to predict the line strengths in an individual AGN using one's best guess for the shape of that AGN's spectrum, it is often of interest to attempt a model for a "typical" AGN; one must then somehow construct a "mean" spectrum. To do this properly, one should choose a well defined sample of AGNs and select a particular method with which to assemble a composite mean continuum shape from the spectra of the objects in this sample.

After the models are computed, the predictions made for the relative line fluxes are then compared to the observed quantities, and a search made for the parameters that yield the best match. In this context, as well as in the context of choosing a continuum shape, it is wise to distinguish between fitting a particular object, and fitting some sample mean. Having made that distinction, it is best to define a quantitative measure of goodness of fit, but one often finds this step omitted in much of the literature. Because of all these problems, and also because certain line ratios are most sensitive to individual parameters, here we will lower our expectations to merely finding approximate ranges for Ξ, n_{H}, and N_{H} in which the predicted line ratios come somewhere near those in table 10.1.

The fact that we see relatively high-ionization stages such as C^{+3} and N^{+4} immediately tells us that the ionization parameter cannot be too small. Its value was first estimated by comparing the relative fluxes of C$\textsc{iii}$] 1909 and C$\textsc{iv}$ 1549 (Davidson 1972) because the ionization balance of C is fairly sensitive to the ionization parameter (see fig. 10.9). However, as figure 10.13 shows, there can be emission regions with Ξ so large at the illuminated edge that C is ionized well past C$\textsc{iv}$, and yet substantial fluxes of both C$\textsc{iv}$ 1549 and C$\textsc{iii}$] 1909 emerge because the optical depth is great enough for first C$\textsc{iv}$, and then C$\textsc{iii}$ to become the dominant ionization stage of C as one looks deeper and deeper inside. Consequently, better constraints are obtained from other line ratios, such as Nv 1240 and Ovi 1034 relative to Lyα, that are more sensitive to the ionization parameter at the illuminated face. Although there are certainly variations from object to object, most estimates of Ξ from single-zone models are in the range 0.1–1.

A number of line ratios constrain the pressure. Perhaps the single most robust conclusion in this entire area stems from the strength of the narrow component of [Oiii] 5007 and its lack of broad wings. This fact is almost certainly explained by a density contrast. [Oiii] 5007 is thermalized (i.e., its flux saturates because the collisional de-excitation rate is as great as the radiative decay rate) when $n_e \gtrsim 3 \times 10^6$ cm^{-3}. Therefore, the density must be less than this in the regions moving with velocities of only several hundred km s^{-1}, and more than this in the faster moving regions.

An upper bound on the pressure/density comes from collisional de-excitation of the semiforbidden line Ciii] 1909. Its critical density is $\simeq 6 \times 10^9$ cm^{-3}. Although its flux can remain significant relative to other lines up to densities a few times this value (particularly if the other lines are thermally limited by a combination of optical depth and collisions), that would not be true if the density were much higher.

A finer indicator comes from a set of neutral He lines. The lower state of HeI 5876 is an excited state that is only populated by electron collisions. Its ratio to HeI 10830 is then a very sensitive indicator of n_e, and also points to densities around a few times 10^9 cm^{-3} (Netzer 1978).

Combining the best-fit ionization parameter and the best-fit density yields an estimate of the distance from the continuum source to the line emission region:

$$r = \left(\frac{L_{\text{ion}}}{9.2\pi c n_{\text{H}} k_B T \Xi} \right)^{1/2}$$

$$= 0.067 L_{\text{ion},45}^{1/2} \Xi^{-1/2} n_{\text{H},10}^{-1/2} \left(\frac{T_4}{2} \right)^{-1/2} \quad \text{pc.} \qquad (10.65)$$

$$= 1.1 \times 10^5 \frac{L_{\text{ion}}}{L_{\text{E}}} L_{\text{ion},45}^{-1/2} \Xi^{-1/2} n_{\text{H},10}^{-1/2} \left(\frac{T_4}{2} \right)^{-1/2} \quad r_g.$$

Thus, the broad line region is quite small compared to the galactic scale but is far outside the relativistic portion of the accretion flow. Because the line ratios change rather little across the entire range of luminosity (§10.1.1.1), the best-fit values of Ξ and n_{H} are almost independent of L, and we expect $r \propto L^{1/2}$.

Now that we have an estimate of the source region size, we can also estimate its filling factor. When Hβ is emitted purely by recombination at a temperature $\sim 10^4$ K, its emission coefficient is $\simeq 1 \times 10^{-25}$ erg cm^{-2} s^{-1}; in practice it is difficult for collisional excitation to increase this by more than a few because the large densities that promote collisional excitation also promote de-excitation (cf. fig. 10.15, displaying the nearly constant excitation temperature of Hα). Thus, given the luminosity and the gas density, we immediately derive an emitting volume. In ratio to the total volume of the source region, the volume filling factor is

$$f = 2 \times 10^{-5} \left(\frac{L_{H\beta}/L_{\text{ion}}}{0.01} \right) L_{\text{ion},45}^{-1/2} n_{\text{H},10}^{-1/2} \Xi^{3/2} \left(\frac{T_4}{2} \right)^{3/2} . \tag{10.66}$$

Clearly, the line-emitting gas occupies a very small portion of the total volume. An immediate corollary is that there must be something else in the region besides line-emitting gas. What that might be is discussed in §10.7.

The mean column density is determined through two lines of argument: the strength of low-ionization lines (e.g., MgII 2800) and the strength of the Balmer lines relative to Lyα. Because the ionization potential of MgII is only 15 eV, it cannot form in substantial abundance unless the Lyman continuum has been weakened. Therefore, substantial Lyman edge optical depth is required, and we can estimate the minimum column density in terms of the incident ionization parameter, as previously argued (eq. 10.32). Assuming $\Xi \simeq 0.3$, this argument gives a minimum column $\sim 10^{22}$ cm^{-2}.

Given parameters chosen in this way, single-zone models for the broad line region do surprisingly well, often predicting relative line strengths within a factor of two or so of those observed. Although it might not be so remarkable that hydrogen recombination lines remain prominent over a wide range of physical conditions—unless elemental abundances are quite different from what we normally see, photo-ionization of H always accounts for a large part of the continuum absorption—it is a genuine surprise that the ionization conditions are so stable that, for example, lines of CIII and CIV are always fairly strong, but CII lines are very weak. Somehow there must be a remarkable degree of self-regulation.

However, these simple models also fail in significant ways. Observations generically show Lyα/H$\beta \simeq 3$–10, much smaller than the factor of 20–30 predicted by the models. In addition, the ratio Hα/Hβ can be anywhere from close to the normal recombination number $\simeq 3$ to several times that. It was to explain these discrepancies that Kwan and Krolik (1979) suggested that much of the Balmer line flux might be created by collisional excitation from $n = 2$ in a very optically thick zone heated by X-rays.

However, later work found that the X-ray/UV flux ratio in typical AGNs (e.g., as in the Elvis et al. mean radio-quiet quasar spectrum) is rather smaller than is required for this explanation to succeed in the generic case, although there are examples in which the X-ray flux is strong enough.

Another—related—difficulty has to do with the FeII emission. If it is as strong as Wills et al. (1985) suggested, its total flux is several times as large as Lyα. Because FeII can only exist in a region where the Lyman continuum intensity is weak, if its energy is derived from photo-ionization, the X-ray flux must then be comparable to or greater than the Lyman continuum flux. This seems unlikely given our present understanding of the global continuum shapes of AGNs. It is possible that these lines are powered by some means other than photo-ionization by the continuum we see (e.g., by Lyα fluorescence: Sigut and Pradhan 1998).

Finally, it is worth pointing out the intrinsic difficulties that stand in the way of inferring elemental abundances from emission line data. Attempts are sometimes made to do this on the basis of single-zone models. Unfortunately, although the existence of strong lines of C, N, O, and others demonstrates that the abundances of the heavy elements cannot be too small, it is very hard to be more quantitative (Davidson 1977). One might wish to begin with the strongest lines, whose fluxes are, of course, the easiest to measure accurately. Because, by construction, these lines account for a large fraction of the total cooling, their fluxes are relatively insensitive to the abundances of their parent elements: these lines must carry a flux equal to the continuum flux absorbed, and a low abundance of the parent species simply leads to a (small) increase in temperature. Weak lines, on the other hand, have excitation rates that really are directly proportional to their abundance, but they are more difficult to measure accurately. In addition, by the argument of §10.3.3.2, those found in the rest-frame ultraviolet are very sensitive to temperature. Because the temperature is controlled by the dominant coolant, this fact introduces an implicit dependence on the abundance of the coolant species; in fact, if the wavelength of the dominant coolant is longer than the wavelength of the line being used as an abundance indicator, the flux of the indicator depends more on the abundance of the dominant coolant than on its own abundance. Temperature can also be sensitive to the shape of the (unobserved) EUV continuum. A further complication comes from the likelihood that several zones of different ionization parameter or pressure are present in the line-emission region. Changes in abundance then trade off against changes in physical conditions when matching the total line strengths to observations. For all these reasons, although attempts have been made to infer deviations from Solar abundance, their reliability has always been questionable.

10.5.3.1.2 multi-zone models: profile modeling and reverberation mapping

Greater resolution in our description of the broad emission line region would, of course, be desirable, but direct imaging will be out of the question until imaging optical interferometry becomes feasible. Using the size scale estimate derived from single-zone photo-ionization modeling, it immediately follows that the angular size scales with the flux from the object:

$$\theta \simeq 15 \left(\frac{F_{\text{ion}}}{10^{-11} \text{ erg cm}^{-2} \text{ s}^{-1}} \right)^{1/2} \Xi^{-1/2} n_{\text{H},10}^{-1/2} \left(\frac{T_4}{2} \right)^{-1/2} \text{ microarcsec.}$$

(10.67)

Even the brightest AGNs in the sky have fluxes only $\sim 10^{-10}$ erg cm^{-2} s^{-1}, so we must seek indirect methods of achieving higher resolution if we are to learn anything about the internal structure of the broad emission line region.

Two ways present themselves: profile modeling and reverberation mapping. They are conceptually similar in that both involve accumulating large quantities of high signal/noise data to enable projection of the line flux onto special surfaces in the source. Although acquisition of the data is much easier for profile modeling, the interpretation is far more direct and far less model-dependent for reverberation mapping.

Consider profile modeling first. Measuring the specific flux as a function of wavelength is equivalent to measuring the flux as a function of line-of-sight velocity. Thus, we can think of a line profile as a mapping of the source emissivity onto its surfaces of constant line-of-sight velocity. If at every location in the source region the line-emitting material has a definite velocity, these surfaces are well defined and infinitesimally thin, and the mapping is well defined. Unfortunately, we don't know a priori where these surfaces are, so any application of this method is highly model-dependent. Worse yet, it is possible that the line-emitting material is effectively collisionless. If so, at each point in the region there is a distribution of velocities, rather than a single unique one. Because f is so small, one could easily imagine the line-emitting stuff being organized into a large number of small clouds that collide with each other only rarely. If this is the case, the wider the distribution, the "fuzzier" the surfaces of constant line-of-sight velocity become, and the less well defined the mapping. For these reasons, little reliable information has been derived from this technique.

Next let us turn our attention to reverberation mapping. The basic idea behind this method is that, by the estimate of equation 10.65, photons require a time

$$t_l \sim 80 \left(\frac{L_{\text{ion},45}}{\Xi n_{\text{H},10} T_4/2} \right)^{1/2} \text{d} \tag{10.68}$$

to travel from the continuum source to the line-emitting gas. By contrast, only minutes elapse while the gas equilibrates to the new continuum level (§10.4). Unless all the gas lies right along the line-of-sight between us and the continuum source, we might therefore expect to see its different components responding to the continuum fluctuations with a range of delays $\sim 100 L_{45}^{1/2}$ d. The resulting line lightcurve would then be a delayed, and smeared, version of the continuum lightcurve. At the order of magnitude level, this delay could be estimated by cross-correlating the continuum and line lightcurves (Bahcall, Kozlovsky, and Salpeter 1972; Gaskell and Sparke 1986).

But a pair of lightcurves contain far more information than a (peculiarly weighted) mean delay. It is possible to disentangle the two to infer the amount of response *as a function of delay* (Blandford and McKee 1982). At the end of this process, because the delays are entirely due to light travel time, we have a map of the emission line region projected onto the surfaces of constant delay with respect to our line-of-sight (fig. 10.19).

More formally, if a line-emitting cloud lies at radius r from the continuum source and polar angle θ with respect to our line of sight, the difference in time of arrival at Earth between a straight flight from the continuum source and a trip that stops first at the cloud is

$$\tau = \frac{r}{c} \left(1 - \cos\theta \right). \tag{10.69}$$

The relation between emissivity j_l in a given line and the continuum flux F_c incident upon it can be quite complicated and nonlinear. However, if the change in the continuum is relatively small, we can describe the response as quasi-linear by expanding about the mean:

$$j_l(F_c) = j_l(\langle F_c \rangle) + \frac{\partial j_l}{\partial F_c} \left(F_c - \langle F_c \rangle \right) + \cdots . \tag{10.70}$$

Taking this approximation permits a simple description of the fluctuating line luminosity:

$$\delta L_l(t) = \int_0^\infty d\tau \, \Psi(\tau) \left[L_c(t - \tau) - \langle L_c \rangle \right], \tag{10.71}$$

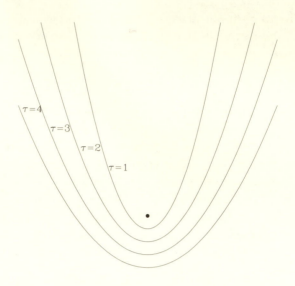

Fig. 10.19   Cross sections of several surfaces of constant delay with respect to an observer located far away and to the top. The delays labeling the curve are in arbitrary relative units and refer to the difference in light travel time between a path directly vertical from the black circle, and a path that starts there, goes to a point on one of the curves, and then turns vertical. Rotated around the line-of-sight to the observer, these curves generate paraboloids.

where the *response* or *transfer function*

$$\Psi(\tau) = c \int_S dA \, \frac{f(\vec{r})}{4\pi r^2} \frac{\partial j_l}{\partial F_c}. \tag{10.72}$$

Here the surface of integration S is the iso-delay surface for delay τ and the quantity $f(\vec{r})$ is the *local* volume filling factor at $\vec{r}$.

We have now reduced the problem to inversion of a convolution relation: given the line and continuum lightcurves, how can we find $\Psi(\tau)$? Although inversion by Fourier transforms is usually impractical (it is characteristically very sensitive to noise), this formal solution does shed light on the issues involved:

$$\Psi(\tau) = \int df \, e^{-2\pi i f \tau} \frac{\hat{L}_l(f)}{\hat{L}_c(f)}, \tag{10.73}$$

where $\hat{L}_{l,c}(t)$ are the Fourier transforms of $L_{l,c}$. In order to have a good description of a response function that varies on the scale τ, one must have

374

good measurements of the lightcurve Fourier transforms at frequencies $\sim \tau^{-1}$. Therefore, the sampling interval Δt must be $< \tau/2$, and the sampling must continue for a duration at least several times τ. If one wishes to search for structure on a range of lengthscales, the minimum number of observations required is $\sim 5 r_{max}/r_{min}$; in practice, given that one performs this exercise because the exact size of the region is not known a priori, the minimum number of observations is likely to be at least 50–100. It is for this reason that the labor involved is so great.

The Fourier description also emphasizes another fact about reverberation mapping: the AGN under study must be "cooperative" in the sense that it varies significantly on these timescales during the monitoring. The effective S/N for an experiment like this is not the ratio of mean flux to *rms* error, but rather the ratio of *rms* fluctuations on the timescales of interest to *rms* error.

There is no reason why this solution cannot be carried out for a number of emission lines simultaneously. One then obtains a mapping of the *marginal* emissivity $\partial j_l/\partial F_c$ (as opposed to the total emissivity j_l) for a number of lines simultaneously, and the techniques of photo-ionization modeling can be applied to the ratios of these marginal emissivities in order to infer mean physical conditions on each isodelay surface.

This procedure can also be generalized to make it sensitive to velocity structure by the simple expedient of replacing the total line flux with the specific flux in each of a number of frequency bins and performing the same solution. The final result is $\Psi(\tau, v)$, a map of the marginal emissivity onto the curves (or belts, if the velocity distribution function has a finite width) which are the *intersections* of the surfaces with delay τ and line-of-sight velocity v.

Although deconvolution can be performed in a numerically stable way by a variety of techniques, interpretation of the response functions that result is ineluctably model-dependent. The paraboloids of constant delay have the admirable property of being located at very well known positions (we are quite confident light travels at c!), but they have the great disadvantage of possessing a symmetry axis that depends on the direction to Earth. Unless by some great good fortune that is the true symmetry axis of the system, some assumption must be made about the true symmetry of the source in order to turn response functions into genuine maps.

Another central difficulty has to do with the fact that this method maps $\partial j_l/\partial F_c$, *not* j_l. There might be large quantities of material whose output is insensitive to F_c that would be entirely missed by reverberation mapping. For example, the line emission from regions optically thin in the Lyman continuum tends to be very insensitive to continuum fluctuations because an increase in ionizing flux causes an (almost) exactly compensating *decrease* in neutral fraction, and hence optical depth. Any line (e.g., H recombination lines, CIV 1549) that draws its energy primarily from the Lyman continuum would, in these circumstances, hardly vary at all, despite large continuum changes. In the same vein, any line whose output is especially sensitive to continuum changes tends to be overemphasized.

In addition, the whole scheme rests upon the validity of three assumptions about the continuum radiation. The first is that the lightcurve we follow must be a fair indicator of the shape of the luminosity history throughout the ionizing continuum and in all directions:

$$\frac{dL_\nu(\nu; \hat{n}; t)}{d\Omega} = L_\nu(\nu_*; \hat{n}_*; t) K(\nu; \hat{n}), \tag{10.74}$$

where $L_\nu(\nu_*; \hat{n}_*; t)$ is the specific luminosity per solid angle at the observed frequency ν_* radiated in our direction $\hat{n}_*$, and $K(\nu; \hat{n})$, the function that relates the continuum we observe to the continuum at other frequencies as seen from other directions, is independent of time. For reverberation mapping to be meaningful, we need to be confident that the continuum we track, which is generally at frequencies below the Lyman edge in the AGN rest frame, can tell us the time variation of the *ionizing* continuum that powers the emission lines. It is equally important that any line-emitting material on other lines of sight sees fluctuations that are proportional to the ones seen in our direction. Note that $\partial K/\partial \Omega$ need not be zero; it just needs to be constant in time. The second assumption is that the time for light to travel across the continuum source is substantially smaller than the minimum resolution scale of the monitoring. If this were not so, each region within the line source would see a *different* continuum lightcurve, and each of them would be different from the one we observe. The third assumption is that continuum photons reach the line-emitting region without any additional pathlength imposed by scattering.

These assumptions can be (approximately) verified post hoc by one's ability to find a function $\Psi(\tau)$ that relates the continuum lightcurve to the line lightcurve(s) through a convolution. If any of them were false, it would be extremely fortuitous that such a solution could be found.

Thus, with a solution in hand, it is possible to use the reverberation mapping technique to make some fairly robust statements, provided one is careful about the degree of model dependence introduced by one's choice of geometrical symmetry. For example, if $\Psi(0) \neq 0$, one of two statements must be true: either there is material closer to the continuum source than $c\Delta t$, or at least some material is located on the line-of-sight, and this material is capable of radiating line photons out its far side. Unless our line-of-sight is special, having material along the line to us also suggests that the line-emitting stuff is distributed fairly isotropically around the continuum source. Similarly, if there is response distributed over a range of timescales, there must be a correspondingly large range of lengthscales within the emission line region.

In the low-luminosity AGNs, whose timescales are consistent with human scheduling constraints, it has been possible to use reverberation mapping to reach just such conclusions. In at least one case (the type 1 Seyfert galaxy NGC 5548), the broad emission line region stretches over possibly as much as two decades in radius, from $\simeq 1$ to > 60 lt-d (data: Clavel et al. 1991, Peterson et al. 1992; Korista et al. 1995; analysis: Krolik et al. 1991; Horne, Welch, and Peterson 1991; Wanders et al. 1995; Done and Krolik 1996). Within this broad range, there is a grading of ionization levels, with the highest ionization material closest in. The luminosity of this Seyfert galaxy is $\sim 10^{44}$ erg s^{-1}; the corresponding generic single-zone radius estimate is $\sim 65 L_{44}(\Xi/0.3)^{-1/2} n_{H10}^{-1/2}(T/2 \times 10^4 \text{ K})^{-1/2}$ lt-d. The rough agreement with observations indicates that the single-zone approximation is not bad as an estimator of the mean conditions, but, of course, it is by its nature incapable of revealing the broad spread in conditions. The ionization grading also means that the entire single-zone procedure is suspect: there is no point in using many different lines to constrain the physical conditions in the source if some of those lines are made in locations entirely disjoint from where the others are produced.

Profile-resolved reverberation mapping indicates that the kinematics of the broad line region are rather complicated (Wanders et al. 1995; Done and Krolik 1996; Ulrich and Horne 1996). The response functions show an overall red-blue symmetry, but (in at least one case, again NGC 5548) the red wing responds noticeably more quickly than the blue wing. The simplest, but by no means the only, interpretation of these facts is that the predominant motion of the radiating material is not radial, but there is a net flow inward.

10.5.3.2 narrow line region

From the prominence of forbidden lines in the narrow line spectrum we can immediately infer that the characteristic densities are at least several orders of magnitude lower than in the broad line region. Because it is impossible for line-emitting gas to exist at ionization parameters several orders of magnitude higher than prevalent in the broad line region, the narrow line region must therefore lie substantially farther from the nucleus than the broad line region. In fact, as we shall see shortly, the most likely distance is so great that it would be extremely surprising if the integrated narrow line flux changed appreciably during a human lifetime. Reverberation mapping is therefore out of the question as an analytic tool for this region. Fortunately, the narrow line region is so big that in the brighter AGNs it can actually be resolved in images taken by optical/UV telescopes in space. In addition, the contrasts in line profile from line to line are sufficiently great that some of the ambiguity is removed from profile modeling.

Just as for the broad line region, inferring densities, temperatures, and so on from individual line ratios is often an easier path to estimated conditions than formal model fitting. However, as we will see shortly, even more than for the broad line region, there is good reason to think that diverse conditions contribute to the total line flux, so it is very important to recognize the limitations of single-zone thinking.

The lower densities that permit forbidden lines to be strong also simplify the atomic physics of line production. Fewer important lines are optically thick, and excited state populations are generally much smaller. Unknown dust extinction can sometimes be a problem, but many of the line diagnostics involve combinations of lines close enough in wavelength that this is not an issue. In fact, the comparative simplicity of the line emission physics allows us to use lines *distant* in wavelength to estimate the extinction by the difference between their measured relative strength, and what we would expect it to be if we had a clear line-of-sight.

The temperature, for example, can be inferred by comparing the flux in two lines within the same ion whose excitation rates differ only because of the energy required to reach their upper states. Figure 10.20 shows a simplified energy level diagram for the ground configurations of two C-like isoelectronic ions, OIII and NII. All the radiative decays are forbidden and sometimes multiply so. Nonetheless, provided $n_e \ll 7 \times 10^5$ cm^{-3} (the critical density for [OIII] 4959,5007; the critical density for [OIII] 2321,4363 is ~ 100 times larger), any ion excited into the ^1D term will de-excite by emission of either λ 4959 (6548) or λ 5007 (6583), and any ion excited to the ^1S term will de-excite by emission of either λ 4363 (5755), branching ratio 0.9, or λ 2321 (3063), branching ratio 0.1. Therefore, the ratio of the fluxes in λ 4363(5755) to λ 5007 + λ 4959 (λ 6583 + λ 6548) depends only on the temperature, due to the contrast in excitation energy. In AGNs, this method works well on the OIII lines (except when Hβ is so broad that it is blended with λ 4959); however, the two members of the NII doublet

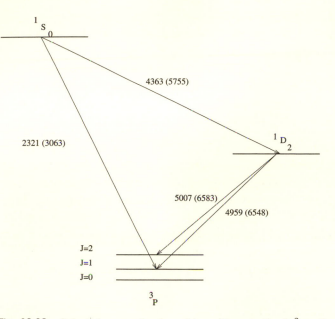

Fig. 10.20 Selected energy levels and transitions in the $2p^2$ config-uration of O III and N II. The numbers by each transition refer to the wavelengths in Å of the associated lines, with the numbers in paren-theses referring to N II. Although the energy differences between the terms are drawn to scale, the fine-structure splitting is exaggerated.

are both so close to Hα that it is very difficult to disentangle them to the necessary accuracy.

Density indicators can be obtained equally simply. For a start, if the ratio of the flux in [O III] 4363 to the sum of the fluxes in [O III] 5007 and [O III] 4959 (or their isoelectronic N II analogs) rises above 0.025, it signals the onset of collisional de-excitation of the ^1D state and a density greater than 10^5 cm^{-3} (this is the way in which the absence of broad wings on [O III] 5007 demonstrates that the broad line region must have a density $\gg 10^6$ cm^{-3}). A more useful density indicator comes from the lines of the ions O II and S II, which share the same p^3 ground-state configuration, but with different principal quantum numbers. In these ions, the ground state is a $^4S_{3/2}$ level, connected by a pair of lines to the $J = 3/2$ and $J = 5/2$ levels of a ^2D term. In the low-density limit, the flux ratio between the two lines (λ 3726, 3729 in O II, λ 6716, 6731 in S II) is simply the ratio of their statistical weights, because the line energies are nearly identical and the collision strengths depend only on the terms, and not on

379

J. On the other hand, as the density rises above 10^3 cm^{-3}, the growing importance of collisions relative to radiative decays leads to a situation in which the ratio of the two upper states' *populations* is the ratio of their statistical weights, while the ratio of the two lines' *fluxes* is the product of the population ratio and the ratio of the lines' radiative decay rates. Thus $\lambda\, 3729(6731)/\lambda\, 3726(6716)$ falls from 1.5 when $n_e \ll 10^3$ cm^{-3} to 0.3 (0.4) when $n_e \gg 10^4$ cm^{-3}.

Just as for the broad line region, the ionization parameter is indicated best by the highest ionization stages present. Ratios of the strength of different ions within the same element (e.g., OIII and OII) are helpful, but there is a degeneracy between lower ionization parameter and greater optical depth. Unfortunately, the weakness of the narrow lines in high-luminosity objects means that we can't make use of a redshift to bring OVI 1034 into a more easily observable band, so in many instances we must make do with (blended) NV 1240, or CIV 1549 which, as already seen in the case of the broad lines, really only provides a lower bound to the ionization level.

Two line ratios can be used to bound the column density of the emitting material: Hβ compared to either HeII 1640 or (preferably) HeII 4686 is a sensitive measure of the column density when $N_{\rm H}$ is small enough that the gas never becomes mostly neutral, and [OI] 6300/Hα helps constrain $N_{\rm H}$ when it is large enough to form a neutral zone.

If a region is optically thick at both the HeII and Lyman edges, and HeII dominates the opacity for photons between $\simeq 55$ and $\simeq 100$ eV, while HI dominates the opacity from 13.6 to $\simeq 25$ eV (the usual state of affairs), the ratio of HeII recombination photons to HI recombination photons is determined entirely by the continuum shape. However, this simple relationship may be broken when one of these conditions is not met. As can be seen in figure 10.16, the HeII edge becomes optically thick well before the HI edge does (this is why the electron fraction declines from 1.2 at the illuminated edge to 1.1 just before the H ionization transition). Consequently, if the total column density of the region is too small for the Lyman edge to become very optically thick, the ratio of HeII recombination photons to HI recombination photons will be greater than the value predicted by the ratio of ionizing photons in the incident continuum. Large ratios of HeII 1640 (the Hα analogue) or HeII 4686 (the Paschen α analogue) to either Lyα or the Balmer lines may therefore signal small column densities (e.g., Binette, Wilson, and Storchi-Bergmann 1996).

On the other hand, OI and HI have nearly identical ionization potentials, so both continuum opacity and charge exchange link their ionization balance very tightly. Therefore, there can be no OI line emission unless the material is thick enough for most H to recombine. Thus, the strength of [OI] 6300 is a measure of both the column density of the matter and the flux in the high-energy continuum whose absorption heats the mostly neutral region.

Applying these simple diagnostics to narrow line spectra immediately reveals that single-zone models fail (Stasińska 1984). For example, the ratio [OIII] 4363/([OIII] 4959 + [OIII] 5007) generally falls in the range 0.005–0.1. The low end of this range is consistent with a temperature $\simeq 11,000$ K and $n_e \ll 10^5$ cm^{-3}. Such a temperature is about what would be expected when the ionization balance contains substantial OIII (see fig. 10.16). However, the high end of this line ratio range must be coming from regions with density high enough to de-excite the λ 4959 and λ 5007 lines, that is, in these cases n_e must be at least 10^5 and possibly as high as 10^6 cm^{-3}. This would not be so bad if the ratio [SII] 6716/[SII] 6731 were not always in the range 0.8–1.5, requiring that the density where those lines be made be less than 10^3 cm^{-3}.

Another indication of inhomogeneity comes from (the small number of) examples in which we can contrast the ionization as estimated from [OIII] lines with OVI 1034/Lyα. For example, in NGC 1068, the [OIII] line ratio would suggest $T \simeq 11,000$ K, or $\Xi \simeq 0.03$; the ratio of [OIII] to Hβ is consistent with about the same Ξ. On the other hand, OVI 1034/Ly$\alpha \simeq 0.3$, requiring $\Xi \gtrsim 1$.

A similar problem arises contrasting the HeII 4686/Hβ ratio with the ratio [OI]6300/Hβ. The former ratio can be anywhere from 0.05 to 0.5, with the upper end of the range especially well populated. By contrast, likely EUV continuum shapes would predict 0.1–0.2 in the optically thick limit. Thus, one might infer optically thin regions are common. At the same time, [OI] 6300 is very often fairly strong (see table 10.2); the material responsible for this line *must* be optically thick at the Lyman edge, and therefore also at the HeII edge.

In addition to all this evidence pointing to a narrow line region comprising a broad range of physical conditions, there appears to be a correlation between line width and either critical density for collisional de-excitation or ionization energy required to reach the parent species (Pelat, Alloin, and Fosbury 1981; Filippenko and Halpern 1984; Filippenko 1985). If the true correlation is with the former quantity, then the fastest-moving gas has the highest density; if with the latter, the fastest-moving gas is subjected to the highest ionization parameter. If both are genuine correlations, the fastest-moving gas must be the closest to the nucleus, and the density must decrease outward, but more slowly than $\propto r^{-2}$.

Nonetheless, despite the fact that single-zone pictures of the narrow line region are very poor portraits, certain order of magnitude estimates can be made in a fashion parallel to those made for the broad line region. For example, the rough scale of the narrow line region is

$$r \sim 500 L_{\text{ion},45}^{1/2} n_{\text{H},4}^{-1/2} \left(\frac{\Xi}{0.03}\right)^{-1/2} \text{ pc}, \qquad (10.75)$$

and its filling factor is as at least as small as the broad line region's:

$$f \sim 4 \times 10^{-6} \left(\frac{L_{H\beta}/L_{\text{ion}}}{0.001}\right) L_{\text{ion},45}^{-1/2} n_{\text{H},4}^{-1/2} \left(\frac{\Xi}{0.03}\right)^{3/2} T_4^{3/2}. \qquad (10.76)$$

These estimates demonstrate that there is one crucial respect in which the narrow line region differs from the broad line region: it is potentially resolvable with ordinary optics, at least in the brighter objects. The angular size that corresponds to the radial scale estimated above is

$$\theta \sim 0.1 \left(\frac{F_{\text{ion}}}{10^{-11} \text{ erg cm}^{-2} \text{ s}^{-1}}\right)^{1/2} \left(\frac{\Xi}{0.03}\right)^{-1/2} n_{\text{H},4}^{-1/2} T_4^{-1/2} \text{ arcsec;} \tag{10.77}$$

HST imaging can therefore reveal internal structure in the narrow emission line regions of the brighter AGNs.

Just this program has been (and is continuing to be) carried out. The type 2 Seyfert galaxy NGC 1068 is so close and so bright that it allows particularly detailed study (plate 5). What the image reveals is that the line emission structure is both very clumpy and very complex. Some of the brighter emission line features are found near the edges of bright spots in the radio map. It is possible that shocks driven into interstellar gas by the radio jet may enhance line emission by compression, and possibly by heating or local ionization (see §10.6). At the same time, the zone of line emission seems to be contained within an envelope with very smooth boundaries. This fact will turn out to be very important in the context of unified models, treated in Chapter 12.

The kinematics of the narrow line gas appear, at zeroth order, to be dominated by ordinary orbital motion in the host galaxy. Nelson and Whittle (1996) have shown that there is a very good correlation between the width of the [OIII] 5007 line and the stellar velocity dispersion in the host's bulge. On the other hand, there is also a correlation between those cases in which λ 5007 is wider than than the bulge dispersion would predict, and particularly strong radio emission, thus suggesting that interaction with the radio jet may be a supplementary force.

10.6 Shocks

10.6.1 Internal shock structure

Although the evidence is very strong that most AGN line emission is powered by absorption of the AGN continuum, there may well be some regions in which the primary energy source is the mechanical energy of shock waves. Because we can directly observe the continuum (albeit with a big spectral gap in the EUV), we can estimate comparatively easily how much energy it makes available for line production; by contrast, our evidence for shock waves is more indirect, and therefore more uncertain. For example, we see radio jets terminating, presumably in shocks, but it is difficult to infer securely much quantitative information about them (§9.3.3). Thus, in practice we are often reduced to the logically shaky procedure of guessing that shocks of a certain character exist, attempting to predict the line emission that would result, and then seeing whether these predictions are better or worse at replicating the observations than a competing photo-ionization model with the same number of free parameters.

Let us begin with the basic physics of shock waves. A short summary of the relations between preshock, or upstream, and postshock, or downstream, properties, is given in Appendix C. Specializing to the case of a purely fluid high-Mach number shock wave striking a gas in which H and He are already fully ionized, we can write the postshock conditions in terms of those prevailing preshock (cf. eqs. C.19):

$$r = 4,$$

$$T_s = 1.38 \times 10^7 \left(\frac{v_s}{1000 \text{ km s}^{-1}} \right)^2 \text{ K}, \tag{10.78}$$

$$\mathcal{M}_2 = \frac{1}{\sqrt{5}},$$

assuming that the electron and ion temperatures equilibrate instantly (see eq. C.21). If the upstream gas had been entirely neutral, the postshock temperature would be a factor of two greater. An immediate consequence of the factor of 4 jump in density is that the postshock velocity is 1/4 the shock speed.

As the fluid flows further downstream, the newly heated ions and electrons collisionally ionize atoms that were not already fully stripped in the upstream state. Estimating the collisional ionization cross section in the Born approximation, we find it takes a time

$$t_{\text{ion}} \simeq 4 \times 10^7 T_{e7}^{1/2} \left(\frac{n_e}{1 \text{ cm}^{-3}} \right)^{-1} \text{ s} \qquad (10.79)$$

for electrons of temperature $T_{e7} 10^7$ K to accomplish this job. This process of collisional ionization consumes some of the gas's heat content; if even H had been neutral in the preshock gas, the temperature, which would have begun a factor of two higher than indicated by equation 10.78, falls by $\simeq 200,000$ K (this figure includes, in addition to the H I ionization potential, the excess energy given freed electrons over and above the ionization potential, and the energy expended on collisional ionization of He). When the electrons are heated up to the ion temperature, the temperature is brought to the value predicted by equation 10.78.

Because the ratio between the rate of collisional excitation to excited bound states and the rate of collisional ionization depends just on atomic physics and the temperature, there is a characteristic spectrum of H and He lines radiated by the ionization zone (when it exists). This line output, dominated by Lyα, can roughly double the energy cost of this zone (Cox and Raymond 1985). Similarly, because the heavier elements are often underionized relative to their ionic balance in collisional ionization equilibrium at this temperature, the cooling rate due to collisional excitation of their lines is enhanced in this region.

Over a longer span of time, the gas continues to cool by emitting radiation. All the mechanisms already described in this chapter can contribute: bremsstrahlung, radiative recombination, and collisional excitation are all possible. The mix of line and continuum output, and the particular selection of lines emitted, depends strongly on the local ionization state of the gas, which continuously changes as the density and temperature of the gas change. Because the ionization equilibration time is often longer than the timescale for changes in the state of the gas, the approximation of local ionization balance is frequently invalid. Photoionization also plays a role in the ionization equilibrium because the postshock temperature in many cases is high enough for the gas to radiate a substantial flux in ionizing radiation. When this locally generated continuum passes through cooler regions, it can help maintain a higher ionization state there than would be supported by the local rate of collisional ionization.

The duration of the cooling can be estimated easily given two reasonable approximations: that the postshock gas pressure p is constant, and that the cooling function $\Lambda(T) \propto T^a$. As shown in figure 10.6, in the vicinity of 10^5 K, a depends strongly on the ionization state of the gas, but above $\sim 10^6$ K, where cooling is dominated by bremsstrahlung, $a \simeq 0.5$.

To make this estimate, we begin with the energy equation, abbreviated to eliminate any compressive work:

$$\frac{d(3k_\mathrm{B}T)}{dt} = -n_e \Lambda(T). \tag{10.80}$$

Use of our two approximations allows us to rewrite equation 10.80 as

$$\frac{d(3k_\mathrm{B}T)}{dt} = -\frac{p}{2k_\mathrm{B}T} \Lambda(T_s) \left(\frac{T}{T_s}\right)^a, \tag{10.81}$$

which can be integrated from $t = 0$ (at the shock front) to t_cool (the time at which the temperature has fallen practically to zero) to yield

$$t_\mathrm{cool} = \frac{6}{2-a} \frac{(k_\mathrm{B}T_s)^2}{p\Lambda(T_s)}. \tag{10.82}$$

Because $\Lambda(T_s) \propto T_s^a$, $t_\mathrm{cool} \propto T_s^{1-a}/n_o \propto v_s^{2(1-a)}/n_o$, for n_o the initial density.

It is often convenient to rephrase this estimate of the duration in terms of the cooling column density:

$$\frac{d(3k_\mathrm{B}T)}{dN} = -\frac{n_e}{n_\mathrm{H}} \frac{\Lambda(T)}{v}. \tag{10.83}$$

The fluid velocity v is related to the temperature through the constancy of the mass flux:

$$v = (n_o v_o)/n = 2n_o v_o k_\mathrm{B} T/p. \tag{10.84}$$

In this form, the cooling column is

$$N_\mathrm{cool} = \frac{3^{2+a}}{4^{3+a}(2+a)} \frac{\bar{m}^{1+a} v_s^{3+2a}}{k_B^a \Lambda_o}, \tag{10.85}$$

for $\Lambda(T) = \Lambda_o (T/T_o)^a$. When collisional ionization equilibrium prevails,

$$N_\mathrm{cool} \simeq 3 \times 10^{17} \left(\frac{v_s}{100 \text{ km s}^{-1}}\right)^4 \text{ cm}^{-2}, \tag{10.86}$$

for $60 \leq v_s \leq 150$ km s^{-1} (Draine and McKee 1993) because $a \simeq -0.5$ in that range of temperatures.

In a more precise treatment, several additional effects should be included. These include possible time variation of the conditions in the shock, compressive work done on the postshock gas, and magnetic pressure. To incorporate them, we return to the conservation equations developed in Appendix C, which we write here in one-dimensional form and assuming infinite electrical conductivity:

$$\frac{\partial \rho}{\partial t} + \frac{\partial}{\partial z}(\rho v) = 0,$$

$$\frac{\partial}{\partial}(\rho v) + \frac{\partial}{\partial z}\left(p + \rho v^2 + \frac{B_t^2}{8\pi}\right) = 0,$$

$$\frac{\partial}{\partial t}\left(u + \frac{3}{2}p + \frac{1}{2}\rho v^2 + \frac{B_t^2}{8\pi}\right) + \frac{\partial}{\partial z}\left[v\left(u + \frac{5}{2}p + \frac{1}{2}\rho v^2 + \frac{B_t^2}{8\pi}\right)\right] = H - C.$$
$$(10.87)$$

These must be supplemented by an equation of state,

$$p = k_{\mathrm{B}}T \sum_Z \sum_{j=0}^{Z} n_{Z,j}(1 + j), \qquad (10.88)$$

the flux-freezing equation,

$$\frac{\partial B}{\partial t} = -\frac{\partial}{\partial z}(vB_t), \qquad (10.89)$$

the equations describing the ionization balance (eq. 10.1 in time-dependent form), and a description of the local radiative heating and cooling (as in §10.3.3). Here B_t is the magnitude of the magnetic field perpendicular to z. Whether the shock can be treated as being in a steady state depends on whether the rate at which the shock changes character (e.g., through a change in the density of material it encounters, or a change in its driving pressure) is faster or slower than the time taken by a fluid element to pass all the way through.

Having written equations 10.87 and 10.89 in conservation form, it is very easy to recast them in a form appropriate to a time-steady situation. First, we define the frame of reference as the one in which the shock front is at rest. Then we set all partial derivatives with respect to time to zero. The first two equations can then be trivially integrated with respect to z, yielding two integration constants set at the shock front: the mass flux and the momentum flux. The time-steady flux-freezing equation likewise leads immediately to a conserved flux, the ratio of magnetic flux to density. The only equations remaining are the energy equation and the ionization balance equations. The latter all have the form $v(\partial n_{Z,j}/\partial z) = \ldots$, where the right-hand sides are all the ionization and recombination rates affecting the ion (Z, j).

In this light, we can now evaluate the quality of one of the approximations made to estimate the cooling time and column density. Because the postshock flow is always subsonic, supposing that the pressure is constant amounts to supposing that the magnetic pressure is never significant, that is, that the compression relative to the preshock density never approaches v_s/v_{A1}, the shock's Alfvénic Mach number. In a completely radiative shock, this approximation must ultimately break down. However, because the cooling time is predominantly spent at the highest temperatures, where the magnetic pressure is least important, the attendant correction is rarely significant to predictions of the primary light output of a shock.

10.6.2   Self-generated photo-ionization

The importance of photo-ionization of postshock material by locally generated photons can be gauged by estimating the ionization parameter due to these photons:

$$\Xi_{\mathrm{post}} = \frac{p_{rad,ion}}{p} \sim f_{\mathrm{ion}} f_{\mathrm{rad}} \frac{v_s}{c}, \tag{10.90}$$

where $p_{rad,ion}$ is the ionizing radiation pressure, f_{ion} is the fraction of the total radiated energy flux in ionizing photons, f_{rad} is the fraction of the work done by the shock that is radiated, and we have estimated the energy flux in the shock by ρv_s^3. f_{ion} and f_{rad} are, of course, at most unity, and $v_s/c \ll 1$ in a nonrelativistic shock. Thus, photo-ionization due to locally generated photons, although playing a role in the postshock ionization and heat balance, cannot be too strong. Its importance can be enhanced, however, in the coolest regions of magnetized shocks, where the gas pressure falls well below the prevailing postshock pressure due to the substitution of magnetic for gas pressure support. Depending on circumstances, photo-ionization by external radiation sources may be either negligible or dominant.

On the other hand, photo-ionization of the *preshock* gas can be very important.[*] In this case, the ionization parameter is

$$\Xi_{\mathrm{pre}} \sim f_{\mathrm{ion}} f_{\mathrm{rad}} \frac{v_s}{c} \mathcal{M}_1^2. \tag{10.91}$$

---

[*]Shocks of this variety are sometimes called, confusingly, "auto-ionizing" shocks; they have nothing to do with auto-ionizing atomic states

Because the postshock gas pressure is always greater than the preshock pressure, the preshock ionization parameter is correspondingly larger than the postshock ionization parameter. In fact, if a shock has been running long enough to be time steady, is fully radiative, and its electrons and ions quickly equilibrate in temperature, H and He are fully ionized by the time they enter the shock whenever $v_s > 110$ km s^{-1} (Shull and McKee 1979). The ionizing photons produced behind the shock drive an ionization front into the gas ahead of the shock front that travels at a speed (in the preshock gas frame)

$$v_i \sim c\Gamma \sim v_s f_{\mathrm{ion}} f_{\mathrm{rad}} \frac{\bar{m} v_s^2}{\langle \epsilon_{\mathrm{ion}} \rangle}, \qquad (10.92)$$

where $\langle \epsilon_{\mathrm{ion}} \rangle$ is the mean energy of the ionizing photons. Thus, when $f_{\mathrm{ion}} f_{\mathrm{rad}} \ll 1$ (true for $v_s < 100$ km s^{-1}), $v_i < v_s$, and the ionization front remains pinned to the shock front. On the other hand, when f_{ion} approaches unity, we can expect that $\langle \epsilon_{\mathrm{ion}} \rangle \lesssim \bar{m} v_s^2$, so if f_{rad} is not too small, v_i can exceed v_s, and the ionization front moves well ahead of the shock front. More detailed calculations (Dopita and Sutherland 1996) show that in fact the scaling can be somewhat stronger, with $v_i/v_s \simeq (v_s/175 \text{ km s}^{-1})^{1.4}$ for $200 < v_s < 500$ km s^{-1}.

Moreover, if the preshock gas is optically thick to the ionizing photons, the radiation produced *ahead* of the shock is comparable to the amount produced behind it because roughly half the photons radiated behind the shock are reprocessed in the upstream material. Thus, the light emitted in the "ionized precursor" can be an important part of the total.

10.6.3 Emergent spectrum

Combining all these segments, we see that there are several characteristic kinds of radiation produced by shocks. In the ionized precursor (should it exist), the gas is in photo-ionization equilibrium, sometimes at moderately high ionization parameter, and produces the sorts of lines discussed elsewhere in this chapter. If there is a region behind the shock hot enough to thoroughly strip all the abundant elements (i.e., T at least several times 10^7 K or more), much of the heat dissipated in the shock will be radiated as X-rays created by the bremsstrahlung process. Somewhat lower maximum temperatures ($T \sim 10^6$–10^7 K) create instead a spectrum rich in soft X-ray lines. Still slower shocks concentrate their emission in EUV lines and radiative recombination continua. In shocks that have accumulated a cool, optically thick postshock tail, much of the energy in ionizing photons produced immediately postshock may be reprocessed in lower energy photons that cool the tail.

Because different shock models produce different balances between these different segments, the range of emergent spectra that can be produced by shocks is quite broad. Where in this range a particular model falls depends on explicitly tunable free parameters (the shock speed, the magnetic field strength, the column density of upstream material), qualitative model choices (shock geometry, whether or not the shock is time steady), and guesses about unknown physics (whether T_e equilibrates rapidly to T_i). Certain features of the output can be identified with just one or two of these choices.

The luminosity radiated in high-energy photons depends most strongly on the shock speed, but just how much of this radiation emerges to distant observers depends on the shock's boundary conditions and geometry. Consider first a shock that is truly plane-parallel. Much (and possibly all) of the ionizing continuum directed upstream can be reprocessed into optical and ultraviolet lines if the column density of material waiting to be shocked (the "ionized precursor") is great enough. Similarly, if the shock has already swept up enough matter, a very large quantity of cool, recombined gas rides along at the tail end of the postshock region and can absorb ionizing photons directed that way. Sideways escape is also uncertain. The opacity of the hottest postshock region can be very small, but if the shock's width is large enough (that is, the shock really is effectively infinitely wide), nonetheless photons may be prevented from escaping in that direction. Thus, the sideways flux depends on the (usually highly uncertain) aspect ratio of the shock. Because of these uncertainties, most published calculations of predicted shock continuum spectra present the flux that crosses the shock front in the upstream direction; this will be approximately correct if either the precursor is not optically thick, or the width of the shock is not too great.

Resonance line production is influenced by whether or not the incoming material is already ionized (i.e., this is controlled by a combination of the initial state of the upstream gas, and the shock speed), by the strength of the ionizing continuum radiated in the hot regions of the shock (this is most directly affected by the shock speed), and by the column densities of cool, neutral matter both before and after the shock. In addition, resonance line photons face transfer problems of the sort discussed in §10.3.4.2. A truly plane-parallel shock tends to focus its line output parallel to its direction of travel. If the optical depth and density are large enough, resonance line photons can be thermalized before escaping (§10.3.5). What sets this situation apart from the situation discussed previously in this chapter is that the optical depth that counts is the depth across a *velocity-coherent* layer; that is, once a photon travels to a region where the local line center is displaced by a Doppler shift by more than a few thermal widths, it no longer sees any significant opacity.

Forbidden line emission is sensitive to the same free parameters as for resonance lines. However, their prediction is rather less uncertain because transfer effects for them are trivial: their small oscillator strengths generally guarantee that their optical depths are negligible.

Having said all that, it is possible to draw certain generic conclusions about the output spectra to be expected from shocks. Shock continuum spectra (if they can be seen) tend to be rather hard up to a maximum photon energy, and then decline rapidly at higher energies. When the maximum postshock temperature is in the range $\sim 10^6$–10^7 K, they are dominated by a pseudo-continuum formed by the blending of large numbers of soft X-ray lines. The equivalent width of these lines is so large because the cross section (per H atom) for excitation of a resonance line in an ion with abundance X_i relative to H is $\sim \alpha_{\rm fs}^{-3} X_i$ times greater than the effective cross section for radiation of equivalent energy photons by bremsstrahlung (compare eqs. 8.1 and 10.18). The resulting blend is very hard because most of the energy tends to come out at photon energies not much greater than $k_{\rm B}T$, and the exponential cutoff in the electron distribution function is replicated in a sharp cutoff in the emitted photon spectrum.

Shocks having ionized precursors exhibit a component very much like that emerging from a photoionized region (§10.5). The only difference lies in the shape of the exciting spectrum. As just discussed, shock continua tend to have a broad, but definite, maximum when plotted in terms of νL_ν, whereas AGN continua are more nearly flat. In addition, if the shocks are fully radiative, they will produce lower ionization, but relatively high temperature, emission lines from the cooled postshock gas. This means, for example, that [OIII] 4363/[OIII] 5007 tends to be higher in shocks than in purely photoionized emission line regions (§10.5.3.2). It also means that lines such as CIII 977, which are generally quite weak in photoionized gases, can be much stronger behind shocks.

10.7 Line-Emitting Gas Dynamics and Life Cycle

What we know for certain about the history and dynamics of AGN emission line matter is very limited: It exists; it moves with a range of line-of-sight velocities generally a few times 10^3 km s^{-1} in the broad line region, and a factor of ten smaller in the narrow line region; and in a few cases reverberation mapping studies provide constraints on the kinematics of the broad line region. All else is speculation.

The scale of the problem is set by the mass budget. In the broad line region, the total mass of emitting material is approximately

$$M_{\text{BLR}} \sim 2 \left(\frac{C}{0.1}\right) \left(\frac{N_{\text{H}}}{3 \times 10^{22} \text{cm}^{-2}}\right) L_{\text{ion},45} \Xi^{-1} n_{\text{H},10}^{-1} M_{\odot} \qquad (10.93)$$

for covering factor C, while in the narrow line region it is much larger:

$$M_{\text{NLR}} \sim 1 \times 10^{7} \left(\frac{C}{0.1}\right) \left(\frac{N_{\text{H}}}{3 \times 10^{21} \text{ cm}^{-2}}\right) L_{\text{ion},45} \left(\frac{\Xi}{0.03}\right)^{-1} n_{\text{H},4}^{-1} M_{\odot}. \qquad (10.94)$$

We do not know the lifetime of emission line material, either in a form suitable for producing emission lines, or in the region. If we estimate this lifetime by the crossing time for matter moving at a velocity comparable to a typical profile width, the implied resupply requirement is substantial, and not very different in the two regions (given the uncertainty in these estimates):

$$\dot{M}_{\text{BLR}} \sim 0.3 \left(\frac{C}{0.1}\right) \left(\frac{N_{\text{H}}}{3 \times 10^{22} \text{ cm}^{-2}}\right) L_{\text{ion},45}^{1/2} \Xi^{-1/2} n_{\text{H},10}^{-1/2} M_{\odot} \text{ yr}^{-1}, \qquad (10.95)$$

and

$$\dot{M}_{\text{NLR}} \sim 0.2 \left(\frac{C}{0.1}\right) \left(\frac{N_{\text{H}}}{3 \times 10^{21} \text{ cm}^{-2}}\right) L_{\text{ion},45}^{1/2} \left(\frac{\Xi}{0.03}\right)^{-1/2} n_{\text{H},4}^{-1/2} M_{\odot} \text{ yr}^{-1}. \qquad (10.96)$$

These numbers are comparable to the mass accretion rate,

$$\dot{M}_{\text{acc}} = 0.18 L_{45} \frac{L_{\text{bol}}}{L_{\text{ion}}} \left(\frac{e}{0.1}\right)^{-1} M_{\odot} \text{yr}^{-1}. \qquad (10.97)$$

It seems unlikely, however, despite the apparent similarity in these estimates, that mass flow through the emission line regions is directly connected to the accretion rate because they scale differently with luminosity.

One immediate issue of concern is the form of the emission line gas. We already know that its volume filling factor is very small. One arrangement that might be possible for the broad-line gas is that it all lies in a single sheetlike structure with very large internal velocity gradients. A possible realization of this might be the surface of an accretion disk. The gas in this sheet might be stationary, but it might also have a steady flow, either inward or outward, glowing in emisison lines only at some favored location within the flow. We already know from images such as Plate 5 that the narrow line gas does *not* form a single coherent structure. Another arrangement, which might work for either the broad or the narrow line gas, is to break up the material into a very large number of small clouds, each having the estimated column density, but otherwise unspecified in shape. Presumably the internal velocity differences within each of these clouds would be subsonic, while cloud-to-cloud velocity differences would be very large. Such clouds might or might not be sufficiently long-lived to cross the region. A third possibility is that the gas is divided into a large number of different pieces, but each piece has no genuine structural integrity; rather, in each piece, there is a flow of gas, supplied by some long-lived source such as a star, that generates lines only where the conditions are just right. All of these suggestions present difficulties.

The most obvious criticism of identifying the broad emission line region with the surface of an accretion disk is that if the disk itself is the source of the ionizing radiation, the disk must flare very strongly in order to intercept enough ionizing flux in physical conditions capable of radiating lines. Given our ignorance of disk structure, this is not impossible, but it does appear to require some special pleading, especially in those objects where the emission line equivalent widths are particularly large. In principle, the flare could be either hydrostatic or due to a wind driven off the disk surface. Quiescent disk models present another problem: Keplerian disks generically produce double-peaked line profiles, which have only been observed in a few lines in a few objects (Eracleous and Halpern 1994). The line profile to be expected from a disk wind depends strongly on the specific dynamics of the wind (Murray and Chiang 1997). Any strongly flattened geometry also suffers from another criticism: we would expect to see many objects face-on, and these would have very small line widths.

The final problem with using (quiescent) disks as the broad line source is not specific to disk models but in fact applies to all models in which the motion of the gas arises from gravitational forces. Photoionization equilibrium arguments give an estimate of the distance from the continuum source to the line-emitting gas (eq. 10.65). Combining this distance with the gravitational speed implies a central mass, and therefore a ratio to Eddington luminosity. Specifically, we find

$$\frac{L}{L_E} = 2 \times 10^{-3} \left(L_{45} \Xi n_{H,10}\right)^{1/2} v_4^{-2} \left(\frac{v}{v_{ff}}\right)^2, \qquad (10.98)$$

where v_4 is the velocity projected on the line-of-sight in units of 10^4 km s^{-1} and v_{ff} is the local free-fall speed. It is important to recall in evaluating this expression that the velocity on the line-of-sight is likely to be rather smaller than free-fall because we are only seeing one of three components. How much smaller depends on the shape of the gas's orbits. The important point is that *any* picture in which the motions are gravitationally driven requires a rather small luminosity relative to Eddington (particularly in low-luminosity AGNs), and as a corollary, very large central (and remnant) masses.

If the emission line gas is instead divided into a large number of independently moving clouds, other problems must be faced. The most pressing of these is their degree of integrity (disks "confine" their emission line gas either by gravity in a hydrostatic atmosphere, or by "inertial confinement" in a supersonic wind). If the rest of the volume has a pressure less than that in the clouds, they will freely expand. In a time very short compared to a crossing time, such transient clouds would self-destruct:

$$\frac{t_{\exp}}{t_{\text{cross}}} \simeq \begin{cases} 0.025 N_{\text{H},23} v_4 \left(\frac{\Xi}{L_{45} n_{\text{H},10}} \right)^{1/2} & \text{broad line gas} \\ 7.5 \times 10^{-3} N_{\text{H},21} (v/300 \text{ km s}^{-1}) \left(\frac{\Xi}{L_{45} n_{\text{H},4}} \right)^{1/2} & \text{narrow line gas.} \end{cases} \tag{10.99}$$

In order to maintain the observed inventory of clouds, there would have to be a very large rate of resupply (by whatever the sources of clouds may be).

Another consequence follows from rapid cloud destruction: the creation of a high-pressure intercloud medium (Krolik, McKee, and Tarter 1981). As the clouds expand, their pressure falls. Because this happens in a time very short compared to a radial crossing time, the ionizing flux hardly changes, so Ξ rises. When it reaches the critical value around 10, the temperature of the gas rises sharply. The only limit to its upward rise in temperature is the achievement of Compton equilibrium, which for typical AGN spectra is 10^7 K or more. If some other sort of heating is present (e.g., stimulated absorption of radio frequency photons from a nearby high brightness temperature radio source, or a strong relativistic particle flux, or frequent shocks, or ...), the temperature might rise above the Compton temperature. To estimate the pressure of the gas created as a result of evaporation of the emission line clouds, we first estimate its equilibrium density from

$$\frac{dn_h}{dt} = \frac{3C N_{\text{H}}}{t_{\exp} r} - \frac{n_h v_h}{r}, \tag{10.100}$$

where n_h is the density of the hot, evaporated gas and v_h is the radial velocity of the hot gas. Then the ratio of the hot intercloud pressure to the cool cloud pressure is

$$\frac{p_h}{p_c} = 3C\mathcal{M}_h^{-1} \left(\frac{T_h}{T_c}\right)^{1/2}, \tag{10.101}$$

where $\mathcal{M}_h$ is the Mach number of the hot gas's radial velocity, T_h is the temperature of the hot gas, and T_c is the temperature of the cool gas in the clouds.

We do not know what T_h is. However, if it is close to the Compton temperature of the AGN continuum, the pressure ratio would be

$$\frac{p_h}{p_c} = 3\mathcal{M}_h^{-1} \left(\frac{C}{0.03}\right)\left(\frac{T_h/T_c}{10^3}\right)^{1/2}, \tag{10.102}$$

intriguingly close to unity. Examination of figure 10.10 then shows that this mechanism holds out the prospect of a genuine equilibrium: When $p_h = p_c$, Ξ is the same in both the clouds and the hot gas; if the continuum shape creates a range of Ξ in which a photoionized equilibrium and a near-Compton balance equilibrium can both exist, clouds and hot phase can both persist for long periods of time.

All these arguments apply with almost equal force to both the broad and the narrow line regions. Transonic motion might be expected for the intercloud gas in either location from a variety of considerations; if so, destruction of some clouds creates a high enough external pressure to confine the rest. In the event that ordinary gas pressure is unable to confine the clouds, other sorts of forces, for example, magnetic fields (Rees 1987), might do the job.

The dynamics of photoionized material can be quite complex. If L/L_E is small enough, gravitational forces could dominate the motion of emission line gas in the broad line region. In the cloud picture, a wide variety of possible orbital shape distributions is possible. As we have already noted, there is some empirical evidence suggesting that gravitational forces due to the host galaxy's mass dominate the dynamics of the narrow line region. In addition to gravity, however, several other forces can be important: radiation pressure, drag against an intercloud medium, and possibly magnetic effects.

To understand the effect of radiation forces, we will focus here on regions with only small internal velocity gradients. The opposite limit (i.e., winds subject to radiation pressure) will be analyzed in §11.4.3.

First, consider a parcel of photoionized gas thin enough to stay mostly ionized, but thick enough that strong resonance lines like Lyα are optically thick. Such a parcel feels radiation force primarily as a result of photo-ionization. In a state of ionization balance, the radiative acceleration is

$$g_{r,\text{ion}} = \frac{\langle \epsilon_{\text{ion}} \rangle n_e \alpha_{\text{rec}}}{c \mu_e}, \tag{10.103}$$

where μ_e is the mean mass per electron and the average over ionizing photon energy is weighted by ionization rate. That is, the radiative acceleration is proportional to the electron density. To gauge its importance, we can compare it to the acceleration of gravity:

$$\frac{g_{r,\text{ion}}}{g} \simeq 75 \Xi^{-1} T_4^{-1.7} \frac{L_{\text{ion}}}{L_{\text{E}}} \frac{\langle \epsilon_{\text{ion}} \rangle}{I(1,1)}. \tag{10.104}$$

Clearly, even for relatively small values of L/L_{E}, the acceleration due to absorbing ionizing radiation can be competitive with gravity. For fixed L/L_{E}, photo-ionization acceleration *increases* in importance as Ξ *decreases* because decreasing Ξ leads to increasing neutral fraction; it is only the neutral atoms, of course, that are available for ionization. For this reason, if the emission line gas is optically thin in the Lyman continuum, and the host galaxy's potential in the narrow line region is not too deep, radiation force may actually be more important in the narrow line region than in the broad line region.

A side effect of photo-ionization-driven radiative acceleration of pressure-confined clouds is a potentially very damaging hydrodynamic instability. Nearly every Lyman edge photon absorbed leads to the creation of a Lyα photon. Because the Lyman edge opacity is $\sim 10^{-4}$ times the line center Lyα opacity, the optical depth to these line photons is likely to be very large even when the continuum is optically thin. As a result, the pressure of Lyα photons can build up to a level that, even though small compared to the gas pressure (see eq. 10.44), can still have dynamical consequences. In particular, ruffling of a cloud's edges can grow exponentially at a rate faster than the acceleration timescale because the easier escape for Lyα photons through the shorter optical paths drains the Lyα pressure from nearby regions, which in turn leads to further thinning in those locations as the external pressure pushes those sections of the cloud boundary inward (Krolik 1979).

The radiative acceleration of optically thick regions is

$$g_{r,\text{thick}} = \frac{L}{4\pi r^2 c N_{\text{H}} \mu_{\text{H}}}, \tag{10.105}$$

or, referred to gravity,

$$\frac{g_{r,thick}}{g} = \frac{L/L_E}{\tau_T},\qquad(10.106)$$

where the Thomson optical depth in the final expression refers to all the electrons present, whether bound or not. The best guess estimates of line-emitting column densities suggest $\tau_T \sim 10^{-3}$–0.1 (§§10.5.3.1, 10.5.3.2), so radiation pressure on optically thick gas can be quite important, though, naturally enough, not quite as effective as for optically thin material.

Drag against an external medium (whatever it may be) could also affect cloud motions. Viewed on the large scale, the momentum of a cloud can be significantly changed in its lifetime if, during that period, it sweeps up an amount of external mass comparable to its own. To see when that might occur, we scale the intercloud density in terms of its pressure relative to the cloud confinement pressure:

$$\frac{N_h}{N_c} \simeq 30\frac{p_h}{p_c}\left(\frac{T_c/T_h}{10^{-3}}\right)\left(\frac{L_{ion,45}n_{H,10}}{T_{c4}\Xi}\right)^{1/2} N_{H,23}^{-1}.\qquad(10.107)$$

If the intercloud pressure is anywhere near the cloud pressure, drag against the external medium is likely to be important, particularly in the broad line region.

Drag also has an immediate effect on the internal structure of clouds. The ram pressure it creates can easily be comparable to the internal cloud pressure:

$$\frac{\rho_h\Delta v^2}{p_c} = \gamma(\Delta\mathcal{M}_h)^2\frac{p_h}{p_c},\qquad(10.108)$$

where Δv and $\Delta\mathcal{M}_h$ are the velocity and Mach number (with respect to the sound speed in the hot phase) of the *relative* motion between the cloud and the external medium. Because ram pressure is greatest at the front edge of the cloud, and decreases along the sides and back by factors of order unity, it can create sizable pressure differentials within a cloud that could lead to destructive motions on a timescale almost as short as the free expansion time if the ram pressure is in fact comparable to the gas pressure.

The origin of clouds is at least as problematic as their dynamics once created. Interstellar conditions in the narrow line region are close enough to "normal" (although the pressure is four orders of magnitude higher than in the interstellar medium near the Sun) that the clouds in that region may be explained by whatever processes produce dense lumps of interstellar gas in ordinary galaxies.

Conditions in the broad line region, however, are sufficiently exotic that other mechanisms are sought. One thought that has been cursorily explored is that clouds might be created by condensing hot intercloud gas through some sort of thermal instability (§6.3.4). For example, some have envisioned injecting gas at a temperature and ionization parameter along the unstable portion of the radiative equilibrium curve (fig. 10.10). However, it is very unclear what sort of dynamics would create gas in such a state; the very fact that it is thermally unstable means that it would be quite difficult to place gas in that position in the diagram. Others have suggested injecting gas at high temperature, but on the net cooling side of the equilibrium curve (Krolik 1988). In this case, finite amplitude perturbations would be necessary to initiate condensation because growth occurs only algebraically in time, not exponentially. Of course, this idea also begs the question of why the initial state of the gas should be in that portion of the diagram.

Frustration with the difficulty of finding reasonable schemes to create (and preserve) clouds in the broad line region has led some to suggest that a dense star cluster might be the source of the broad line gas (because we know interstellar clouds exist, even though we don't know how they form or survive, the existence of clouds in the narrow line region is seen as much less problematic). In the most cogent of these suggestions (Scoville and Norman 1988; Kazanas 1989), the gas source is the slow-moving stellar wind that is a generic feature of the red giant stage of stellar evolution. Unfortunately, this model also has daunting drawbacks. First, only a small fraction (how small depends on the stellar population and evolutionary history) of all the stars at any given time are in the appropriate wind-generating stage. This means that there must be many other stars in the region that don't contribute to line emission but do greatly burden the mass budget. In addition, these stars may participate in a variety of processes, such as stellar collisions, whose presence can already be ruled out. Second, because the characteristic speed of these winds is $\simeq 10$ km s^{-1}, they inject material into the intercloud medium at the free expansion rate of clouds, so that the winds may create a high-pressure medium. If that were to occur, the winds would be self-limiting because they would have difficulty expanding out against the pressure of this medium. Third, the line emission would be dominated by the portions of the winds that occlude the greatest solid angle while still remaining optically thick in the Lyman continuum. For this region to have an ionization parameter in the range identified in §10.5.3.1 requires tuning the wind density, that is, the ratio of mass loss rate to wind velocity, to a value that lies well outside what is normally seen in stellar winds (Alexander and Netzer 1994). For all these reasons, the stellar wind models for emission line clouds have not gained much of a following.

11   Intrinsic Absorption and Outflows

11.1   Phenomenology

In addition to emission lines, atomic absorption features are also commonly found in AGN spectra. The farther away an AGN is located, the greater is the probability that gas along the line-of-sight absorbs or scatters part of its light. When AGNs have redshifts greater than $\simeq 2$, mingled Lyα lines, Ly continuum edges, and heavy element lines can remove significant chunks of flux. There can be so much intervening Lyα absorption that the continuum level immediately blueward of the Lyα emission line can be substantially depressed below the continuum just to the red. Although these features are a primary source of information about the physical conditions in both the intergalactic medium and high-redshift galaxies, they have little to do with the AGN itself.

Here, we will instead limit our attention to those absorption features that appear to be associated with the AGN proper, or its host galaxy. Almost all the varieties of AGNs sometimes exhibit absorption, but the nature of the absorption changes somewhat from one variety to the next. One property, however, is shared by all—to the accuracy with which we can identify the systemic velocity (the uncertainty may be as much as a few thousand km s^{-1} in high-redshift quasars), any velocity shift is always to the *blue* of the systemic velocity. Because the absorbing gas is, after all, *between* the continuum source and us, it must therefore be moving *away* from the AGN.

Several tens of percent of type 1 Seyfert galaxies exhibit narrow (several hundred km s^{-1}) absorption in the C IV 1548,1551 doublet, usually offset $\sim$ 1000 km s^{-1} to the blue of line center (fig. 11.1). In some cases absorption in O VI 1032,1038, Si IV 1394,1403, Lyα, or lower ionization lines such as Mg II 2796,2804, C II 1335, or Si II 1193 can also be seen.* In a few cases (like NGC 4151, the object shown in fig. 11.1), there is also absorption at the Lyman edge.

A similar fraction have soft X-ray features indicating that the light has passed through an ionized gas (these are the "warm absorbers" mentioned

---

*For the remainder of this chapter we will resume the use of the doublet naming convention we first adopted in §10.1.1, so that C IV 1548,1551 will be called C IV 1549, etc.

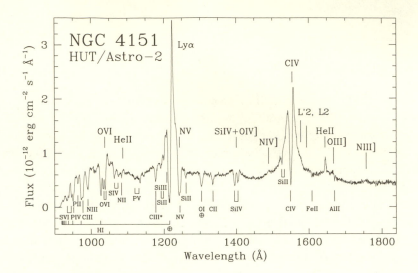

Fig. 11.1   Far UV spectrum of NGC 4151 from *Hopkins Ultra-violet Telescope* data (Kriss et al. 1995). Note the rich absorption spectrum, including many higher series Lyman lines.

in §8.8.1). Ionization edges of OVII and OVIII are the most commonly identified soft X-ray features, but others are also possible (fig. 11.2). In both the UV and X-ray bands, the strength of these features varies on timescales of days to years.

At higher luminosities and redshifts, absorption can become more prominent. Many (most?) radio-loud quasars show CIV 1549 absorption similar to that seen in Seyfert galaxies, but stronger: widths of $\sim$ 1000 km s^{-1} are frequent, and the offset relative to the emission lines can be as much as $\simeq$ 5000 km s^{-1} (Anderson et al. 1987). In fact, some velocity offsets are so large that it is ambiguous whether the absorbing matter is truly associated with the quasar.

A rather different kind of absorption can be seen in BL Lac objects, from which a substantial slice ($\Delta\epsilon/\epsilon \sim 0.1$) of the soft X-ray band is sometimes removed. This absorption might be due to the Lyα line of OVIII formed in material with an internal velocity range $\sim 0.1c$ and blue-shifted with respect to the systemic velocity, or it might be a photo-ionization edge in highly ionized oxygen shifted by a rather smaller amount (Canizares and

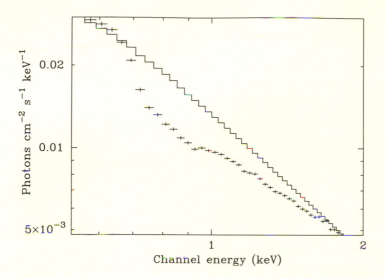

Fig. 11.2 The soft X-ray spectrum of the Seyfert galaxy MCG - 6-30-15 from data in the *ASCA* archive reduced and plotted by K. Weaver. The upper, smoother histogram is the spectrum fitted by a simple power law; the associated χ^2 is far too large to be acceptable. The lower histogram gives the spectrum when the model is a power law plus two absorption edges, one at the energy of the OVII edge, 720 eV, the other at the OVIII edge, 870 eV. This model fits the data well.

Kruper 1984; Sambruna et al. 1997)

The most spectacular absorption is found in radio-quiet quasars. Approximately 10% of the quasars found in samples selected by optical color and emission line strength display broad absorption lines extending anywhere from 2,000 to as much as 60,000 km s^{-1} blueward of line center in the quasar rest frame (Weymann et al. 1991). In fact, so much flux can be removed from the (rest-frame ultraviolet, observed frame optical) continuum that some quasars with these features drop out of flux-limited samples. The true incidence may therefore be as high as 12–15%. Although the line profiles are, by definition, "broad," there is great variation in the specific forms that are seen (fig. 11.3). Some absorption troughs are smooth; some quite bumpy; some begin in the middle of the associated absorption line; some begin several thousand km s^{-1} away.

The list of lines seen in absorption in "BAL" quasars is almost as

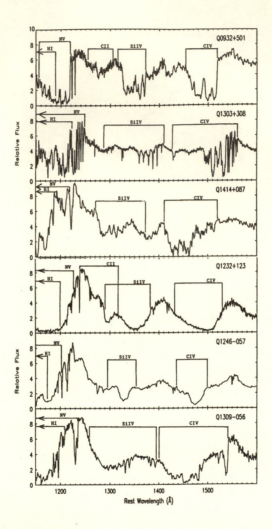

Fig. 11.3   Rest-frame ultraviolet spectra for several broad absorption line quasars. The profile shapes are clearly quite diverse, but CIV 1549, SiIV 1397, NV 1240, and Lyα are all present (Turnshek 1988).

stereotyped as the list of broad emission lines. Whenever there is any detectable absorption, it can usually be seen in CIV 1549, SiIV 1397, NV 1240, Lyα, and OVI 1034, provided the wavelength coverage is adequate to look at all these wavelengths. In addition, $\simeq$ 10–20% of the time there is also broad absorption in the lower ionization lines MgII 2800 and AlIII 1857.

When the S/N and wavelength coverage are especially good, weaker absorption can also be seen in lines belonging to a much longer list of ions: CIII, NIII, NIV, OIII, OIV, SIV, SV, SVI, PV, and FeIII (Korista et al. 1992). The shapes of the absorption profiles for the higher ionization lines typically seem to be crudely consistent with each other, but the low-ionization lines are systematically narrower than the high-ionization lines.

A glance at figure 11.3 shows that sometimes the absorption is "black," that is, there are frequencies at which it removes essentially *all* flux from the spectrum, whereas elsewhere the absorption is only partial. Naively, one might interpret the depth of absorption simply as an indicator of optical depth: where the absorption is black, the optical depth must be at least several; where it is only partial, the optical depth is smaller. Real life is more complicated, and in ways that typically cause the naive estimate to be an *underestimate* of the true column density between us and the continuum source. The absorbing matter may not have the same depth along the lines of sight to all parts of the continuum emission region. If this is the case, partial absorption can be the result of very large optical depths along some rays, while others are entirely transparent. The correct mean column density, averaged over the projected area of the continuum source, would then be seriously underestimated by the naive estimate. Another possibility is that the optical depth as a function of frequency varies more rapidly than can be distinguished given the resolution of the spectrum. In this case, too, what appears to be partial absorption and modest optical depth could in truth be very large column densities, but only at some velocities. Still another source of uncertainty arises from the possibility of scattering redirecting photons from other lines of sight into ours. This process also has the effect of mimicking the appearance of small column densities along the direct path. For all these reasons, then, measurement of the column density in the absorbing matter is a subtle business.

Like other quasars, BAL quasars may also vary in flux. Generally speaking, when the continuum flux varies, the optical depth of the absorption changes, but not its velocity structure (Barlow 1994). That is, the *relative* optical depth, comparing one velocity to another in the absorption trough, changes less than the absolute level of absorption. However, there are instances in which sizable continuum fluctuations are not accompanied by any perceptible change in the absorption, and significant changes in the absorption occur without any change in the continuum level.

The fact that there are no truly low-luminosity AGNs with genuinely broad absorption lines immediately indicates that there must be some connection between this phenomenon and either luminosity or redshift. However, the incidence of broad absorption in modest luminosity, modest redshift quasars (i.e., $L_{opt} \sim 10^{45}$ erg s^{-1}, $z \sim 0.2$) is roughly as great as in higher luminosity, higher redshift objects. This suggests that the dependence is not a smooth one.

As we have seen, (modest) absorption occurs very often in type 1 Seyfert galaxies and radio-loud quasars, but we discover strong broad absorption in only $\sim 10\%$ of radio-quiet quasars. The question immediately arises as to whether gas capable of creating broad absorption lines exists in all quasars but is found on our line of sight only 10% of the time, or such gas can be found on all lines of sight in a special 10% of the quasar population. One way to approach this question is to ask whether there are any isotropic characteristics of quasars by which broad absorption line quasars can be distinguished from the others. If there were, that might indicate a true physical difference between those with and without broad absorption lines. At least with respect to the shape of the optical/ultraviolet continuum, or the character of the broad emission lines, the answer to this question is "no"—any differences are at most very subtle (Weymann et al. 1991).

There are, however, other features that distinguish BAL quasars from normal radio-quiet quasars. Extremely weak soft X-ray flux appears to be a hallmark of this class. Taken as a group, their flux in the 1 keV band is an order of magnitude weaker compared to their UV flux than in average quasars (Green and Mathur 1996). One possible explanation for this weakness is that a large thickness of X-ray absorbing matter (at least $\sim 10^{22}$ cm^{-2} of unstripped gas with Solar abundances) lies between us and their nuclei. In fact, a spectrum of the apparently brightest known BAL quasar, PHL 5200, shows evidence for an absorbing column density of just this magnitude (Mathur, Elvis, and Singh 1995). Because there are no examples of non-BAL quasars (at least among those selected by emission line equivalent width, colors, radio flux, and so on) with absorbing columns anywhere close to this figure, it is a reasonable speculation to suppose that the X-ray absorption is associated with the material responsible for the UV absorption. If so, its presence does not help us determine whether BAL quasars are a special population.

The most striking correlation between other properties and the presence of broad absorption lines has to do with radio emission (Stocke et al. 1992): broad absorption line quasars are *almost never* radio-loud (see fig. 11.4). Genuine radio-loud quasars (in the sense that $R > 100$: §9.1.1), with only very rare exceptions, seem to show at most only the more modest form of absorption we have already described. Thus, broad absorption line quasars *can* be distinguished on the basis of another property from *some* quasars, but not from the majority population.

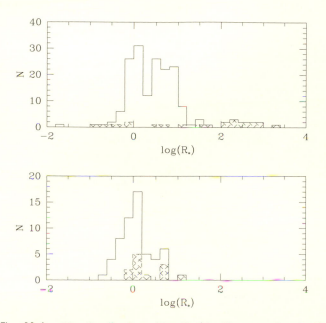

Fig. 11.4   The distribution of $\log R_*$ (the K-corrected ratio of F_ν at 5 GHz to F_ν at 2500 Å) for quasars in the samples studied by Stocke et al. (1992). (Top) Non-BAL quasars are shown, (bottom) BAL quasars. Those actually detected at 5 GHz are shown by crosshatch, bounds by the outlined histogram. No broad absorption line quasars are truly radio-loud. Data tables provided courtesy of S. L. Morris.

Another way to try to answer the question of whether broad absorption line quasars are truly different from ordinary (radio-quiet) quasars is a corollary of the nature of resonance line scattering. When line photons are removed from one observer's line-of-sight by scattering, they are not destroyed; they are merely redirected to some other observer. If the mean optical depth for all lines of sight were the same as the optical depth along ours, we would expect to see an *emission* component with as many photons as have been removed in the absorption feature. If our line-of-sight also gives a fair sample of the mean velocity distribution, the scattered photons we see in emission should be spread as far to the red as the absorption stretches to the blue. In addition, of course, photons created by collisional excitation in either the broad emission line region or the broad absorption line gas should add to the emission component. The results of this test indicate that, if anything, the number of photons removed by absorption typically *exceeds* the number seen in emission by factors of a few. On this basis, we expect that the average covering fraction C_{abs} is significantly less than unity, so that the fraction of quasars with absorbing matter on some lines of sight is rather larger than the fraction in which broad absorption lines are actually observed.

Further support to the inference that the covering fraction is less than unity, but the phenomenon common, is given by the polarization properties of BAL quasars. Whereas most quasars have polarizations of 1% or less, the typical continuum polarization of those with broad absorption lines is several percent. Even more strikingly, the polarization within the absorption trough can rise to $\simeq 10\%$; in other words, the polarized flux exhibits a much weaker absorption feature than is seen in total flux (Goodrich and Miller 1995; Cohen et al. 1995). Perhaps the simplest explanation of these facts is that some of the continuum light arrives at Earth having taken a detour around the thickest part of the absorption; the scattering that deflects its path may also imprint a polarization if the scattering geometry is not circularly symmetric on the sky (see §12.4.1). There is no shortage of scattering mechanisms available to do this: electron scattering can reflect light in the continuum, while the same resonance line opacity that obstructs photons headed toward us can also scatter photons from other lines of sight into ours. The fact that such alternate paths are available also suggests that C_{abs} is well less than unity: after all, if some photons are able to pass through less optically thick regions than others, there cannot be very opaque matter on all lines of sight.

Thus, the balance of the evidence suggests that broad absorption line gas exists in all quasars, but any particular observer can see it in only some of them. Assuming that this is true, we could identify the BAL fraction in quasar samples with C_{abs} if quasars radiated isotropically. However, there are numerous mechanisms (see Chap. 12) that are likely to make them quite anisotropic. Correlations between the directions in which absorbing matter is found and the directions in which quasars are intrinsically bright or dim could complicate the relationship between C_{abs} and the observed BAL fraction. For example, if the absorbing matter were always in the intrinsically dim direction, the very steep quasar counts distribution (fig. 3.4) would ensure that BALs would be severely underrepresented in flux-limited samples. In that case, the true C_{abs} might be several times greater than the simple estimate (Krolik and Voit 1998).

11.2   Physical Conditions in the Absorbing Gas

Given our usual guess about the shape of the ionizing continuum in the EUV, we can use the techniques of Chapter 10 and the information provided by the the observed ionization states to constrain the ionization parameter in the absorbing matter.

Such a wide range of ionization stages is seen in the Seyfert absorption features that they almost certainly come from a diverse range of circumstances. Even in a single object, the UV absorption lines in Seyfert galaxies can require ionization parameters Ξ ranging from ~ 0.01 to ~ 1 (e.g., Kriss et al. 1996), while their soft X-ray absorption indicates a higher level of ionization, $\Xi \sim 10$ or more.

In BAL quasars, the substantial abundance of N^{+4} and O^{+5} indicates that Ξ is at least ~ 1–10. On the other hand, in order to preserve C^{+3} and not ionize away the N^{+4} and O^{+5}, the ionization parameter should not be too much greater than this. It is also possible that, as in the Seyfert case, a range of conditions may be found.

In the last chapter we learned how to use collisional deexcitation rates to estimate the density in emission line regions. Those arguments do not apply here, but inferences based on variability may be made instead. Imagine that a certain ion's column density changes due to a shift in the ionization balance. From equation 10.1 we see that the fractional rate of change in the ionic abundance is

$$\frac{\partial \ln n_{Z,j}}{\partial t} = R_+ - R_-, \tag{11.1}$$

where

$$
\begin{aligned}
R_+ = \frac{n_{Z,j-1}}{n_{Z,j}} & \left[\int_{I(Z,j-1)} d\nu \, 4\pi \frac{J_\nu}{h\nu} \sigma_{ph,Z,j-1}(\nu) \right. \\
& \left. + \int_{I_K(Z,j-1)} d\nu \, 4\pi \frac{J_\nu}{h\nu} Y_{fl}(Z,j-1)\sigma_{K,Z,j-1}(\nu) \right] \\
& + \frac{n_{Z,j-2}}{n_{Z,j}} \int_{I_K(Z,j-2)} d\nu \, 4\pi \frac{J_\nu}{h\nu} Y_A(Z,j-2)\sigma_{K,Z,j-2}(\nu) \\
& + \frac{n_{Z,j+1}}{n_{Z,j}} n_e \left[\alpha_{rec,Z,j}(T) + n_e \alpha_{3,Z,j} \right] \\
& + \frac{n_{Z,j-1}}{n_{Z,j}} n_e C_{Z,j-1}(T)
\end{aligned}
\tag{11.2}
$$

and

$$R_- = \int_{I(Z,j)} d\nu \, 4\pi \frac{J_\nu}{h\nu} \sigma_{ph,Z,j}(\nu) + \int_{I_K(Z,j)} d\nu \, 4\pi \frac{J_\nu}{h\nu} \sigma_{K,Z,j}(\nu)$$
$$+ n_e \alpha_{rec,Z,j-1}(T) + n_e C_{Z,j}(T) + n_e^2 \alpha_{3,Z,j-1}. \qquad (11.3)$$

We have neglected charge exchange reactions, which are usually unimportant for highly ionized species. As equation 11.3 shows, R_-, the fractional rate at which an ionization stage's abundance falls, is independent of the density of the ion and depends only on the underlying physical conditions: the ionizing intensity, the electron density, and the temperature. On the other hand, the replenishment rates contributing to R_+ all depend on ratios of population between other species and the ionic species of interest. R_-, therefore, is more easily used to infer the physical conditions.

Decreases in the absorbing column density could be due either to ionization of the ion in question to higher stages, or recombination to lower stages. Ordinarily, valence shell photo-ionization dominates collisional ionization and inner-shell photo-ionization, and radiative recombination dominates three-body recombination. If the column density decreases while the luminosity falls, we might reasonably suppose that recombination dominates ionization. Using the radiative recombination rate as estimated in equation 10.5, we then find that

$$n_e \simeq 10^4 \left(\frac{\Delta t_{\rm rec}}{1 \text{ yr}}\right)^{-1} \left(\frac{Z_{\rm eff}}{4}\right)^{-2} \text{cm}^{-3}, \qquad (11.4)$$

where Z_{eff} is the charge of the ion and Δt_{rec} is the time for an order unity change in the ionic abundance due to recombination.

Alternatively, if we see the column density decreasing simultaneously with an increase in the ionizing flux, we have reason to believe that it is ionization causing the change. Because the absorbing matter must be on the line-of-sight, the luminosity we measure is exactly the same as the luminosity incident on the absorbing matter. We can then use our measured luminosity to estimate the distance from the continuum source to the absorber:

$$r \simeq 500 L_{ion,45}^{1/2} \left(\frac{\Delta t_{\rm ion}}{1 \text{ yr}}\right)^{1/2} \left(\frac{Z_{\rm eff}}{4}\right)^{-2} \text{pc}, \qquad (11.5)$$

where the ionizing luminosity is defined as the luminosity in the continuum corresponding to the ion in question, an order unity change occurs over a time $\Delta t_{\rm ion}$, and we have used the approximate photo-ionization cross section given in equation 7.95. Combining this distance estimate with an estimate of the ionization parameter yields a new estimate of the density:

$$n_H \sim 10^3 \Xi^{-1} T_4^{-1} \left(\frac{\Delta t_{\text{ion}}}{1 \text{ yr}} \right)^{-1} \left(\frac{Z_{\text{eff}}}{4} \right)^2 \text{ cm}^{-3}. \tag{11.6}$$

Further constraints on the distance from the source to the absorption region come from the fact that the absorption often eats well into the associated broad emission line. When this occurs, it follows that the absorption region must lie *outside* (or be contiguous with) the broad emission line region. If for some reason line-emitting and line-absorbing gas are perfectly correlated—that is, there is absorbing matter on every line-of-sight passing from the central source through line-emitting gas—the absorber could lie anywhere on this line. However, in the limit that there is no such correlation, substantial absorption of the line emission implies that the absorbing gas must come in chunks whose transverse dimensions are at least as large as the broad emission line region. If there are N chunks with a total covering factor C_{abs}, the typical distance from the center to the absorbing gas must be at least $\sim (N/C_{\text{abs}})^{1/2}$ times the size of the broad emission line region.

11.3   The Magnitude of the Problem

Whatever the nature of the material producing the absorption, its motion conveys considerable quantities of mass, momentum, and energy. Consider first the matter responsible for UV line absorption. Imagine that the absorption takes place with maximal efficiency, that is, that if the mean optical depth over some span of velocities Δv is unity, then the optical depth at each frequency is also unity. Then the column density of C^{+3} ions needed for each thermal Doppler width is $\sigma_l^{-1} = 2.5 \times 10^{12} \tau_{\text{CIV}} T_4^{1/2} \text{ cm}^{-2}$, where τ_{CIV} is the actual mean optical depth in the CIV 1549 line.

To cover the entire span of velocities where there is line absorption, the total column density of H atoms is

$$N_H = 2 \times 10^{20} \tau_{\text{CIV}} \left(\frac{X_{\text{CIV}}}{10^{-4}} \right)^{-1} \left(\frac{\Delta v/c}{0.1} \right) \text{ cm}^{-2}, \tag{11.7}$$

where X_{CIV} is the abundance of C^{+3} ions relative to the abundance of H. If the C abundance is Solar, $X_{\text{CIV}} = 10^{-4}$ means that about 1/3 of all C atoms are in that ionization stage. We might naively expect that

409

the characteristic column density in Seyfert galaxy UV line absorbers is $\sim 10^{-2}$ times the corresponding column density in quasars because their lines are roughly a factor of 100 narrower. The column density associated with line absorption is independent of temperature because the number of thermal widths required to cover Δv scales with temperature in a fashion identically opposite to the line cross section. More fundamentally, the column density is independent of temperature because in this maximal efficiency (zero shielding) limit, each atom is an independent absorber, so the total column density is independent of the local atomic velocity distribution whatever it might be.

As we have already remarked, there are several reasons why the estimate of equation 11.7 is quite uncertain, with most of these reasons tending to make it too low. A further uncertainty (of potentially either sign) is whether the abundances of the observed ionization stages of C, N, and O are actually $\sim 10^{-4}$. If, somehow, the elemental abundances of the medium-Z elements were elevated (as e.g., happens in a young type 2 supernova remnant), the H column density would be considerably smaller than estimated here.

We also do not know whether the material is clumped or smoothly distributed. Writing the volume filling factor along covered lines of sight as f, the mean density in absorbing gas is

$$n_{\rm H} = \frac{N_{\rm H}}{rf} = 70 f^{-1} \tau_{\rm CIV} r_{\rm pc}^{-1} \left(\frac{X_{\rm CIV}}{10^{-4}} \right)^{-1} \left(\frac{\Delta v/c}{0.1} \right) {\rm cm}^{-3}. \qquad (11.8)$$

Comparing this estimate to the ones made in equations 11.4 and 11.6, we see that f may well be considerably less than unity.

Estimating the ionization parameter in the same fashion, we find

$$\Xi = 3 \times 10^7 f L_{46} r_{\rm pc} \tau_{\rm CIV}^{-1} \left(\frac{X_{\rm CIV}}{10^{-4}} \right) \left(\frac{\Delta v/c}{0.1} \right)^{-1} T_4^{-1}. \qquad (11.9)$$

Clearly, something is seriously wrong. There are (at least) three possible ways to reconcile the estimate of equation 11.9 with the ionization stages we see. The first is that $f \ll 1$. This choice raises all the difficulties of cloud confinement discussed in §10.7. The second is that the naive estimate of the column density is a gross underestimate. This alternative has the disadvantage of drastically increasing the mass, momentum, and energy fluxes associated with the absorption (see eqs. 11.11, 11.12, and 11.13). The third is that, relative to the near-UV portion of the continuum that we observe, the continuum is dramatically suppressed at energies above

~ 100 eV, so that the ionization stages we see may be retained despite the very large value of Ξ. This choice is supported by the observed weakness of BAL quasars at 1 keV, but there must be some special mechanism to create this "filtration."

Using equation 11.7 as a fiducial point, we find that the associated rate of mass loss (assuming a steady state) is

$$\dot{M} = 4\pi C_{\mathrm{abs}} r \mu_{\mathrm{H}} \langle v \rangle N_{\mathrm{H}}(r/\Delta r)$$

$$= 0.04 r_{\mathrm{pc}} \left(\frac{r}{\Delta r}\right) \left(\frac{C_{\mathrm{abs}}}{0.1}\right) \tau_{\mathrm{CIV}} \left(\frac{X_{\mathrm{CIV}}}{10^{-4}}\right)^{-1} \left(\frac{\langle v/c \rangle}{0.05}\right) \left(\frac{\Delta v/c}{0.1}\right) M_\odot \ \mathrm{yr}^{-1},$$

$$(11.10)$$

where μ_{H} is the mean mass per H atom, Δr is the radial thickness of the absorbing gas, and $\langle v \rangle$ is the mean outflow velocity. If the absorption really is done with maximal efficiency, and the optical depth is constant across the range of velocities with absorption, $\langle v \rangle = \Delta v/2$. In quasars especially, $\dot{M}$ can be an interesting fraction of the accretion rate:

$$\frac{\dot{M}c^2\eta}{L} = 0.025 r_{\mathrm{pc}} \left(\frac{C_{\mathrm{abs}}}{0.1}\right) \tau_{\mathrm{CIV}} \left(\frac{X_{\mathrm{CIV}}}{10^{-4}}\right)^{-1} \left(\frac{\langle v/c \rangle}{0.05}\right) \left(\frac{\Delta v/c}{0.1}\right) \left(\frac{\eta}{0.1}\right) L_{46}^{-1}.$$

$$(11.11)$$

Outward-directed momentum is also carried away with the flow. A useful standard of comparison for the momentum flux is the momentum flux in radiation:

$$\frac{\dot{p}c}{L} = 0.01 r_{\mathrm{pc}} \left(\frac{C_{\mathrm{abs}}}{0.1}\right) \tau_{\mathrm{CIV}} \left(\frac{X_{\mathrm{CIV}}}{10^{-4}}\right)^{-1} \left(\frac{\langle (v/c)^2 \rangle^{1/2}}{0.05}\right)^2 \left(\frac{\Delta v/c}{0.1}\right) L_{46}^{-1}.$$

$$(11.12)$$

Finally, the kinetic energy flux can be compared to the radiative luminosity:

$$\frac{\dot{E}}{L} = 3 \times 10^{-4} r_{\mathrm{pc}} \left(\frac{C_{\mathrm{abs}}}{0.1}\right) \tau_{\mathrm{CIV}} \left(\frac{X_{\mathrm{CIV}}}{10^{-4}}\right)^{-1} \left(\frac{\langle (v/c)^3 \rangle^{1/3}}{0.05}\right)^3 \left(\frac{\Delta v/c}{0.1}\right) L_{46}^{-1}.$$

$$(11.13)$$

Thus we see that the outflowing line-absorbing gas can take away almost as much mass as the accretion flow brings in, and almost as much momentum as the radiation carries out. Only if the abundances of the medium-Z elements are substantially greater than Solar can the impact of this outflow be mitigated; conversely, if the abundances of the absorbing species are substantially *less* than the Solar abundance of the corresponding element, the outflow becomes that much more massive. In addition, the mass, energy, and momentum efflux all increase $\propto r$, so that the associated dynamical problems become harder and harder the farther away the absorbing matter is located.

The column densities associated with X-ray absorption tend to be several orders of magnitude greater than the column estimated in equation 11.7 for the UV absorbers assuming maximal efficiency. Consequently, if they are moving with the same speed as the UV line absorbers, the associated rates of mass, momentum, and energy loss are greater still.

11.4 Dynamics

We will now explore what sorts of forces might possibly drive winds of such magnitude. Although we discuss them separately, it is quite possible for all three mechanisms to act in concert.

11.4.1 Thermally driven winds

One way to drive an outflow is simply to heat gas to temperatures comparable to the *escape temperature*, the temperature at which the gas's thermal energy matches its gravitational binding energy. Matter is injected at some radius (a source must, of course, be found); exposed to the radiation from the AGN, its temperature rises if its ionization parameter is greater than the critical value; if it reaches the escape temperature, out it goes. In a time-steady state with spherical symmetry, such a wind obeys equation 6.30.

Certain properties of such a wind can be estimated at the order of magnitude level. First, whether such a wind can be created depends on whether the gas can be given enough heat to escape within a free-fall time (Krolik and Begelman 1986). Many different mechanisms might contribute to heating: a few of the many that have been considered are photo-ionization, Compton recoil, and collisions with energetic particles (e.g., as in Begelman, de Kool, and Sikora 1991). Here we will parameterize the heating rate by that due solely to Compton scattering. Photoionization (or other) heating will supplement that rate, and radiative cooling will diminish it (particularly if the gas is not far from the radiative equilibrium line in the Ξ–T plane); Compton heating alone may therefore not be too bad an approximation. The heating criterion is then

$$\frac{L\sigma_T}{4\pi r^2} \frac{4k_B T_C}{m_e c^2} \frac{\phi}{\mu_e} > \frac{(GM)^{3/2}}{2r^{5/2}}, \tag{11.14}$$

where we suppose that the gas is injected with the kinetic energy corresponding to a circular orbit at radius r, and ϕ describes how different the true net heating rate is from the purely Compton rate. This criterion can be conveniently rewritten as

$$\frac{L}{L_E} > \frac{1}{8\phi}\frac{m_e c^2}{k_B T_C}\frac{v_{\rm orb}}{c} \simeq 0.07\phi^{-1}T_{C7}^{-1}\left(\frac{v_{\rm orb}/c}{0.001}\right),\qquad (11.15)$$

where T_{C7} is the Compton temperature in units of 10^7 K, and $v_{\rm orb}$ is the speed of a local circular orbit. This form shows that a wind may be driven off for modest values of L/L_E, but only from relatively shallow portions of the gravitational potential.

Once a wind takes off, its temperature may not ever rise all the way to the Compton temperature because adiabatic expansion cooling can compete with radiative heating. This balance can be approximated by supposing that the wind achieves an asymptotic Mach number of order unity. Continuously heated thermal winds rarely achieve Mach numbers much greater than this. In that case, the Compton heating rate is balanced by an effective cooling per unit mass $\sim (k_B T/\bar{m})^{3/2}r^{-1}$ for a mean mass per particle $\bar{m}$; the temperature achieved at that balance is

$$T \sim \frac{\bar{m}}{k_B}\phi^{2/3}\left(\frac{4k_B T_C}{m_e c^2}\right)^{2/3}\left(\frac{L\sigma_T}{4\pi r\mu_e}\right)^{2/3} \sim 5\times 10^6 T_{C7}^{2/3}L_{45}^{2/3}r_{\rm pc}^{-2/3} \text{ K.}$$
$$(11.16)$$

Here $r_{\rm pc}$ is the radius in units of parsecs. Requiring this temperature to be greater than the escape temperature is exactly equivalent to the criterion of equation 11.15.

The electron scattering optical depth can also be readily estimated. If gravity were entirely negligible, the momentum flux in the outgoing wind would be equal to its initial momentum flux, that is, the pressure at the base of the wind. Using this relation, the greatest feasible electron scattering optical depth is

$$\tau_{T,\rm max} \sim \Xi^{-1}\phi^{-2/3}\left(\frac{4k_B T_C}{m_e c^2}\right)^{-2/3}\left(\frac{l/4\pi}{\mu_e/m_e}\right)^{1/3} \sim \Xi^{-1}\left(\frac{v_{\rm orb}}{c_s}\right)^2\frac{L}{L_E}.$$
$$(11.17)$$

Because Compton heating is the driving force, it should be no surprise that the key scaling parameters determining the maximum optical depth are the Compton temperature and, once again, the compactness. However, the second form also affords useful insight. In a thermally driven wind, the ratio of orbital speed to sound speed is automatically order unity; therefore, because the critical ionization parameter for evaporation is ~ 10 (§10.5.1), we can expect Thomson depths $\sim 0.1(L/L_E)$ whenever L/L_E is large enough to create a wind at all. Note that we have used v_{orb} and L_E, which are both $\propto M$, simply for normalization; $\tau_{T,max}$ is independent of the central mass because we have assumed that gravity is negligible.

Comparing these inferred conditions—$\Xi \sim 10$, $T \sim 10^6$ K, $N_H \sim 10^{23} L/L_E$ cm^{-2}—to those found in AGN absorbing systems, we see that they match fairly well those found in the warm X-ray absorbers associated with Seyfert galaxies (and also to other inferred properties of Seyfert galaxies: see §12.6.1). If the terminal Mach numbers are a few, these winds can be expected to flow outward at several hundred km s^{-1}, similar to the velocities seen in the Seyfert UV line systems, but far slower than in the quasar broad absorption systems.

11.4.2 Magnetically driven winds

Another possibility is the sort of magnetically driven wind discussed in §7.2.2. The same questions apply here as in the context of removing angular momentum from accretion disks: What generates the field structure and makes it strong enough to force matter into corotation? How is the mass outflow rate determined?

11.4.3 Radiation pressure driven winds

When the absorption covers a range of velocities $\sim 0.1c$, and its optical depth is at least ~ 1, we automatically know that the very act of scattering photons out of our line-of-sight has transferred a fraction ~ 0.1 of the momentum in radiation from photons to the absorbing gas. At the same time, we also know that if the angle-averaged momentum flux in the absorbing gas is $\sim 1\%$ of the momentum flux in radiation, and the covering fraction is ~ 0.1, then the momentum flux in the flow is ~ 0.1 the momentum flux in radiation along the lines of sight where the gas is located. That these two estimates of the momentum flux in the absorbing gas should (crudely) agree may be no coincidence.

In Chapter 10 we discussed how continuum radiation pressure can accelerate photoionized gas. In this case, because the column density of neutral hydrogen is much smaller ($\sim 10^{16}$ cm^{-2} if there is no self-shielding), the Lyman edge is optically thin, and line pressure is relatively more important.

11.4.3.1 the force due to a single line

Because absorbing gas in quasars can move at speeds that are interesting fractions of c, it is necessary to describe the radiative acceleration with a proper respect for relativistic effects, as treated, for example, in §9.3.2. Initially restricting our attention to the impact of a single line, we can rewrite equation 9.48 in the form

$$g_{rl} = c\frac{d\beta}{dt} = \frac{\kappa_l}{c\gamma} \int dx\, \phi(x) \int d\mu\, I_x(x, \mu)(1 - \beta\mu)(\mu - \beta), \qquad (11.18)$$

where we suppose that the intensity is independent of azimuthal angle around the radial direction, κ_l is the opacity per unit mass averaged over the line profile, $\mu = \hat{n}_i \cdot \hat{r}$ (for photons traveling in the direction $\hat{n}_i$), x is the frequency offset from line center in units of the local thermal width $\Delta\nu_D$, and $\phi(x)$ describes the line profile.

Here we expect that the intensity could change considerably from place to place, for we see optical depths that are at least order unity. To evaluate g_{rl}, we must then solve the transfer equation determining I_x as a function of position. Following Castor (1970), we write it in the lab frame, using a cylindrical coordinate system (fig. 11.5), and integrating over azimuthal angle:

$$\frac{\partial I_x}{\partial z} = \kappa_l \rho \phi(x)(s_x - I_x). \qquad (11.19)$$

Here s_x is the source function with respect to length (we will write the source function with respect to optical depth as S_x), and x is the offset of a fixed frequency in the lab frame *relative to the resonant frequency as Doppler shifted by the local velocity*. The path of integration is at fixed impact parameter p, so that the polar angle of the ray of interest (parameterized by $\mu = z/r$) changes as a function of z along the path.

If the properties of the flow are spherically symmetric, the formal solution to the transfer equation is

$$I_x = I_x^{(o)} e^{-\tau_x(z)} + \int_{-\infty}^{\tau_x} d\tau_x'\, S_x(r) e^{\tau_x' - \tau_x}, \qquad (11.20)$$

for

$$\frac{d\tau_x}{dz} = \kappa_l(r)\rho(r)\phi(x). \qquad (11.21)$$

The boundary condition at either $z = -\infty$ or the place where the ray intersects the continuum source is $I_x^{(o)}$. Note that $r = \sqrt{p^2 + z^2}$.

Because the opacity is resonant, an approximation of great simplifying power—the *Sobolev approximation*—is available to help us create a genuine solution from this formal one. It is valid provided only three rather unrestrictive conditions are met: The velocity of the flow $\vec{\beta}c$ is purely radial, it increases monotonically with radius, and the state of the absorbing

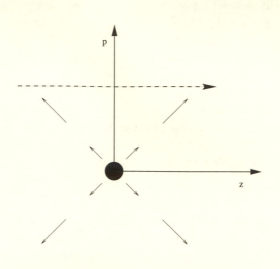

Fig. 11.5 The black circle at the center of the coordinates is the origin of both the flow and the radiation. The thin arrows represent the velocity field; the dotted line with an arrow-head represents the path of integration for the transfer equation. An observer is located at $+\infty$ along the z-axis.

gas varies only slightly in a distance large enough for the bulk velocity to change by an amount much greater than the thermal speed. When these conditions are met, we can consider all the interesting transfer effects as occurring only in the immediate vicinity of our point of interest (Sobolev 1960; Castor 1970). Due to a combination of spherical divergence and the velocity gradient, the Doppler shift to distant points is so great that line photons that interact in one locale find any other spot to be completely transparent (barring accidental coincidences with other resonance lines: §11.4.3.4).

Taking that approximation, the optical depth integrated along an entire ray is the same as the optical depth integrated across the single region where it is resonant:

$$\tau(p) = \int_{-\infty}^{+\infty} dz\, \kappa_l(z,p)\rho(z,p)\phi(x)$$

$$= \kappa_l(r)\rho(r) \int_{-\infty}^{+\infty} dx\, \frac{\phi(x)}{dx/dz}.$$

(11.22)

416

Because the local frequency offset is

$$x = \frac{1}{\nu_l \Delta v_{\rm th}/c} \left[\nu - \frac{\nu_l}{\gamma(1 + \beta z/r)} \right] \qquad (11.23)$$

for a line whose rest-frame frequency is ν_l and occurs in atoms moving with thermal speed $\Delta v_{\rm th}$, we rename $\tau(p)$

$$\tau_\mu = \frac{\kappa_l(r)\rho(r)r(\Delta v_{\rm th}/c)\gamma(1 + \beta\mu)^2}{\beta\left[1 + \mu^2(\partial \ln \beta/\partial \ln r - 1) + \beta\gamma^2\mu(1 + \beta\mu)\partial \ln \beta/\partial \ln r\right]}. \qquad (11.24)$$

At the resonant point, the local direction cosine of the integration path is indeed μ, so it is appropriate to interpret this optical depth as the one pertaining to photons passing through radius r with direction μ. Paths with other values of p intersect the spherical surface with radius r at other locations and at different angles; by spherical symmetry, $\tau_\mu(r, \theta) = \tau_\mu(r, \theta')$, so we are able to build up the set of optical depths for all photon directions by computing them one at a time at different values of p. The effective optical depth depends on direction in the way shown because the velocity gradient in a radial wind is due to a mix of acceleration and spherical divergence whose relative proportions depend on direction. However, to order of magnitude, the optical depth in any direction can be estimated by its value in the radial direction

$$\tau_1 = \frac{\kappa_l(r)\rho(r)(\Delta v_{\rm th}/c)}{\partial\beta/\partial r(1 + 2\beta\gamma^2)}. \qquad (11.25)$$

That is, the optical depth relevant to photons whose frequencies are all within one thermal width of each other is the optical depth across the distance required for the flow to accelerate one thermal width. This length is sometimes called the *Sobolev length* and defines the size scale of regions that are coherent with respect to the transfer of line radiation.

A second application of the same local approximation to the formal solution (eq. 11.23) reduces it to

$$I_x = I_x^{(o)}e^{-\tau_\mu\Phi(x)} + S_x(r)\left[1 - e^{-\tau_\mu\Phi(x)}\right], \qquad (11.26)$$

where

$$\Phi(x) = \int_{-\infty}^{x} dx' \, \phi(x'), \qquad (11.27)$$

and x is a function of z through the dependence of velocity on position. By the normalization of $\phi(x)$, $\Phi(+\infty) = 1$. As is usually the case, to transform a formal transfer solution into a genuine solution, we must specify the source function. When the source function is due solely to conservative scattering of a flux $\mathcal{F}_x$ of initially monodirectional photons, the source function in the fluid frame can be approximated in a very simple way: $S'_{x'} = \mathcal{F}'_{x'}/2$. To find first the flux in the moving frame, and then the source function in the lab frame, we make use of the familiar invariant I_ν/ν^3:

$$\mathcal{F}'_{x'} = \frac{F_x}{\gamma^3(1+\beta)^3}, \tag{11.28}$$

so that

$$S_x = \frac{F_x}{2\gamma^6(1+\beta)^3(1-\beta\mu)^3}. \tag{11.29}$$

We now see that all the frequency dependence in the integral (eq. 11.21) for the radiative acceleration has one of two forms, either $\int dx\, \phi(x) \ldots$ or $\int \phi(x)\exp[-\tau_\mu\Phi(x)]\ldots$. Both integrals are very simple: the former is simply unity, while the latter, often called the *Sobolev escape probability*, is just $[1 - \exp(-\tau_\mu)]/\tau_\mu$. The radiative acceleration is then

$$g_{rl} = \frac{\kappa_l\rho}{c\gamma}\mathcal{F}_x^{(o)}\left[(1-\beta)^2\left(\frac{1-e^{-\tau_1}}{\tau_1}\right) + \right.$$
$$\left. \int d\mu \frac{\mu-\beta}{2\gamma^4(1+\beta)^3(1-\beta\mu)^2}\left(1 - \frac{1-e^{-\tau_\mu}}{\tau_\mu}\right)\right], \tag{11.30}$$

where

$$\tau_1 = \tau_\mu(\mu=1) = \frac{\kappa_l\rho(\Delta v_{\text{th}}/c)\gamma(1+\beta)^2}{\partial\beta/\partial r\,[1+\beta\gamma^2(1+\beta)]}, \tag{11.31}$$

and $\mathcal{F}_x^{(o)}$ is the flux that would have obtained in the absence of optical depth.

Thus, the net radiative acceleration is due to two pieces: a push, whose strength is due both to the direct flux and to the part of the diffuse intensity with $\mu > \beta$, and a drag, whose strength is proportional to the intensity with $\mu < \beta$ (cf. §9.3.2). When the scattering is conservative, the diffuse intensity is always comparable to the direct flux. In the limit of small β, the diffuse intensity's net contribution to the force is identically zero; however, when β is significant, this is not necessarily so. The reason is that when β is not negligible, the scattered photons are "carried along" with the flow due to relativistic beaming, so that their net momentum transfer to the gas is reduced. In the limit of $\beta \to 1$, the effects of both the direct flux and the diffuse intensity diminish $\propto \gamma^{-4}$.

If the flow is clumpy, a small reinterpretation allows almost identical expressions to apply. In this case, τ_μ is no longer directly related to the global velocity gradient or spherical divergence; it is simply the optical depth across a clump in the appropriate direction.

11.4.3.2 the force multiplier

In the wavelength range we observe, there are anywhere from one or two to perhaps a dozen lines in which significant absorption can be seen, depending on the object and the S/N of the observations. If we had access to wavelengths shorter than the Lyman edge, it is likely that we would see a great many more lines in absorption, for most resonance lines in ionized heavy elements lie in the EUV. Therefore, in order to gain a realistic estimate of the total radiation force from lines, it is necessary to sum over a large number of lines. Exactly which lines are important, and their opacities, depends on the spectrum of the ionizing continuum and the ionization parameter. For the purpose of argument, let us suppose that the ionizing spectrum is known and its shape is not affected by intervening absorption. After much use of photo-ionization codes, it is possible to compute the distribution of opacities $dN_l/d\kappa(\Xi)$, that is, the number of lines N_l with line center opacities κ in units of cross section per H atom. Making use of the fact that the radiative acceleration due to Thomson scattering sets a natural unit for the problem, we use the distribution of line opacities to write the total radiative acceleration in terms of the Thomson acceleration and a "force multiplier" $M(N_H, \Xi, \beta)$:

$$g_r = \frac{\mathcal{F}\sigma_T}{c\mu_e} M(N_H, \Xi, \beta), \qquad (11.32)$$

where

$$N_H = \frac{\rho \Delta v_{th}/c}{\partial\beta/\partial r} \qquad (11.33)$$

and we suppose that $\mathcal{F} = \nu \mathcal{F}_\nu$ doesn't change much over the frequency band where most of the lines are located. In the limit of $\beta \ll 1$, M becomes nearly independent of β.

The most obvious interpretation of the force multiplier is that it gives the ratio by which line opacity increases the radiation force relative to that provided by Thomson scattering alone. However, by taking various limits of M, we see that it has several other interpretations as well. If all the lines were optically thin,

$$M(N_H, \Xi) = \frac{\Delta v_{th}/c}{\sigma_T(n_e/n_H)} \frac{(1-\beta)^2}{\gamma} \int d\kappa \, \frac{dN_l}{d\ln\kappa}. \qquad (11.34)$$

In the very optically thin limit, therefore, M is independent of N_H. Note that even when $\gamma \gg 1$, the force due to the diffuse intensity is negligible when the lines are very optically thin because, of course, there are few scattered photons. In the opposite limit, in which all the lines are optically thick, $M \propto N_{\mathrm{H}}^{-1}$ because all the photons' momentum is used, and the acceleration declines with increasing column mass. The typical situation is in between, $-1 < \partial \ln M / \partial \ln N_{\mathrm{H}} < 0$. If Ξ is near the range in which the observed lines are strong and the CIV 1549 line is marginally optically thick, M is in the range 10–1000 when $\beta \ll 1$ and generally decreases with increasing Ξ.

Further physical insight into the meaning of the force multiplier can be gleaned by recalling that when thermal Doppler broadening defines the line profile, line cross sections can be written in the form

$$\sigma_l(\nu) = \frac{\sqrt{\pi} e^2}{m_e \nu \Delta v_{\mathrm{th}}} f_{lu} e^{-x^2}, \tag{11.35}$$

where f_{lu} is the absorption oscillator strength. If the line is optically thin even at line center, the acceleration due to an individual line (cf. eqn. 11.33) is then

$$g_{rl} = \frac{\pi e^2}{m_e c^2 \mu_{\mathrm{H}}} \frac{(1-\beta)^2}{\gamma} f_{lu} X_{\mathrm{H}} \mathcal{F}_\nu(\nu_l), \tag{11.36}$$

where X_{H} is the abundance of the species in which the line is found relative to the abundance of H nuclei. From this expression it is easy to find the ratio of the line acceleration to the Thomson opacity in this optically thin limit

$$\frac{g_{rl}}{g_{re}} = \frac{3}{8} \frac{\mathcal{F}}{\int d \ln \nu \, \mathcal{F}} f_{lu} \frac{\lambda}{r_e} \frac{\mu_e}{\mu_{\mathrm{H}}} X_{\mathrm{H}}, \tag{11.37}$$

for $\lambda = c/\nu$. That is, in the optically thin limit, resonance scattering produces a force $\sim f_{lu}(\lambda/r_e) X_{\mathrm{H}}$ times the electron scattering force, where $r_e = e^2/(m_e c^2)$ is the electron classical radius, $\simeq 3 \times 10^{-5}$ Å. Thus, one strong resonance line in a moderately abundant species, if optically thin, can produce an acceleration ~ 100 times the electron scattering acceleration. Because in any particular ionization state there are only a few strong resonance lines in abundant species, the maximum $M \sim 1000$ in the low N_{H} (and low β) limit, and M decreases as the strong lines become optically thick with increasing N_{H}, or the relativistic corrections diminish the efficiency of scattering as β increases.

11.4.3.3 dynamics under the influence of radiation pressure

If the system is time steady, the equation of motion is then

$$v \frac{dv}{dr} = \frac{GM}{r^2} \left[-1 + \frac{L}{L_{\mathrm{E}}} M(N_{\mathrm{H}}, \Xi) \right]. \tag{11.38}$$

Thus, when L/L_E is not too small, the radiative acceleration can exceed gravity, and without too much strain could in principle accelerate gas in this state to speeds of order what is seen in quasar broad absorption line systems.

Several complications arise when trying to solve equation 11.38 for the flow pattern in a radiatively driven wind. Suppose, for initial simplicity, that the outflow is smoothly distributed. In this case (as eqn. 11.12 shows), something special must be added to the picture in order to find the observed ionization stages, which are also the ionization stages in which strong resonance lines couple the outgoing radiation flux to the gas. One possibility (Murray et al. 1995) is that virtually all photons energetic enough to K-shell ionize C, N, and O (i.e., energies greater than 200 eV) are removed from the continuum before it strikes the absorbing gas. This has the effect of making the He-like ionization stages of these elements the most abundant forms. Because the Li-like ions whose UV lines we see represent only a fraction of their parent elements' abundances, the total column density is increased by several orders of magnitude.

A secondary problem afflicting any flow accelerated by optically thick line radiation pressure is that they are dynamically unstable (see especially Owocki and Rybicki 1985, Gayley and Owocki 1995). Imagine that a region the size of a Sobolev length is split in two, with the outer half given a small addition to its speed so that it travels δv faster than the inner half. Because (by assumption) $\tau_l \gg 1$, the total force on the region increases by $\sim \delta v/\Delta v_{\rm th}$, while its total mass, of course, is unchanged. Thus, small accelerations create positive feedback, and the perturbation grows. Clearly, the characteristic growth time is $g_r/\Delta v_{\rm th}$. Because the dynamical time is $\sim v/g_r$, the number of growth times available is $\sim v/\Delta v_{\rm th} \gg 1$. Consequently, this instability can be expected to cause major disruption to the flow. Indeed, in O stars, where line radiation pressure is a well established explanation for the observed winds, there is much evidence for the shocks that result from this instability (e.g., Lamers 1994).

Consider next the possibility that $f \ll 1$ so that the absorbing gas is divided into a very large number of small clouds. If the clouds are optically thick, they are subject to the dynamical instability just described. In addition, no matter what their individual optical depth is, many of the same dynamical difficulties apply to these clouds as to the broad emission clouds discussed in Chapter 10. The first problem, of course, is that of confinement. Just as for the broad line clouds, if they were able to expand freely, they would dissipate in a time very much less than a crossing time. In order for a hot phase to confine these clouds by normal gas pressure, its column density through the region would be

$$N_h = 6 \times 10^{23} L_{46} r_{\mathrm{pc}}^{-1} \Xi^{-1} T_4^{-1} T_{h8}^{-1} \ \mathrm{cm}^{-2}, \qquad (11.39)$$

where T_{h8} is the temperature of the hot phase in units of 10^8 K. Comparing this quantity to the minimum total column density of all the clouds on a line-of-sight, $\sim 10^{20}$ cm^{-2}, we see that such a confining medium would present a tremendous drag load to the outflow unless either there is much hidden absorbing material, or the hot phase temperature were at least $\sim 10^{11}$ K, that is, even the ions in the hot phase would need relativistic thermal speeds! Even a modest amount of drag also entails the same problems with hydrodynamic effects described in Chapter 10. Drag against an external medium is a much more severe problem for absorption line clouds than for emission line clouds because their column densities are so much smaller.

11.4.3.4 line-locking

We close this chapter with discussion of a property of outflows driven by line radiation pressure that might be described either as a complication or as a distinctive signature. *Line-locking* is a mechanism that causes the force due to scattering in one line to be affected by either emission or absorption in a different, but not too distant, line. Two varieties of line-locking exist: "absorption-absorption" and "emission-absorption."

Consider the absorption-absorption variety first. In an accelerating wind, the observer-frame frequency range scattered in a particular line by an individual fluid element moves to higher and higher frequencies as the fluid element moves out and the redshift it imposes on the continuum grows. Suppose, then, that there are two lines with rest-frame frequencies $\nu_1 > \nu_2$ and $\nu_1/\nu_2 < \sqrt{(1 + \beta_{\mathrm{max}})/(1 - \beta_{\mathrm{max}})}$, where β_{max} is the terminal speed of the wind. Lyα and Nv 1240 are good examples for line 1 and line 2, respectively; their frequency ratio corresponds to a velocity difference of 5900 km s^{-1}, well within the range of velocities seen in many BAL quasars. If material close to the continuum source is optically thick in transition 1, material farther out, moving faster by a velocity difference $[1 - (\nu_2/\nu_1)^2]/[1 + (\nu_2/\nu_1)^2]$, will feel a diminished force because the continuum that should be available for scattering in transition 2 has been diminished. The amount of material with velocity near any particular value is inversely proportional to the acceleration at that point, so the optical depth at that velocity difference might be enhanced.

Emission-absorption line-locking works in the opposite sense. Suppose that the spectrum incident upon the innermost part of the broad absorption line region contains a strong emission feature in transition 1 (if line 1 is Lyα, this supposition should apply almost universally in these objects). When absorbing material is accelerated to a velocity equal to the splitting (5900 km s^{-1} for the Lyα–Nv 1240 example), the force due to scattering in transition 2 increases because F_ν in a strong line feature is, by definition, greater than in the continuum. In the case of Lyα, the peak in F_ν is often several times F_ν in the neighboring continuum. If line 2 scattering provides a significant part of the total force, the radiative acceleration of material moving with that speed is substantially increased. As a result, there is a comparative *deficit* of material with velocities in this range. Because all the material is affected, echoes of this effect can be seen in a diminution of optical depth at this velocity in all the other absorption lines.

Because strong absorption features are often found immediately to the blue of strong emission features, the two line-locking mechanisms can both operate. Where these effects are important, the optical depths of all lines would be reduced over a range of a few thousand km s^{-1} centered on a blue-shift equal to the line 1–line 2 velocity separation, and enhanced over a rather wider range of greater blue-shifts. In fact, evidence for Lyα–Nv 1240 line-locking of just this sort has been reported on the basis of features in the Civ 1549 absorption found in several BAL quasars (Arav et al. 1995).

12 Anisotropic Appearance and Unification of Disparate AGN Varieties

12.1 General Principles

The simplest astronomical objects, like stars, appear very much the same from whatever direction they are viewed. In keeping with the natural procedural prejudice of avoiding adding complications to models until they are necessary, we have so far (mostly) avoided asking whether AGNs should likewise look the same from any point of view. We have now reached the stage, however, where it has become essential to ask, and to answer, that question.

The previous chapter has already broached this issue. There we argued that the most likely interpretation of the fact that ~ 10–15% of all quasars show broad absorption lines is that broad absorption takes place in all quasars, but is only visible on certain lines of sight. In this chapter we will show how arguments like this can be generalized and then used to great effect in simplifying the complicated picture of subvarieties set out in Chapter 1.

The converse of the statement that some properties of AGNs are dependent on viewing angle is that AGNs with different properties may, in fact, be intrinsically identical but appear different only because we see them from different sides. It is for this reason that the whole subject of angle-dependent appearance of AGNs is often called *unification schemes*: it is by ascribing contrasting appearance to the mere accident of viewing angle that we are able to unite different classes of AGNs that otherwise seem to be quite distinct.

Put more formally, the program on which we are about to embark is to choose several sets of apparently different AGNs and claim that the contrasts in their observed properties can all be explained on the basis of specific physical models rotated in polar angle with respect to a single axis. Many different mechanisms might create this sort of anisotropy, but it is convenient to divide them into two categories: those in which the radiation is intrinsically anisotropic, and those in which it is collimated by external obscuration. As we shall show in this chapter, both mechanisms are at work in AGNs.

The most natural candidate for the defining axis is, of course, the rotation axis of the central black hole. This is both the symmetry axis for the accretion flow and the likely orientation of any jet. Whether this special direction correlates in any fashion with a symmetry axis of the host galaxy is another matter. Searches for such correlations (after allowance for the "uninteresting" effect of extinction in the ordinary interstellar medium of the host) have generally come up empty-handed (§13.5).

A surprisingly large number of methods are available for testing assertions about anisotropic appearance of AGNs. It is convenient for conceptual purposes to separate these tests into two groups: those that directly test for anisotropy in individual objects, and those that examine a population for statistical consistency with the version of anisotropy proposed. The first sort are, of course, logically much more compelling. Unfortunately, they are not always feasible, and when they are, it is typically only for special cases, rather than whole samples. Consequently, we must often make do with the second class.

At first, one might think it an impossible task to determine what any individual object looks like from a different point of view. After all, the nearest AGNs are ~ 10 Mpc away, and many are a great deal farther, so it would be very difficult (to say the least!) for us to move to a significantly different vantage point. However, we can get around this difficulty if there are structures in the AGN itself, or its host, upon which we can see AGN light projected. These may be reflection nebulae, or reprocessing regions, or even masers; the key point is simply that they should send some photons our way that can be tagged as having been generated by radiation from the nucleus that began life headed in a different direction. Another way to search individual objects for anisotropic appearance is to observe them in multiple wavelengths, for the degree of anisotropy may depend strongly upon wavelength in a recognizable way. Both dust extinction and photoelectric opacity can easily create this sort of signal.

In the absence of any trick by which we might "turn an AGN on its side," we can instead check whether the statistical properties of the set we are proposing to unify agree with those predicted by the model. Although such a procedure cannot result in a positive proof of the model in question, the more the model predicts correctly, and the less it relies on tuning unknown parameters, the greater our confidence that it is correct. A simple and well known example of this technique is the study of the ellipticity of disk galaxies. We believe that disk galaxies are intrinsically round because the eccentricity distribution of their images is consistent with what we would expect if a set of round disks had been projected on the sky at randomly chosen angles*.

The viewing angle effects we expect in AGNs are more complicated than the simple projection of disks, but the basic procedure for testing the hypothesis of viewing-angle dependence is conceptually the same: We define the set of objects we think might be intrinsically the same, and the physical mechanism that creates the anisotropy. Next we identify those observable properties that should depend on direction to the observer. Third, we compute *within the model* what that dependence should be. Fourth, we use the computed angle dependence to predict what the distribution of these properties should be *among the objects in our samples*, assuming, of course, that the intrinsic orientations are distributed randomly with respect to our line-of-sight. Frequently the ways in which samples are selected introduce strong biases with respect to these angle-dependent properties, so this last step can be very important. Finally, we check the degree of statistical consistency of these predictions with our samples.

Several generic sorts of statistical tests can be applied to check the anisotropic appearance model: Does the distribution of the angle-dependent observables follow the predicted distribution? If there are several angle-dependent observables, do they correlate in the expected way? For that matter, is the distribution of isotropic properties indeed the same in the classes to be united? Although this last test does not really speak to the question of whether we correctly understand the objects' possible anisotropic appearance, it can be turned around in a fruitful way: A good way to begin the search for classes to match up is to cull the known varieties for those whose distributions of isotropic properties are similar.

An important possible complication to these schemes is the role of unknown parameters in influencing the nature of the anisotropic appearance. For example, if there is dust extinction on some lines of sight to an AGN, its flux at Earth will depend very differently on direction depending on the optical depth of that dust, or its distribution of grain sizes. The power of unification schemes is enhanced to the degree that the underlying model depends on such additional parameters (i.e., other than viewing angle) only weakly, or that Nature has arranged for the distributions of those other parameters to be narrow. However, even when there is an additional

---

*To demonstrate even this simple fact quantitatively actually requires careful sample selection. Subtle orientation biases (e.g., extinction in the disk plane) can be surprisingly difficult to eliminate.

parameter that strongly influences the character of the anisotropy, and the distribution of its values is quite wide, we may still be able to test whether viewing angle is an important variable if it is possible to select objects for comparison that all possess the same value of that additional parameter.

12.2 Mechanisms for Intrinsically Anisotropic Emission

There are numerous ways to produce intrinsically anisotropic emission in nonspherical systems, and as we have already seen, AGNs are quite likely to contain strongly nonspherical structures, whether they be accretion disks or jets. Here we will concentrate on two mechanisms of particular importance to AGNs: nonspherical relativistic motion and limb darkening from flat surfaces.

12.2.1 Nonspherical relativistic motion

A spherical relativistic outflow looks just as spherically symmetric as a static spherical surface. Directed relativistic outflow, however, as in the jets frequently observed in radio-loud AGNs, can appear very differently depending on the angle of view. Not surprisingly, the more narrowly collimated the flow, the more its appearance changes with viewing angle. When the opening angle is very small ($\ll 1/\gamma_r$, where γ_r is the Lorentz factor of the radiating plasma), the outflow might as well be unidirectional as far as relativistic effects are concerned.

As already discussed in §9.3.4.3, Doppler shifting and relativistic beaming both act so as to enhance the flux seen in the direction of motion and diminish the flux in the opposite direction. Consequently, if our sample is selected (as usual) on the basis of observed flux, two countervailing tendencies compete to determine how many beamed objects we see. On the one hand, only those objects that we see within an angle $\sim 1/\gamma_r$ of the favored axis are boosted; on the other hand, when our line-of-sight does lie in that direction, they are much brighter and therefore much more likely to appear in a flux-limited sample. Both effects can be strong, even for motion that is only mildly relativistic, but the latter usually outweighs the former. The favored fraction of solid angle is only $\sim \gamma_r^{-2}$ of the whole; but (as shown in §9.3.4.3), when viewing a smooth, time-steady jet at fixed observed frame frequency, the contrast in F_ν between perfect alignment and perfect counteralignment is $[(1 + \beta_r)/(1 - \beta_r)]^{2+\alpha}$, for $F_\nu \propto \nu^{-\alpha}$. When γ_r is a few or more, the aligned/counter-aligned flux ratio is $\simeq (2\gamma_r)^{4+2\alpha}$, which can easily be a large number.

In this context it is important to note that although beaming certainly alters the *number* of sources found at a given flux level, and, as we shall see in detail in §12.5.1, the shape of the observed luminosity function, it does *not* affect the arguments of §3.1 relating the *shape* of the source count distribution to the luminosity function and the geometry of the Universe. In a homogeneous Euclidean Universe, the source count distribution is always $N(> F) \propto F^{-3/2}$, whether the sources are beamed or not. Beaming may change the proportionality constant, but not the shape of the curve itself.

Nonspherical motions (not necessarily relativistic) can also alter the image of a source on the sky. When we observe in the radio domain, we are fortunate to have the extraordinary angular resolution given by VLBI techniques, a resolution far outstripping anything currently available in any other band. With that resolution, we are able to obtain morphological information about radio jets in AGNs on very small scales. We have already seen (§9.3.3.2) that relativistic motion can substantially alter the apparent speed of transverse motion for discrete blobs of radiating stuff, making the apparent transverse speed $\beta_{\mathrm{app}} = \beta_p\sqrt{1 - \mu^2}/(1 - \beta_p\mu)$, where β_p is the speed of the moving *pattern* defining the surface brightness peak and μ is the cosine of our viewing angle with respect to the direction of motion. It is important to remember that β_p is not necessarily the same as β_r.

If the radiating matter is better described by a continuous flow, it exhibits no outflowing blobs. Instead, its transverse length is constant in time and depends on viewing angle in a much simpler way that has nothing to do with relativistic effects. Suppose, for example, that the radiation is produced by a jet moving through a fixed channel, and it is all emitted within a distance from the origin that is fixed in our frame. Then (if its outgoing intensity distribution is isotropic) the projection of the jet length on the sky is just its intrinsic length times $\sqrt{1 - \mu^2}$. The only complicating factor in this case is a possible distribution in the intrinsic length (or anistropy in its emission).

12.2.2 Angle-dependent disk emission

In §7.5.5 we have already discussed the possible angle dependence of accretion disk emission. Although this angle dependence is likely to be less dramatic than that due to relativistic beaming, it can still be appreciable enough to influence the statistics of flux-limited samples. Even the simple classical projection effect that acts on optically thick disks would depress the flux we receive, relative to what is seen by an observer at the pole, by a factor $\leq \mu$ for a fraction μ of all disks. Because the impact of local limb-darkening and general relativity can be at least as great, and both of these are strongly model-dependent, there can be no general predictions of the angle dependence expected in accretion disk spectra (§7.5.5). Instead, any particular disk model entails its own specific prediction for angle dependence of the flux.

X-ray emission may also come from a flattened configuration, possibly associated with an accretion disk (Chap. 8). It is possible that the X-ray source is optically thin in all directions, so that its emission is isotropic, or nearly so. However, this is not necessarily always the case. A flat corona can be optically thin in one direction, but quite optically thick in the perpendicular directions. If so, it will be limb-darkened in a way qualitatively much the same as a classical electron scattering atmosphere. X-ray reflection from an optically thick, cool disk (§8.7.4) also tends to be strongly limb-darkened because of the same effects of slab geometry: the absorption optical depth is much smaller for directions parallel to the disk axis than for more equatorial directions. Just as for the ultraviolet continuum, these angular distributions, too, can be bent toward the equatorial plane by relativistic effects.

If, as is quite likely, electron scattering opacity is important in the atmospheres of accretion disks, inclination-dependent polarization can also be expected, but it, too, is highly model-dependent (§7.5.6).

12.3 Angle-Dependent Obscuration Mechanisms

Well outside the region(s) where the intrinsic angle dependence of the radiation is determined, a variety of mechanisms can further modify the angular distribution of escaping light. Perhaps the simplest geometry capable of creating anisotropy in the appearance of an AGN is that of a large number of small clouds, distributed with statistical spherical symmetry. In this case, the AGN would look different depending on line-of-sight, but there would be no correlation between viewing angle and appearance.

A slightly more complicated, but also perhaps more interesting, geometry is one in which the obscuration is azimuthally symmetric but does depend on polar angle. For example, there might be large obscuration near the equatorial plane, but little near the poles, so that observers close to the axis are able to see the AGN clearly, but those observers near the equator might find their views thoroughly blocked.

Several different physical mechanisms can provide this obscuration. A flaring accretion disk (§7.3.4.2) is one possibility. At larger distances, interstellar clouds might be placed in such an arrangement. Exposed on their inner surfaces to the AGN continuum, they will be ionized to a rather greater degree than normal interstellar clouds and can present a significant free-free opacity (eq. 7.92) to radio frequency photons. Dust can be quite effective in absorbing and scattering photons all the way from the mid-infrared through the extreme ultraviolet (§10.3.4.1). Depending on the grain size and composition, and the photon wavelength, the balance between absorption and scattering can vary over a fairly wide range. That part of the photon energy that is absorbed must, in the steady state, be reradiated. Because dust cannot survive at temperatures much above $\simeq 1500$ K, thermal dust reradiation is inevitably in the infrared. Consequently, significant dust obscuration automatically entails significant infrared reradiation.

When the neutral fraction in the gas is greater than $\sim 10^{-3}$, H photoelectric opacity dominates dust opacity for photons in the Lyman continuum. At still higher energies, absorptive opacity is almost entirely due to photo-ionization whether or not there is dust present (§8.7.1). At the highest energies (above ~ 10 keV), photoelectric opacity becomes less important than Compton opacity; at energies that high, the fact that an electron is bound in a (low-Z) atom has no bearing on its Compton scattering efficiency. As figure 8.10 shows, a column density $\sim 10^{21}$ cm^{-2} suffices to block 1 keV X-rays; a column density $\sim 10^{24}$ cm^{-2} or greater is Compton thick. Obscuration that thick deflects all photons from ~ 10 to ~ 100 keV; at still greater energy, the cross section enters the Klein-Nishina regime. There the opacity falls, scattered photons increasingly tend to go in the forward direction, and the process acquires some absorptive character due to recoil losses.

Those high-energy photons that scatter, but avoid being absorbed, eventually find their way out of the obscuration. The exact angular distribution of the escaping photons depends on the aspect ratios of the system. Imagine, for example, that the obscuration has the shape of a torus. To the extent that its optical depth in the equatorial plane exceeds that in the vertical direction, more of the escaping flux emerges in the polar direction than in the equatorial direction. Thus, even though scattering does not destroy photons, but only redirects them, viewers in the disfavored directions still find that they are deprived of flux from the nucleus.

The overall effectiveness of the obscuration depends only on its optical depth and shape, not its location. In principle, if these quantities are preserved, it could be located anywhere in the galaxy. What most distinguishes different locations as likely sites for obscuration is the local *dynamical* environment, for that controls how geometrically thick the obscuration can be.

In fact, the support of obscuring material against gravity presents a difficult dynamical problem. Consider, for example, toroidally distributed obscuration. As was shown in §7.1, in the thin disk approximation the angle subtended by a hydrostatic disk is $h/r \simeq c_s/v_{\rm orb}$. In the simplest model of accretion disks, one expects rather little flaring (§7.3.4). Out in the host galaxy, typical rotational speeds are ~ 100 km s^{-1}, so to block a significant solid angle, if the torus were to be hydrostatically supported, the sound speed in the gas would have to be at least a few tens of km s^{-1}, corresponding to a temperature at least $\sim 10^5$ K. However, such a high temperature would lead to rapid destruction of any dust grains that might be present.

Several solutions have been proposed for this problem. One is that the gas is highly clumped. The clumps' velocities can then be written as the sum of a fixed orbital velocity $\vec{v}_{\rm orb}$ and a second part $\Delta\vec{v}$, where $\vec{v}_{\rm orb}$ is the velocity corresponding to circular orbits in the equatorial plane, and $\Delta\vec{v}$, whose magnitude $|\Delta\vec{v}| \simeq (h/r)|\vec{v}_{\rm orb}|$, may point in other directions. If $\Delta\vec{v}$ is randomly directed, the collision rate between clumps is $\sim (h/r)C\Omega$, where C is the mean number of clumps intersected by a line of sight in the equatorial plane (also known as the "covering factor"). Unless somehow clump collisions are highly elastic, the energy in their random motions will rapidly dissipate, causing them to settle into the equatorial plane after just a collision or two. Because the collision time is not much longer than an orbital time if h/r is not too small, and the orbital time is much smaller than the canonical AGN lifetime of 10^8 yr or more (§3.5), unless these clouds are more than 1 kpc from the galactic nucleus, such transient obscuration would be difficult to reconcile with the large fraction of AGNs in which obscuration is present (§§12.6.4, 12.7.3).

Thus, appealing to highly supersonic clump motions merely shifts the question to how those motions are sustained (or organized). In this regard, it may be significant that when geometrically thick obscuration occurs, it is often located near r_*, the point at which the black hole's potential is comparable to the galactic potential (§4.4). The rate of orbital shear $(d\ln\Omega/d\ln r)$ changes there, with consequences that may be favorable for the creation of toroidal belts of obscuration (Krolik and Begelman 1988; Yi, Field, and Blackman 1994).

It may also be possible that $\Delta\vec{v}$, although changing in direction from place to place, may exhibit strong local correlations, as, for example, in a warped disk (Pringle 1997). Correlations in the random velocities strongly suppress collisions. Put another way, correlations in $\Delta\vec{v}$ can cause the disk to be *locally* thin, even while its global orientation changes by a large angle. Spreading out the change in orientation over a large dynamic range in radius allows the warp to appear locally smooth.

Another possibility is that the gas is supported by radiation pressure (Pier and Krolik 1992a). Most of the energy striking the obscuration is converted into infrared photons. If the dust is close enough to the nucleus that the typical wavelength of these photons is a few tens of microns or less, and the dust opacity per unit gas mass is similar to its local value (fig. 10.7), the opacity mid-IR photons face is $\sim 1 - 100\kappa_T$. Because the radiation force is proportional to the flux times the opacity, it is $\sim (\kappa/\kappa_T)(L/L_E)$ as strong as gravity. If L/L_E isn't too small, and the dust isn't too far away, radiation pressure could be the obscuration's support. To the degree that the obscuration is flattened, and therefore the flux escaping from it is predominantly in the vertical direction, the cancellation of gravity will be most effective in the vertical direction.

A third possibility is that the obscuring matter is simply not in hydrostatic equilibrium: it could be continually streaming in, or streaming out. Matter might be expelled and carried to high altitude by a magnetized wind of the sort described in §7.2.2, for example (Königl and Kartje 1994).

12.4 Mirrors and Screens:or, How to See AGNs from Another Side

Even if an AGN appears to be very faint as seen from our point of view, it still might be much brighter seen from other vantage points. If, in the directions where the AGN shines brightly, there is matter to "light up," and we are able to see the illuminated matter, we can be given a telltale signature of the AGN in reflected light or fluorescence. All we need to do is have some way of separating the light due to the nucleus from that produced in the "mirror" (in the case of reflection) or the "screen" (in the case of fluorescence or some other form of reradiation). Fortunately, there are several distinguishing marks of reflection or reprocessing that we can use.

Because the distnction between "mirror" and "screen" is one of function rather than structure, there are materials that can act simultaneously as both. In fact, because the opacity of matter can be such a strong function of photon energy, there are also structures that serve simultaneously as obscuration (for some kinds of light) and mirrors or screens for other kinds.

Other physical relationships can also exist between mirrors, screens, and obscuring matter in AGNs. Mirrors can exist *because* obscuration is present. At the inner edge of the obscuration, the matter is subjected to the full strength of the radiation from the central nucleus. If the ionization parameter there is greater than the critical value for maintenance of a cool equilibrium (see §10.5.1), the matter at the inner edge of the obscuration is continuously evaporated into a much more highly ionized, much hotter state. This evaporated gas can escape and flow out the hole of the torus, thus providing a source of matter for a mirror in just the location where it can reflect light to observers whose views of the nucleus would otherwise be blocked. There it can be driven out to larger radii by any of the mechanisms discussed in §11.4.

12.4.1 Polarization

The most powerful marker of reflected light is the polarization induced by many kinds of scattering. The existence (indeed the prevalence) of polarization-dependent scattering follows immediately from the fundamental physics of photon scattering. When a small object (i.e., smaller than the wavelength associated with the photon) responds to a photon by creating a temporary electric dipole (examples of small objects commonly encountered in astronomical contexts include dust particles, atoms with resonance transitions, or free electrons), the direction of the temporary dipole is linked to the original photon polarization. The polarization of the scattered light is determined by the polarization of the temporary dipole, whence the polarization dependence of the scattering process.

In macroscopic materials (like dust), the linkage between the incident polarization and the temporary dipole is given by

$$\vec{d} = V\,(\underline{\epsilon} - \underline{I}) \cdot \vec{E_i}, \tag{12.1}$$

where V is the volume of the dust grain, $\underline{\epsilon}$ is the dielectric tensor, $\underline{I}$ is the identity matrix, and $\vec{E_i}$ is the electric field of the incident wave. In the simplest case (isotropic polarizability, i.e., the eigenvalues of $\underline{\epsilon}$ are all equal), the temporary dipole is exactly parallel to the initial photon polarization. In other cases (materials that polarize more readily in some directions than others), the temporary dipole need not be parallel to the initial polarization.

The partial cross section for scattering is then

$$\frac{\partial \sigma}{\partial \Omega} \propto |\hat{e}_f^* \cdot (\underline{\epsilon} - \underline{I}) \cdot \hat{e}_i|^2, \tag{12.2}$$

433

where $\hat{e}_{i,f}$ are the initial and final photon polarization unit vectors.

The quantum mechanical analog (relevant to atomic or molecular resonance scattering) is that the transition amplitude for scattering is the product of two pieces: the matrix element

$$\langle E|\hat{e}_i \cdot \vec{d}|G\rangle$$

for a photon of the incident polarization to raise the atom from the ground state (quantum numbers G) to the excited state (quantum numbers E), and the matrix element for the reverse process via emission of a photon with the new polarization

$$\langle G|\hat{e}_f^* \cdot \vec{d}|E\rangle.$$

The scattering cross section is proportional to the magnitude squared of the transition amplitude, or

$$\frac{\partial \sigma}{\partial \Omega} \propto |\hat{e}_f^* \cdot \langle G|\vec{d}|E\rangle\langle E|\vec{d}|G\rangle \cdot \hat{e}_i|^2. \tag{12.3}$$

The formal similarity of the quantum mechanical and classical cross sections is obvious. The most important change in point of view is simply transforming the temporary electric dipole $\vec{d}$ from a function to an operator. In fact, the analogy carries further—when the ground state of the atom is completely spherically symmetric (total angular momentum quantum number $J = 0$, or free electrons in the Thomson limit) $\partial \sigma/\partial \Omega \propto |\hat{e}_f \cdot \hat{e}_i|^2$, but less symmetric ground states lead to cross sections with more complicated relations between $\hat{e}_i$ and $\hat{e}_f$ (Lee, Blandford, and Western 1994).

Because photons traveling in a given direction must be polarized perpendicular to that direction, the correlation between initial and final photon polarizations produces strong polarization in the scattered light (see fig. 12.1). Consider, for example, a photon initially traveling in the $\hat{z}$ direction, so that its polarization could lie anywhere in the x–y plane. If it is deflected into the $\hat{x}$ direction by a scatterer with an effectively isotropic dielectric tensor, the possible new polarizations could in principle be anywhere in the y–z plane, but because $\partial \sigma/\partial \Omega \propto |\hat{e}_f^* \cdot \hat{e}_i|^2$, none of the photons in the initial beam can be polarized parallel to $\hat{z}$. Therefore, those photons scattered in the $\hat{x}$ direction are 100% polarized parallel to $\hat{y}$. Anisotropic dielectric tensors lead to more complex (and typically weaker) polarization in the scattered light.

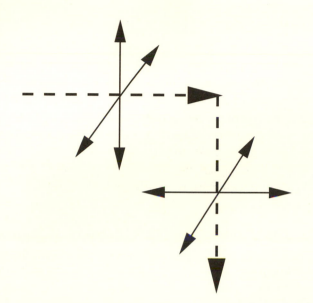

Fig. 12.1 Polarization-dependent scattering geometry. The dashed lines represent the initial and final photon directions, the solid lines with double-headed arrows are a particular choice for the two linear polarization basis states. When the photon changes direction by 90°, only one of the polarizations overlaps with a possible polarization direction in the final state.

If all the light striking the mirror were moving in exactly the same direction, the polarization observed perpendicular to that direction (assuming a scatterer with approximately isotropic polarizability) would approach 100%. However, such simple geometry rarely occurs. In any realistic situation, the mirror extends over a finite size, and therefore light striking it, even from a point source, has a finite range of initial directions. If the source size is comparable to the size of the mirror, the range of directions may be still broader. Any single observer then sees light scattered through a variety of scattering angles, and consequently diluted polarization. In the limit that the initial ray directions of the scattered light are circularly symmetric on the sky, no net polarization emerges. Thus, our ability to observe substantial polarization depends on some combination of resolving the mirror, and a break in its azimuthal symmetry when projected on the sky plane.

When those conditions obtain, it is possible to distinguish scattered light from locally produced light because the locally produced light generally has at most very weak polarization. A number of useful inferences can be obtained from the spectrum of the polarized light.

Scatterers effective over a wide bandwidth provide information on the shape of the spectrum before scattering. When electrons are the scattering agent, scattering is completely independent of photon energy (provided $\epsilon \ll m_e c^2$), so the spectrum of polarized light is (almost; see §12.6.1) simply proportional to the spectrum illuminating the electrons. When dust is the scattering agent, the relation between the two spectra is more complicated. The absorptive part of the dust opacity can redden both the illuminating beam and the reflected light, whereas the general trend for the scattering cross section to increase with decreasing wavelength tends to "blue-en" the polarized reflected light (§10.3.4.1). Inference of the intrinsic spectrum from a dust-reflected spectrum can be further complicated by the possible existence of quasi-resonant effects (e.g., the 2200 Å graphite absorption feature) that may, depending on the grains' chemistry and structure, interrupt the global tendency of scattering efficiency to increase with decreasing wavelength.

Atomic resonance lines, of course, scatter only those photons that match the resonance frequency to within the line width. If the atoms are at rest, the polarized spectrum is then simply a series of spikes at the resonance frequencies. When they have an organized bulk motion, however, the combined correlations of polarization and Doppler shift can yield intereesting constraints on the material's kinematics.

Although reflection polarization is a very powerful signature when it is present, there are several important limitations on when it can be effective. On the one hand, the optical depth for scattering can't be too small, or the reflected flux will be too weak to detect. On the other hand, if the optical depth for scattering is great enough that the average reflected photon has been scattered more than once, the tight correlation between scattering angle and polarization is lost. Which of these limitations is the more serious depends, of course, on the scattering agent. In a fully ionized plasma with Solar abundances, the electron scattering cross section per unit mass is the usual Thomson opacity $0.36(n_e/n_H)$ cm^2 gm^{-1}; the Galactic dust scattering cross section per unit mass is $\simeq 170$ cm^2 gm^{-1} at 5000 Å, and increases (though not necessarily monotonically) toward shorter wavelengths (fig. 10.7). Atomic resonance scattering cross sections are typically $\sim 10^{-13} T_4^{-1/2}$ cm^2, so the opacity at line center is $\sim 10^{11} X_H$ cm^2 gm^{-1}, where the abundance of the ground state relative to the number of H nuclei is X_H. These different mechanisms are thus sensitive to quite disjoint ranges of column density. The optimal column density for atomic lines is very small: $\sim 10^{12} X_H^{-1}$ H cm^{-2}; that for dust is somewhat larger: $\sim 10^{20}$ H cm^{-2} if the dust/gas ratio is similar to that in our local interstellar medium; electron scattering mirrors work best at still larger columns of ionized atoms: $\sim 10^{23}$ cm^{-2}.

12.4.2 Reflection without dilution

Although starlight is often a serious dilution of the near-infrared to ultraviolet spectrum in even the smallest of apertures, in other bands, for example, X-rays or radio, an electron-scattering mirror can produce a far less encumbered reflection of the nuclear spectrum. The great advantage of diminishing stellar dilution is, of course, that one no longer needs to throw out a large fraction of the photons (by considering only the polarized flux) in order to gain a clear view of the reflected AGN light.

Even within the radio and X-ray bands, certain frequency ranges are less likely to be contaminated by non-AGN emission than others. Because they are optically thin to synchrotron radiation, while the core sources in AGNs are often optically thick, supernova remnants may weaken relative to AGN emission at higher radio frequencies. In the X-ray band, the flux from supernova remnants and low-mass X-ray binaries falls rapidly above a few keV, leaving high-mass X-ray binaries as the most likely sources of extraneous light at higher energies. Thus, spectra of galactic nuclei at high radio frequency (more than ~ 1 GHz) and hard X-rays (more than a few keV) can potentially yield nearly pure reflection spectra.

12.4.3 Emission lines

Gas capable of producing emission lines acts like a "screen"—it absorbs light of one variety, and reradiates in a different form. The extensive discussion in Chapter 10 of the physics of emission line production illustrates how complex the relationship can sometimes be between incident ionizing continuum and emergent lines. However, in the right circumstances relatively secure inferences can be made.

In this context, the most robust tie is between H and He recombination lines and the ionizing continuum. If these lines come from a region well outside the nucleus, the density in the line-emitting matter is generally low enough that few of the complications discussed in §10.3.5 can interfere with simple recombination cascade. We can then use the observed flux in recombination lines such as Hβ to infer the rate at which Lyman continuum photons are absorbed, and hence a lower bound on the rate at which they are produced:

$$N(H\beta) = CY(H\beta)N_{\text{ion}}, \tag{12.4}$$

where $N(H\beta)$ is the number of Hβ photons produced per unit time, N_{ion} is the rate at which H ionizing photons are generated, C is the fraction of solid angle around the central source that is covered by material optically thick at the Lyman edge, and $Y(H\beta)$ is the yield (in photons/photon) of Hβ photons for each Lyman continuum photon absorbed. $Y(H\beta)$ is a slowly varying function of temperature (and is also somewhat sensitive to whether the Lyman edge is optically thick or thin), but for temperatures $\sim 10^4$ K it is $\simeq 0.1$ (see, e.g. Osterbrock [1989] for tables from which $Y(H\beta)$ can be derived). The inferred N_{ion} is a lower bound because $C \leq 1$.

The Fe Kα line (see §8.7.2) can also provide a relatively unsullied measure of the strength of the X-ray continuum incident upon gas in the host galaxy. Just as $Y(H\beta)$ is a slow function of temperature, the proportionality constant between the number of Kα photons produced and the number of K-edge photons absorbed depends only weakly on Fe's ionization stage. Only one complication may confuse the issue: in those ionization stages in which there are vacancies in the L-shell (FeXVIII–FeXXVI), photons can be resonantly scattered in the Kα transition. In the stages having between one and seven L-shell electrons (FeXVIII–FeFeXXIV), the temporary return to a multiply excited state during the scattering event can lead to line photon destruction via Auger ionization. On the other hand, if the medium is optically thin to Compton scattering, the apparent intensity of the line in the scattered spectrum can be enhanced by continuum photons scattered in the resonance (Band et al. 1990).

A warm, electron-scattering mirror can also serve as a very bright Fe Kα screen. If the luminosity radiated toward the mirror is $\int d\Omega \, dL/d\Omega$, the reflected luminosity (in the optically thin limit) is simply

$$L_{\text{refl}} = \int d\Omega \, \tau_{\text{T}}(\Omega) \frac{dL}{d\Omega}. \tag{12.5}$$

Because the Fe Kα luminosity is similarly

$$L_{K\alpha} = \int d\Omega \int d\epsilon \, \tau_K(\epsilon, \Omega) Y(K\alpha) \frac{\epsilon_{K\alpha}}{\epsilon} \frac{dL_\epsilon}{d\Omega}, \tag{12.6}$$

its equivalent width is

$$W(K\alpha) \simeq \frac{\langle \tau_K \rangle}{\tau_{\text{T}}} \frac{Y}{\alpha} \left(\frac{\epsilon_{K\alpha}}{\epsilon_K} \right)^\alpha \epsilon_{K\alpha}$$
$$\simeq 1 \frac{[\text{Fe/H}]}{[\text{Fe/H}]_\odot} \text{ keV}, \tag{12.7}$$

438

where the numerical evaluation has set the continuum power-law slope $\alpha = 0.7$. Thus, for Solar Fe abundance, a very large equivalent width can be expected; its actual value can be used to measure the Fe abundance in the mirror (Matt, Brandt, and Fabian 1996).

Screens can also radiate emission lines at the opposite end of the energy spectrum, in the infrared and millimeter bands. When the matter is cool enough for dust to survive (as in dusty obscuring material), atoms in the gas phase are more likely to be in molecules than atoms, and the ionization fraction is likely to be rather low. In these circumstances, photoelectric events create fast electrons, and the ratio of free electrons to atoms and molecules is so small that these fast electrons give up more energy by inelastic scattering with the atoms and molecules than by Coulomb scattering with other free electrons. As a result, a significant part of the absorbed energy is used for direct excitation of a variety of atomic and molecular electronic transitions. Because the photons produced when these excited states decay mostly lie in the ultraviolet, where dust extinction is large, the energy in this channel is largely deposited in the dust grains, which transform it into infrared photons. Because of this local extinction, only a small fraction of the energy emerges in lines; the strongest tend to be either H_2 vibrational transitions between 2 and 10 μ, or very high quantum number CO rotational transitions (Krolik and Lepp 1989).

Although their total luminosity may not be large, the extremely high brightness temperature found in masers makes them a potentially very useful tool for revealing hidden AGNs. Where dusty molecular material is exposed to the X-ray continuum radiated by an AGN, there is generally a layer in which the optical depth is large enough to permit molecules to form, but not so great as to quench population inversions caused by scattering with other molecules. In particular, strong masing action can occur in the 22 GHz 6_{16}–5_{23} rotational line of H_2O. The surface brightness of the maser tends to be fixed at $\sim 100\, L_\odot\, \mathrm{pc}^{-2}$ (relative to the X-ray-illuminated surface area, not the projected masing area) because the conditions for its existence are limited to a specific range in the ratio of X-ray flux to gas pressure (Neufeld, Maloney, and Conger 1994). If the radiating surface is a few square parsecs or more, masers this bright can be seen from very large distances.

12.4.4 Infrared continuum

Although dusty molecular gas does release some energy in molecular emission lines, the bulk of its radiation is in an infrared continuum.

Because dust grains emit quasi-thermally, and the obscuration is often optically thick even in the mid-infrared, the characteristic wavelength of this reradiated continuum is simply determined by the effective temperature at the location of the obscuration: $\lambda_c \simeq 4(T_{\text{eff}}/1000 \text{ K})^{-1} \; \mu$. Often the column density is great enough that the optical depth is substantial even in the infrared. Reference to figure 10.7 shows that if the obscuration is Compton-thick, and has a dust opacity per unit mass like that in the local interstellar medium, it is optically thick even to wavelengths as great as $\sim 100 \; \mu$. Consequently, even though dust grains are inefficient thermal emitters (because they are considerably smaller than the wavelengths they are trying to radiate), the output spectrum from a single portion of the obscuration may closely approach a truly thermal shape.

Its specific character, however, depends on details such as the magnitude of the column density and the geometry of the obscuring matter. Because obscuration that collimates the emergent radiation is the interesting case, and the optical depth therefore depends strongly on polar angle, in general we expect the infrared continuum to depend strongly on polar angle at all wavelengths for which it is optically thick. The character of this anisotropy depends on the detailed configuration, but in plausible geometries there can easily be order of magnitude contrasts between polar and equatorial views (Pier and Krolik 1992b; Granato and Danese 1994; Efstathiou, Hough, and Young 1995). The predictions of a particularly simple model are shown in figure 12.2 to illustrate these trends. For the parameters chosen in this model, the infrared continuum is strongly anisotropic for all wavelengths shorter than several tens of microns.

12.4.5 Spectra versus images

Still further information can be gleaned if either a "screen" or a "mirror" is resolved on the sky. Its surface brightness will always be jointly controlled by intrinsic physical conditions (gas pressure, dust content, etc.) and the strength of its illumination by the nucleus. However, certain aspects of its shape can be interpreted directly. For example, if (as is often seen: plate 7) an emission line region has sharp edges consistent with the outline of a cone projected on the sky, we can naturally infer that we are seeing photo-ionization due to a collimated beam. Similarly, if we see a region reflecting polarized light (e.g., plates 6 and 8), all the local $\vec{E}$-vectors must be perpendicular to the line toward the source; their intersection, therefore, gives an excellent measurement of the source's position even if the direct line-of-sight from us to it is blocked.

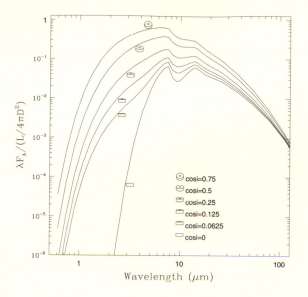

Fig. 12.2   Predicted infrared continuum from a very simple model: toroidal obscuration with constant density inside a figure having a square cross section (from Pier and Krolik 1992b). The half-opening angle of this torus is 30°, and its total column density is 1.6×10^{24} H cm^{-2}, corresponding to a total extinction $A_V \simeq$ 900 mag. The sharp contrast between the different curves, each corresponding to a different inclination angle i, demonstrates the strong angle dependence of the reradiated infrared continuum.

12.5   Unification of Flat-Spectrum Compact Radio Sources with Steep-Spectrum Extended Sources

We are now ready to apply these principles to some actual examples of AGN classes. To start, we will examine a case in which the anisotropy is entirely due to the strong intrinsic angle dependence of radiation when it emanates from relativistically moving matter.

As we discussed in Chapter 9, one of the numerous ways in which radio-loud AGNs can be classified is according to whether most of their flux comes from very extended regions and the spectrum is relatively steep, or most of their flux comes from a compact region at the nucleus and the spectrum is relatively flat. Although these classes appear to be quite distinct, it is in fact quite likely that they are intrinsically the same, but for the accident of our viewing angle. All that is required to make the connection is an allowance for relativistic beaming of the core component.

That relativistic beaming might explain the distinctions between class-es of AGNs separated on the basis of their radio properties was suggested early on. However, it took a while for the right partners to be matched up. Originally, it was hoped that relativistic beaming would explain the difference between radio-loud and radio-quiet quasars (Scheuer and Read-head 1979). A few years later Orr and Browne (1982), and in a less well known paper Kapahi and Saikia (1982), suggested that there would be bet-ter agreement with the source statistics if only core- and lobe-dominated quasars were considered. Finally, Scheuer (1987) raised the possibility that radio galaxies could also be included. It is this last version that now seems to work the best.

Because it is hard to find either "mirrors" or "screens" that are effective at radio wavelengths, the tests of this unification scheme are all of the statistical sort. Following the procedure set out in §12.1, we will work out the predicted distributions of several observed properties supposing that lobe- and core-dominated radio-loud AGNs are really all members of the same family of objects but differ in appearance to us because the core radiation comes from material in relativistic motion. We will then discuss how well these predictions agree with observations.

12.5.1 Luminosity function of compact radio sources

The most obviously angle-dependent property is the core luminosity L_b. In the most powerful version of this unification scheme, one posits a unique (or at least tight) relation between the unbeamed luminosity from the jet L_* (i.e., the luminosity as seen at an angle such that the Doppler boost factor $\delta = 1$; §9.3.4.3) and the extended isotropic luminosity L_o. The simplest version of such a picture is one in which they are directly proportional: $L_* = qL_o$.

More flexible pictures might allow for complicating effects such as envi-ronmental influences on the lobe luminosity (as emphasized in §9.3.4, we do not have a firm grasp on what controls how brightly jets shine), or the time delay between the epoch at which the energy that is now in the extended lobes was released, and the date at which the energy of the plasma now in the radio core was generated (e.g., Gopal-Krishna, Kulkarni, and Wiita 1996). Given typical lobe sizes ~ 100 kpc and typical core sizes ~ 10 pc, the current state of the extended emission is telling us about the state of the nucleus $\sim 10^6$ yr ago (or more), while the core reflects conditions only a few decades in the past. The state of the nucleus could well have changed greatly in that interval. Other plausible extensions include a distribution of q values, a nonlinear (but still unique) relation between L_o and L_*, or comparing L_b to the optical luminosity of the nucleus instead of L_o (as advocated by Wills and Brotherton 1995).

To start, let us explore the simplest, most powerful, version of the relativistic beaming unification model. Suppose then, that the radio emission comes from a steady relativistic flow moving with Lorentz factor γ_r. The core radiation we see has luminosity per unit frequency $L_b = \delta^{2+\alpha} L_*$, so $L_b = \delta^{2+\alpha} q L_o$. From this relation, two direct predictions follow, using only the *observed* luminosity function dN_e/dL_o of extended emission and a guess about the distribution of jet Lorentz factors $f(\gamma_r)$: the luminosity function dN_c/dL_b for the cores of radio-loud AGNs, and the luminosity function dN_c/dL_o of core-dominated sources with respect to their extended luminosity.

The second prediction is actually easier to carry out (Antonucci and Ulvestad 1985). Suppose that to appear in our sample as core-dominated, $L_b \geq L_{b,\min}$, and that all jets have Lorentz factor γ_r. Then we see all core-dominated objects in the sample from a direction such that $\mu \geq \mu_{\mathrm{crit}}$, where

$$\mu_{\mathrm{crit}} = \frac{1}{\beta_r} \left[1 - \frac{1}{\gamma_r} \left(\frac{q L_o}{L_{b,\min}} \right)^{1/(2+\alpha)} \right]. \tag{12.8}$$

Unless γ_r is large enough, of course, no viewing angle will provide a big enough boost. In the limit of $\gamma_r \gg 1$, this means that

$$\gamma_r > (1/2)(L_{b,\min}/q L_o)^{1/(2+\alpha)}.$$

Knowing μ_{crit}, we can immediately find the number of core-dominated sources, for

$$\frac{dN_c}{dL_o} = \frac{1}{2} \left(1 - \mu_{\mathrm{crit}} \right) \frac{dN_e}{dL_o}. \tag{12.9}$$

There should be at least $\sim \gamma_r^2$ times as many lobe-dominated sources as core-dominated ones at fixed L_o. Because this ratio is ~ 100 in, for example, the 3C sample, the relativistic beaming model works nicely for Lorentz factors ~ 10.

To find the luminosity function with respect to L_b, we begin with the fact that, for each L_o, the range of beamed luminosities runs from $q L_o [\gamma_r(1+\beta_r)]^{-(2+\alpha)}$ to $q L_o [\gamma_r(1-\beta_r)]^{-(2+\alpha)}$. Over that range, the number of sources with core luminosity L_b and extended luminosity L_o is

$$
\begin{aligned}
\frac{\partial^2 N_c}{\partial L_b \partial L_o} &= \frac{\partial^2 N_c}{\partial \mu \partial L_o} \frac{d\mu}{dL_b} \\
&= \frac{dN_e}{dL_o} \frac{1}{2(2+\alpha)\beta_r \gamma_r} L_*^{1/(2+\alpha)} L_b^{-1-1/(2+\alpha)}.
\end{aligned}
\tag{12.10}
$$

The factor of 2 in the denominator is simply due to the isotropic orientation assumption, $i.e.$ $dN_c/d\mu = 1/2$. Outside this range of luminosities there should be no objects. The contribution to the core luminosity function from each L_o and γ_r has a shape that depends on α but not γ_r; the only influence of γ_r is to set the range of observed luminosities, and therefore the number density per unit luminosity.

The next step is to find the observed core luminosity function given a spread in intrinsic luminosity and Lorentz factor:

$$
\frac{dN_c}{dL_b} = \frac{q^{1/(2+\alpha)}}{2(2+\alpha)} L_b^{-1-1/(2+\alpha)} \int d\gamma_r \frac{f(\gamma_r)}{\beta_r \gamma_r} \int_{L_-}^{L_+} dL_o\, L_o^{1/(2+\alpha)} \frac{dN_e}{dL_o},
\tag{12.11}
$$

where

$$
L_{\pm} = [\gamma_r(1 \pm \beta_r)]^{2+\alpha} L_b/q.
\tag{12.12}
$$

Suppose, for example, that $dN_e/L_o = (N_e/L_e)(L_o/L_e)^{-s}$ between two cut-offs $L_{o,\min}$ and $L_{o,\max}$ (the measured luminosity function at $z = 0$ can be approximated in this fashion with $s \simeq 5/2$; see fig. 3.9). In that case, if $1 - s + 1/(2 + \alpha) > 0$, the upper limit of the L_o integral dominates; if that combination has the other sign (as is more likely, for example, when $s \simeq 5/2$), the lower limit dominates. The physical meaning of these limits is obvious: whether most of the beamed sources we see correspond to un-beamed sources at the high or the low end of the luminosity distribution. Taking the more likely case, the core luminosity function becomes

$$
\begin{aligned}
\frac{dN_c}{dL_b} \simeq \frac{q^{1/(2+\alpha)}}{2(2+\alpha)} \frac{N_e/L_e}{s - 1 - 1/(2+\alpha)} &\left[L_{o,\min}^{-s} \left(\frac{L_{o,\min}}{L_b}\right)^{1+1/(2+\alpha)} \int_{\gamma_*}^{\infty} d\gamma_r \frac{f(\gamma_r)}{\beta_r \gamma_r} \right. \\
&\left. + q^{s-1-1/(2+\alpha)} L_b^{-s} \int_1^{\gamma_*} d\gamma_r\, f(\gamma_r)\, (\beta_r \gamma_r)^{(1-s)(2+\alpha)} \right],
\end{aligned}
\tag{12.13}
$$

where γ_* is defined by

$$
\gamma_*(1 - \beta_*) = \left(\frac{qL_{o,\min}}{L_b}\right)^{1/(2+\alpha)}.
\tag{12.14}
$$

Thus, at the low-luminosity end, the luminosity function is stretched out with logarithmic slope $-1 - 1/(2 + \alpha)$, while at the high-luminosity end it follows the shape of the underlying unbeamed luminosity distribution. If $f(\gamma_r) \simeq \delta(\gamma_r - \gamma_o)$, the transition comes at

$$L_{\text{break}} \simeq \frac{qL_{o,\min}}{[\gamma_o(1 - \beta_o)]^{2+\alpha}} \simeq (2\gamma_o)^{2+\alpha}\, qL_{o,\min}, \qquad (12.15)$$

where the second approximate equality obtains when $\gamma_o \gg 1$. When $f(\gamma_r)$ is broad, the transition is smoothed out. Note that if $1 - s + 1/(2 + \alpha)$ had the opposite sign, the sense of curvature in the luminosity function would be reversed.

This predicted luminosity function must now be compared with the observed. Unfortunately, this procedure is complicated by two problems. First, there is only a finite range over which we can establish the unbeamed luminosity function. When the boosting is strong, the counterparts to the observed beamed sources have very low extended luminosity. This means that, constrained as we always are to flux-limited samples, we may not have a good empirical measure of the unbeamed luminosity function in the range of luminosities we need. Second, at the end of §12.1 we alluded to the possibility of additional parameters that might complicate the construction of unified schemes; the proportionality constant q (or, more generally, the specific relation between L_o and L_*) and the Lorentz factor distribution $f(\gamma_r)$ are prime examples. Indeed, one might well imagine that these two relations are coupled in some way. However, in most attempts to test this unification scheme, these relations are chosen to be as simple as they can be while still resulting in an acceptable fit to the observed core luminosity function after adjusting the free parameters. Such a procedure clearly suffers from a measure of subjectivity, but it appears to be the best we can manage for the time being.

The results of performing this exercise are nonetheless quite encouraging. One version (Urry, Padovani, and Stickel 1991; Padovani and Urry 1992) is displayed in figure 12.3. In this work, instead of simply predicting the core luminosity function from the extended emission luminosity function, a slightly more complicated argument was followed in which the luminosity functions for defined AGN categories were used. In addition, the "parent" (i.e., unbeamed) population was split in two, separating FR1 (low-luminosity) from FR2 (high-luminosity) radio galaxies. Steep-spectrum radio-loud quasars were included in the latter group. The beamed populations were taken to be BL Lac objects (for the FR1 galaxies) and flat-spectrum radio-loud quasars (for the FR2 galaxies). This division was motivated by the observation that the extended emission in most (but not all: Kollgaard et al. 1992) BL Lac objects has FR1 morphology, while in radio-loud quasars it is generally FR2. In order to compensate for the relatively small numbers of sources in the samples, cosmological evolution in the luminosity functions was (approximately) removed by modeling. After adjustment of parameters, the best agreement for the low-luminosity segment was found for $q = 0.01$ at 5 GHz and $f(\gamma_r) = 380\gamma_r^{-4}$ over the range $5 \leq \gamma_r \leq 35$, while for the high-luminosity sources the best fit is obtained for $q(2.7 \text{ GHz}) \simeq 5 \times 10^{-3}$ and $d\ln f/d\ln \gamma_r \simeq -2.3$. In both cases, the scheme fails if only a single value of γ_r is permitted.

Because the quality of the fit is not terribly sensitive to the exact value of $d\ln f/d\ln \gamma_r$, these results indicate both that relativistic beaming can indeed unite these subclasses of radio-loud AGNs, and that both the distribution of Lorentz factors and the relation between extended and compact luminosity may be only weak functions of luminosity. On the other hand, if the distinction in $f(\gamma_r)$ between the FR1s and the FR2s is real, it might explain why counterjets are often seen in FR1s but almost never in FR2s (§9.1.2). The range of Lorentz factors inferred from matching the luminosity functions is also pleasantly consistent with the range required to explain both superluminal expansions on the VLBI scale (§§9.3.3.2, 12.5.2) and the escape of high-energy γ-rays (§8.8.2). This consistency lends strength to the belief that the statistical concurrence is real.

These results reinforce the comments made earlier (§12.2.1) about the way in which beaming can distort the observed source counts. Although, for any given intrinsic luminosity, only $\sim 1\%$ of all sources are substantially boosted in strength as seen from our direction, beamed sources are comparable in number to unbeamed sources at high observed luminosities. Consequently, in many flux-limited surveys beamed sources are as common as unbeamed sources.

12.5.2 Jet length and speed

If the core emission does come from a jet, and the length of that jet is a fixed quantity l_o, then the distribution of jet lengths projected on the sky should be

$$\frac{dN}{dl} = \frac{dN}{d\mu}\frac{d\mu}{dl} = \frac{N}{2l_o}\frac{l/l_o}{\sqrt{1 - (l/l_o)^2}}. \tag{12.16}$$

In practice, this distribution is difficult to verify because it is so easy for there to be a broad range of intrinsic lengths l_o, as well as complicating factors due to uneven surface brightness along the jet, anisotropic emission, edge-darkening, and so on.

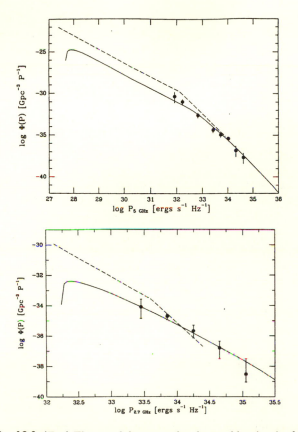

Fig. 12.3 (Top) The match between the observed luminosity function of BL Lac objects and the luminosity function predicted from the FR1 luminosity function by the relativistic beaming model described in the text (Urry et al. 1991). (Bottom) The same match is shown for FR2 galaxies and flat-spectrum radio-loud quasars (Padovani and Urry 1992) . In both panels, the filled circles are the measured luminosity function for the beamed population, the solid line is the model prediction for their luminosity function, and the dashed line is the measured luminosity function for the unbeamed population.

On the other hand, if the core emission is irregular enough to display identifiable moving spots, it is possible to measure the distribution of transverse velocities for these spots in a sample of objects selected on the basis of isotropic properties (e.g. extended radio luminosity). In the simplest picture, all the spots in a given AGN move in the same direction and with a Lorentz factor γ_p that may vary across the sample but is fixed in each individual object. Their apparent speed on the sky is then $\beta_{\mathrm{app}} = \frac{\beta_p\sqrt{1-\mu^2}}{1-\beta_p\mu}$ (§9.3.3.2). Other, more complicated relations between β_{app}, β_p, and μ can also be constructed, but almost any model of superluminal motion that makes use of relativistic kinematics shares the same qualitative properties: $\beta_{\text{app}} \sim \beta_p$ in most directions, but rises to $\sim \gamma_p$ when seen within an angle $\sim \gamma_p^{-1}$ from the direction of motion. Because the distribution of angles $dN/d\mu = N/2$, the observed distribution of apparent transverse speeds should be

$$\frac{dN}{d\beta_{\text{app}}} = \int d\gamma_p \, \frac{1}{2} f(\gamma_p) \frac{d\mu}{d\beta_{\text{app}}}, \qquad (12.17)$$

for a distribution f of pattern Lorentz factors γ_p.

Because β_{app} usually varies more slowly with angle than does the apparent luminosity, if the sources are intrinsically bright enough, it is possible to measure substantial transverse velocities even in sources whose direction of motion lies across, or possibly even opposite to, the line of sight. In fact, there are a number of examples of lobe-dominated radio galaxies in which β_{app} (slightly) greater than 1 has been seen.

Testing measured distributions of β_{app} for consistency with the predictions of the relativistic jet model is made tricky both by ambiguities of the models and limitations of the data. However, it is satisfying at the qualitative level that β_{app} does seem to be largest in the most core-dominated objects (Vermeulen and Cohen 1994).

12.5.3 Depolarization

Until now we have supposed that there was a perfect vacuum between the radio source and us. That is not always so, and, in fact, the greater pathlength between us and the jet on the far side of the source can sometimes produce detectable effects. The most common ways for intervening material to affect radio frequency signals are free-free absorption and Faraday rotation of the polarization. The former requires relatively high-density gas to be effective (eq. 7.92); the latter scales with the column density of electrons (§7.5.6), so that lower density gas can still be effective. Scaling the amount of likely Faraday rotation to fiducial numbers characteristic of the regions surrounding radio sources, we estimate that the rotation of the polarization angle is

$$\Delta\phi \simeq 3.6 \times 10^4 \tau_{\text{T}} \nu_{\text{GHz}}^{-2} \left(\frac{\langle B_z \rangle}{10^{-6} \text{ G}} \right) \text{ radians.} \qquad (12.18)$$

Thus, Faraday rotation (and depolarization if the beam we receive passes through inhomogeneous regions) can be quite substantial. Although we have no a priori way of knowing what the *original* plane of polarization was, the strong dependence on frequency allows us to infer the amount of rotation by measuring how much it changes as a function of ν. Once this has been done, we have, of course, also inferred the original plane of polarization.

In §9.1.2 we remarked on the fact that jets are frequently seen on kiloparsec scales. In FR2 radio galaxies, they are almost always seen only on one side of the nucleus, but in FR1s, frequently on both sides. If we make the supposition that the contrast in jet strength from one side to the other is due to relativistic beaming, we can also expect a contrast in Faraday rotation between the two: The path to the jet pointing away from us is likely to traverse a greater integrated electron column density and parallel magnetic field than the path to the jet pointing toward us. Thus, we can (at least statistically) identify which one is which on the basis of the contrast in Faraday rotation.

As first found by Laing (1988), the Faraday rotation of the radio emission from the side with the weaker (or absent) jet is almost invariably greater than the Faraday rotation of the radio emission from the jet. Although this test cannot be clearly quantified (we have no way of predicting a priori how much Faraday rotation there should be, or how smoothly it is distributed across the central parts of the host galaxy), and other interpretations might also be possible, its results are clearly in excellent agreement with the idea that the strength of small-scale jet features is strongly modulated by relativistic beaming.

12.6   Unification of Radio-Quiet Broad Line and Narrow Line AGNs via Toroidal Obscuration

Our next study of the consequences of anisotropic appearance has to do with the majority of AGNs, those that are radio-quiet. They, too, can be divided in a natural way into two subclasses—those with and those without broad optical and ultraviolet emission lines (§10.1.3). The presence of a strong nonstellar continuum, either in the optical/ultraviolet band, or the soft X-ray band, is also closely linked with the existence of broad emission lines (§1.3). At low luminosities, those radio-quiet AGNs possessing broad emission lines and a strong nonstellar continuum are called type 1 Seyfert galaxies, while those with only narrow emission lines and a weak nonstellar continuum are dubbed type 2 Seyfert galaxies. At luminosities high enough that the host galaxy is difficult or impossible to see, the terminology switches from Seyfert galaxies to quasars. Although a small number of "type 2 quasars" have been discovered, the vast majority of known quasars are type 1. To what degree that is a physical effect, and to what degree a selection effect, is controversial (§12.6.5).

Be that as it may, there is now excellent evidence—particularly for the Seyfert galaxies, but also for quasars—that these two subclasses of AGNs are intrinsically identical but for the accident of our viewing angle. The primary reason for their very different appearance is a toroidal belt of very optically thick obscuration.

Unlike the situation for radio properties, there are a great many mechanisms providing "mirrors" and "screens" in the higher frequency bands from the infrared through X-rays. Thanks to them, and to the existence of strongly energy-dependent obscuration, there are many instances in which we can directly demonstrate that the nucleus looks different on lines of sight other than our own.

12.6.1 Mirrors

As we discussed in §12.4.1, a polarizing mirror can be a very powerful tool for revealing AGNs otherwise hidden behind thick obscuration. Indeed, just this sort of mirror has been seen in quite a number of type 2 Seyfert galaxies (fig. 12.4).

Because it is by far the nearest and brightest, the best observed of them is NGC 1068. It is so close that the mirror, even though it extends only $\simeq$ 50–100 pc from the nucleus, can actually be resolved (plate 6). Although far outside the continuum and broad line production regions, the mirror coincides with the inner parts of the narrow line region.

The optical/ultraviolet total flux spectrum of the region containing the mirror is dominated by narrow emission lines and starlight. In striking contrast, its polarized flux spectrum looks just like a classic type 1 Seyfert galaxy: strong broad emission lines and a clearly nonstellar continuum (fig. 12.5; Antonucci and Miller 1985; Miller, Goodrich, and Mathews 1991; Kriss et al. 1992; Antonucci, Hurt, and Miller 1994). The polarization fraction of the nonstellar light is large (16%) and independent of wavelength. Strikingly, the direction of the polarization $\vec{E}$-vector is very nearly perpendicular to the (weak) radio jet found in this AGN. As Antonucci and Miller (1985) stressed, the *only* way to understand the contrast between the polarized and total flux spectra is if there is a normal type 1 nucleus hidden behind very optically thick obscuration, and electron scattering centered on the jet axis manages to reflect a small fraction (perhaps $\sim$ 1%) of the light around the obscuration.

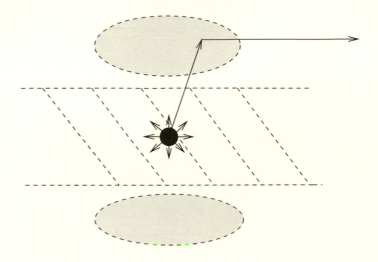

Fig. 12.4   Schematic view of a polarizing mirror (shaded ovals) near an AGN with toroidal obscuration (broken lines). There is a hole in the obscuration that permits light directed toward the mirror to leave, but in all other directions light from the center strikes the obscuring matter and is absorbed. An observer far to the right can receive light reflected from the mirror if it follows a path such as the one shown by the arrow-tipped line.

In the happy case of NGC 1068, there is also a *second* mirror, far enough from the nucleus ($\sim$ 300 pc) that it can be easily separated from the first (Miller et al. 1991). Contrasting the spectra from the two mirrors gives us the information to draw a still more detailed picture. Because the polarized flux from the outer mirror rises very steeply toward the blue, this mirror is almost certainly composed of dust, not ionized electrons. Moreover, the Balmer lines in the polarized spectrum of this mirror are blue shifted by 400 km s^{-1} relative to the corresponding lines in the polarized spectrum of the inner mirror, and somewhat narrower. Because the lines in the outer mirror have a redshift close to the systemic recession velocity of the host galaxy, the shift is naturally interpreted as due to a bulk outflow in the inner mirror:

$$\frac{\Delta\nu}{\nu} = \left(\frac{\vec{v}}{c}\right) \cdot (\hat{n}' - \hat{n}), \qquad (12.19)$$

where $\vec{v}$ is the velocity of the mirror ($|\vec{v}| \geq 400$ km s^{-1}), and $\hat{n}, \hat{n}'$ are the initial and final photon directions (cf. eqn. 10.39). Similarly, the broader lines in the spectrum of the inner mirror are most likely due to the broadening that accompanies electron scattering. The inferred temperature is $\sim 3 \times 10^5$ K.

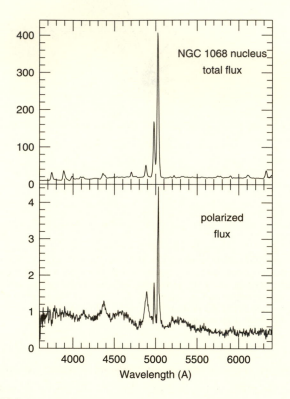

Fig. 12.5 (Top) The total flux spectrum of NGC 1068. (Bottom) The polarized flux spectrum. Both are taken from Miller et al. (1991). Note how the Hβ and HeII 4686 lines appear with broad profiles in the polarized flux, and the continuum rises in strength relative to the [OIII] 5007 line.

The fraction of the light directed at the inner mirror that is scattered toward us (f_{refl}) can also be estimated using the spectrum of the outer mirror. Because the inner mirror reflects by electron scattering, we can use any wavelength to define this quantity, for example, $f_{\text{refl}} \equiv F_{\text{refl}}(\text{H}\beta)/F_{\text{tot}}(\text{H}\beta)$. This ratio can be rewritten as

$$\frac{F_{\text{refl}}(\text{H}\beta)}{F_{\text{tot}}(\text{H}\beta)} = \frac{F_{\text{pol}}(\text{H}\beta)/P}{F_{\text{tot}}([\text{OIII}]\ 5007)} \frac{F_{\text{tot}}([\text{OIII}]\ 5007)}{F_{\text{tot}}(\text{H}\beta)}, \tag{12.20}$$

where P is the polarization fraction of the light scattered from the inner mirror and F_{tot}([OIII] 5007) is the total flux in the [OIII] 5007 narrow emission line. Because the polarization properties of the narrow lines are quite different from those of the broad lines, the great majority of the narrow line flux must be generated either near or outside the inner mirror. The second ratio in equation 12.20 can be estimated by the line ratio *in the reflected light from the outer mirror* because it has a clear view of both emission line regions and is so far away that it sees both as essentially points. With this substitution, all the ratios are measurable quantities, and the reflected fraction can be estimated. In this case, it is $\simeq 0.016$. From the opening angle of the emission line cone, we can also estimate the solid angle of the mirror ($\sim \pi$ sterradians). The optical depth is therefore ~ 0.06, nicely in line with the estimate of equation 11.17 if L/L_{E} is a few tenths (as is confirmed by an independent argument: Greenhill et al. 1996).

These inferences of optical depth and temperature for the inner mirror also call to mind the properties of the X-ray "warm absorbers" (§§8.8.1, 11.1). These, too, are media with high, but not quite complete, ionization, and Thomson depths of $\sim 10^{-3} - 0.1$ that occupy the line-of-sight to many type 1 Seyfert nuclei. It is possible that "warm absorbers" and electron scattering mirrors are the same structures viewed from different directions (Krolik and Kriss 1995).

That we can resolve these two mirrors in NGC 1068 is due to our good fortune in being so close. In other, more distant, AGNs there may well be two mirrors, but unresolvable by us. Diagnostic information in these cases is therefore more difficult to obtain. Analysis of these reflection regions must also be sensitive to the possible complications due to mixing light from two (or more) different mirrors. More distant cases are also, of course, harder to interpret because apertures of fixed angular size enclose greater amounts of starlight. To measure the polarized fraction of the reflected light, one must subtract off the starlight, and this becomes progressively more difficult.

Despite these difficulties, the polarization spectra of numerous type 2 Seyfert galaxies have been measured. In virtually every case where the signal/noise ratio is large enough to see them, broad emission lines appear in the polarization spectrum, confirming the presence of a hidden type 1 nucleus. In similar confirmation of the picture inferred from NGC 1068, the direction of polarization is nearly perpendicular to the radio axis, at least to the accuracy with which that direction can be determined (Antonucci 1983). Generally speaking, the polarization after subtraction of that fraction of the continuum that is obviously starlight is significant, but not large, perhaps a few percent (Tran 1995). Interestingly, the polarization fraction is often larger in the emission lines than in the continuum, suggesting that there is some other source of unpolarized continuum (dubbed "FC2" for "featureless continuum #2"). This additional component may be due to a stellar population (e.g., a "starburst" where the star formation rate is particularly high) without the strong absorption lines that allow easy identification in the spectrum (Heckman et al. 1995).

12.6.2 Screens

Many Seyfert galaxies also possess screens on which light from the central nucleus may be projected, whether or not we have a direct view of the nucleus. For example, most display elongated regions of line emission (an example is shown in plate 7), just as one would expect in disk galaxies illuminated by a collimated beam of ionizing radiation (Wilson 1994).

Likewise, their near-infrared spectra rise very steeply toward longer wavelengths, suggesting thermal emission (Edelson, Malkan, and Rieke 1987). NGC 1068 is so close that the region of infrared reradiation can actually be partially resolved at 10 μ (Cameron et al. 1993) and imaged at very small physical scale at 2 μ (fig. 12.6). Unfortunately, the predicted angular dependence of the infrared spectrum depends so strongly on details of the obscuring matter's shape and optical depth that there are too many extraneous parameters to apply the sort of statistical test to the Seyfert galaxy infrared luminosity function that is possible with relativistic beaming models applied to radio galaxies' radio luminosity function. Nonetheless, the obscuration model does explain one important fact—why it is that in many Seyfert galaxies, the bolometric luminosity is dominated by the mid-infrared (as shown in fig. 1.5).

Numerous H_2O masers have also been found, but only in type 2 Seyfert galaxies (Braatz, Wilson, and Menkel 1994). One might expect that the direction of greatest amplification would lie in the equatorial plane of the molecular gas, so the apparent angle dependence of maser emission is certainly in keeping with this unification model, even in the absence of detailed predictions.

12.6.3 Energy-dependent obscuration

Yet another indication of anisotropic emission in the infrared through soft X-ray bands is that type 2 Seyfert galaxies are often seen with hard X-ray spectra that appear to be very similar to those of type 1 Seyferts but for very large column densities. These are sometimes as little as $\sim$

10^{22} cm^{-2} but in other cases can exceed $\sim 10^{24}$ cm^{-2} (fig. 12.7). Relative to the (obscured) nuclear optical luminosity, the hard X-ray luminosity is very large; relative to the bolometric luminosity (in these obscured cases, dominated by the infrared), the X-ray luminosity contributes more or less its usual significant, but minority, share.

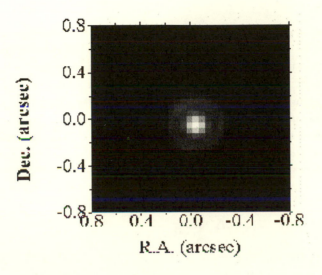

Fig. 12.6   A speckle interferometric K-band (2 μ) image of the nucleus of NGC 1068 (Thatte et al. 1997). Essentially all the flux is contained within a 1 pc radius of the nucleus. If the dust is optically thick, its flux is so small that we can be seeing only a small part of the hot inner surface of the toroidal obscuration. Such a partial view is what one might expect, given the nearly equatorial view implied by the fact that NGC 1068 is a type 2 Seyfert.

12.6.4   Relative numbers of type 1 and type 2 Seyfert galaxies

Another way to develop (or test) this sort of unification scheme is to proceed along a track parallel to the one we have already traveled when we examined the unification of flat-spectrum compact radio sources with steep-spectrum extended sources. There, given one or two adjustable parameters, the unification model successfully predicted the core luminosity function on the basis of the measured lobe luminosity function. Here, we can do the same thing: choose an isotropic property of Seyfert galaxies, measure its distribution, make a parameterized prediction of the distribution of the related nuclear property, and see whether a successful fit can be found.

Unfortunately, truly isotropic components are difficult to find in Seyfert galaxies. As discussed in §12.4.4, the obscuration can be so optically thick that even the infrared continuum can be substantially anisotropic, although this anisotropy can be expected to diminish at longer wavelengths where the optical depth is smaller. So far, three choices have been pursued, none of them ideal.

The first is, in fact, not so much a test of the model as a way to measure one of its parameters, the solid angle $\Delta\Omega$ through which clear lines of sight to the nucleus exist. At low nuclear luminosities, a relatively good (but inefficient) selection criterion is the flux from the *host* galaxy. Except for high inclination angles, when interstellar extinction makes it especially difficult to see type 1 nuclei (Keel 1980; Maiolino and Rieke 1995), this selection method introduces little directional bias. With the further caveat that there is no correlation between the host luminosity and the transparent solid angle, the ratio of type 1 Seyfert galaxies to type 2's in a sample selected solely on the basis of host flux should determine $\Delta\Omega$ directly:

$$\Delta\Omega = 4\pi \frac{n_1}{n_1 + n_2}, \tag{12.21}$$

where $n_{1,2}$ are the space densities of the two types. However, this method's applicability is somewhat limited: once the nuclear luminosity becomes too high, it easily contaminates measurements of the total galaxy luminosity and thus biases the sample selection. It is for this reason that Huchra and Burg (1992) regarded their 1:1 ratio of type 1 Seyferts to type 2's as an upper bound to the true ratio. If the nuclear luminosity were separated from the host luminosity, the sample could be adjusted to provide a fair measure of $\Delta\Omega$, as well as the optical luminosity function for type 1 Seyfert galaxies.

The other two methods are both closer in spirit to the radio galaxy luminosity function studies. In both cases, a component of the AGN emission is sought that is both isotropic and bears (within the model) some definite relation to the anisotropic components.

The first candidate for an isotropic radiation component in Seyfert galaxies is a narrow emission line, for example, [OIII] 5007 or the narrow core of a Balmer line. As demonstrated in §12.4.3, the luminosity L_l in such a line is simply proportional to the intrinsic nuclear luminosity times the covering fraction of radiating material. However, there are two subtleties

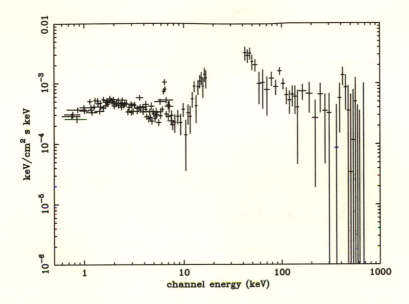

Fig. 12.7 The X-ray spectrum of the type 2 Seyfert galaxy NGC 4945 (Done et al. 1996). Although very weak at soft X-ray energies, the flux rises extremely steeply at $\simeq$ 10–20 keV, just as would be expected if a column density $\sim 10^{24}$ cm^{-2} were interposed between us and a nucleus with "normal" type 1 Seyfert X-ray emission. We see X-rays at lower energies predominantly through reflection.

here. First, there is some hazard of anisotropic emission if a part of the emission line region is subject to obscuration. Second, in those directions where there is obscuration between the nucleus and the emission line region, there can be no photoionized line emission (this is, of course, the "ionization cone" effect illustrated in plate 7). Consequently, the isotropic part of the line luminosity we see is better described by

$$L_l = C_i \frac{\Delta\Omega}{4\pi} Y_l L_{\text{nucl,intr}}, \qquad (12.22)$$

where C_i is the covering fraction of emission line material averaged over the illuminated solid angle, Y_l is the yield in energy units, and $L_{\text{nucl,intr}}$ is the true, intrinsic luminosity of the nucleus, than by the simpler expression of equation 12.4. Consequently, although the line luminosity can be isotropic, its strength is proportional to the principal parameter describing the degree of anisotropy of the nuclear emission. Unless $\Delta\Omega$ varies over only a narrow range, any sample selected on the basis of emission line flux is biased toward large $\Delta\Omega$, and the fraction of type 1 Seyfert galaxies found in it is larger than in a fair sample.

Nonetheless, given enough data, this bias can be overcome, at least in a model-dependent fashion. If the space density per unit nuclear luminosity and clear solid angle is $n_{\text{tot}} f(\Delta\Omega, L_{\text{nucl}})$, the observed line luminosity function is

$$\frac{dn_{\text{tot}}}{dL_l} = \int d(\Delta\Omega) \, f\left(\Delta\Omega, L_{\text{nucl}} = \frac{4\pi L_l}{C_i Y_l \Delta\Omega}\right) \frac{4\pi n_{\text{tot}}}{C_i Y_l \Delta\Omega}, \qquad (12.23)$$

while the observed line luminosity function for type 1 Seyfert galaxies alone is

$$\frac{dn_1}{dL_l} = \int d(\Delta\Omega) \, f\left(\Delta\Omega, L_{\text{nucl}} = \frac{4\pi L_l}{C_i Y_l \Delta\Omega}\right) \frac{n_{\text{tot}}}{C_i Y_l}. \qquad (12.24)$$

Given a parameterized description of $f(\Delta\Omega, L_{\text{nucl}})$, these expressions may be fitted to the data and the best-fit parameter values found. Note that in this approach n_{tot} and C_i cannot be determined separately; only their ratio can be found. Because $C_i \leq 1$, this ratio gives a lower bound on the total space density.

It is difficult in practice to construct a sample that is genuinely selected solely on the basis of emission line flux. Because a line must have a minimum equivalent width even to be detectable, there is almost inevitably some dependence on equivalent width, and therefore on the host plus nuclear continuum luminosity, as well.

Two such surveys have been performed to date. Using a search method based purely on emission line equivalent width, and supposing that there was only a narrow distribution of $\Delta\Omega$, Osterbrock and Shaw (1988) found that there are between four and nine times as many type 2 Seyferts as type 1 (the former number applies if one counts types 1.8 and 1.9 as type 1, the latter if one groups them with the type 2's). Similarly, Salzer (1989), using a sample sensitive to both the flux and equivalent width of [O III] 5007 and Hβ, discovered roughly five times as many type 2's as type 1's.

The second candidate for an isotropic AGN component is the long-wavelength infrared flux. At long enough wavelength, the obscuration must eventually become optically thin, and its infrared continuum therefore isotropic. If, for example, the parameters of the model shown in figure 12.2 are typical, the nuclear infrared continuum is reasonably isotropic for $\lambda > 50$–$75\ \mu$, but is still substantially anisotropic at shorter wavelengths. In many cases, it is also possible that the host luminosity becomes comparable to or greater than the nuclear infrared luminosity at wavelengths longer than 50–75 μ. Thus, the far-infrared flux is likely to be a genuinely isotropic property, but one that may depend on both the nucleus and the host.

One example of the use of infrared selection is the study of Keel et al. (1994), who chose a sample based on both 25 and 60 μ flux with the idea that substantial 25 μ flux distinguishes AGNs from normal galaxies. They found a ratio of type 2 to type 1 Seyferts of only $\simeq 2$, which one might expect to be an underestimate if type 1 Seyferts are generally brighter at 25 μ than type 2 Seyferts of the same intrinsic luminosity.

Although there are imperfections in all these studies, it is clear that our view of the nucleus is unobscured in only a minority of Seyfert galaxies. In other words, the opening angle of the obscuration is relatively small, probably as little as only a few tenths of 4π.

12.6.5 Extrapolation to higher luminosity?

The very strong evidence for the importance of obscuration in Seyfert galaxies naturally suggests that it might be equally important in their higher luminosity cousins, the radio-quiet quasars. Unfortunately, although it is relatively easy to find nearby narrow line AGNs (e.g., select galaxies on the basis of their total luminosity, and look for narrow emission lines in their spectra), it is much harder to find them far away. At large distances, we are often only able to find objects when their total luminosity is greatly enhanced by the unobscured luminosity of their nuclei. Consequently, objects whose nuclear continua are blocked from our line-of-sight are difficult even to find, much less examine spectrally. They are passed over by flux-limited optical continuum surveys, especially those keyed to nonstellar colors (§2.1). They are missed by emission line surveys (§2.2) both because the equivalent widths of the broad lines are generally much larger than the equivalent widths of the narrow lines and because (if we are interested in high-redshift objects) there are fewer strong forbidden lines in the rest-frame ultraviolet. They are invisible to soft X-ray searches (§2.4) because the optical depth to K-shell photo-ionization of elements like C, N, O etc. is large whenever there is optically thick dust extinction. Almost the only detection methods unaffected by obscuration are those based on radio flux, and even these can be hampered by free-free opacity associated with the obscuration. And, of course, most AGNs are radio-quiet.

Despite these difficulties, a few examples of narrow line quasars have been found. Strong mid-infrared emission has proven a particularly fruitful search key. Confirmation can be drawn from either hard X-rays, which shine through when the obscuration is optically thin to Compton scattering, or screens (e.g., emission lines) and mirrors (extended polarized flux) that may exist in these objects. By these means we have found at least a few objects of genuinely quasar-class luminosity that have very strongly anisotropic appearance due to toroidal obscuration (e.g., Ogasaka et al. 1997, Young et al. 1996). What we do not know, due to the unsystematic character of the searches so far, is how common they may be, and whether *all* quasars possess such structures.

12.7 Unification of Radio-Loud Quasars and Radio Galaxies via Toroidal Obscuration

Broad- and narrow line radio galaxies are also very likely the same creatures viewed from different perspectives (Barthel 1989). Indeed, we have already tacitly drawn this connection when we united flat-spectrum core-dominated radio-loud AGNs with their steep-spectrum lobe-dominated brethren; counting radio-loud quasars on an equal footing with radio galaxies was essential to making that brand of unification work.

Although the evidence that radio-loud quasars and radio galaxies are fundamentally the same is not nearly so strong and complete as it is for type 1 and type 2 Seyfert galaxies, it is still very persuasive. In fact, because radio galaxies are easier to find at high redshift, and the reflection signals tend to stand out better in the rest-frame ultraviolet, in some ways it is *easier* to see the telltale signs of hidden nuclei in radio galaxies than in type 2 Seyfert galaxies. In addition, it is, of course, much easier to determine the direction of the radio axis in radio galaxies than in Seyfert galaxies, whose radio emission is much weaker and much less clearly structured.

12.7.1 Mirrors

Antonucci (1984) was also the first to discover a narrow line radio galaxy—3C 234—whose polarization spectrum mimics that of a quasar. The polarization spectrum of this $z \simeq 0.2$ radio galaxy is very similar to the polarization spectrum of NGC 1068: a strong blue continuum, broad Balmer lines, and an $\vec{E}$-vector perpendicular to the radio axis. Since that time, more have been identified, including several at $z \simeq 1$ (di Serego Alighieri et al. 1994). In the latter group, the redshift is large enough that we observe the rest-frame ultraviolet. Because starlight is quite weak at short wavelengths, broad MgII 2800 is visible in the total flux spectrum, despite the predominance of narrow forbidden lines.

Just as in the case of the Seyfert galaxies, it is almost impossible to imagine any other explanation for the presence of broad emission lines in the polarized spectrum when they are absent in the total flux spectrum. There must be thick obscuration on the direct line-of-sight, and a polarizing scatterer on the axis, so that we see nuclear light only via polarizing reflection. Thus, the same picture appears to apply to radio-loud AGNs of high luminosity as to the radio-quiet Seyfert galaxies: toroidal obscuration and a mirror on the axis.

Despite the many similarities with type 2 Seyfert galaxies, the polarization properties of narrow line radio galaxies differ significantly in one important respect: the dominant reflection agent in the NLRGs is more likely to be dust than electrons. Unlike Seyfert galaxies, in which the reflection is confined to a very small region near the galactic nucleus, in radio galaxies, the region emitting polarized light covers much of the galaxy. Imaging polarimetry then allows a direct measurement of the opening angle of the mirror. The reflected fraction can be estimated by comparing the mean reflected optical luminosity seen in the radio galaxies with the mean optical luminosity of radio-loud quasars with the same extended radio luminosity (di Serego Alighieri et al. 1994 find $f_{\mathrm{refl}} \simeq 0.02$ for their three radio galaxies). The optical depth through the mirror is then $f_{\mathrm{refl}}/\Delta\Omega$, and its mass is $A_{\mathrm{refl}} f_{\mathrm{refl}}/(\kappa\Delta\Omega)$, where A_{refl} is the cross-sectional area of the mirror and κ is its opacity per unit mass. When the reflection is spread over the entire volume of the host galaxy, the relatively small opacity of electron scattering would require an extremely large total mass; dust is relatively much more efficient.

The scattering efficiency of dust is also, of course, strongly wavelength-dependent (as illustrated in fig. 10.7). Dust-reflected light should therefore show a filtered version of the intrinsic spectrum. Because reddening inside the "mirror" plays off against the "bluening" effect of scattering, the shape of the filter depends on the geometry and optical depth of the reflection region, as well as on the size distribution and composition of the grains.

Because the polarized light from radio galaxies is distributed over a large part of the host galaxy, it is relatively easy to image. If the collimation by obscuration model is correct, we would expect the polarized image to have the same (bi-)conical outline as did the emission line images of type 2 Seyfert galaxies. The axis of this cone must be perpendicular to the polarization electric field (by definition). Because the polarization direction is generally perpendicular to the radio axis, we expect that these cones should be aligned with the radio axis. Moreover, in the rest-frame ultraviolet (which is shifted into the visible band when $z > 1$ or so), because the reflected light can be stronger than the starlight of the host galaxy, the *total flux* image should have the same shape.

Total flux images are very much consistent with this prediction. As first pointed out by McCarthy et al. (1987) (using 3CR galaxies with $0.5 < z < 2.0$) and Chambers et al. (1987) (using 4C ultra-steep radio spectrum galaxies without redshifts determined in advance), the continuum images of high-redshift radio galaxies are elongated in the direction of the radio axes. This tendency has been dubbed the *alignment effect*. While often due to reflection, there may be exceptions (Dey et al. 1997).

Polarization images are harder to obtain with good S/N, but in at least some cases (e.g., 3C 321; see plate 8), the polarized flux is clearly elongated in the direction of the radio axes, the $\vec{E}$-vectors of the polarization are perpendicular to the direction toward the nucleus, and the polarized flux is so strong that at 3200 Å observed wavelength (rest-frame 2900 Å) the mean polarization is $\simeq 9\%$.

12.7.2 Screens: infrared reradiation

Because the brightest known radio galaxies are rather fainter than the brightest Seyfert galaxies, their infrared spectra are significantly harder to measure. In addition, the host contamination problem is likely to be equally severe. Consequently, the number of radio galaxies with secure detections in the infrared, much less well defined spectra, is rather small.

Nonetheless, it is possible to attempt a stripped-down version of the luminosity function test that we earlier applied to radio galaxies. In this case, one compares the mid-infrared flux to a quantity that one expects to be proportional to the bolometric luminosity and isotropically radiated, for example, the extended radio flux. Using samples of 3CR galaxies matched in low-frequency radio power and redshift, Heckman et al. (1992, 1994) found that the broad line radio galaxies (at higher luminosity this means radio-loud quasars) were, on average, four to five times brighter than the corresponding narrow line galaxies across the range of mid- to far-infrared wavelengths. If the narrow line galaxies generated infrared emission only by conventional stellar processes, even this much power in the mid-infrared would be very surprising. On the other hand, although the obscuration model would predict similar luminosities for the two classes in the far-infrared, as we discussed in §12.4.4, they do predict a contrast of this sort in the mid-infrared. These results therefore give at least partial support to the idea that narrow line radio galaxies in general contain genuine active nuclei, but they are hidden by very optically thick dusty obscuration.

12.7.3 Relative numbers of radio-loud quasars and radio galaxies

Because the extended radio emission must escape freely in all directions, the relative numbers of obscured and unobscured nuclei can be measured from any complete sample selected on the basis of low-frequency radio flux. Fortunately, the 3CR survey is entirely identified and so is perfectly suited to this task. Overall, the ratio of obscured to unobscured is about 2.3, but this appears to be a function of radio power. At the highest radio luminosities, the numbers are about even, while at the lowest there are hardly any unobscured nuclei (Lawrence 1991). This trend may also explain the apparent rarity of broad-line nuclei in FR1 radio galaxies (Falcke, Gopal-Krishna, and Biermann 1995). When the radio luminosity is low, the opening angle of the obscuration may be so small that it is no wider than the beaming angle for its relativistic jet. Any unobscured view of an FR1 nucleus would then show only a BL Lac object.

13   Properties of AGN Host Galaxies

AGNs do live, after all, in galaxies, and so it is a very natural question to ask what effect the character of the host galaxy has on the nature of the nucleus, and vice versa. One might easily imagine that the properties of the host influence which variety of AGN is created at its center, its luminosity, or even whether an active nucleus may exist in that galaxy at all. At the same time, it is equally easy to imagine that an object at the center of a galaxy that irradiates its host with a luminosity comparable to or greater than that of all its stars, and possibly also pushes a rapidly moving jet of plasma through the host's interstellar medium, could greatly alter the course of the host galaxy's evolution.

Because galactic evolution is extremely complex, we are far from being able to analyze any of these possible host-nucleus interactions on the basis of first principles. Our primary tools for investigating these questions are therefore empirical. We can measure a variety of nuclear properties (AGN type, luminosity in various bands, etc.) and attempt to relate them to measurable properties of the host galaxy (morphology, luminosity, size, rotation velocity or dispersion, gas mass, etc.). The host's environment (nearby galaxy density, presence of close companions) may also be relevant. In this chapter we will discuss the issues involved in constructing an experiment to look for connections between active nuclei and their hosts.

13.1   Methodology

There are two sorts of methodological issues (one having to do with imaging, the other having to do with how to construct appropriate statistical samples) that, while applicable to many areas in astronomy, rise to particular prominence here. So that we can evaluate the quality of proposed answers to the questions raised in the previous paragraphs, we first digress to consider these methodological issues.

13.1.1   Imaging the host

If we are to study host galaxies, then we must be able to separate them from their active nuclei. Plate 1 is a pretty picture of an AGN embedded in a bright, well resolved galaxy. Unfortunately, for many AGNs it is very difficult to make such a picture. As was briefly discussed in §1.2.1, in many cases, particularly the more distant ones, numerous difficulties stand in the way of successfully imaging the host galaxy. Moreover, although they are most easily described in terms of images, all of these problems apply equally strongly to attempts to obtain the host's spectrum.

First, the angular scales are often not much bigger than an atmospheric seeing disk. At an angular diameter distance D_A (see Appendix F), the physical scale in an image is $19.4(D_A H_o/c)(h/0.75)^{-1}$ kpc arcsecond^{-1}. Now, as figure 13.1 shows, in any cosmology with $\Omega_M > 0$ and $\Omega_\Lambda = 0$, as the redshift increases, the angular diameter distance rises to a maximum (which is always $< 0.5c/H_o$), and declines at larger redshift. Thus, imaging of host galaxies is significantly hampered by seeing once the redshift exceeds ~ 0.03.

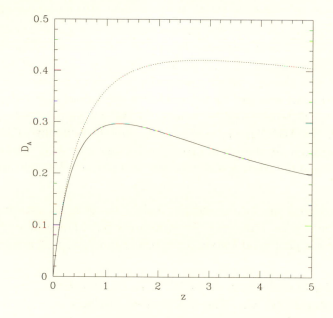

Fig. 13.1   Angular diameter as a function of redshift z in units of c/H_o. The solid curve corresponds to $\Omega_M = 1.0$ and $\Omega_\Lambda = 0$; the dotted curve to $\Omega_M = 0.1$, $\Omega_\Lambda = 0$.

Moreover, this problem is exacerbated by two others. First, the host's bolometric surface brightness is dimmed by cosmological expansion $\propto (1+z)^{-4}$ (eq. F.26). Consequently, for clear imaging of distant galaxies, a very dark background (and an ability to subtract it accurately) are necessary.

Moreover, even the flux from an unresolved source decreases rapidly with increasing z; if $F_\nu \propto \nu^{-\alpha}$, $\nu F_\nu \propto (1+z)^{-3-\alpha}$.

Second, there is the problem of dynamic range. Frequently the AGN is far more luminous than the inner part of its host and sometimes outshines even the integrated light of its host by a large factor. Fiducial numbers for angular resolution generally refer to a characteristic width θ_* that describes the central peak of the point spread function $I(\theta)$, but when a very bright point source lies close to a much fainter object, one must also worry about how much of the point source's light is distributed over the wings of the point spread function. In diffraction-limited images produced by a circular aperture of radius a in light of wavelength λ, $I(\theta) = [2J_1(x)/x]^2$, where J_1 is the first-order Bessel function and $x = 2\pi\theta a/\lambda$ (Born and Wolf 1980). For $x \gg 1$, $I \simeq 4/(\pi x^3)$. Thus, even in a diffraction-limited image, although half the light is contained within an angle $\theta \lesssim (1/4)(\lambda/a)$, there is still 1% outside $\simeq 10\lambda/a$. Because quasars can easily produce 100 times the luminosity of their hosts, light in the far wings of the point spread function can be very significant. Similar problems obtain in images whose quality is controlled by seeing, where I generically declines as a Gaussian out to a few times θ_*, but switches to diffraction-limited behavior at larger angles (King 1971). If the point spread function is well known, it may be possible to subtract the AGN light by model-fitting, but the reliability of the result depends strongly on the luminosity ratio between the AGN and its host, the accuracy of the point spread function, and the signal/noise ratio of the data.

These effects are further complicated by the possible presence of extinction due to dust in the host. An awkwardly placed dust lane can create a spurious morphological feature, or hide a real one, as well as tilt a spectrum.

In order to study the hosts of AGNs, one must therefore exploit whatever advantages one can. One mechanism, of course, is to restrict the study to relatively nearby galaxies, whose larger angular sizes and surface brightnesses make it relatively easy to separate galactic light from nuclear light. For obvious reasons, this is not altogether satisfactory. Similarly, one might also limit oneself to AGNs whose nuclei are obscured, such as type 2 Seyfert galaxies and radio galaxies. However, this alternative introduces two other disadvantages—it is hard to say whether the hidden AGNs really are equivalent to the visible ones, and it is very hard to find type 2 radio-quiet AGNs at large redshift. Yet another method is to focus on wavelengths where the nucleus/galaxy luminosity contrast is not so unfavorable, such as the near-infrared (the effects of extinction are also reduced by moving from the optical into the infrared). As we have seen, there is often a local minimum in the spectra of AGNs near wavelengths of a few microns, but many galaxies are quite bright there. Finally, there is the technical solution—to use an optical system with the greatest achievable angular resolution, a well known point spread function, and a detector with low background and a response that remains linear over a very wide dynamic range. These properties maximize one's ability to subtract the nuclear point source accurately.

13.1.2 Sample definition

To establish any empirical connection we must construct samples and analyze them statistically. Most astronomical samples are, in one way or another, flux-limited. Some of the special problems raised by working with flux-limited samples are discussed in Appendix D. In this subsection we concentrate on how to construct appropriate samples for studying galaxy-nucleus interactions.

Suppose first that we can be confident that any possible connection between nuclear property X and host property Y is completely independent of anything else going on in the active nucleus or its host. Then to construct an ideal sample in which to search for a statistical connection between them, we would optimize three different characteristics: the sample size, its effective signal/noise ratio, and its bias. The ideal sample size is one large enough to avoid small number fluctuations, but not so big that it consumes unaffordable amounts of telescope time, human time, and other resources. The effective signal/noise ratio is the ratio between the range of one of the variables found in the sample and its measurement error; ideally we would like to maximize this for both X and Y. Finally, any bias should, of course, be minimized. In this context that means avoiding any selection criteria that might create an implicit link between the values of X and Y in the sample's members.

Now, if we are looking at active nuclei and their hosts, we have only two choices for how to begin sample selection: we can initially select on a nuclear property, or we can initially select on the basis of a galaxy property. We'll call the first choice Method A and the second Method G.

In Method A, we might prepare our initial list based on some purely nuclear property such as hard X-ray flux or variability, and then measure the distribution of properties X and Y. It is not necessary to avoid selecting on the basis of X, provided there is no connection to the value of Y. However, in an X-flux-limited sample, most of the members will tend to have values of X near the minimum, and so such a sample's effective signal/noise ratio in X might not be very great. The most common problem with Method A samples is that because they tend to be selected on the basis of nuclear brightness, there may be many objects in which we find only an upper bound for Y, or it may even be undefined. For example, in most quasar samples, there are at best upper bounds for the hosts' optical fluxes, and the hosts' Hubble types are completely unknown.

Selection in Method G might use an indubitably host property such as far-infrared flux, or the host's Hubble type. With regard to selecting on the basis of Y flux, the same comments that we made in the previous paragraph with regard to X-flux-selection apply—it is not necessarily to be avoided, but it may be limiting. Similarly, because AGNs can be found in only a minority of galaxies, in most of such a sample we will find at best an upper bound for X, and often we obtain no information at all. Radio flux, for example, can be bounded, but radio morphology is completely undefined if there is no detectable radio flux to map.

The problem gets a bit more complicated when we recognize the possible influence of other factors. The conceptually clearest way to deal with this complication (and also some other issues of sample selection; see Appendix D) is to work with the multivariate probability distribution $p(X, X', X'', \ldots; Y, Y', Y'', \ldots)$, where X', X'', ... are the other nuclear properties of interest, and Y', Y'', ... are the other host properties. This probability distribution is the probability density per unit spatial volume and per unit volume in multidimensional parameter space of finding an AGN with that particular list of characteristics.

In terms of this probability density, we can search for a special tie between X and Y in either of two ways: by taking a cut (or cuts) in this multi-dimensional space at one or several sets of fixed values for all the other parameters; or by integrating over all the other variables to examine only the bivariate probability density $p(X, Y)$. In either case, if there is a relationship between X and Y it will show itself as a non-uniform probability density in the X–Y plane. To the degree that there is a definite functional relation between X and Y that is independent of other variables, the probability density will be confined to a narrow band in that plane.

Which variables we plan to fix, and which to integrate over, has implications for whether Method A or Method G works best. If we wish to fix any of the AGN variables such as X', etc., it is best to build that cut into the original selection. This choice therefore entails use of Method A.

Conversely, fixing any of the host variables Y', etc. is best done through Method G.

13.2 Host Morphology

One of the first questions we might ask about AGN hosts is whether there is anything unusual about their morphology. After all, one of the basic questions about AGN evolution is whether they occur only in certain special galaxies, or can be ignited in any galaxy, but last for only a fraction of a Hubble time (§3.5). Beyond that question, we might wish to hunt for clues in the morphology of the host that could tell us about the dynamics that funnel accretion fuel into the nucleus.

13.2.1 Sample selection

The first step in attempting to answer the question, "Is there a connection between host morphology and the existence or nature of an AGN?" is to construct an appropriate sample of galaxies. There are many possible realizations of Method A suitable to this investigation. X-ray flux, for example, is an efficient way of finding radio-quiet AGNs (although it does miss "type 2" objects), whereas radio flux obviously selects for radio-loud AGNs.

The most natural way to implement Method G in this context is to construct a galaxy catalog selected on the basis of host galaxy flux and study the subset in which AGNs are found. This method is free of morphological bias to the degree that the luminosity functions (and distributions of all other properties affecting selection, such as surface brightness) for different Hubble types are identical in shape. Whether this is true is controversial. Analysis of the Center for Astrophysics redshift survey (§2.2), an example of this technique, suggests that the luminosity function of galaxies depends at most weakly on morphology: In the band of the survey, elliptical galaxies are on average about 0.5 mag brighter than spirals, and the fraction of very faint galaxies is significantly higher among Sm-Im galaxies than for the other types, but the shapes of the luminosity functions for all the ordinary spiral types are very similar (Marzke et al. 1994). On the other hand, Binggeli, Sandage, and Tammann (1988) maintain that there is a clear offset between the different varieties of spirals, with Sa and Sb galaxies on average $\simeq 1.5$ mag brighter than Sc spirals, and the mean Sc $\simeq 1.5$ mag brighter than the mean Sd or Sm.

13.2.2   Host morphology and existence of an AGN

Assuming we have been able to assemble a suitable sample, the next step in the study of a possible correlation between host morphology and character (or existence) of an AGN is to contrast the distribution of AGN host morphology with the analogous distribution for normal galaxies. For the purpose of describing Hubble-type distributions, it is useful to define a numerical variable that maps to morphological type. The standard coding defined by de Vaucouleurs, de Vaucouleurs, and Corwin (1976) is displayed in table 13.1.

Table 13.1: Morphological Type and the T Parameter

cE	E	cD	SO$^-$	SO	SO$^+$	SO/a	Sa	Sab	Sb	Sbc	Sc	Scd	Sd	Sdm	Sm	Im
−6	−5	−4	−3	−2	−1	0	1	2	3	4	5	6	7	8	9	10

Because only those surveys that reach out to very great distance achieve fair samples of the Universal galaxy morphology distribution, the comparison of AGN host morphologies to normal galaxy morphologies is done most cleanly with an AGN sample selected from the same region as the normal galaxies. The RSA sample of Ho et al. (1997a) is a good example of such a survey, even though it found mostly rather low-luminosity AGNs.

Using that sample, Ho et al. found that most AGNs are in early-type disk galaxies. By comparing the distribution of AGN host Hubble types to the distribution of Hubble types for all galaxies in the survey, they also found that the probability of supporting an AGN is greater in this same range of early-type disk galaxies than in other galaxy varieties (fig. 13.2). Because only a small minority of AGNs are radio-loud, essentially all the AGNs in this sample are radio-quiet.

Farther away, and in AGNs of higher luminosity, the situation is not so clear. Part of the problem is that the very distance of these AGNs makes the morphology of their hosts harder to perceive. A first suite of *HST* observations of low-redshift quasars and their hosts (shown in plate 9) illustrates

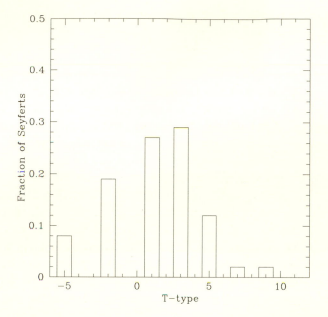

Fig. 13.2   The fraction of a given Hubble type containing a Seyfert nucleus. Unclassifiable and peculiar hosts have been omitted. Data are from Ho et al. (1997a).

this point. In no case is the host image clear enough to show much detail of its morphology. Nonetheless, there is enough information in these images to demonstrate that there is considerable diversity in the character of low-redshift quasar host galaxies. In some cases they are normal-appearing spirals and ellipticals; in others, they are strongly disturbed, or interacting with other galaxies.

13.2.3   Host morphology and radio loudness

As we have seen, the relative strength of radio emission is a fundamental divide between the different AGN categories (§1.3, Chap. 12). This division in AGN phenomenology is strikingly replicated in a division between the sorts of hosts containing radio-loud and radio-quiet AGNs: radio-loud AGNs are associated with elliptical galaxies, whereas radio-quiet AGNs are found predominantly in disks (as first noticed by Adams 1977).

Because radio-loud AGNs are rare, only the very largest AGN catalogs selected without regard to radio flux contain objects spanning a wide range of radio strength. The overwhelming majority of objects found that way are radio-quiet. Consequently, if we wish to study the dependence of relative radio power on host morphology, we need to divide the problem into two parts, one study directed specifically at radio-loud AGNs, the other directed at all the rest.

Method A is automatically entailed for the first study, of course, because radio flux is itself a nuclear property. In one such effort, Martel et al. (1998) took *HST* R-band images of nearly all the extragalactic 3C radio sources, some 273 galaxies. To study their morphology distribution, it is best to consider only those at relatively small redshift: on the one hand, only for these is the image quality good enough to permit reliable determination of the morphology; and on the other, this restriction also eliminates any effects due to cosmological evolution (of either the population of radio galaxies or the nature of their hosts). Another advantage of using the 3C catalog is that its members were found on the basis of their low-frequency flux, an emission component that should be very nearly isotropic. The result of this survey is very clear: the overwhelming majority, more than 80%, are in elliptical galaxies, and the remainder have hosts whose morphologies are, for one reason or another, difficult to determine. A sampling of 3C host morphologies is displayed in plate 10.

It is hard to imagine a selection effect that would exclude spiral galaxies from this survey, and yet they are entirely absent. Because 3/4 of all galaxies are spirals, this is a very strong (and surprising) result. In the language of §13.1.2, let us define the bivariate probability distribution $p(L_r/L_{bol}, T)$ with T the standard morphological type displayed in table 13.1. We need make no "effective volume" correction for morphological type because all varieties could be identified in these images. Put in this language, there is a clear peak in $p(L_r/L_{bol}, T)$ for values of T squarely in the range of elliptical morphology when L_r/L_{bol} is in the radio-loud range (i.e., $L_r/L_{bol} \sim 10^{-3}$ at $\nu_r = 5$ GHz).

Next we turn to the opposite end of the relative radio power spectrum, the radio-quiet AGNs. These are efficiently discovered either in the course of general purpose AGN surveys, or as a by-product of ordinary galaxy surveys. However, there is a difficulty in comparing the distribution of host morphologies for radio-quiet AGNs with that of radio-loud objects: The rarity of radio-loud AGNs means that to accumulate a reasonable sample we need to look out to redshifts of a few tenths, where the AGNs we discover have intrinsically very large luminosities. Even though the bolometric luminosities of these AGNs are very high, most are obscured (§12.7), so that in the majority, there is no bright nucleus to overwhelm images of the host galaxy. However, it is hard to find radio-quiet AGNs of this bolometric luminosity when they are obscured, while unobscured AGNs this luminous far outshine their hosts. Consequently, most studies of radio-quiet AGN host galaxy morphology have concentrated on lower luminosity objects. In these AGNs, intrinsically lower optical luminosity and obscuration combine with better spatial resolution (because they are nearer) to permit relatively easy determination of the host morphology. Thus, the best studies of radio-quiet AGN host morphology are not directly comparable to studies of radio-loud host morphology because the mean bolometric luminosities of the two groups are offset by at least one, and sometimes several, orders of magnitude.

Nonetheless, a good idea of the morphological distribution of the host galaxies of radio-quiet AGNs can be gleaned from the Seyfert galaxies discovered in the CfA redshift survey (Huchra and Burg 1992), a group whose bolometric luminosities are merely a factor of 10 or so smaller than those of the low-redshift 3C radio galaxies. This sample is, of course, an example of Method G. These AGNs are largely in early-type (S0–Sb) disk galaxies (fig. 13.3). To the extent that the redshift survey itself did not select in a way biased with respect to Hubble type, this is a fair representation of the distribution of host Hubble types for Seyfert galaxies. It is starkly different from that of the radio galaxies, which would peak in the range $-5 \leq T \leq 0$.

13.2.4   Fine points of host morphology: bars and other disturbances

Having established the zeroth-order morphological classification distribution for AGNs, it is also of interest to ask whether any deviations from "classical" morphology can be correlated with nuclear activity. As discussed at greater length in §14.3.1, it is possible, for example, that nonaxisymmetric structures in the host's gravitational potential create the torque that drives accretion fuel inward. If replicated in the starlight of the host, features of this sort might present an easily legible signature of activity in the nucleus. Because radio-loud and radio-quiet AGNs are almost entirely found in elliptical and disk galaxies, respectively, and the ordinary shapes of these two galaxy classes are quite different, it is best to treat searches for abnormal morphology in the two varieties separately.

Even in inactive disk galaxies, there is quite a range of possible morphology. Spirals may be more or less tightly wound, more or less prominent in different kinds of light, and so on. As a result, it is difficult to define the normal range of variation, much less departures from normal behavior. Consequently, despite the obvious interest of the subject, it is very difficult to establish robust quantitative results.

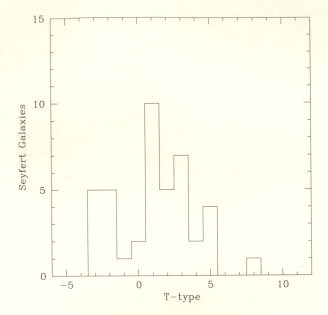

Fig. 13.3 The distribution of morphological types for the CfA Seyfert sample (data from McLeod and Rieke 1995). Eleven unclassifiable and peculiar hosts have been omitted.

For example, there is a long history of claims and counterclaims that AGNs are preferentially located (or absent) in barred galaxies (e.g., Simkin, Su, and Schwarz 1980; Mackenty 1990). Some of this debate has been fueled by theoretical speculation that bars may be responsible for driving interstellar gas toward the centers of galaxies with active nuclei (§14.4.3). To decide definitively whether bars are especially common in AGN hosts, numerous difficulties must be overcome. For a start, quantitatively defining when a "bar" exists is a subtle matter: Exactly what shapes are included? How large must they be? How much contrast must there be between the bar and the smooth disk? In addition, serious biases are introduced by the wavelength of observation. Bars are phenomena involving the gravitational potential of a galaxy. The best image of a bar is therefore made in the frequency band that most closely traces the mass distribution. This is usually the near-infrared. Unfortunately, most galaxy morphological classifications are made from blue light images. Because relatively small numbers of hot stars can greatly alter the appearance of a galaxy at short wavelengths, there is great uncertainty in the actual incidence of bars in the population of normal galaxies, not to mention among those hosting active nuclei. For all these reasons, this is an area in which clear-cut answers are very hard to come by. However, the most careful recent work (Ho, Filippenko, and Sargent 1997b; Mulchaey and Regan 1997) indicates that, at least, for low-luminosity AGNs, the existence of a bar seems to be uncorrelated with the existence of an AGN.

Similarly, there are many suggestions that a large fraction of quasar hosts are interacting with neighboring galaxies (e.g., Hutchings, Crampton, and Campbell 1984). Once again, it is a difficult matter to quantitatively define what constitutes evidence for interaction. "Tidal tails," long, thin structures extending out from the main body of the galaxy are a strong indication that the galaxy has been recently damaged by a transient gravitational tide (Toomre and Toomre 1972). However, particularly for high-redshift AGNs, the available imaging data (e.g., plate 9) are often too indistinct to unambiguously identify such features, although there are a few cases (e.g., PG 1012+008 in plate 9) in which the symptoms of interaction are quite clear. In an attempt to cope with this problem, substitute criteria, easier to recognize in blurry images, have been used instead of true tidal tails. These include low surface brightness "bridges" between the host and another galaxy, irregularities that appear with common orientation in both galaxies, or simple deviations from smoothness in the host's isophotes. To what degree these surrogate symptoms reliably signal real dynamical disturbance remains to be solidly established. Given the complexity of the structure(s) that might arise from a genuine interaction (see §14.4.4), this is not an easy job.

On the other hand, the appearance of normal elliptical galaxies is rather simpler, so deviations are easier to detect. "Normal" ellipticals have radial surface brightness profiles well approximated by the de Vaucouleurs "$r^{1/4}$ law," that is, $S \propto \exp[-(r/r_e)^{1/4}]$, where the effective radius r_e is conventionally chosen to be the (projected) radius within which half of the light is contained. With that definition, $S(r) = S(r_e) \exp\{-7.67[(r/r_e)^{1/4} - 1]\}$. In addition, they exhibit a strong correlation between total luminosity L_{tot} (or equivalently r_e), velocity dispersion σ, and central surface brightness S_o: $L \propto \sigma^{3.45} S_o^{-0.86}$ (Dressler et al. 1987; Djorgovski and Davis 1987). This relationship has become known as the "fundamental plane" for elliptical galaxies.

However, even among inactive ellipticals there are counterexamples. Some ellipticals have surface brightness profiles that are shallower than the $r^{1/4}$ form; the most distended of these are called cD galaxies, which are almost always located in the centers of rich galaxy clusters. On the other hand, there are also ellipticals whose stellar surface brightness is more sharply peaked at the center than predicted by the $r^{1/4}$ law; in the most extreme cases, the light from the central cusp can dominate the total output of the galaxy.

As measured with respect to this axis (peaky profiles vs. shallow profiles), the hosts of radio-loud AGNs depart strongly from normal elliptical galaxies and do so in a way that is correlated with their radio morphology. Roughly a quarter of FR1 radio galaxies are cD's versus less than 1% of all ellipticals. By contrast, few FR2 radio galaxies are cD's, but 40% are considered nuclear-dominated galaxies, a category at least as rare as cD's. In evaluating the prevalence of bright nuclei in the hosts of FR2s, it is important, however, to recall that some of those classified as nuclear-dominated may acquire that label because of the AGN contribution (cf. §12.7). This uncertainty is, of course, yet another realization of the angular resolution problem (§13.1.1).

More radio galaxies than normal ellipticals also seem to show disturbances, but it is harder to be quantitative, and the variety of disturbance seems to vary according to Fanaroff-Riley class. For example, roughly a quarter of all FR1s have double nuclei, while a similar fraction of FR2s show other sorts of irregularities in their surface brightness (Zirbel 1996b).

13.3   Host Luminosity

It is natural to ask whether there is any tie between the host luminosity L_h and the luminosity L_n. In a simple-minded way, one might expect that a bigger galaxy could provide more fuel and therefore feed a brighter active nucleus. For either the host or the nucleus, this question might be raised with respect to the luminosity in a particular band (blue light, X-rays, radio ...) or the bolometric luminosity. No matter which sort of luminosity is at issue, the best way to answer this question is to measure the bivariate probability function $p(L_n, L_h)$.

In addition to the usual problems of such an enterprise (§3.3, Appendix D), here there is a special one to overcome: how to separate cleanly the nuclear from the host luminosity (§13.1.1). Because high-luminosity AGNs dramatically outshine their hosts, it is only possible to measure the host luminosity in high-luminosity AGNs when very high angular resolution imaging data are available. Conversely, at the low-luminosity end, considerable care must be taken when measuring the nuclear optical or ultraviolet luminosity to distinguish truly nuclear light from stellar light, even when the aperture used is small compared to the size of the galaxy.

Two approaches may be taken to solve this problem. One is to use an imaging system with the greatest possible angular resolution. From that data, one can then attempt to separate the nuclear light from the total by fitting a point-source contribution to the galaxy's image. This method suffers from the obvious drawback of depending on the resolution of the image (and the quality of one's knowledge of the point spread function), problems that are particularly acute if there is a high surface brightness stellar region close to the galaxy's center.

The second approach is to measure the luminosity of the nucleus in a band where it dominates, and the luminosity of the host in a band where it is strongest. For example, one might measure the hard X-ray luminosity of the nucleus and the near-infrared luminosity of the host. However, this device does not entirely solve the problem, for the hard X-ray (or radio or ...) luminosity may not be a fair gauge of the nuclear bolometric luminosity, while the nucleus may contribute some light (or in the most luminous cases, continue to dominate the total) even in the near-infrared.

Because neither of these approaches is foolproof, this is also an area in which few strong results have been obtained. Although indications have been found that there may be correlations between the near-infrared luminosity of the host and the optical luminosity of its active nucleus (e.g., McLeod and Rieke 1994; Kotilainen and Ward 1994), this topic is a very controversial one.

The search for a correlation between radio luminosity and host optical luminosity is a happy exception to many of these difficulties. On the one hand, the AGN radio luminosity, except in the weakest cases, nearly always greatly dominates any intrinsic host radio luminosity; on the other, there are numerous AGNs with measurable radio power but either relatively faint nuclear optical emission, or even essentially none at all (§12.7).

Exploitation of these advantages has led to the discovery of a number of intriguing correlations. As we have already mentioned (§9.1.2), the radio power division between FR1 and FR2 radio galaxies depends on the optical luminosity of the host galaxy. Because the luminosity of the dividing line increases in proportion to the R-band luminosity squared, the mean optical luminosity of the hosts to FR1 sources is rather greater than the optical power of FR2 sources with the same radio luminosity. Moreover, the probability that an elliptical galaxy is a radio source increases with increasing optical luminosity up to a maximum $\simeq 1/3$, reached when $L_R \geq 10^{44}$ erg s^{-1} (Ledlow and Owen 1996; Fig 13.4). Reflecting this fact, powerful radio galaxies (i.e., FR2 radio galaxies and the more luminous FR1s), although associated with hosts having a wide range of optical luminosities, are on average brighter than typical elliptical galaxies, with a mean luminosity comparable to the brightest ellipticals in loose groups (Zirbel 1996a).

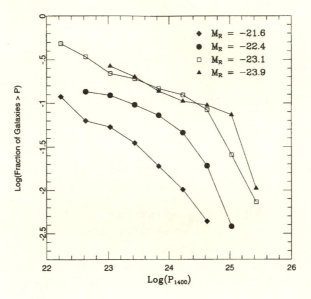

Fig. 13.4   Fraction of elliptical galaxies with a radio source brighter than a given luminosity (at 1.4 GHz) for several different optical luminosities (Ledlow and Owen 1996). The absolute optical magnitudes are in the R band.

These radio power correlations can also be extended across the radio-loud–radio-quiet divide. Nelson and Whittle (1996) found an apparent correlation between the core radio luminosity and the optical luminosity of the host's bulge (fig. 13.5). That is, just as brighter elliptical galaxies are more likely to create brighter radio sources, so too do brighter bulge components of disk galaxies (the portion of disk galaxies most similar to elliptical galaxies) seem to engender brighter radio sources, albeit on a much smaller scale.

Although this conclusion is based on a luminosity-luminosity plot, it is unlikely to suffer the more common ills of this sort of correlation. Relatively little in the way of effective volume correction is required because the Seyferts were selected in a way wholly independent of either radio or

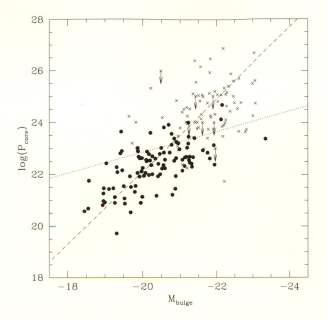

Fig. 13.5 Radio core luminosity (in $W\,Hz^{-1}$) versus absolute B magnitude of the host's bulge for samples of Seyfert galaxies (filled circles) and radio galaxies (crosses). Data are from Nelson and Whittle (1996) and Zirbel and Baum (1995). Although there is no apparent correlation for the radio galaxies taken alone, there is a weak correlation for the Seyfert galaxies by themselves, and a much stronger one for the two populations combined. Note that although this plot does not make any effective volume correction, the apparent correlation is much steeper than the one (shown by the dotted line) that would be induced by a spurious distance-distance correlation—the slope of the dashed line is $\partial\ln L_{\mathrm{radio,core}}/\partial\ln L_{\mathrm{opt}} \simeq 3.5$. The effect is therefore likely to be real, although this plot may not give a picture that is quantitatively accurate.

bulge luminosity, and every Seyfert in the sample has a measured value for each of these quantities. Similarly, the radio galaxies were selected on the basis of low-frequency extended radio flux, not 1.4 GHz core flux, yet in only a few cases was no core radio emission detected.

13.4   Host Interstellar Medium

Somewhere in the host there must be a reservoir of accretion fuel from which the nucleus draws its supply. A natural place to look for it is the host's interstellar gas. One might imagine, for example, that those galaxies with active nuclei are those with unusually large amounts of interstellar gas. To test that idea, one could look for correlations between the mass of gas in the host, or its relative proportion, and the luminosity of the nucleus.

13.4.1 Signals

As always, there are complications in the actual performance of this experiment. To start, what is the best measure of the total gas mass in the host? H may exist as H_2, atomic H, or ionized H^+, and each of these has its own spectroscopic signature.

13.4.1.1 the 21 cm line

The strength of the H$_I$ 21 cm hyperfine line can give a very direct measure of the associated mass in atomic H. Because this transition involves only a very small amount of energy, for almost any conceivable excitation temperature the population ratio between its upper and lower states is almost exactly the ratio of statistical weights, $g_u/g_l = 3$. Moreover, because the matrix element linking the two states is very small (the spontaneous decay rate A_{21} is only 2.85×10^{-15} s^{-1}), the photon absorption cross section σ_{21} is also very small, only $1.42 \times 10^{-18} T_4^{-1/2}$ cm^2 at line center when the velocity distribution is thermal. Therefore, when the column density is not too large, the luminosity per unit area that we see is simply

$$\frac{L_{21}}{A} = \frac{3}{4} N_{HI} A_{21} h\nu_{21} = 2.70 \times 10^{-32} N_{HI} \text{ erg s}^{-1} \text{ cm}^{-2}, \qquad (13.1)$$

where N_{HI} is the column density of H atoms in units of cm^{-2}.

Unfortunately, the measurement of this flux is complicated when there is a strong radio continuum source within the telescope beam. This is the case, of course, whenever we are interested in radio-loud AGNs. The difficulty is due to the fact that radio spectra are synthesized from finite duration integrations of the signal. The signal autocorrelation functions for each integration are then Fourier transformed to find an estimate of the power spectrum. However, a square wave window function of width ΔT in the time domain does not produce a square wave in the frequency domain— rather, it produces a central peak whose width is $\sim (\Delta T)^{-1}$ accompanied by subpeaks spaced every $(\Delta T)^{-1}$ apart whose amplitudes fall $\propto (\Delta f)^{-2}$. In other words, the frequency bins have sizable side lobes. These side lobes mix power from different regions in frequency space; when the continuum is strong relative to the line emission, the measured spectrum that results is relatively smooth (though curved) in the middle of the frequency band but becomes increasingly bumpy toward the edges; this is a realization of the well known Gibbs phenomenon. If the line emission does not stand up well above the continuum, it can easily be lost in this "ringing."

Self-absorption of the HI hyperfine transition sets in when the optical depth

$$\tau_{21} = \sigma_{21} N_{\mathrm{HI}} \left[1 - \exp\left(-\frac{h\nu_{21}}{k_B T_{21}} \right) \right]$$

$$\simeq \sigma_{21} N_{\mathrm{HI}} \frac{h\nu_{21}}{k_B T_{21}}$$

(13.2)

becomes significant. Here T_{21} is the excitation temperature of the transition. For the same reason that the populations are very nearly in statistical balance, the stimulated emission correction is always very important, and the optical depth is $\propto T_{21}^{-1}$. T_{21} is usually very close to the kinetic temperature of the gas, but intense background radiation at either 21 cm or Lyα can also influence it (Field 1958). When the atomic velocity distribution is thermal, and the excitation temperature is equal to the gas kinetic temperature,

$$\tau_{21} = 9.6 \times 10^{-21} T_{100}^{-3/2} N_{\mathrm{HI}},$$

(13.3)

where T_{100} is the gas temperature in units of 100 K. Thus, substantial column densities are needed to produce self-absorption, particularly if the velocity distribution is broader than thermal.

However, such large column densities do sometimes occur. When they are present, and the atomic H gas itself is the only source of radiation at 21 cm, the brightness temperature saturates at the gas temperature. In such a case, the 21 cm flux underestimates the true mass of atomic H. On the other hand, when there is a compact radio source in the nucleus (and it is bright enough), 21 cm *absorption* along the line-of-sight can be a powerful probe of the host's interstellar medium. For these measurements, the quantity of interest is not the radio flux, but the optical depth in the line. If there is some way to estimate the gas temperature, equation 13.3 transforms the measured optical depth into a column density of H atoms along the line-of-sight. In addition, galactic orbital speeds are generally several hundred km s^{-1}, much greater than atomic thermal speeds, so the optical depth as a function of projected velocity provides information about the gas's kinematics as well as its mass.

13.4.1.2 line transitions in H_2

The mass of molecular gas is more difficult to measure. Because H_2 is a symmetric diatomic molecule, it has no electric dipole moment in its equilibrium state. It immediately follows that the matrix element for E1 transitions between different states of nuclear motion (i.e., vibrational or rotational motion) is identically zero. To see H_2 molecules in emission, therefore, one must look at E2 transitions, which generically have very small A coefficients (these are compiled in Turner, Kiby-Docken, and Dalgarno 1977). When they take place within the ground vibrational state, only rotational transitions with $\Delta J = 2$ are permitted for this mechanism; thus, the lowest-lying upper state is $J = 2$. Because H_2 has a relatively small moment of inertia, the energy required to reach this state is comparatively high; in temperature units $\simeq 510$ K. The surface brightness to be expected from the 28 μ $J = 2 \rightarrow 0$ line is then

$$\frac{L_{\rm rot}}{A} = 2.1 \times 10^{-24} \frac{5 e^{-510/T_{\rm rot}}}{Z_{\rm H}} N_{H_2} \ {\rm erg \ s^{-1} \ cm^{-2}}, \qquad (13.4)$$

where $T_{\rm rot}$ is the excitation temperature of the $J = 2 \rightarrow 0$ transition, $Z_{\rm H}$ is the molecular hydrogen partition function, and N_{H_2} is the column density of H_2. Note that the statistical weight of each rotational state with even J (para-H_2) is $2J + 1$, whereas the statistical weight of each odd J (ortho-H_2) state is $3(2J + 1)$. The tiny A coefficient (2.94×10^{-11} s^{-1}) again makes it very likely that $T_{\rm rot}$ is equal to the kinetic temperature of the gas; however, UV fluorescence can perturb the population, particularly for higher J states.

Just as for the 21 cm line, the small matrix element for the H_2 rotational transitions also ensures that they stay optically thin even for relatively large column densities. In this case, the $J = 0 \rightarrow 2$ line center optical depth is

$$\tau_{\rm rot} = 7.9 \times 10^{-25} T_{100}^{-1/2} N_{H_2} \qquad (13.5)$$

for a thermal velocity distribution and $T_{\rm rot} = T \ll 500$ K. This optical depth falls with rising temperature both because of the diminishing fraction of H_2 molecules in the ground state and because of the increasing importance of the stimulated emission correction to the optical depth (which is, of course, also a consequence of greater population in excited states). Thus, self-absorption in H_2 rotational transitions requires column densities roughly 10^4 times greater than does self-absorption in the hyperfine transition.

As a result, inferring the mass of H_2 directly from its own spectrum is fairly difficult. To see it in emission requires sensitivity at 28 μ, a wavelength where the atmosphere is quite opaque. In addition, inference of the column density requires a significant adjustment that depends on the temperature; any signal that one does see comes preferentially from regions where $T \gtrsim 500$ K. On the other hand, very large column densities are necessary in order to create an absorption line against a bright background source.

13.4.1.3 CO and other molecules

Because it is so hard to see H_2 directly, lines from other molecules, which do have permanent electric dipole moments, are used as surrogate indicators of the total gas mass. The most commonly used of these is the $J = 1 \rightarrow 0$ transition of CO. This line (at 2.6 mm or 115 GHz) has a comparatively large A coefficient, 7.17×10^{-8} s^{-1}. Its optically thin surface brightness is found in a way directly analogous to that for H_2 transitions (eq. 13.4) with one technical difference: because the transition energy (equivalent to 5.5 K) is so small compared to any likely temperature, the partition function is well approximated by an integral:

$$Z_{CO} \simeq \int_0^\infty dJ(2J+1) \exp\left[-\frac{J(J+1)E_o}{k_B T}\right]$$
$$\simeq \frac{T}{2.76 \text{ K}},$$
(13.6)

where E_o is half the energy of the $J = 1 \rightarrow 0$ line.

A corollary of larger matrix elements is greater optical depth. For example, in the case of the CO $J = 1 \rightarrow 0$ line,

$$\tau_{CO} = 6.9 \times 10^{-22} N_H T_{100}^{-2} \left(\frac{X_{CO}}{10^{-4}}\right)\left(\frac{\Delta v}{1 \text{ km s}^{-1}}\right)^{-1},$$
(13.7)

where N_H is the column density of H nuclei (in any chemical state), X_{CO} is the fractional abundance of CO relative to H nuclei, and Δv is the velocity spread of the gas. This estimate assumes that the rotational state populations are in thermal equilibrium.

As we discussed in the context of HI 21 cm emission, optically thick lines are ordinarily not the most suitable for measuring the column density of the associated atom or molecule; in fact, the intensity of an optically thick line is usually at most a very weak function of the column density (i.e., the flat part of the curve of growth extends over a wide range of optical depths). Because H column densities in excess of 10^{21} cm^{-2} are, if anything, the norm in molecular regions, saturation by self-absorption is an important effect for the CO $J = 1 \rightarrow 0$ line.

Before discussing how to use optically thick emission lines to estimate the column density of molecular gas, it is necessary to digress for a moment to explain the peculiar units in which these quantities are usually quoted. Because radio astronomers frequently measure intensity in terms of brightness temperature and (relative) frequency in velocity units, the frequency-integrated intensity observed in molecular emission lines is often given in units of K km s^{-1}. Assuming isotropic emission, the conversion factor to luminosity per unit area is then

$$\frac{L}{A} = 3.48 \times 10^{-7} T v \lambda_{mm}^{-3} \text{ erg s}^{-1} \text{ cm}^{-2}, \qquad (13.8)$$

where T is measured in K, v in km s^{-1}, and the wavelength λ of the line in mm. This unit also has a natural physical interpretation when the line is optically thick: The intensity of such a line is Planckian at its excitation temperature across the optically thick bandwidth. Viewed in this light, the conventional units are quite suitable. A measurement of the line width then immediately yields an estimate of the transition's excitation temperature. In most instances (see below), the excitation temperature is very close to the kinetic temperature of the gas.

Surprisingly, it has been found empirically that in our own Galaxy there is a good correlation between the intensity of the CO $J = 1 \to 0$ line and other measures of the total gas column density (e.g., reddening, or the column density of molecules such as ^{13}CO with optically thin transitions). The proportionality coefficient is uncertain at the level of a factor of two but is $\simeq 3 \times 10^{20}$ cm^{-2} K^{-1} (km s^{-1})$^{-1}$ (Scoville and Sanders 1986).

In our Galaxy, the success of this correlation may be due to the fact that many molecular clouds are self-gravitating. In that case, the line width v relevant to equation 13.8 is proportional to $N_{\text{H}} \rho^{-1/2}$, where ρ is the gas density. If $T \propto \rho^{1/2}$, we would then find that L/A for optically thick lines like CO $J = 1 \to J = 0$ is $\propto N_{\text{H}}$. This sort of dependence of temperature on density would not be too surprising if all molecular clouds were formed from gas of similar specific entropy and lost some of their heat by radiation as they were compressed. Whether molecular clouds in AGN hosts share these properties is very uncertain. Nonetheless, CO luminosity is often the only indicator of molecular gas we have.

When there is a bright compact radio source, the optical depth of CO absorption (eq. 13.7) can also be used to measure the column density of molecular gas. However, it can sometimes be possible that the radio continuum source is bright enough to alter the population balance on the CO rotational ladder (Maloney, Begelman, and Rees 1994). The expression given in equation 13.7 must then be modified in two ways: the fractional population in $J = 0$ changes, and the stimulated emission correction $1 - \exp(-h\nu/k_{\text{B}}T) \simeq h\nu/k_{\text{B}}T$ must also be adjusted appropriately.

13.4.1.4 H recombination lines

Finally, ionized H can be detected through its recombination line emission (§10.3.5.2). The luminosity per unit area in, say, Hβ is given by

$$\frac{L_{\mathrm{H}\beta}}{A} = \int dz\, n_e n_p \alpha_{rec,1,1} \phi_{\mathrm{H}\beta} h\nu_{\mathrm{H}\beta}, \tag{13.9}$$

where $\phi_{\mathrm{H}\beta}$ is the fraction of all recombinations resulting in emission of an Hβ photon (see Osterbrock 1989 for tables of $\phi_{\mathrm{H}\beta}$, and the same quantity for other recombination lines). If the density were constant, the surface brightness in the line would be $\propto n_e N_{\mathrm{HII}}$, that is, it depends on the product of the ionized H column density and the electron density. Thus, to infer the H$\textsc{ii}$ column density we need an estimate of the electron volume density. However, because the emissivity per unit volume depends on the square of the density, more inhomogeneous regions are also more efficient in producing recombination photons, so to infer the H$\textsc{ii}$ column density from the surface brightness an estimate must also be made of the "clumpiness" of the region. Given the powerful ionizing radiation produced by the nucleus in AGNs, it is very likely, of course, that the quantity of ionized gas in the host is strongly influenced by the presence of the nucleus.

13.4.2 Surveys

All three of these techniques have been utilized in attempts to correlate properties of the host's interstellar medium with the nature of the nucleus. Mirabel and Wilson (1984) surveyed 91 Seyfert galaxies in the 21 cm line and found that the ratio between L_{21} and the total blue light luminosity of the galaxy is usually quite compatible with the normal ratio for hosts of the same morphological type, although occasionally there are anomalies in the shape of the line profile. Of course, in those galaxies with particularly luminous type 1 Seyfert nuclei, the luminosity of the nucleus contributes significantly to the total blue light.

Heckman et al. (1989) similarly compared the CO luminosity of a large sample of Seyfert galaxies to the total blue luminosity of the galaxy, normalizing to the distribution of that ratio for normal galaxies of the same Hubble type. They found that the ratio was a few times larger than normal in type 2 Seyfert galaxies, but about normal in the type 1's. They found a similar result comparing the CO luminosity to the 21 cm luminosity, that the ratio L_{CO}/L_{21} was about five times greater in type 2 Seyfert galaxies than in normal spirals of the same Hubble type, while this ratio was about normal in the type 1 galaxies. Applying the Galactic conversion from L_{CO}

to H_2 mass would imply that the mass of H_2 is about equal to the mass of HI in type 2 Seyferts, but, as discussed above, the use of that conversion factor for active galaxies is dubious. It is possible that the discrepancy between the two types is due to a bias in favor of especially luminous type 2 Seyfert nuclei in their sample; if that is so, there may be no distinction in CO luminosity between the two types when compared at fixed bolometric nuclear luminosity and host Hubble type.

Finally, Baum and Heckman and their collaborators (Baum et al. 1988; Baum and Heckman 1989ab) have studied the incidence and kinematics of ionized H in radio galaxies. Compared to elliptical galaxies of similar stellar luminosity, the radio galaxies are typically about 10 times brighter in $H\alpha$, and the linear extent of that emission is similarly about 10 times greater. Because their observations resolve the emission region, they are able to measure its total volume. Assuming the ionized gas is evenly spread throughout the region then gives (a rather generous) lower bound to its density, hence an (equally generous) upper bound to its mass. For the radio galaxies of their sample, this upper bound is typically $\sim 10^8$–10^9 $M_\odot$.

On the basis of these studies, we arrive at the (weak) conclusion that the total gas mass of galaxies with active nuclei is not grossly different from that found in suitably matched inactive galaxies. However, there may be differences in its state (ionized vs. atomic vs. molecular), its kinematics, and its distribution within the host.

13.5   Orientation

Neither galaxies nor AGNs (as Chap. 12 emphasized) are perfectly round. It is therefore an interesting question to ask how their structures are oriented relative to one another. If the matter of the galaxy is the accretion fuel of the AGN, one might expect that any special orientation of the galaxy (e.g., its angular momentum axis) might be communicated to the AGN (determining, e.g., the direction of its black hole's spin, or the direction of any radio jet that might be produced).

How, then, can we measure the relative angle between the galaxy's orientation and the AGN's? The first thing we must do is define what we mean by the galaxy's orientation. For this purpose, rotation is probably the best measure. In most cases it is relatively well defined, although in elliptical galaxies the speed of organized rotation is generally not very large compared to random motion speeds, and there are disk galaxies with a range of rotation axes (i.e., warps). Absent explicit data about velocities within the galaxy, the direction of the galaxy's minor axis is often used as a stand-in for the rotation axis.

The next question is to identify the AGN axis. Here we have a choice. We might define it by the direction of a radio jet, if present; we might also define it by the orientation of an elongated emission line region, if one of these exists; another alternative is to choose the direction perpendicular to the $\vec{E}$-vector of polarization, if that can be measured. There can be ambiguity in these definitions. In some AGNs none of these signatures may be found; in others, several may exist, with discrepant directions; in still others, there may only be one (the radio jet), but it may change direction as a function of distance from the nucleus. Nonetheless, the fact (§12.6.1) that the directions of polarization in Seyfert galaxies are strongly correlated with the radio jet directions in those galaxies is reassuring evidence that these orientations may be related to more fundamental axes, such as that of the spin of the black hole.

When we measure one of these directions, what we find, of course, is not its absolute direction in space, but only its projection onto the sky plane. Consequently, we cannot find the *true* distribution of relative angles. Instead, we can at best test for consistency between proposed true distributions and the observed distribution. In doing so, one should also keep in mind that the host galaxies of AGNs may not be randomly oriented with respect to our line-of-sight, for there can be orientation-dependent selection effects (e.g., extinction in the plane of disk galaxies) that influence our ability to find AGNs or measure their orientation (e.g., by the directions of elongated emission line regions).

Preliminary attempts to explore this issue have already arrived at some surprising results. In particular, it seems that (at least in Seyfert galaxies) there is *little* correlation between the axis of the AGN (as measured by radio jets) and the axis of the host galaxy (Schmitt et al. 1997).

13.6 Evolutionary State

That the AGN phenomenon was so much stronger at $z \simeq 2$–3 than today (§3.3) has long elicited the hunch that there is a connection between the youth of a galaxy and the likelihood that an AGN forms inside it. The question then naturally arises, "What counterparts to young galaxies might exist now, in which local AGNs may live?" One possible answer is a category of galaxies called *starbursts*.

No single definition of the starburst phenomenon exists. Some call a galaxy a "starburst" when the inferred rate of star formation would use up the galaxy's estimated supply of interstellar gas in much less than a Hubble time; some define this state as occurring when the star formation rate is much higher than the galaxy's average rate over the lifetime of the Universe; some associate "starbursting" with a large ratio of far-infrared to optical luminosity; and some only label galaxies as "starbursts" when they show evidence (such as strong Balmer line emission) for an exceptional number of Hii regions or other symptoms of hot young stars. Fortunately, there are enough galaxies that would be identified as starbursts by all of these criteria that the term is meaningful.

Possible affiliation with AGNs comes through several different sorts of ties. One possibility is that an unusually large fraction of mass in gas rather than stars (whether in the entire galaxy, or just the central region) may permit a strong flow of accretion fuel into the nucleus (see Chap. 14). Another is the basic fact of extraordinary luminosity: in the most powerful starburst galaxies (the "ultraluminous infrared galaxies"), the far-infrared luminosity rivals the luminosity of quasars, rising to as much as $\sim 10^{45}$ erg s^{-1}. In addition, these galaxies have a local space density comparable to that of quasars (Soifer et al. 1987).

If we are to believe that starbursts really promote AGN activity, we need an affirmative answer to either of two questions—"Are starburst galaxies more likely to house AGNs than a matched set of galaxies forming stars at a normal rate?" and "Are the host galaxies of AGNs more likely to be in a starburst state than galaxies of the same variety without AGNs?"

Answering either one requires, of course, a quantitative definition of what constitutes a starburst. If we use the definition of extraordinary far-infrared luminosity, we must confront the ambiguity of whether the warm dust in those galaxies is heated by absorbing starlight or AGN radiation. It may be that the ultra-luminous infrared galaxies are merely deeply obscured quasars (Sanders et al. 1988).

Other symptoms are less ambiguous in interpretation. For example, Markarian and associates tagged some 1500 galaxies on the basis of their unusually blue optical light (§2.1). Most owe their blue tint to young, hot stars and in that respect can be considered starbursts (or at least weak cousins to "true" starbursts). About 10% of the Markarian list are Seyfert galaxies; although some of them may owe their presence in the sample to the AGN light alone, this cannot be true for the type 2 Seyferts. The presence of these type 2 Seyferts in the Markarian sample thus suggests another tie between AGNs and starbursts.

There are also several indications that AGNs are more likely to be found in galaxies with starbursts than if their hosts had been chosen without regard to star formation rate. Heckman et al. (1995), for example, suggested that the ultraviolet continua of type 2 Seyfert galaxies were due to a surrounding starburst. In these AGNs, we can see only reflected light from the nucleus, yet their UV continua are moderately strong, and their UV spectra show none of the signs of being reflected nuclear light.

Finally, there is an especially provocative speculation about the role of galaxy interactions in triggering both sorts of events. Essentially *all* ultra-luminous infrared galaxies are partners in an interaction with another galaxy so strong that it has (or will soon) lead to their merger (Sanders et al. 1988). It is often speculated that galaxy interactions may also create AGNs (§14.4.4). If this speculation is correct, starbursts may then be a part of the AGN-creation process (see, e.g., Norman and Scoville 1988).

13.7   Environment

The possible connection of galaxy-galaxy interactions to AGN activity raises the question of whether there is any correlation between the environment of the host galaxy and its proclivity to house an AGN of any particular sort. Making clear inferences, however, is very difficult because the web of variables involved is especially tangled. To understand why, let us imagine designing a model experiment.

Suppose we wish to determine whether the presence of close companions around a galaxy makes it more likely that it develops an active nucleus. Once again, the two generic sample construction methods of §13.1.2 are available. In this context, if we chose Method A, we would first construct a sample of active galaxies in a way that is independent of their galactic environment and measure their distribution of nearby galaxy density. To examine a correlation with respect to fixed host galaxy properties (i.e., take a "cut" in galaxy parameter space), we would then construct a separate sample of normal galaxies whose distribution of properties is chosen to match as nearly as possible the distribution of properties in the host galaxies of the AGN sample. In this sample we would measure how many close companions the galaxies have and contrast those numbers with the ones derived from the AGN sample.

Implementing Method G in this context means that we assemble a sample of galaxies without regard to the existence of active nuclei, but having a range of nearby galaxy density. We could then measure the fraction of galaxies with active nuclei as a function of local galaxy density. Accepting all active nuclei means that we are effectively integrating over the distribution of different AGN properties; accepting only special kinds restricts the search for possible correlations more narrowly.

Both Method A and Method G have been tried, but, as we shall see, it is extremely difficult to implement either one in a way that is entirely satisfactory.

Both methods share a common difficulty: how to measure the true three-dimensional density. Measuring *projected* density on the sky is an operation requiring some care but can be done; to measure the true *volume* density one must obtain the redshifts of all the galaxies projected nearby. That could require a very large number of spectra, and many of the candidate companions may be quite faint. Rather than do this, one might approximate the volume density by subtracting the mean number of galaxies expected to be seen nearby in projection. To see whether the volume density of galaxies near the AGNs is greater than normal, one can then estimate the expected number of galaxies using an expression for the luminosity function of normal galaxies at the redshifts of the AGNs. This last step is subject to large uncertainties when applied at high redshift.

With these comments in mind, consider Method G first. Most galaxy samples are selected on the basis of optical flux. Because many active nuclei have optical luminosities comparable to those of ordinary galaxies, the frequency of nuclear activity is automatically enhanced in any survey selected in this way. If the luminosity of active nuclei depends on the galactic environment, the degree of enhancement will therefore be a function of the character of the galaxies' neighborhoods, and this method will produce a spurious result. Therefore, in order to make this method succeed, the galaxy sample must somehow avoid this bias. Assuming that this problem has been solved, the next obstacle is to build a large enough sample that it contains both a substantial range in local density and enough AGNs across this range that their frequency can be estimated without being subject to the large errors associated with small number statistics.

Next consider Method A. The first step—assembling an AGN sample without environmental bias—is relatively easy. The second step—measuring their distribution of nearby galaxy density—is subject to all the problems just enumerated. The third step—constructing the comparison galaxy sample—is the trickiest of all. We must first determine which properties must be matched to the AGN host galaxies. Morphological type is one that we must certainly be careful about, for there is a strong correlation for all galaxies between morphological type and nearby galaxy density, in the sense that the mix of types changes from predominantly disks at low density to mostly ellipticals and S0's at very high density (Dressler 1980; Postman and Geller 1984; Santiago and Strauss 1992). Because of this correlation, and the extremely strong connection between radio-loud AGNs and elliptical hosts on the one hand and radio-quiet AGNs and disk hosts on the other, studies of the environments of radio-loud and radio-quiet AGNs require almost completely disjoint comparison samples. Host luminosity is another property that can also have its own correlation with environment, for reasons having nothing to do with the processes creating active nuclei. More subtly, the number of companions discovered can be expected to depend on redshift. If "close" is defined by a fixed physical separation, faint companions will be difficult to detect at large redshift, while the angular size subtended at low redshift will be so large that one would be subject to large statisical fluctuations in the number of background galaxies found. On the other hand, if "close" is defined in terms of galaxy angular diameters, while the latter problem is eliminated, the former certainly remains, and in addition the physical distance encompassed scales with redshift.

Having enumerated all these hazards, it should come as no surprise that many contradictory results have been found, and there is much disagreement about whose techniques are best. A few weak conclusions can be stated, however. Given the strong correlation between galaxy morphological type and environment (ellipticals strongly favor groups and clusters, disks prefer low-density surroundings), it should not be surprising that radio-loud and radio-quiet AGNs differ in their environments.

There is a weak tendency for Seyfert galaxies to have more close companions than normal galaxies have box(e.g., Dahari 1984, Laurikainen and Salo 1995), but this correlation may be due to the presence of enhanced star formation in the host, and the more strongly established connection between galaxy interactions and starbursts (Mackenty 1989). There are also hints that quasars (both radio-quiet and radio-loud) favor small group environments at redshifts less than $\simeq 0.5$ (Yee and Green 1984), but at higher redshifts ($z \simeq 1$), there is no signal of any correlation between radio-quiet quasars and neighboring galaxies (Boyle and Couch 1993).

At low redshift, radio-loud objects seem, if anything, to be less associated with clusters of galaxies than normal elliptical galaxies. Moreover, the detailed character of their preferred neighborhoods depends on radio morphology, at least for the more luminous radio galaxies: FR1 sources tend to be in richer groups than FR2, and the groups in which FR1s are found generally have greater luminosity contrast between their brightest member (almost always the host of the FR1) than do the groups in which FR2s are located (Zirbel 1996a). Curiously, earlier in the history of the Universe (by $z \simeq 0.5$), FR2 radio galaxies and radio-loud quasars are often found in clusters (Yee and Green 1987; Hill and Lilly 1991).

13.8 Impact of the Nucleus on the Host

Remarkably, despite the immense outpouring of ionizing photons and relativistic plasma produced by AGNs, they appear to inflict remarkably *little* damage on their host galaxies. In fact, the interstellar gas content of most AGNs is not far from the normal range and deviates, if at all, in the sense of being *richer* (§13.4).

One quantitative measure of the effect that we might have expected an active nucleus to have on the interstellar medium of its host is the ionization parameter one would predict on the basis of the luminosity of the nucleus and the interstellar pressure of a typical spiral galaxy:

$$\Xi \sim 10^3 L_{\text{ion},46} r_{\text{kpc}}^{-2} p_4^{-1}, \tag{13.10}$$

where r_{kpc} is the distance from the nucleus to any particular point in kpc and p_4 is the interstellar gas pressure in units of 10^4 K cm^{-3}. An ionization parameter as small as a few tens (§10.5.1) suffices to heat (static) gas up to the Compton temperature. Because the Compton temperature for most AGN continua is comparable to or greater than the escape temperature from ordinary locations in galaxies ($\sim 10^6$–10^7 K), we might therefore expect that many AGNs could drive away any interstellar gas present in their hosts.

So the question is: "Why doesn't this happen?" There are three answers.

The first is that we do not know the lifetime of AGNs. It is possible that the duration of any one episode of luminous nuclear activity is shorter than the time it would take to expel interstellar gas from the host. Characteristic dynamical times in the outer parts of galaxies are generally several hundred million years, and it is well within the realm of possibility that active nuclei shut off more quickly than that.

492

The second is that, as we have seen in Chapter 12, the output of AGNs can be highly directional. In some cases the radiation is confined to narrow beams by relativistic effects; in others, very optically thick obscuration collimates the emerging radiation so that only within a pair of cones does it emerge into the outer galaxy. Where the radiation travels freely outward, it does "light up" interstellar gas, generating the narrow emission line spectrum, but where it is weak due to relativistic effects, or on lines of sight outside of thick obscuration, any effect is weak. Because obscuration often blocks the majority of solid angle, if its orientation is uncorrelated with the orientation of structure in the host, on average the majority of the host is shielded.

Even if much of the host is effectively protected by relativistic beaming or obscuration, the portion exposed to the nuclear radiation ought to be destroyed as thoroughly as predicted by our naive estimate. Only if its pressure were substantially greater than the fiducial local Solar neighborhood pressure to which we scaled would it be able to radiate away the energy it must absorb through photo-ionization. Examination of the narrow line spectrum (§10.5.3.2) indicates that just this happens (this is the third answer to our question). Like the interstellar gas in the inner few hundred pc of other galactic bulges (Helfer and Blitz 1997), the pressure in the narrow line region is elevated by several orders of magnitude relative to the pressure farther out in the galactic disk. At this higher pressure, the photoionized gas can efficiently radiate away the energy it absorbs when exposed to the nuclear continuum.

An AGN can also influence its host in another way, through its contribution to the gravitational potential. In §4.4 we described how massive black holes could be sought in galaxies through observations of the orbits of gas and stars. Now we can turn that around and examine how the alteration of orbits in a host galaxy by a black hole at its center alters the evolution of the host. The basic principle is that when the black hole is sufficiently massive, it can significantly symmetrize the structure of the surrounding galaxy. In disk galaxies, the orbital families supporting bars become chaotic (and therefore fill a spherical region) when the black hole mass is as large as $\simeq 0.05$ of the total galaxy mass (Norman, Sellwood, and Hasan 1996). On the other hand, stars on "box" orbits in triaxial elliptical galaxies (i.e., orbits with segments that pass close to the center) are scattered by close encounters with the central mass, also forming a rounder structure (Norman, May, and van Albada 1985; Gerhard and Binney 1985). These issues will be discussed further in §14.3.1.

14   Onset and Fueling

14.1   Introduction

Although we have learned much about AGNs, we have not yet dealt with some of the most basic questions about them: What sets off nuclear activity in galaxies? How is it maintained? What ends it? Why are certain galaxies active at certain times? How is it that the character of the host and its environment determine the nature of events in the nucleus? Taken as a group, these are perhaps the most fundamental problems pertaining to AGNs, but at present we hardly know how to begin answering them. The most we can do is to enumerate the specific physical issues involved and remark on a few tentative ideas that may be relevant.

In a sense, all these problems are the same as the central problem of Chapter 7, but acting on the galactic scale. In order to create a massive black hole, and then to keep it supplied with fuel, large amounts of matter must be brought very close to the center of the galaxy, but this requires stripping away virtually all the matter's angular momentum. To drive home the scale of the problem, we will begin with some order of magnitude estimates.

As we estimated in §4.2, if AGN light is made by reasonably efficient accretion onto black holes, the remnant mass per L_* galaxy is $\sim 10^7 \, M_\odot$. If only a minority of galaxies ever housed AGNs, the remnant mass per participating galaxy is even greater. By comparison, the total instantaneous gas inventory in a typical disk galaxy is only $\sim 10^8 \, M_\odot$; that is, the amount that must be brought into the center over the lifetime of the AGN is a substantial fraction of the amount available at any one time. Similarly, the *rate* at which mass must be brought to the center is an interesting fraction of the total rate at which gas is cycled through a galaxy's interstellar medium. In a typical contemporary disk galaxy, interstellar gas is converted into stars at a rate $\sim 1 \, M_\odot \, \mathrm{yr}^{-1}$. If half of this were diverted into the center of the galaxy and accreted onto a black hole with a radiative efficiency ~ 0.1, the resulting luminosity would be $\simeq 3 \times 10^{45} \, \mathrm{erg \, s^{-1}}$. In other words, it would be enough to fuel an AGN in the middle of the observed low-redshift luminosity distribution (e.g., as shown in fig. 3.7); still more is required to support a truly high-luminosity AGN. We conclude that when

494

an AGN is active, its gas consumption represents a major part of the gas consumption of the entire galaxy.

Removing the angular momentum from this gas is a serious problem. At a typical point in a disk galaxy, the angular momentum per unit mass of orbiting material is $\sim 10^{29}$ cm^2 s^{-1}; this is $\sim 10^5$ times greater than the specific angular momentum of material in a marginally stable orbit around a black hole of $10^8\ M_\odot$. As most of this book has been devoted to showing, the physical conditions of matter change drastically as one moves from the edge of the black hole outward into the host galaxy. Whatever mechanism is responsible for the torque close to the black hole, it seems unlikely that it also operates on material so far away, and in such different circumstances. Thus, we can expect that there is a hierarchy of mechanisms, handing off to one another as the material slips inward. At some levels the hand-off may be immediate, but there may also be points along the path better visualized as slowly filling reservoirs that impose a delay on the fuel before it can proceed farther. There may also be cases in which the fuel goes part way in, only to find itself stopped in a cul-de-sac.

If such a hierarchy of torquing mechanisms exists, it may correspond to the hierarchy of nested structures we have already discovered within AGNs. As shown schematically in figure 14.1, if we were to examine a garden-variety AGN by a series of successive steps in magnification, we would first see the global structure of the host. After a single enlargement, the narrow line region would be revealed. Inside the narrow line region we could see the obscuring torus, which surrounds the broad emission line region, which is itself far outside the part of the accretion disk where most of the energy is released. We can expect that as we discover the mechanisms responsible for removing angular momentum from material in each zone, we will also discover why each zone takes on its specific structure. In previous chapters, we have dealt with each of the inner regions in turn; in this chapter, the primary emphasis will be on mechanisms acting in the host proper, or in the narrow line region.

14.2 The Timescale Problem

14.2.1 Creation and growth of the black hole

The most natural place to begin is the issue of how the first massive black holes were assembled. Their raw materials were presumably drawn from their host galaxies, which themselves must have been relatively early

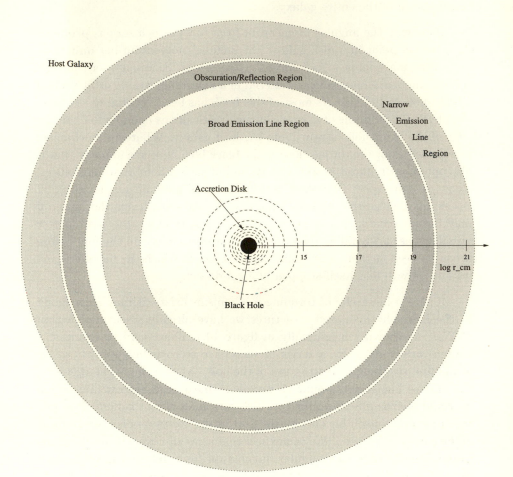

Fig. 14.1 The "onion-skin" model of AGNs. The distances are scaled logarithmically, not linearly, but should not in any event be taken as more than indicative. Many are rough estimates, and almost all scale with the luminosity of the AGN. Note in particular the several orders of magnitude outside the accretion disk that are unlabeled. Because we have yet to identify any photons as coming from this range of radii, there is very little we can say about it.

specimens, considering that we know of quasars at redshifts as early as $\simeq 5$. Unfortunately, we know little about how galaxies were made, except that it is likely that they were put together piece by piece, as gravitational condensations having a wide range of masses agglomerated. This ignorance of galaxy formation is, not surprisingly, a tremendous handicap for the study of AGN formation. Numerous points about the early history of galaxies—how rapidly they accumulate mass; the balance between gas, stars, and nonbaryonic matter; the rate at which new stars are formed— are all of central interest for AGN evolution, and our knowledge of all of them is extremely limited.

Certain things can be said with confidence, however. Most notably, the creation of massive black holes must have occurred just about as fast as it possibly could, for the Universe was still very young (fig. 14.2) when the earliest quasars appeared. Depending on the values of Ω_M and Ω_Λ, the age of the Universe at $z \simeq 5$ was only $\simeq (0.05\text{--}0.1)H_o^{-1}$, or $\sim 5\text{--}10 \times 10^8(h/0.75)^{-1}$ yr. This is not a very long time (as stressed, e.g., by Turner 1991), whether one measures it in units of galactic free-fall times [$\sim 10^8(r/10 \text{ kpc})$ yr in fully grown medium-sized galaxies] or black hole growth times [$\sim 4 \times 10^7(\eta/0.1)(L_E/L)$ yr; eq. 6.25].

Generically, there are three possibilities for the initial creation of the black hole: collapse of an isolated star, merger of two neutron stars, or collapse of a gas cloud, although numerous complicated steps can lead to any one of these (see, for example, fig. 4.2). The first process almost certainly occurs soon after the formation of the first stars in a galaxy, for stars whose initial mass is several tens of Solar masses run through their lives in $\sim 10^7$ yr, only to collapse at the end into black holes.

The second possibility is also quite likely, after a short delay. Stars somewhat less massive than those forming black holes live only slightly longer, and then collapse, leaving a neutron star remnant. Some fraction of these are in close binaries with another massive star that soon after is transformed into a second neutron star. If common envelope evolution or some other strongly dissipative mechanism brings the stars close enough before the second star collapses, and the second supernova does not destroy the binary, gravitational radiation eventually brings them together (cf. eq. 14.4). The shortest delay in plausible models is $\sim 10^7$ yr (Portegies Zwaart and Spreeuw 1996). Although this number is very uncertain, it seems quite likely that there are enough double neutron star binaries to make the steady state merger rate at least $\sim 10^{-6}$ yr^{-1} per galaxy (Narayan, Piran, and Shemi 1991; Phinney 1991).

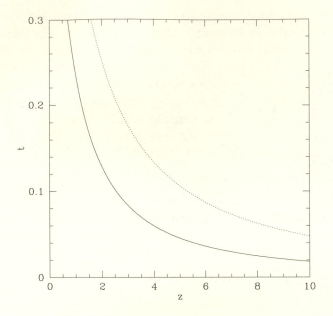

Fig. 14.2 The age of the Universe as a function of redshift z in units of H_o^{-1}. The solid curve corresponds to $\Omega_M = 1.0$ and $\Omega_\Lambda = 0$, the dotted curve to $\Omega_M = 0.1$, $\Omega_\Lambda = 0$.

The stumbling block that lies between the creation of a stellar mass black hole and the ignition of an AGN is, of course, the many e-foldings of mass that lie between a black hole whose original mass is $\sim 10\,M_\odot$ and achieving a final mass $\sim 10^8\,M_\odot$. Sixteen e-folds (as would be required to reach $10^8\,M_\odot$) take $\sim 6 \times 10^8(\eta/0.1)(L_{\rm E}/L)$ yr if accomplished through radiative accretion. It is important to note the significance of the radiative efficiency in this estimate. Low-efficiency accretion (§7.3.5) permits much more rapid growth of the black hole in its early stages, although the arguments of §4.2 require high efficiency during the last few e-foldings if the ultimate mass budget is not to become prohibitive.

Whether the third possibility occurs in Nature is more speculative, but it may offer a more direct pathway to creation of AGNs. There is no easy way to estimate the initial mass of a black hole formed by direct collapse of a gas cloud, or the rate of such events. When a cloud collapses under its own self-gravity, very large internal density contrasts form, so it is unlikely that the entire mass of the cloud would be contained within the initial event horizon; however, just how much of the initial mass *is* contained is likely to depend strongly on initial conditions. Whatever the initial black hole mass is, once it is formed, its further growth is subject to many of the same constraints that apply to black holes formed initially by stellar events. In only two ways do the situations differ: cloud collapse can create a black hole with an initial mass that could (in principle) be considerably greater (or smaller) than $\sim 10\,M_\odot$; and one might expect that a new black hole formed directly from the collapse of diffuse gas might be embedded in a region of above-average gas density.

14.2.2 Heavy element abundances

In addition to the creation and growth of the black hole in the central engine, another timescale problem confronts models of AGN evolution: the timescale for creating medium-Z elements in the host's interstellar medium, so that the quasar at the center can generate the familiar repertory of C, N, O, Mg, Si, and Fe emission lines that is so consistently seen. That N and Fe are part of this list is a particular problem. Although the other elements with strong lines are the products of reactions in which α-particles are added to nuclei, and are therefore quickly synthesized in short-lived massive stars, most N and Fe is believed to be made more slowly. N is largely a product of H-burning on the CNO cycle and therefore requires some prior synthesis of C and O. Because the CNO cycle dominates energy production for main-sequence stars above $\simeq 2\,M_\odot$ when there is at least a trace amount of C and O, most of the N injected into the interstellar medium arrives when intermediate-mass stars eject planetary nebulae at the ends of their ~ 1 Gyr lives. Fe is mostly a product of type 1 supernovae, events that require a similarly long wait from the initial creation of their progenitor stars (Arnett 1996).

The fact that we see quasars with apparent abundances not far from Solar therefore means that the time interval between the first generation of stars in their hosts and the ignition of the quasar cannot have been too short. As figure 14.2 demonstrates, this is a serious constraint for high-redshift quasars.

14.2.3 Black hole centralization and multiple black holes

Finally, there is the question of how long it takes the growing black hole to find its way to the center of the galaxy. Although the limits on high-z AGNs in this regard are very poor, AGNs in all *nearby* galaxies are located at the center of the galaxy to within a very small margin of error.

Perhaps the single most plausible mechanism to accomplish this centering is dynamical friction against the stars in the host. A black hole of mass M moving with speed v_M through a number of stars with mean mass m pulls the slower-moving stars into a density concentration behind itself. Deceleration results, at a rate

$$\frac{dv_M}{dt} = -16\pi^2 \ln \Lambda \, G^2 m(M + m) v_M^{-2} n_*(< v_M), \qquad (14.1)$$

where $n_*(< v_M)$ is the local density of stars moving slower than v_M, and Λ is a ratio of greatest to least impact parameters that is a direct analogue to the argument of the logarithmic enhancement to the Coulomb scattering cross section (Binney and Tremaine 1987).

This steady deceleration drives a secular loss of the black hole's orbital energy E at a rate

$$\frac{d\ln|E|}{dt} \sim 6 \times 10^{-20} \left(\frac{m}{M_\odot}\right) \left(\frac{M}{10\,M_\odot}\right) \left(\frac{n_*(< v_M)}{1\,\mathrm{pc}^{-3}}\right) \left(\frac{v_M}{100\,\mathrm{km\,s^{-1}}}\right)^{-3} \mathrm{s}^{-1},$$
$$(14.2)$$

where we have made the assumption that the orbital evolution is slow enough for the virial theorem to apply. Thus, a black hole created by stellar collapse in the outer portions of a galaxy would take many Hubble times to lose any significant energy by dynamical friction, but one created close to the center, where the stellar density is several orders of magnitude greater, can sink relatively rapidly. Note, however, that the expression used above for deceleration by dynamical friction is not entirely appropriate when the mass of the object becomes comparable to the mass of the stellar system with which it is interacting, as can occur when the black hole is approaching its final mass.

Dynamical friction becomes an especially important consideration if a galaxy possesses not one massive black hole, but two (Begelman, Blandford, and Rees 1980). Two massive black holes in a single galaxy may not be a terribly rare occurrence. After all, if the probability per galaxy of creating a single black hole is not too small, the probability of creating two may not be negligible. In addition, galaxy mergers may bring two massive black holes into the same system.

When there are two massive black holes in a single galaxy, dynamical friction helps bring both to the dynamical center of that galaxy at a rate determined just as in equation 14.2. They become mutually bound gravitationally when their separation s is small enough that the galaxy's tidal field cannot separate them, that is, when

$$s < \left(\frac{M_1 + M_2}{n_*}\right)^{1/3}, \qquad (14.3)$$

500

where the two masses are $M_{1,2}$. If s is greater than the distance r_B from the galaxy's center to their center of mass, this is also the point at which their mass dominates the local potential. If $s < r_B$, the potential might still be dominated by stars at this stage.

Once the two black holes become bound to each other, dynamical friction acts simultaneously in two different ways: it continues to remove energy from the orbit of their mutual center of mass through the galaxy, and it also removes energy from their orbits around their mutual center of mass. If $s > r_B$, although dynamical friction does continue to bring the black hole's center of mass closer to the galaxy's, this motion is not very interesting. However, if $s < r_B$, and there are enough stars in the galactic nucleus that their mass, rather than $M_1 + M_2$, dominates the potential, the rate at which r_B shrinks is accelerated when the two black holes become bound, for dynamical friction is proportional to the total mass. No matter how s compares to r_B, dynamical friction continues to remove energy from the motion of the black holes around their center of mass even after they become bound to one another. The rate is very slightly diminished when stellar mass dominates the potential, for then the orbital period of the black hole binary is shorter than the time it takes for a distant star to pass by one of the black holes. As a result, energy transfer due to distant encounters is very small because of phase-averaging. However, closer encounters remain effective, so the only change in the dynamical friction rate is through a decrease in the argument of the logarithm.

Ultimately, when the black holes are very close, the most rapid rate of orbital energy loss is by gravitational radiation. Loss of energy by this mechanism causes a circular orbit to decay on a timescale

$$t_{gr} = 7.66 \times 10^9 \left(\frac{s}{10^{17} \text{ cm}} \right)^4 \left(\frac{M_2}{10^8 \, M_\odot} \right)^{-3} [q(1+q)]^{-1} \text{ yr}, \qquad (14.4)$$

where $q = M_1/M_2$.

14.3 Basic Fueling Mechanisms

Once a black hole somehow forms, grows to AGN size, and sinks to the center of its host galaxy, its ability to function as an AGN depends on its being fed (or having recently been fed if it is losing energy primarily by spin-down). We can identify a number of mechanisms that might help bring matter toward the center of the host galaxy, but we do not know which of them is most important, or how they relate to one another. In this section, therefore, we will discuss them individually, leaving to the next section some examples of how they might combine to create AGNs.

14.3.1 Gravitational forces

A given gravitational field produces identical acceleration on all matter, whether it is gas, stars, or nonbaryonic dark matter. It is, in that sense, very simple. However, in the simplest pictures of galactic structure, gravity is incapable of pushing anything close to the center—gas, stars, or anything else. The reason is that *simplest* pictures are generally *symmetric* pictures. Because gravity is a central force, if its potential is symmetric around an axis, the component of angular momentum parallel to that axis J is rigorously conserved by all gravitational forces. In a time-steady potential, orbital energy E is likewise conserved, so it is possible to describe orbits in terms of energy-conserving motion in an effective potential. That is, the orbital kinetic energy per unit mass is

$$\frac{1}{2}\left(\dot{\varpi}^2 + \dot{z}^2\right) = E - \Phi(\varpi, z) - \frac{J^2}{2\varpi^2}, \tag{14.5}$$

where distance from the symmetry axis is denoted ϖ, distance along the symmetry axis is z, and Φ is the gravitational potential. For any $J \neq 0$, there is then a distance of closest approach

$$\varpi_{\min}(z) = \frac{J^2}{\sqrt{2\left[E - \Phi(\varpi_{\min}, z)\right]}}. \tag{14.6}$$

Thus, if only gravitational forces act, there is an unbreachable barrier preventing gas with angular momentum from feeding a galactic nucleus in an axisymmetric potential.

Consequently, any model of gravitational forces removing angular momentum from either stars or gas must involve an asymmetric structure. In disk galaxies, axisymmetry is generally a fairly good approximation, so any such asymmetry can be viewed as a perturbation to an underlying axisymmetric potential, except perhaps for transient effects due to strong encounters with other galaxies. In elliptical galaxies, whose equilibrium structure can be intrinsically tri-axial, axisymmetry may not necessarily be a good starting point, but some of the matter may trace orbits with analogous symmetries that prevent close approaches to the center equally effectively.

For simplicity, consider the situation in an infinitely thin disk galaxy. At any given moment, its potential at each cylindrical radius ϖ can be described as a Fourier series in azimuth:

$$\Phi(\varpi, \phi) = \sum_m \Phi_m(\varpi)e^{im\phi}. \qquad (14.7)$$

The z-component of the torque exerted by this potential is

$$\mathcal{G}_g(\varpi, \phi) = -\text{Re}\left[i\sum_m m\Phi_m(\varpi)e^{im\phi}\right]. \qquad (14.8)$$

Clearly, the axisymmetric part ($m = 0$) contributes nothing to the torque. Just as clearly, the magnitude of the torque depends on how large the nonzero $m\,\Phi_m$ are and the phase relationships between the different components. It can also be expected to vary substantially both in ϖ and ϕ.

To compare torques of different origin, it is convenient to normalize their magnitude to the fractional rate of change per orbital period of the angular momentum of orbiting matter. We note parenthetically that even to define an approximate orbital period requires that the axisymmetric part of the potential dominate the others. With this assumption, an instantaneously nonaxisymmetric potential alters angular momentum at a normalized rate

$$\frac{d\ln J}{d(\Omega t)} = -\text{Re}\left[\frac{i}{\partial\Phi_o/\partial\ln\varpi}i\sum_m m\Phi_m(\varpi)e^{im\phi}\right]. \qquad (14.9)$$

Thus, to order of magnitude, $d\ln J/d(\Omega t) \sim m_*\Phi_{m_*}/\Phi_o$, where m_* is that $m_* \neq 0$ for which $m\Phi_m$ is greatest.

The distinction between *instantaneous* torque and *time-averaged* torque is an important one because the phases of the Φ_m in general vary in time, that is, the pattern formed by the nonaxisymmetric part of the potential can both rotate around the galaxy, and change form as time goes on. Consider, for example (as did Lynden-Bell and Kalnajs 1972), what happens when the nonaxisymmetric portion of the potential is a long-lived linear feature (a "bar") rotating with a fixed pattern speed Ω_p that is independent of radius (as is necessary for it to retain its straight shape for more than a fraction of an orbit). The orbital frequencies of stellar orbits in the bar region are not necessarily the same as the pattern speed. If the bar is weak enough to be viewed as a perturbation to the underlying axisymmetric potential, the orbital frequency of stars close to the center of the galaxy is considerably greater than Ω_p. In this situation, as a star approaches the potential minimum created by the bar, it is accelerated; as it leaves, it is decelerated. Averaged over many passages, these effects nearly cancel. The cancellation is not complete, however, for while the star is close to the potential minimum, it is deflected in such a way that it spends a little extra time near that azimuthal angle. This deflection produces a net torque in the sense of slowing down the star's orbital motion. That is, (an externally imposed) bar removes angular momentum from those stars traveling at small enough radii that their orbital frequencies exceed Ω_p. This is, of course, the sense desired in order to send matter inward. The sense of the effect is opposite for stars with $\Omega < \Omega_p$. However, if $\Omega \simeq \Omega_p$, the force seen by the star is nearly constant for long periods of times. Consequently, stars near resonance with the bar can be very strongly affected by the bar potential. Note, however, that this entire argument depends crucially on the lifetime of the nonaxisymmetric feature. If it is short-lived, none of these frequency comparisons is relevant, for there is no well defined pattern speed.

The degree to which these comments can be carried over to elliptical galaxies depends on their degree of symmetry, and on the populations of stars in different orbital families. With respect to the sort of argument presented in the previous paragraphs, there is no essential difference between axisymmetric elliptical galaxies and disk galaxies. However, not all elliptical galaxies are axisymmetric. In tri-axial elliptical galaxies, none of the components of a star's angular momentum are necessarily conserved, but there can nonetheless be other conserved integrals of the motion that constrain the orbits (Binney and Tremaine 1987). In some cases, the effect of these conserved quantities is to create sizable forbidden regions around one of the axes; these are called *tube orbits*. In others, called *box orbits*, stars can pass arbitrarily close to the galactic center, but the actual pericenter distance varies substantially from one passage to the next. Stars (or gas clouds) traversing box orbits have no difficulty reaching very small radii; instead, if we are concerned with supplying accretion fuel, the question is how to stop them before they swing back out again.

When an elliptical galaxy is strongly centrally concentrated, still further complications appear. Those orbits passing especially close to the center lose (at least) one of their conserved quantities and become chaotic. A large point mass (e.g., a black hole) located at the galactic center can produce similar effects (Gerhard and Binney 1985; Norman et al. 1985). If many stars are put on chaotic orbits, their time-averaged spatial distribution is relatively round, and the associated gravitational potential rounder yet. As a result, close to the center of an elliptical galaxy in this condition, angular momentum may once again be a conserved quantity.

14.3.2   Hydrodynamic dissipation

By definition, hydrodynamic effects act only on gas, and not on stars. Highly clumped gas is a marginal case—when the clumps' collision frequency is greater than the orbital frequency, the hydrodynamic picture is a better approximation, but when the the clumps' collision frequency is smaller than the orbital frequency, they behave more like a collisionless system (see, e.g., Goldreich and Tremaine 1978 on how the effective viscosity depends on the ratio of collision frequency to orbital frequency).

Absent global torques (e.g., of the sort described in the previous subsection), angular momentum in a fluid element is conserved. Consequently, in order to alter a gas parcel's angular momentum by hydrodynamic effects, it must be mixed with other gas having different specific angular momentum. Mixing of this sort, is, of course, intrinsically dissipative. It can be accomplished slowly by conventional atomic diffusion in fluids with gradual contrasts in specific angular momentum, or rapidly by bulk motions (e.g., shocks) that create very sharp gradients. Turbulent motions can also enhance mixing. In fact, because *all* purely hydrodynamic torques can be thought of as due to a thermodynamically irreversible mixing process, they can equally well be considered as the product of one variety or another of viscosity.

Because hydrodynamic torques can all be described in the language of viscosity, we already have a convenient formulation for their description: the "α" model of accretion disks (§7.2.1). That formalism makes two implicit assumptions that often, but not always, apply to the galactic context: that the turbulence is quasi-isotropic, and that the gas is confined to a geometrically thin disk. When those assumptions are appropriate, it is easy to see (cf. eq. 7.80) that

$$\frac{d \ln J}{d(\Omega t)} \sim \alpha \left(\frac{h}{r}\right)^2 , \qquad (14.10)$$

where α is the magnitude of the stress due to turbulent viscosity in units of the gas pressure, and h is the vertical thickness of the disk. Thus, unless α is near unity, and the disk is not terribly geometrically thin, isotropic turbulent mixing in a thin disk cannot explain the inflow we see occurring in AGNs. However, in geometrically thicker systems (e.g., elliptical galaxies), or in cases with strongly anisotropic peculiar velocities (e.g., the perturbations due to strongly nonaxisymmetric potentials in disk galaxies), these effects might be more important.

Another possible exception to this (not-so-general) argument has to do with singular events. The accretion disk analogy presupposes a time-steady flow. Although such an assumption is appropriate there, it is not necessarily appropriate to gas flows on the galactic scale. Individual events, such as galaxy-galaxy encounters (§14.4.4), might temporarily produce very large-scale gas motions in which enough gas mixing occurs to convey inward the fuel to power an AGN for a very long time.

14.3.3 Magnetic forces

As discussed in §§7.2.2 and 7.2.3, there are two ways magnetic forces can remove angular momentum from a rotating system: they can increase the rotation rate of matter that then flows away, or magnetic fluctuations can carry off the angular momentum directly. The two mechanisms may be linked, of course, if angular momentum is carried part way in the field, and then delivered to material, or vice versa.

First, consider magnetized winds. Just as with accretion disks, they can in principle extract angular momentum very quickly, but their efficacy depends critically on unknown considerations such as the geometry and strength of the magnetic field, and the rate at which mass can be injected into them.

Consider next the case in which the angular momentum transport is entirely in the field. The torque can then be described in terms of the Maxwell stress tensor. For example, the z component of the inter-ring magnetic torque (at radius r) in a thin disk of plasma is given by equation 7.18, which we repeat here in slightly different notation:

$$\mathcal{G}_m = \frac{1}{2}\varpi^2 \int dz\, B_\varpi B_\phi. \tag{14.11}$$

In fact, as also remarked in the discussion following equation 7.18, if the field is very tangled, its effect can be described in terms of the α-formalism. We might then simply substitute $\alpha_M = \langle B_\varpi B_\phi \rangle/(4\pi p)$ for α in equation 14.10 and conclude that magnetic forces—unless the energy density in field is much greater than the gas pressure—would be as ineffective as hydrodynamic forces in promoting inward gas flow in galaxies.

Thus, in time-steady, geometrically thin systems, we find that one magnetic mechanism (magnetized winds) may be active, but any conclusions we draw about it are extremely model-dependent, while the other mechanism (direct Maxwell stress) is unlikely to be important. However, our analysis of both magnetized winds and turbulent magnetic stresses has implicitly assumed a time-steady inflow. Just as for hydrodynamic mechanisms, there may be one-time occurrences in which magnetic forces are

506

much stronger than normal, and at that special time manage to push large amounts of gas toward the galactic nucleus.

14.3.4 Poynting-Robertson drag

Another mechanism special to gas is the loss of angular momentum due to radiation or scattering of photons. Consider first the case of radiation. If, as is usually the case, the photons are isotropically radiated in the fluid frame, in the frame of the entire system they carry a mean specific angular momentum equal to that of the fluid. So, for example, the photons that carry off the heat dissipated in an accretion disk also remove angular momentum from the disk. However, compared to the accretion torque, the rate of angular momentum loss is quite small at all radii far from the edge of the black hole: generically, it is a fraction $\sim r_g/r$ of the accretion torque.

The origin of the torque due to scattering is very similar. Photons initially having zero angular momentum (as, e.g., from a source at the center), are given angular momentum as a result of scattering against orbiting matter. The torque due to scattering can, however, be much greater than that due to intrinsic radiation if the incident flux is sufficiently great. The reason, of course, is that the amount of energy available in the *total* luminosity of the system can be much greater than that available at a particular radius far from the central black hole. To illustrate this point, let us again consider the example of gas orbiting far outside an accreting black hole, but close enough that AGN light dominates the local radiation intensity. In this case, the ratio between the Poynting-Robertson torque and the local accretion torque is ηC, where η is the radiative efficiency of the accretion in rest-mass units, and $C \sim h/r$ is the angular covering factor of the gas around the central source. Although C is also unlikely to be as large as unity, it can be much larger than r_g/r at large r.

Let us now place this mechanism in the context of the galactic host of an AGN. Suppose that the local intensity (which may have a significant contribution from starlight) is $L'/(4\pi r^2)$, and the scattering opacity κ. Angular momentum is lost by Poynting-Robertson drag at a rate per orbital period

$$\frac{d\ln J}{d(\Omega t)} = \frac{1}{\sqrt{2}} \frac{\kappa}{\kappa_{\mathrm{T}}} \frac{L'}{L_{\mathrm{E}}} \left(\frac{M_{\mathrm{AGN}}}{M(<r)}\right)^{1/2} \left[\frac{r_g(AGN)}{r}\right]^{1/2}, \tag{14.12}$$

where κ_{T} is the Thomson opacity, L_{E} is the Eddington luminosity of the AGN, and $M(< r)$ is the total mass (galactic plus AGN) interior to radius r. Whether this mechanism is important depends on whether the opacity can be enough larger than Thomson to outweigh the small ratio $[r_g(AGN)/r]^{1/2}$.

14.3.5 Stellar-interstellar mass exchange

Mass exchange between the gas phase and stars can be of fundamental importance in governing mass inflow in galaxies. It plays a basic role in regulating the interstellar pressure, of course. In addition, it determines how much mass is available for those dynamical mechanisms that act only on gas.

Mass can be exchanged in both directions between the host's interstellar medium and its stars: Interstellar gas can condense into stars, and stars can return some of their mass to the interstellar medium by ejecting winds, or exploding as supernovae . We have excellent data tabulating how much mass a star having a given initial mass eventually returns to the interstellar medium, and how long it takes to do so. Stars with initial mass greater than $\simeq 8\,M_\odot$ return all but $1.4\,M_\odot$, part in a wind and part as a result of a supernova explosion (Arnett 1996), and do this within their lifetime of $\sim 10^7$ yr. Lower mass stars live longer (lifetime $\propto M^{-3}$ roughly) but still eventually restore a large fraction (50–80%) of their initial mass (Weidemann 1987). If the density of stars is known, and the distribution of initial masses specified, it is easy to use tabulations of this sort to calculate the rate at which a population of stars re-injects mass into the surrounding interstellar medium.

These two prerequisites are not easily satisfied. Our understanding of star formation is not adequate to the job. Consequently, it is very difficult for us to predict how many stars are made where, at what time, and with what distribution of masses. As a result, we cannot have much confidence in our estimates of either the mass dropout rate due to star formation, or the mass return due to subsequent stellar evolution. Several reasons, both theoretical and experimental, explain why it is that we understand what happens to stars after they are born much better than we understand how they are created. Once a star is made, it remains very nearly in hydrostatic equilibrium, and all we need to do is track the rate at which it cooks light elements into heavier ones, and how fast the heat that is released makes its way to the surface. By contrast, star formation is a fundamentally dynamical process. In addition, our empirical measures of the dynamical elements of stellar evolution, for example, the mass loss rate in stellar winds, are rather better tied to the underlying physical quantities than are the corresponding empirical measures of star formation rates. For all these reasons, considerable controversy attaches to this subject.

To within uncertainties at the factor of a few level, we do possess empirical measures of the integrated gas consumption rate in normal disk galaxies. Typically we see this gas being consumed for star construction over timescales of a few times 10^9 yr (Kennicutt, Tamblyn, and Congdon 1994). The situation is rather murkier with regard to the questions of why this rate varies from galaxy to galaxy, and how it depends on location *within* a single galaxy. Schmidt (1959) made the very plausible suggestion that the star formation rate increases in proportion to a power (between 1 and 2) of the gas surface density Σ_g. In order to better fit such data as Hα surface brightness distributions, gas/star ratios as functions of radius, and heavy-element abundance gradients, others have elaborated on this idea by suggesting that multiplicative correction factors of various forms should be applied to the "Schmidt law." For example, Kennicutt (1989) suggested that star formation can proceed only when Σ_g exceeds a threshold whose value (which depends on the galactic rotation curve) signals the onset of local gravitational instability (see §14.4.3), while Wyse and Silk (1989) proposed instead a correction factor proportional to the local orbital frequency. Given the uncertainty prevailing in this subject, it is hard to construct models for gas flows in galaxies harboring AGNs whose description of star formation can be relied upon.

If dense stellar clusters exist very close to the AGN, a number of exotic mechanisms may also tear mass from stars. For example, when the orbital speeds of the stars exceed the mean escape speeds from within them (~ 600 km s^{-1} for normal main sequence stars), stellar collisions can liberate much of the colliding stars' mass (Spitzer and Saslaw 1966). Red giant stars present much larger cross sections for physical collisions than do main sequence stars, and their surface escape speeds are also much lower, but the amount of mass contained in their diffuse envelopes is typically a small fraction of a Solar mass, and so their presence does not greatly enhance the overall rate of mass loss due to stellar collisions.

Stars venturing extremely close to the AGN might find their outer layers heated so strongly by absorption of the AGN continuum radiation that they lose mass at a significant rate. The key criterion for determining whether the irradiation is strong enough to drive a wind is how the gas temperature compares to the escape temperature $T_{esc} \equiv GM_s m_p/(r_s k_B)$, where M_s and r_s are the mass and radius of the star.. Using the typical escape speeds just cited, we immediately find that T_{esc} for a main sequence star is $\simeq 4 \times 10^7$ K, whereas it is only $\simeq 1 \times 10^4$ K for a red giant. If most of the heat is absorbed above the photosphere, the temperature of any gas whose pressure is low enough to make its ionization parameter

Ξ greater than the maximum value permitting a cool equilibrium ($\simeq 10$: §10.5.1) rapidly rises toward the Compton temperature, which is generally $\sim 10^7$–10^8 K. Such a high temperature is comparable to T_{esc} for a main sequence star, and far above it for a red giant.

On the other hand, if most of the heat is absorbed below the stellar photosphere, the energy is thermalized. Only if the new effective temperature of the stellar atmosphere is comparable to or greater than T_{esc} can much mass be removed from the star. This criterion establishes how close to the center stars must venture if subphotospheric heating is to damage them: within $\sim 1 \times 10^{16} L_{45}^{1/2} T_{esc,4}^{-2}$ cm. That is to say, only red giants straying very close to the center of an AGN can lose much mass by irradiation if the energy penetrates below the photosphere.

14.3.6 Gas heating and cooling

The gas's hydrodynamic behavior depends, of course, on its pressure, which, in turn, depends on its density and temperature. Although its density is determined by a combination of the sources and sinks just listed and advection of material from one place to another, its temperature is determined by a balance between heating and cooling. In this regard it is important to note that on small scales what we mean by temperature is the usual measure of the energy in random atomic motions; on larger scales, however, the strong clumping that frequently characterizes interstellar gas means that the more relevant measure of "temperature" is the energy per unit mass of random clump motions. When this latter definition is the relevant one, "cooling" refers not directly to radiative or conductive losses, but to inelastic clump collisions.

Conventional temperature balance, when dominated by radiative processes, is relatively easy to determine. Calculations like the ones described in §10.3.3, although perhaps tedious, are conceptually straightforward. Photoionization by the continuum generated by either the AGN or hot stars may be readily balanced by molecular, atomic, or ionic line emission at low temperatures, or by bremsstrahlung (§8.2.1) and inverse Compton scattering (§8.2.2) at higher temperatures.

Difficulties arise when nonradiative processes become important. Possibilities include cosmic ray heating, or shock waves driven by a potential plethora of pistons: stellar winds, supernovae , or jets issuing from the AGN. Because relativistic nuclei are extremely inefficient at generating photons directly, it is very hard for us to obtain a direct measure of their density almost anywhere (the only exception to this rule is the γ-rays produced by pion decay following the reaction $p + p \rightarrow p + p + \pi$). Although knowledge of the stellar population allows us to predict how many shock wave sources there are, and the momentum flux driving each shock, their efficiency in heating the gas depends on both the stars' spatial arrangement and the distribution of gas density. In particular, clustered hot stars can lead to strongly nonlinear effects such as interconnected supernova remnants (Cox and Smith 1974; McKee and Ostriker 1977), and venting of hot gas out of the galactic plane (McCray and Snow 1979; Tomisaka and Ikeuchi 1987; Norman and Ikeuchi 1989).

Because random clump motions bear almost no direct relation to radiative processes, these difficulties are accentuated when we attempt to understand either clump stirring or the dissipation of clump motions. In addition to heating the gas they strike, shock waves also accelerate it. Stellar winds and supernovae may therefore be as central to clump "heating" as to conventional heating of interstellar gas. Of course, all the problems that arise in the quantitative study of conventional heating apply with equal force in the quantitative study of the origin of clump random motions. Clump "cooling" would be simple to compute if the motions were truly random—the rate per unit volume would be $\epsilon_{cc} n_c^2 \sigma_{cc} (\Delta v_c)^3$, where n_c is the number density of clumps, σ_{cc} their mutual collision cross section, Δv_c their *rms* random speed, and ϵ_{cc} is an efficiency factor that lumps together the inelasticity of clump collisions and appropriate geometrical averages. However, many of the stirring processes may be coherent on scales large compared to the mean separation between clumps. If so, clump motions would be correlated, and the actual cooling rate rather smaller than that given by this simple estimate.

14.4 Triggers and Scenarios

We now raise the question of what sorts of events might bring these mechanisms into play. Although we have thus far discussed them separately, in many plausible scenarios (use of this overworked term is appropriate here, for ideas about AGN creation and fueling are as yet little more than stories) several act in synergistic fashion. Here we present several examples of how the basic mechanisms described in the previous section may combine to help fuel an active galaxy's nucleus. These should be viewed more as illustrative examples of how these mechanisms might behave than as descriptions of reality. They are ordered by increasing complexity, not plausibility. Much of the discussion will be framed in terms of a very useful tool, the tensor virial theorem; readers not familiar with this device are advised to read Appendix E.

14.4.1 Cooling flows

Bright X-ray haloes surround many elliptical galaxies. Although there is much scatter around the trend line, the X-ray luminosity L_X for isolated elliptical galaxies increases rapidly with the blue light luminosity of the galaxy L_B: $L_X \sim 10^{40}(L_B/10^{44})^{1.8}$, for both luminosities measured in erg s^{-1} (Fabbiano, Kim, and Trinchieri 1992). Those elliptical galaxies located in rich clusters of galaxies may be embedded in even brighter—and much larger—X-ray haloes: luminosities as high as $\sim 10^{45}$ erg s^{-1} are sometimes seen (Sarazin 1988). In both cases, the origin of these X-rays is almost certainly bremsstrahlung by hot gas trapped in the gravitational potential.

From equation E.24 (the time-steady scalar virial theorem with negligble surface terms), we know that gas supported by its own pressure in hydrostatic equilibrium against gravity has a mean pressure comparable to its mean potential energy density. In a point mass potential, the mean kinetic energy (both atomic and bulk) is exactly half the mean potential energy; when the potential is shaped otherwise, the ratio differs, but it is always order unity. Because both the potential energy and the pressure are proportional to the gas density, the mean temperature of gas that supports itself against gravity by pressure gradients is of order the potential energy per unit mass; i.e., it must be approximately the virial temperature, which is $\sim 10^7$ K in elliptical galaxies. For gas bound in a rich cluster it can be an order of magnitude hotter. Thus, X-ray emission is exactly what we would expect from such hot gas.

Because the bremsstrahlung cooling time for gas in the inner parts of these haloes is shorter than a Hubble time, if there is no source of energy to restore the radiating gas's heat, it must fall inward. The associated rate of mass influx can then be estimated crudely by supposing that the entire X-ray luminosity comes from compressive work done by the gravitational potential on the settling gas:

$$\dot{M} \sim \frac{m_p L_X}{(5/2)k_B T} \sim 0.7 \left(\frac{L_X}{10^{41} \text{ erg s}^{-1}} \right) \left(\frac{k_B T}{1 \text{ keV}} \right)^{-1} M_\odot \text{ yr}^{-1}. \quad (14.13)$$

In other words, these galaxies may be examples of spherical accretion akin to the picture envisaged in Chapter 6. Unlike in almost every other accretion scenario that has been suggested, angular momentum may not play much of a role here because elliptical galaxies are, typically, rather slow rotators. If this picture applies, the estimate of equation 14.13 shows that the associated accretion rate can be quite substantial.

However, in evaluating the possible importance of cooling flows in galactic accretion models, several important technical points should be borne in mind. First, although the specific angular momentum of this gas *on the scale where we observe it* can easily be much smaller than would support it against gravity (in the language of the virial theorem, the integrated bulk kinetic energy $K \ll W$, the integrated pressure), angular momentum conservation guarantees that as it flows inward, rotation will eventually increase in relative importance.

Second, what we see on the galactic halo scale is not the entire spherical accretion process described in Chapter 6; it is only the outer, subsonic part. The characteristic accretion radius r_s defined in §6.1 is here $\simeq 200 M_{11} T_7^{-1}$ kpc, for M_{11} the mass of the galaxy (or mass of the cluster within r_s) in units of $10^{11} M_\odot$ and T_7 the temperature in units of 10^7 K.

Finally, there is reason to believe (Murray and Balbus 1992) that these flows *never* achieve a time-steady state. Instead, cooling-provoked collapse spreads steadily outward from the center in such a way that the accretion rate increases with both radius and time. If there is no luminous object at the center, the temperature of the innermost gas plummets; if there is a bright AGN, Compton scattering of its continuum can fix the temperature of the gas on the inside of the flow to the Compton temperature (Fabian and Crawford 1990). For a conventional AGN continuum shape, this may be $\sim 10^7$ K or perhaps a few times greater. However, if there is optically thick obscuration that converts much of the high-energy portion of the spectrum to infrared photons, those portions of the flow in the shadow of the obscuration equilibrate via the Compton process at a much lower temperature. On those sides, Compton scattering has the effect of accelerating the sort of thermal collapse that occurs without a central AGN.

14.4.2 Axisymmetric collapse enhanced by magnetic braking

Next consider a slightly more complex picture, in which the host is no longer thought of as spherical but is at least roughly axisymmetric. Although this symmetry must be broken for gravitational torques to work, other, noncentral, forces can still transport angular momentum. In fact, in a fashion that emphasizes the importance of synergism between the different dynamical mechanisms separately enumerated in the previous section, the magnetic braking picture is one in which MHD effects catalyze a process that eventually leads to gravitational collapse.

It is convenient to analyze the equilibrium of such a system by means of the virial theorem. Consider, for example, a fluid equilibrium with total mass M, mass per particle m, characteristic linear scale R, mean temperature T, total magnetic flux $\mathcal{F}_M$, total angular momentum J, and no nonrotational internal bulk motions. If this equilibrium is constrained to maintain a constant shape (i.e. a fixed ratio of equatorial to axial scale length), we can, on the basis of dimensional analysis alone, rewrite equation E.24 in terms of globally conserved quantitites:

$$a_1 \frac{M}{m} k_\mathrm{B} T(R) + a_2 \frac{\mathcal{F}_M^2}{R} + a_3 \frac{J^2}{MR^2} = a_4 \frac{GM^2}{R} - M\Phi_\mathrm{ext}(R). \qquad (14.14)$$

Here $a_{1,2,3,4}$ are all dimensionless numbers, generally speaking of order unity. T is implicitly a function of R through the gas's equation of state. The right-hand side of equation 14.14 summarizes the gravitational binding energy; to achieve equilibrium, the integrated pressure, magnetic energy, and rotational kinetic energy must add up to equal the gravitational binding energy. The characteristic size of the system in equilibrium is found by solving equation 14.14 for R. Given the nonlinearity in R that is present in this equation, there can be zero, one, or several solutions.

Let us suppose that at least one solution does exist, and that we have chosen a particular one. The next question to investigate, since we are interested in how to collect material at the center of a galaxy, is what must happen for the system to shrink to smaller R. This might happen in either of two ways: if the equilibrium is stable to small perturbations in R, there might be a secular change in some of the conserved quantities; alternatively, an unstable equilibrium will destroy itself, either expanding or contracting until nonlinear effects (e.g., approach to a stable equilibrium) limit further changes in size.

Use of the virial theorem also aids in testing for stability. The easiest way to apply the test is to compute the logarithmic derivatives with respect to radius for each of the terms in the equation expressing virial balance. If the equation of state can be expressed in terms of an adiabatic index γ (i.e., $p \propto \rho^\gamma$), then spherical compression gives a logarithmic derivative $-3(\gamma-1)$ for the pressure term. The logarithmic derivatives for the magnetic and rotational energies are, of course, -1 and -2, respectively. On the right-hand side, the logarithmic derivative of the self-energy is -1, while for the external potential, a wide range is possible in principle, but in galaxies, where the potential is often quasi-logarithmic, typical numbers are small in absolute value and negative.

Because growth in the left-hand side of equation 14.14 without compensating growth in the right-hand side leads to expansion, instability with respect to (shape-preserving) small perturbations occurs when the total logarithmic derivative of the left-hand side is larger (less negative) than the right-hand side. For example, if the primary support is from pressure, and self-gravity is the dominant confining force, instability results when $\gamma < 4/3$.

Absolute stability cannot be guaranteed by this test because we are considering only a very special class of perturbations, those that preserve the shape of the system and are nondissipative and so conserve J and $\mathcal{F}_M$. However, with respect to this narrow class of perturbations, we can readily see from the form of equation 14.14 that support by rotation is almost always stable. On the other hand, support by magnetic forces against self-gravity is neutrally stable, for both logarithmic derivatives are -1. This neutral stability means that there is a critical ratio of magnetic flux to mass $(\mathcal{F}_M/M)_{\mathrm{crit}} \sim G^{1/2}$; greater magnetic flux per unit mass prevents any self-gravitating equilibrium, while smaller magnetic flux per unit mass can never on its own provide complete support.

Let us suppose for the time being that the equilibrium is genuinely stable (collapse would not be a problem if it were unstable). Further compression then requires a change in one of the conserved quantities. Thus, we expect that many collapses proceed secularly through a succession of near-equilibria. For instance, if the equation of state is nearly adiabatic, there may be slow cooling via radiation; alternatively, there may be some small magnetic diffusivity that allows magnetic flux to escape gradually. Because radiation is relatively easy to accomplish, and, as we have just seen, magnetic flux alone cannot resist self-gravity, removal of angular momentum often ends up as the controlling factor regulating secular collapse. This is, of course, yet another version of the point emphasized in §14.1.

In this connection, it is important to recognize that the assumption of fixed shape in the scaling arguments of the previous paragraphs can be somewhat misleading. Both magnetic forces and rotational support are highly directional. Magnetic stress does not resist collapse along the field lines, while rotation does not prevent compression parallel to the rotation axis. Thus, even in the presence of substantial magnetic flux or rotation, collapse in one direction can be relatively easy. Indeed, that is clearly the case in disk galaxies: rotation supports them against radial collapse, but they are very thin in the axial direction.

Let us now focus on how a rotationally supported axisymmetric structure can lose angular momentum by magnetic torques. Imagine that the structure we are following has already contracted (by radiative cooling or the loss of some magnetic flux) to the point where it is supported primarily by rotation. The ease of contraction along field lines means that it is likely to remain connected by poloidal field to the gas remaining outside. At the same time, however, the rotation rate of the inner structure is considerably greater than the outer material, precisely because it has been compressed while conserving angular momentum. Consequently, the field lines with footpoints in the inner, rapidly rotating material, will begin to spin up the outer matter. This happens as fast as a torsional Alfvén wave can propagate outward, and continues until the wave reaches such large distances that the Alfvén speed falls below the speed of co-rotation.

An alternative view of the same process can be had from the vantage point of the tensor virial theorem (eqs. E.17, E.21). The net magnetic torque can be described completely by the antisymmetric part of the magnetic surface integral:

$$
\begin{aligned}
\frac{dJ_z}{dt} &= \frac{1}{8\pi} \oint dA \left(\hat{n} \cdot \vec{B} \right) \hat{z} \cdot \vec{r} \times \vec{B} \\
&= \frac{1}{8\pi} \oint dA \, \frac{\varpi B_\varpi + z B_z}{\sqrt{1 + (z/\varpi)^2}} B_\phi,
\end{aligned}
\tag{14.15}
$$

where ϖ is the cylindrical radius, and $B_{\varpi,z,\phi}$ are the components of the magnetic field in cylindrical coordinates. If the structure is geometrically thin, this expression reduces to the familiar statement that the torque is $(1/2)\varpi^2 \int dz \, B_\varpi B_\phi$ (eq. 14.11). As time goes on, the faster rotation of the central structure relative to distant material causes B_ϕ to grow and grow; this is the expression of the outward motion of the Alfvén wave described in the previous paragraph.

As the cloud shrinks, the importance of its self-gravity grows relative to the importance of external gravity (self-gravity is $\propto R^{-1}$; galactic potentials are often roughly logarithmic). When the equilibrium is best described by rotation resisting self-gravity, non-shape-preserving perturbations become important, and stability of the cloud depends on the relative importance of thermal and rotational support. This is the topic of the next subsection.

14.4.3 Bar-driven inflow

If the gravitational field can develop some asymmetric components, then there is no need to wait for (intrinsically weaker) magnetic forces to act. Gravitational torques alone can do the job. The question is, "How can they arise—and how strong can they be—when the underlying galaxy is fundamentally axisymmetric?"

First consider the case in which rotation is the dominant support against gravity, that is, a disk galaxy. Bars, spiral arms, and other non-axisymmetric features are often observed in these galaxies, although the surface brightness amplitude of these features may be considerably exaggerated relative to the amplitude of their associated gravitational potential due to the nonlinear relationship between star formation and potential depth. The creation and maintenance of these features is as yet only incompletely understood, but certain qualitative results appear to be well supported. In particular, there appears to be a close connection between the criterion for *local* stability against self-gravity and the ultimate creation of *global* features. Here we will provide only a brief account of this aspect of galactic dynamics; a much fuller treatment can be found in Binney and Tremaine (1987).

We begin by examining what happens when small perturbations are introduced into an infinitely thin fluid disk whose equilibrium is axisymmetric. The full equations of motion (in cylindrical geometry) are

$$\Sigma \left(\frac{\partial v_\varpi}{dt} + v_\varpi \frac{\partial v_\varpi}{\partial \varpi} + \frac{v_\phi}{\varpi} \frac{\partial v_\varpi}{\partial \phi} - \frac{v_\phi^2}{\varpi} \right) = -\Sigma \frac{\partial \Phi}{\partial \varpi} - c_s^2 \frac{\partial \Sigma}{\partial \varpi} \qquad (14.16)$$

and

$$\Sigma \left(\frac{\partial v_\phi}{dt} + v_\varpi \frac{\partial v_\phi}{\partial \varpi} + \frac{v_\phi}{\varpi} \frac{\partial v_\phi}{\partial \phi} + \frac{v_\phi v_\varpi}{\varpi} \right) = -\frac{\Sigma}{\varpi} \frac{\partial \Phi}{\partial \phi} - \frac{c_s^2}{\varpi} \frac{\partial \Sigma}{\partial \phi}, \qquad (14.17)$$

where Σ is the surface density, $v_{\varpi,\phi}$ are the velocity components in the (cylindrical) radial direction and the azimuthal direction, and we assume a fluid equation of state in which

$$\frac{\partial}{\partial \Sigma} \int dz\, p = c_s^2. \qquad (14.18)$$

The continuity equation in these coordinates has the form

$$\frac{\partial \Sigma}{\partial t} + \frac{1}{\varpi} \frac{\partial}{\partial \varpi} (\varpi v_\varpi \Sigma) + \frac{1}{\varpi} \frac{\partial}{\partial \phi} (v_\phi \Sigma) = 0. \qquad (14.19)$$

In equilibrium, all fluid motions within an axisymmetric disk must be circular, and there is no dependence of any quantity on ϕ. In the limit that

517

the fluid's support against gravity (whether intrinsic or external) is entirely due to rotation, it follows that the orbital frequency is given by

$$\Omega^2 = \frac{1}{\varpi}\frac{d\Phi}{d\varpi}. \tag{14.20}$$

Next suppose that the equilibrium is perturbed by fluctuations that vary $\propto e^{i(k\varpi+m\phi-\omega t)}$. The Fourier representation is clearly appropriate for the ϕ- and t-dependence (the disk's equilibrium is, of course, periodic in ϕ, and time-independent), but it is only valid for the ϖ-dependence when $k\varpi \gg 1$ because parameters such as Ω and Σ can vary significantly on the scale of ϖ. After linearization, equations 14.16, 14.17, and 14.19 become

$$i\left(m\Omega - \omega\right)\delta v_\varpi - 2\Omega\delta v_\phi = -ikc_s^2\delta\Sigma/\Sigma_o - ik\delta\Phi, \tag{14.21}$$

$$\left(2\Omega + \frac{d\Omega}{d\ln\varpi}\right)\delta v_\varpi + i\left(m\Omega - \omega\right)\delta v_\phi = -\frac{im}{\varpi}\left(\delta\Phi + c_s^2\delta\Sigma/\Sigma_o\right), \tag{14.22}$$

and

$$k\delta v_\varpi + im\delta v_\phi = \left(\omega - m\Omega\right)\delta\Sigma/\Sigma_o. \tag{14.23}$$

In the final equation, we have made use of the fact that $k\varpi \gg 1$ to eliminate several terms.

To find the perturbation to the gravitational potential due to the changes in surface density requires a bit more care. The linearized Poisson equation,

$$\left(\frac{ik}{\varpi} - k^2\right)\delta\Phi - \frac{m^2}{\varpi^2}\delta\Phi + \frac{\partial^2\delta\Phi}{\partial z^2} = 4\pi G\delta\Sigma\delta(z), \tag{14.24}$$

cannot be trivially reduced to two-dimensional form. Instead, we first impose what is called the "tight-winding" limit. That is, not only do we require $k\varpi \gg 1$; we also require that $k\varpi \gg m$. If we are interested in simple nonaxisymmetric structures such as bars, $m \sim O(1)$ anyway, so this is not a significant additional assumption. At this level of approximation, there is only one form for the z-dependence of $\delta\Phi$ that will be consistent with the δ function on the right-hand side: $\delta\Phi \propto e^{-|kz|}$. This z-dependence makes the left-hand side automatically zero for $z \neq 0$, while its discontinuity in slope at $z = 0$ creates a singularity in the second derivative at $z = 0$ that also matches the behavior of the δ function. Assuming that form for the z dependence of $\delta\Phi$, we can then integrate equation 14.24 in the vertical direction over a short distance that includes $z = 0$:

$$\lim_{\Delta z\to 0}\int_{-\Delta z}^{+\Delta z} dz\left(-k^2\delta\Phi + \frac{\partial^2\delta\Phi}{\partial z^2}\right) = 4\pi G\delta\Sigma,$$

$$\lim_{\Delta z\to 0}\left\{-2k^2\Delta z\delta\Phi + \left[\frac{\partial\delta\Phi(+\Delta z)}{\partial z} - \frac{\partial\delta\Phi(-\Delta z)}{\partial z}\right]\right\} = 4\pi G\delta\Sigma, \tag{14.25}$$

$$-2k\delta\Phi = 4\pi G\delta\Sigma.$$

Using the perturbed Poisson and continuity equations to solve for $\delta\Phi$ and $\delta\Sigma$ in terms of δv_r, we can then solve the ϖ and ϕ equations of motion. The result is most transparently posed in terms of the natural units of the problem. Guided by the fact that we are interested in self-gravity, we choose as the characteristic lengthscale the Jeans length, $\lambda_J \equiv c_s^2/(\pi G\Sigma_o)$. The orbital frequency Ω defines the primary characteristic timescale, but here there is also another, closely related, characteristic frequency, the epicyclic frequency

$$\kappa = \left(4\Omega^2 + \Omega\frac{d\Omega}{d\ln\varpi}\right)^{1/2}. \tag{14.26}$$

The epicyclic frequency (also called the "Lindblad frequency") is the frequency at which a particle following a circular orbit [with orbital frequency $\Omega(\varpi)$] oscillates in the radial direction if perturbed. In terms of these characteristic scales, the dispersion relation takes the form

$$(\omega/\Omega - m)^2 = \left(\frac{\kappa}{\Omega}\right)^2 \left\{1 + \left[(k\lambda_J)^2 - 2k\lambda_J\right]\right\} \left(\frac{\pi G\Sigma_o}{\kappa c_s}\right)^2. \tag{14.27}$$

The dimensionless ratio

$$Q \equiv \frac{\kappa c_s}{\pi G\Sigma_o} \tag{14.28}$$

has a special role in determining the character of these oscillations. First identified by Toomre (1964), it expresses the ability of the fluid to resist self-gravity. When it is large, only neutrally stable standing wave oscillations with $\omega = m\Omega \pm \kappa$ exist. However, when it is small (but nonzero), there is a range of wavelengths

$$1 - \sqrt{1 - Q^2} < k\lambda_J < 1 + \sqrt{1 - Q^2} \tag{14.29}$$

over which perturbations with $m = 0$ grow exponentially (Goldreich and Lynden-Bell 1965 showed that $m \neq 0$ modes are secularly damped by shearing). Just how small Q must be for an instability to exist is clear: because k is purely real by definition, there are at least some unstable wavelengths when $Q < 1$. Because the unstable wavelengths are all such that $k\lambda_J \sim 1$, the growth rate is always $\sim \kappa$.

Although its derivation (and also its final form) are a bit more complicated, the dispersion relation for a collisionless system (e.g., stars) is very similar to the one just found for a fluid. In the collisionless case, the criterion for instability is

$$Q \equiv \frac{\sigma_\varpi \kappa}{3.36 G\Sigma_o} < 1. \tag{14.30}$$

Here σ_ϖ is the radial velocity dispersion of the stars. This result, first found by Toomre (1964), is known as the *Toomre criterion*. When the gas is so highly clumped that the cloud-cloud collision time is longer than the dynamical time, it is effectively collisionless, and the clouds' contribution can be folded into the stars'.

Toomre's criterion makes a test for *local* stability very easy—all one need do is check whether the local random speeds and epicyclic frequency are great enough to balance the self-gravity of the local surface density. At the same time, it is easy to see that the "softer" the equation of state, that is, the smaller the random speeds of the system, the more susceptible it is to self-gravitating instability.

Although Toomre's criterion is most often applied in the context of a galactic disk, the argument behind it is equally valid when applied to other disks, for example, accretion disks. Consider for example, the gas-pressure-dominated disks discussed in §7.3.4.2. If the only heat source is local dissipation and most H remains ionized even at relatively large distance from the black hole, they become unstable to self-gravity outside a radius $x \sim 5000\alpha^{16/25}(L/L_{\rm E})^{17/25}L_{46}^{-1}$. In more realistic treatments, there are numerous effects that might alter this estimate quantitatively, such as other heat sources or recombination (Shlosman and Begelman 1989). Nonetheless, this simple estimate does indicate that accretion disk physics acquires a new wrinkle when disks extend relatively far from the black hole.

Whether local instabilities can merge together into a globally growing disturbance is another matter. To examine this question, we must now consider what happens not just in one locale, but across the entire galaxy. Suppose, for example, a tightly wound wave packet is created near a particular radius. It propagates radially with the group speed

$$
\begin{aligned}
v_g &= \text{sgn}(k)\pi G\Sigma_o \frac{k\lambda_{\rm J} - 1}{\omega - m\Omega} \\
&= \text{sgn}(k)\frac{k\lambda_{\rm J} - 1}{\omega - m\Omega}\frac{\Sigma_o}{\bar{\Sigma}}v_{\rm orb},
\end{aligned}
\tag{14.31}
$$

where $v_{\rm orb} \equiv \varpi\Omega$ and $\bar{\Sigma} \equiv \varpi\Omega^2/(\pi G)$ is a measure of the *mean* surface density of gravitating mass inside ϖ. Thus, these waves can propagate either in or out, and (for $k\lambda_{\rm J} \sim 1$), they do so at a speed that is smaller than the orbital velocity by the ratio of the local surface density to the equivalent "smoothed-out" surface density of the mass creating the galaxy's total potential.

To find the long-range propagation of such a wave packet, we fix the frequency ω and require that the wave satisfy appropriate boundary conditions at $\varpi = 0$ and the outer edge of the disk. As a first approximation to that solution, consider the WKB approximation, in which k is allowed to vary as a function of ϖ. This approximation reveals that several different sorts of barriers exist to long-range wave propagation.

Solving the dispersion relation for k in terms of ω gives

$$k\lambda_{\mathrm{J}} = 1 \pm \sqrt{1 + Q^2\left[\left(\frac{\omega - m\Omega}{\kappa}\right)^2 - 1\right]}. \qquad (14.32)$$

In order for the wave to propagate, k must be real. Where k is complex, the wave is evanescent, and its amplitude decays exponentially. This occurs wherever the argument of the square root is negative, which happens when the wave frequency is sufficiently close to the local orbital frequency:

$$\left(\frac{\omega - m\Omega}{\kappa}\right)^2 < 1 - \frac{1}{Q^2}. \qquad (14.33)$$

When $Q \gg 1$ (the disk is very stable locally), this band is quite wide; when Q is just slightly greater than unity (marginal stability), the evanescent zone is narrow.

Moreover, the WKB approximation must break down when $k = 0$, and this, too, can happen. As shown by the $\pm$ in equation 14.32, there are two possible wavelengths for each frequency ω. The longer wavelength mode can (formally) stretch to infinite length when

$$\omega - m\Omega = \pm\kappa. \qquad (14.34)$$

That is, when the wave frequency (as measured in a frame moving with the orbital frequency) matches the epicyclic frequency, the long wavelength mode must fail to satisfy the requirement $k\varpi \gg 1$.

A particularly interesting phenomenon, called *swing amplification* (Goldreich and Lynden-Bell 1965, Julian and Toomre 1966) can occur at such a resonance when the mode is a "leading wave" ($k < 0$). Because $d\Omega/dr < 0$, regions with enhanced density at smaller radius catch up in phase with high-density regions farther out. This causes a feature that may have initially been truly tightly wound to stretch out and unwind; that is, k decreases. When the wave frequency matches the epicyclic frequency, orbiting matter stays in phase with the wave for a significant time and can be strongly affected. If Q is not much greater than unity, the matter is susceptible to self-gravity, and the wave feature can be strongly amplified.

Thus, the parameter Q appears in two different guises: as a strict indicator of *local* stability, and as an approximate indicator of susceptibility to the global process of swing amplification. It can also be interpreted in yet a different vein by averaging it over the galaxy and applying the scalar virial theorem. At the order of magnitude level, the mean velocity dispersion can be identified with $[W/(3M)]^{1/2}$, the mean orbital frequency with $[|U_{\text{int}}|/(MR^2)]^{1/2}$, and the mean surface density with $U_{\text{int}}/(GMR)$. Here W is the volume-integrated pressure and U_{int} is the potential energy due to self-gravity. Combining these, we find that

$$\langle Q^2 \rangle \sim \frac{WM}{|U_{\text{int}}|}. \tag{14.35}$$

In the absence of magnetic forces or any external potential, the scalar virial theorem reduces to $2K + 3W = |U_{\text{int}}|$. Therefore, we can estimate

$$\langle Q^2 \rangle \sim 1 - \frac{MK}{|U_{\text{int}}|}. \tag{14.36}$$

That is, as the galaxy becomes increasingly supported by organized motions rather than random motions, in some sense its mean Q diminishes. Thus, we might expect that sufficiently flat galaxies would become unstable to large-scale nonaxisymmetric wave modes.

This is, in fact, observed in numerical simulations of galaxies. In that context, the relative proportion of support through organized motions is commonly measured in terms of the ratio of rotational kinetic energy to total potential energy (commonly called "T/W" in that literature, after its first use by Ostriker and Peebles 1973). When that ratio is more than $\simeq 0.14$, simulated galaxies develop strong bars. It was because our Galaxy is *not* dominated by an extremely strong bar that Ostriker and Peebles first suggested that a significant part of the Galaxy's mass is in some form, other than visible stars, whose velocity dispersion is much greater than the dispersion of stars in the Solar neighborhood. In our context, where bar creation may be desirable, the criterion may be interpreted in the opposite sense—a large ratio of rotational energy to potential energy may be conducive to fueling active nuclei.

A relatively large ratio of gas mass to stellar mass may also be favorable to bar formation. The velocity dispersion of interstellar clouds in our Galaxy is a factor of 3–4 smaller than the velocity dispersion of Population I stars. Consequently, for equal surface densities in the same location, the self-gravitating response of gas is stronger by that same factor, that is, the effective Q is rather smaller than the value given by the stars alone. If this contrast in characteristic random speeds is a generic feature of galaxies, a higher proportion of gas (as might have been the case when galaxies were first assembled) would make bar formation easier.

When present (see §13.2.4 for a summary of the observational evidence on whether the hosts of AGNs are especially likely to be barred), bars can be effective agents for driving matter inward. As we have already discussed, their direct gravitational torque tends to push matter at radii inside corotation further inward. In addition, the noncircular orbital motions induced by a bar can drive fluid mixing by supporting strong shock waves along almost its entire length (Roberts, Huntley, and van Albada 1979; Athanassoula 1992: fig. 14.3). The dissipation that creates new entropy in the shock wave simultaneously mixes the angular momentum of the pre- and post-shock gas. As a result, in a strong shock the specific angular momentum of the gas can easily change by order unity as it passes through the shock front, and bars tend to drive the gas they encounter inward (Schwarz 1984).

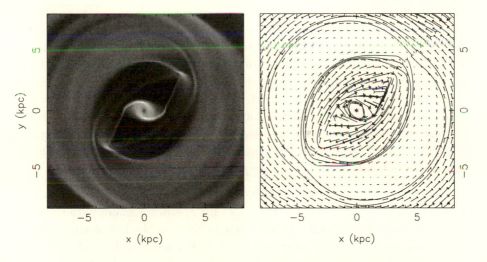

Fig. 14.3   Gray-scale representation (in the bar frame) of gas flow in a bar potential (from Athanassoula 1992). (Left) Lighter tone indicates higher density. Note the pair of standing shock waves and the scouring of gas from much of the rest of the bar's body. (Right) The arrows represent gas velocity (again, with respect to the bar reference frame), with the arrow length proportional to the speed.

For several reasons, the gas inflow triggered by a bar may enhance the importance of gas self-gravity in the inner portions of a galaxy. First, shocks are intrinsically compressive, so the local density of the gas grows relative to the stars. In addition, the decrease in the gas's specific angular momentum caused by the bar's shock wave preferentially concentrates the gas in the inner part of the galaxy. Finally, although the shock initially heats the gas, much of that heat is likely to be radiated away; by contrast, the increase in stellar random speeds caused by an encounter with a bar is never lost (Norman et al. 1996). Thus, the net result of bar action may be to create a special region in the inner region of the galaxy where the conditions are especially propitious for self-gravitating collapse of interstellar gas (see, e.g., Shlosman, Frank, and Begelman 1989).

14.4.4 Encounters

Several lines of thought all suggest that galactic encounters may help ignite nuclear activity in galaxies. On the empirical side, there have long been hints that the hosts of AGNs are especially likely to have near neighbors, or to be somehow "disturbed" in morphology (see §§13.2.4 and 13.7 for critical discussions of the observational evidence). Then there is the obvious point that an encounter with another galaxy breaks the axisymmetry of the gravitational potential in a galaxy, permitting strong gravitational torques to be exerted. Finally, there is the fact that the AGN phenomenon was so much stronger at redshifts of a few than it is today. While the peculiar velocities of galaxies may not have changed drastically since that epoch, their density has diminished by an order of magnitude or more. The encounter rate must therefore have been much greater then. If galaxies formed by a process of hierarchical merging from small proto-galactic fragments, the density of such fragments must have been very high, and the associated collisions similarly frequent.

Detailing the events of a galaxian encounter is a complicated effort. Simply defining an example is complex, for there are many parameters that influence the outcome. Before even beginning, the initial structures of the interacting galaxies must be specified. For a disk galaxy, the list of parameters begins with the total mass of the system, and the three ratios that define the relative mass contained in gas, disk stars, bulge stars, and dark matter. The surface density of disk stars is well approximated by an exponential; the scale length of this exponential is another free parameter. Conventionally, the dark matter in disk galaxies is thought to have an isothermal velocity distribution, so that it is distributed spherically with a density that falls $\propto (r^2 + r_c^2)^{-1}$. The core radius r_c of the dark matter defines the asymptotic rotation speed of the galaxy. Given a ratio of luminosity to mass, the empirical Tully-Fisher relation can then be used to relate r_c to the total mass. Stars in the bulge are thought to be similarly isothermal, but with a smaller core radius. The thickness of the stellar disk is set by choosing a *rms* random speed for the disk stars, and supposing that their ellipsoid of peculiar velocities is spherical. Finally, the thickness of the gaseous disk is determined by the gas temperature or the *rms* random speeds of interstellar clouds (depending on whether the interstellar medium is pictured as more nearly continuous or clumped). Although it is simplest to suppose that both the random speeds of the stars and the temperature of the gas are independent of radius, one could easily imagine making both functions of position. Thus, eight free parameters are required merely to define the initial state of each interacting disk galaxy. Specifying the initial state of an elliptical galaxy is far more complicated because to do so requires defining the entire stellar distribution function.

After defining the initial states of the galaxies, characterizing the collision itself requires more choices: the relative speed of the collision (or alternatively, the kinetic energy of the collision); the (projected) impact parameter (or alternatively, the orbital angular momentum); and four angles to describe the orientation of the two galaxies relative to the orbital angular momentum of the encounter. With such a large universe of possibilities for the initial conditions, it should come as no surprise that a large number of outcomes are possible.

If we knew everything there is to know about mass and energy interchange between stars and interstellar gas (see §§14.3.5, 14.3.6), we would be able to compute the dynamics of interstellar gas from first principles. As was made clear in those earlier sections, we are far from being able to do so. This means that more choices must be made that have far-reaching implications: how star formation depends on local conditions, and the effective equation of state of the gas. The latter function hides all the complications of mass dropout and re-injection, as well as all the heating and cooling processes at work in the gas.

Even without the subtleties of the interstellar equation of state, the mechanics of galaxy encounters would be far too complicated for analytic approaches. Consequently, this is an area where simulation techniques reign supreme. Because numerical simulations involve great quantities of a rare commodity (time on large, fast computers) and, as we just demonstrated, there are many different varieties of galaxy encounters, results are largely anecdotal. Nonetheless, certain features stand out as relatively insensitive to detailed parameter choices.

To highlight these points, we will contrast two encounters, one weak and one strong. In the interest of specificity (and simplicity), let us assume that the net orbital energy (ignoring internal motions and binding energy) is zero. Suppose first that the mass of one of the galaxies is much smaller than the mass M_g of the other, and the distance of closest approach a is far outside the radius of even the larger galaxy. Relative to the system center of mass (very near the center of the larger galaxy), at pericenter the orbital frequency of the little galaxy is $(GM_g/a^3)^{1/2}$, which is much less than the orbital frequency for any of the stars in the larger galaxy (and very likely much smaller than the orbital frequency of stars in the smaller galaxy around its center of mass). The weak tidal force felt by the stars on the outer fringes of both galaxies can be viewed as perturbations relative to the much stronger force they feel from their own galaxies. Because this perturbation oscillates at a frequency much smaller than the stars' epicyclic frequency, the stars respond only weakly. Deeper inside the galaxies, the tidal force is much weaker and the frequency mismatch even greater, so there is even less of an impact.

Next consider a stronger encounter in which both galaxies have mass M_g, and the distance of closest approach a is now not so large compared to their sizes. As they approach pericenter, the galaxies' angular rotation rate around the system center of mass approaches $\Omega_a = 2(GM_g/a^3)^{1/2}$. Meanwhile, if the orbit could be smoothly extrapolated to pericenter as if the galaxies were point masses, stars inside the galaxies would rotate around the system center of mass at the rate

$$\Omega_* = \frac{\Omega_a}{1 - 2r_f/a}\left[1 - g\left(\frac{fa}{r_f}\right)^{1/2}\right], \tag{14.37}$$

where r_f is the distance from a particular star to its galaxy's center, and the fraction of that galaxy's mass enclosed within r_f is f. The factor g expresses the relative sense of rotation. It is 1 for stars on the near side of the galaxy (the ones that feel the greatest force from the other galaxy) if the galaxy rotates in the same sense as the orbital angular momentum, and -1 for near-side stars in the event of the opposite sense of rotation.

In this much stronger collision, the tidal forces are no longer mere perturbations. As the expression for Ω_* shows, if a galaxy rotates in the prograde sense, the *net* rotation rate of the near-side stars can come close to matching the rotation rate of the collision. This near-coincidence of rotation rate keeps the direction of the tidal force nearly constant for a large part of the collision. Thus, galaxies rotating parallel to the orbital angular momentum are affected the most strongly, but retrograde galaxies can also be significantly disturbed, albeit not quite as much.

When stars feel a consistent strong tidal force, they are drawn together in that direction. This bunching creates dynamical friction much as described in §14.2.3. As a result, the relative speed of the two galaxies diminishes, and they are able to come closer together. At the same time, the bunching forced by the tides creates barlike features (see fig. 14.4). Stronger bars are created in galaxies that (as discussed in §14.4.3) are more susceptible to them even without the stimulus of a collision. These bars cause strong dissipation in the gas of the galaxies, driving it toward their centers. At this stage of the collision, the ability of tidal forces to distort the colliding galaxies can be viewed as an inelasticity mechanism in which orbital energy is transferred into internal energy of the galaxies.

If the two galaxies are brought close enough, the interaction can become much stronger. Each galaxy's collisionless component (stars plus dark matter) now finds itself moving through a relatively high density of background stars and dark matter, and therefore subject to strong dynamical friction. The fluid components of the two galaxies shock against each other and lose much orbital energy as the shock wave converts it into heat, and the heat is radiated away. During this portion of the collision, the gravitational potential is far from symmetric, and strong gravitational torques transfer angular momentum outward. The gaseous component tends to be even more strongly affected than the stars for two reasons. It loses angular momentum more rapidly than the stars because the linear features in the potential generally lag behind the gas. In addition, the gas, is, of course, dissipative, so it loses energy rapidly, which the stars cannot do at all. As a result, the interstellar media of the galaxies can end up strongly concentrated toward the center of the resulting merged galaxies (Barnes and Hernquist 1992, 1996).

Thus, although the initial impact of galaxy encounters is felt only on the outskirts of galaxies, a sequence of events makes it possible for encounters to squeeze gas much closer to the centers of the affected galaxies.

14.5 Some Open Questions

In the previous section we described several mechanisms that might all help to convey material inward from galactic scales by an order of magnitude or two. What befalls the gas there is less clear.

An enhanced rate of star formation is one possibility (§13.6). It would actually be rather surprising if much of the inflowing gas did *not* drop out into stars (§14.3.5). Support for this thought may be found in the fact that

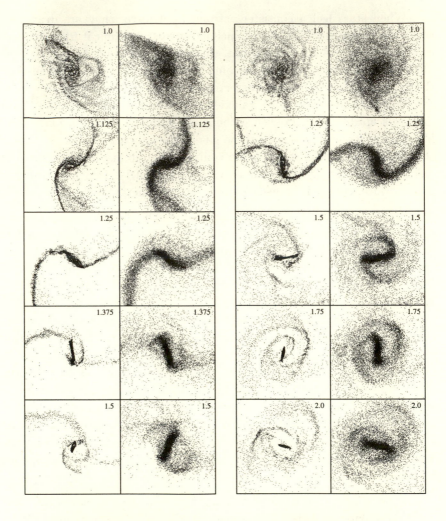

Fig. 14.4   Surface density of stars and gas in two initially identical disk galaxies during part of a close encounter (Barnes and Hernquist 1996). Time (in units of 250 Myr) is shown in the top right-hand corner of each image. After roughly another 200 Myr, these galaxies merge. The initial rotation axis of one of the galaxies (shown in the left-hand column) was parallel to the orbital angular momentum; the initial rotation axis of the other was inclined by 71°. Within each column, the left-hand series shows the gas, the right-hand series the stars. The gas, whose mass was conserved throughout the simulation at 10% of the mass in the stellar disks, was given an equation of state with "realistic" cooling rates that generally kept its temperature near 10^4 K. Note how in both galaxies the gas swiftly becomes far more centrally concentrated than the stars.

there are a number of nearby AGNs in which star formation does appear to be proceeding at an unusually high rate within the central kpc or so (Heckman et al. 1995).

The most dramatic cases of rapid star formation are the ultra-luminous infrared galaxies, present-day galaxies in which the rate of star formation appears to be extremely high. Almost all of them appear to be in the midst of a stong interaction with a neighboring galaxy (Sanders et al. 1988). This fact appears to endorse the prediction made by galaxy encounter simulations that these events do in fact concentrate gas in the centers of interacting galaxies. Whether such an event subsequently leads to creation of an AGN is an important open question. At the very least, the many massive stars created in these events must surely spawn a sizable number of new black holes, one of which may be the seed for an eventual AGN.

As stressed in Chapter 3, the frequency of very luminous AGNs was far greater at redshifts of a few than today. It is possible that in these AGNs the central gas accumulation was the result of merging gas-rich proto-galactic fragments before they form many stars. If so, an enhanced rate of star formation due to such an accumulation could actually be the very process of galaxy formation. In this case, locating high-redshift AGNs might be a fruitful way of finding galaxies in a crucial stage of their lifecycle.

A major uncertainty shared by all these speculations is whether mechanisms that push matter from, say, a few kpc to a few hundred pc, are then able to hand off to new mechanisms that carry the matter to a few tens of pc or closer. As remarked in the context of bar-driven inflow, it may be that concentrating gas near the center of a galaxy by one means enhances the probability that an independent mechanism will be set off to carry it in the next factor of ten in radius. Final collapse by self-gravity may be one such mechanism. There may also be "way-stations" in the form of dense stellar clusters in which matter is stored for a time, but then ultimately discharged into the heart of the galactic nucleus (David, Durisen, and Cohn 1987ab; Norman and Scoville 1988). Perhaps the fraction of galaxies with active nuclei is determined by the probability of success in these hand-offs.

Finally, it is worth pointing out that the converse of not understanding what drives global mass flows in galaxies is ignorance of what stops them. Squelching mechanisms may be as important to the evolution of nuclear activity in galaxies as mechanisms that initiate them. What extinguishes nuclear activity, and whether that extinction is permanent, are major questions that remain entirely open.

15 Where We Stand

Having come this far, it is now appropriate to look back and see what we have accomplished. As in any ongoing field of research, there are certain questions whose answers now seem quite clear, others about which we have some plausible ideas, and still others—some among them extremely important—about which we haven't a clue. This final chapter will present a summary of where the field stands as of the date of publication of this book. For readers not long after, it may be useful as a guide both for attaching confidence levels to ideas, and for selecting new research topics. For readers several years or more in the future, it may serve as an amusing record of a time of ignorance.

15.1 Accomplishments

Arguably the most important question about AGNs for which we have a good answer is the nature of their central engines. The arguments that extremely massive accreting black holes lie at the center of all AGNs are quite strong:

- Without relativistic accretion efficiency it would be hard to satisfy their fuel budgets (§4.2).

- What else could produce Fe Kα line profiles many tens of thousands of km s^{-1} wide (§8.8)?

- When $\sim 10^7 \, M_\odot$ of gravitating mass are found within a fraction of a pc of the center of a galaxy (§4.4), what other dynamically stable form could it take but a black hole?

Many of the radiation mechanisms in AGNs are also now clearly identified:

- No other mechanism competes with the synchrotron process (§9.2) for its ability to produce polarized emission from radio frequencies up to the optical band and beyond.

- Thermal dust radiation (§12.4.4) is far and away the best candidate to generate infrared continua over extended spatial regions, as is seen in some Seyfert galaxies. The similarity of spectral shape seen in quasars suggests that their infrared power is made in the same way.

• Only photo-ionization (Chap. 10) can explain emission lines that vary—after an appropriate time delay—in synchrony with variations in the ionizing continuum. It is also the most natural explanation for high-ionization emission lines generated at relatively low temperatures, and equivalent widths that hardly vary over many orders of magnitude in continuum luminosity.

• Thermal Comptonization (§8.3) by hot electrons of modest Compton optical depth is an excellent explanation for the cutoff power-law shape of the X-ray spectrum in most AGNs.

• Inverse Compton scattering by nonthermal relativistic electrons (also discussed in §8.3) seems equally unassailable as a device for creating the hard X-ray/γ-ray continuum seen in flat-spectrum, core-dominated radio-loud AGNs such as BL Lac objects and optically violently variable quasars.

Third, there can now be little doubt that the appearance of most AGNs depends strongly on the direction from which they are viewed. This anisotropy of appearance accounts for some of the multitude of categories into which AGNs have been historically divided.

• Relativistic kinematics of some sort are absolutely required to explain the superluminal expansion of structure in the cores of radio-loud AGNs (§9.3.3.2). Relativistic beaming is equally necessary if high-energy γ-rays are to emerge from their source regions with the strength that we see (§8.8.2). Directed relativistic motion of a radiating object automatically creates strong anisotropy of emission (§12.2.1). There is strong reason to believe that relativistic beaming is the device that distinguishes flat-spectrum, core-dominated radio-loud AGNs from steep-spectrum, lobe-dominated radio-loud AGNs (§12.5).

• Partial obscuration (§12.3) can also strongly modulate the appearance of an AGN as a function of viewing angle. This effect transforms broad-line AGNs, both radio-quiet (§12.6) and radio-loud (§12.7), into narrow line objects as the observer's line-of-sight rotates.

15.2   The Future

Despite these successes, large gaps definitely remain in what we understand about AGNs.

Perhaps the foremost of these is why they should occur at all, and why they exhibit the distribution of luminosities they do (Chap. 14). Their sources of accretion fuel and the devices that regulate its flow are both very largely unknown.

From today's standpoint, it is a reasonable guess that the strong peak in AGN activity near $z \simeq 2$ is closely related to the fact that most galaxies were still quite young at that time. Just why that is, and what influence the presence of an AGN may have on the subsequent evolution of its host galaxy, are both important unanswered questions.

Another respect in which the nature of the host galaxy somehow influences the character of AGN activity in a fundamental, yet still mysterious, way is the correlation between host morphological type and whether the AGN is radio-loud or radio-quiet (§13.2.3).

Relativistic jets are at the bottom of much of the phenomenology of radio-loud AGNs. Because they are created in the innermost regions of the central engine, the connection with host morphology is especially perplexing. However, we also lack any understanding of the dynamics that accelerate so much matter to such high speeds (§9.3.2). This ignorance of outflow dynamics also includes the slower flows evidenced by intrinsic absorption systems (Chap. 11).

Last, although we are confident we have identified the principal radiation mechanisms employed by AGNs, we do not have any clear idea why most of them should be in operation. Although thermal radiation that peaks in the ultraviolet is a very natural consequence of accretion on this scale, we have no fundamental understanding of why hotter plasma should be generated to inverse Compton scatter photons up to X-ray energies, or why relativistic electrons, the sine qua non of both the synchrotron and inverse Compton mechanisms, should be accelerated in large numbers. Similarly, we do not know the origin of the cooler gas that, when photoionized by the AGN continuum, produces the observed emission lines. It is for this reason that Chapter 7 began with well-motivated theory, and then compared its predictions to observations, while Chapters 8, 9, and 10 all began with phenomenology intended to motivate the theoretical ideas presented later in those chapters. If we only knew that AGNs are accreting massive black holes, we might have predicted that they would be strong UV continuum sources; it is extremely doubtful that we would have been able to predict anything else about their spectra.

With this enumeration of successes and challenges for the future, this book closes.

A   Basic General Relativity

General relativity is the relativistic theory of gravity. In the context of AGNs, it is an essential part of two subjects: defining the large-scale "geography" of the Universe, and the dynamics of material near black holes.

Obviously, a thorough treatment of general relativity cannot fit within the scope of this book. Instead, as results are needed, they are introduced in a manner that is very "once over lightly." This appendix provides a very concise overview and a presentation of the notation that will be standard in this book.

Several good texts on general relativity provide more background: Weinberg (1972) and Misner, et al. (1973) are particularly recommended.

From special relativity, we learned that measurements of space and time depend on one's frame of reference in such a way as to preserve the dynamical laws of physics, for example, the laws of electromagnetism which define the speed of light. As a result, there is no experiment by which one can ascertain the "absolute velocity" of a constant-velocity frame of reference. That is, there is no frame that can be said to uniquely specify the rest frame of the Universe, for all can be seen to be moving if one simply changes frame of reference.

We now ask the question, "Can this equivalence of frames be generalized to frames of reference with changing velocity?" One might answer (too facilely) by saying, "No. One can always distinguish accelerating frames by the forces that appear in them." However, there is a common experience that belies this response, namely that of free-fall. If a frame of reference is freely falling in a uniform gravitational field, the only "fake force" that appears is the one that cancels the apparent gravity. Consequently, the dynamics due to all other forces act in a way that is indistinguishable from a constant-velocity frame. In fact, because the acceleration of the frame identically cancels the gravitational acceleration, mechanics becomes *simpler* than viewed from a constant-velocity frame.

On this basis, let us try the supposition that the laws of physics act in the same way in all freely falling frames. In regions of weak gravity, this definition includes constant-velocity frames. However, there is an important difference between the way these frames are defined and the way constant-velocity frames are defined in special relativity: because gravitational fields always vary from place to place, a single freely falling frame of reference can only be defined locally. In a Universe containing matter that is not spread with perfect homogeneity, the local gravitational acceleration *must* change magnitude and direction from place to place. Consequently, there is no one frame that is freely falling everywhere, so "inertial" (i.e., freely falling) frames in general relativity can only be defined locally.

As we know from special relativity, distances $\vec{x}$ and time intervals t depend on the reference frame in which they are measured, but there is a combination of distance and time, $s^2 = c^2 t^2 - |\vec{x}|^2$, which is invariant— that is, s^2 is the same no matter in what reference frame it is measured. The same statement also holds for many other quantities—momentum and energy, currents and densities, and many more. For this reason, it is convenient to define a new sort of geometric quantity, the four-vector, whose length is defined to be the invariant combination of its components. We can express the existence of the associated invariants in a special notation:

$$s^2 = x^\mu \eta_{\mu\nu} x^\nu. \tag{A.1}$$

Here x^μ is a four-vector, in which the timelike component occupies the first place, and is followed by the three components of the spacelike vector. The indices μ, ν then run from 0 to 3. The matrix η has the form

$$\eta_{\mu\nu} = \begin{pmatrix} 1 & 0 & 0 & 0 \\ 0 & -1 & 0 & 0 \\ 0 & 0 & -1 & 0 \\ 0 & 0 & 0 & -1 \end{pmatrix}, \tag{A.2}$$

and we have adopted the "Einstein convention," in which indices found once in a superscript and once in a subscript are summed over. Viewed in this light, $\eta_{\mu\nu}$ can be reinterpreted as a "metric"—the matrix that tells us how to define the invariant length of a four-vector.

Writing the indices as either subscripts or superscripts is not an arbitrary matter. The reason is that there are two classes of four-vectors, distinguished by their Lorentz transformation properties. Quantities that transform like ordinary spatial four-vectors (e.g., four-momenta or four-currents)) are called "contravariant"; quantities that transform like gradients with respect to spatial four-vectors (i.e., like $\partial/\partial x^\mu$) are called "covariant" (not to be confused with "covariant" equations, those written purely in terms of four-vectors, tensors, and so on).

To see the difference, consider a Lorentz transformation

$$y^\nu = \Lambda^\nu_\mu x^\mu, \tag{A.3}$$

534

where the transformation matrix has the usual form

$$\Lambda^{\nu}_{\mu} = \begin{pmatrix} \gamma & -\gamma\beta & 0 & 0 \\ -\gamma\beta & \gamma & & 0 \\ 0 & 0 & 1 & 0 \\ 0 & 0 & 0 & 1 \end{pmatrix}, \tag{A.4}$$

and we have used the conventional notation, that is, the relative speed between the two frames is β, $\gamma = (1 - \beta^2)^{-1/2}$, and the relative velocity is directed along the first spatial axis. Because the relationship between x and y is linear, the partial derivative matrix

$$\frac{\partial y^{\nu}}{\partial x^{\mu}} = \Lambda^{\nu}_{\mu}. \tag{A.5}$$

This fact tells us how to transform gradient four-vectors:

$$\frac{\partial}{\partial y^{\nu}} = \frac{\partial x^{\mu}}{\partial y^{\nu}} \frac{\partial}{\partial x^{\mu}} = \left(\Lambda^{-1}\right)^{\mu}_{\nu} \frac{\partial}{\partial x^{\mu}}. \tag{A.6}$$

Because $\Lambda \neq \Lambda^{-1}$, we must distinguish these two classes of four-vectors.

With this technical aside, we are now ready to generalize this formalism to include the effects of gravity. The essential statement of general relativity is the Einstein field equation

$$R_{\mu\nu} - \frac{1}{2}R g_{\mu\nu} + \Lambda g_{\mu\nu} = -8\pi G T_{\mu\nu}, \tag{A.7}$$

where the tensor $R_{\mu\nu}$ and the scalar R are both related to derivatives of the metric that describe the intrinsic curvature of space-time, and $T_{\mu\nu}$ is the total stress-energy tensor. This equation tells how, in the presence of gravitating masses and their associated energy, the metric (written typically as $g_{\mu\nu}$) departs from the simple form of $\eta_{\mu\nu}$. In general relativity, the diagonal elements of the metric are not necessarily all 1's and -1's, and the nondiagonal elements can also be nonzero. Metrics are, however, symmetric, and, by the equivalence principle, in any freely falling frame, $g_{\mu\nu}$ does reduce to $\eta_{\mu\nu}$.

Reexamining equation A.1, we now see that metrics have another function: they provide the relationship between contravariant and covariant descriptions of the same physical four-vectors:

$$x_{\mu} = g_{\mu\nu}x^{\nu}, \tag{A.8}$$

so that

$$s^2 = x^{\mu}x_{\mu} = x^{\mu}g_{\mu\nu}x^{\nu}. \tag{A.9}$$

Metrics can also be generalized to operate on higher-rank tensors than vectors. The same basic rules prevail, however: indices repeated in covariant and contravariant locations should be summed over, and one multiplication by the metric transforms one index from contravariant to covariant, or (if the contravariant form of the metric is used) the other way around.

Because the elements of $g_{\mu\nu}$ can be anything, it is no longer necessary that the components of four-vectors be defined with respect to orthonormal bases. Any basis set will do, provided it spans the full four-space. It will therefore be much more important to distinguish between "coordinate" quantities and "proper" quantities or invariant quantities than it normally is in special relativity.

Coordinate quantities are those measured with respect to whatever peculiar axes we choose to label points. They have *no* physical significance beyond that of labels. "Proper" quantities are those measured with respect to certain special frames. The proper time between two events is the invariant interval defined by the magnitude of the four-vector connecting the two events (it is the name in general relativity for the invariant interval s of special relativity introduced in eq. A.1). In particular, it is often of interest to measure the proper time between two events that occur in the frame of motion of a freely falling particle, for example, the emission of two photons. In that case, the proper time is the time between the two events as measured in the rest frame of that freely falling particle. Proper distance, on the other hand, is the spatial length between two events measured in a frame for which the two events are simultaneous. This is, in some sense, the most natural way to describe the length of an object in its own rest frame.

B Basic Magnetohydrodynamics

Astrophysical gases nearly always have ionized fractions large enough that interaction with organized electric and magnetic fields is important. If one were to be complete in the treatment of all these interactions, it would be necessary to follow all the different subvarieties of particles. That is, one would be forced to solve the kinetic equations for all the different sorts of ions as well as for the electrons. The associated "plasma" effects can be extremely complicated, and lengthy textbooks present very partial introductions to the subject (Krall and Trivelpiece 1973 is one good example, but many others also exist).

Fortunately, in many astrophysical contexts, a much simpler picture, known as the *Magnetohydrodynamics* or *MHD* approximation, contains almost all the interesting physics. In this approximation, the gas is treated as a single fluid with macroscopic properties: density ρ, velocity $\vec{v}$, pressure p, and electrical current $\vec{J}$. The primary conditions required for this approximation to be valid are that any gradients have lengthscales that are very long, and any time-variable quantities do so very slowly. These conditions will be clarified below.

The governing equations for fluid dynamics in the MHD approximation are an equation of state (or a heating and cooling equation) and the mass continuity equation

$$\frac{\partial \rho}{\partial t} + \nabla \cdot \vec{v} = 0, \tag{B.1}$$

the force equation:

$$\rho \left[\frac{\partial \vec{v}}{\partial t} + \vec{v} \cdot \nabla \vec{v} \right] = \frac{1}{c} \vec{J} \times \vec{B} - \nabla p + \vec{F}_{\text{ext}}, \tag{B.2}$$

Ohm's law:

$$\vec{E} + \frac{1}{c} \vec{v} \times \vec{B} = \vec{J}/\sigma, \tag{B.3}$$

and Maxwell's equations, of which Faraday's law and Ampère's law are of special interest:

$$\nabla \times \vec{E} = -\frac{1}{c} \frac{\partial \vec{B}}{\partial t} \tag{B.4}$$

and

$$\nabla \times \vec{B} = \frac{4\pi}{c} \vec{J}. \tag{B.5}$$

Almost all the terms in these equations are familiar, but a few require comment. First, from a point of view exhibiting slightly more detail, the current $\vec{J} = |e|(\langle Z\rangle n_i \vec{v}_i - n_e \vec{v}_e)$, where $|e|$ is the absolute value of the electron charge, $n_{i,e}$ and $\vec{v}_{i,e}$ are the density and velocity of the ions and electrons separately, and $\langle Z\rangle$ is the mean charge of the ions in electron units. The total Lorentz force on the fluid is then

$$\vec{F}_{\mathrm{L}} = |e| \left(\langle Z\rangle n_i - n_e\right) \vec{E} + \frac{1}{c}\left(\langle Z\rangle n_i \vec{v}_i - n_e v_e\right) \times \vec{B}. \tag{B.6}$$

Free charges will stream so as to cancel out local net charge densities on any lengthscale large compared to the Debye length

$$\lambda_{\mathrm{D}} = \left(\frac{kT}{4\pi n_{i,e} e^2}\right)^{1/2}. \tag{B.7}$$

Because MHD is a long lengthscale limit, we set $\langle Z\rangle n_i = n_e$, and the Lorentz force reduces to

$$F_{\mathrm{L}} = \frac{1}{c}\vec{J} \times \vec{B}. \tag{B.8}$$

Second, the term $\vec{F}_{\mathrm{ext}}$ in the force equation represents all other possible bulk forces acting on the fluid. This most notably includes gravity but could also include radiation pressure, or inertial forces due to viewing these events in an accelerating frame.

Third, σ in Ohm's law is the electrical conductivity. In general σ depends on the frequency of the electric and magnetic fields; here we evaluate it in the low-frequency limit. In the great majority of astrophysical applications, the conductivity is so large that it is effectively infinite. The appropriate dimensionless quantity for defining the applicability of the infinite conductivity limit is the *magnetic Reynolds number*

$$R_M = \frac{4\pi L V \sigma}{c^2}, \tag{B.9}$$

where L is the characteristic scale of the gradient in magnetic field and V is the characterisic scale of the fluid velocity. This dimensionless ratio compares the time required for magnetic field to diffuse through the gas relative to the characteristic dynamical timescale.

Fourth, Ampère's law in the MHD equations is written without the displacement current $(1/c)(\partial \vec{E}/\partial t)$. This is another expression of the low-frequency limit at the heart of the MHD approximation.

Having made these comments, it is possible to be more precise about just how long the lengthscale and timescale must be for the MHD approximation to apply. The essential point is that they should be long enough to make any quantities dependent on kinetic properties negligible. In addition to $L \gg \lambda_D$, this stipulation requires

$$\left(\frac{r_{Li}}{L}\right)^2 \ll \frac{1}{T\Omega_i} \ll 1, \tag{B.10}$$

where T is the characteristic timescale, r_{Li} is the ion Larmor radius and Ω_i is the ion cyclotron frequency. These conditions ensure that details of the ionic cyclotron motion are immaterial; if $T_i = T_e$, the electron cyclotron motion extends over much smaller lengthscales and much shorter timescales than the ion cyclotron motion, so if L and T are large enough compared to the ion quantities, they are certainly large enough compared to the electron quantities.

An immediate consequence of the large R_M limit of Ohm's law is that the fluid cannot slip across magnetic field lines. This condition is known as *magnetic flux-freezing*. Taking the curl of Ohm's law, and using Faraday's law to substitute for $\nabla \times \vec{E}$ results in

$$\frac{\partial \vec{B}}{\partial t} = \nabla \times \left(\vec{v} \times \vec{B}\right) - \frac{c}{\sigma}\nabla \times \vec{J}. \tag{B.11}$$

The large R_M (or large σ) limit transforms this equation into a kinematic restriction binding the time variation of the magnetic field to the fluid velocity:

$$\frac{\partial \vec{B}}{\partial t} = \nabla \times \left(\vec{v} \times \vec{B}\right). \tag{B.12}$$

To interpret equation B.12 geometrically, imagine that we wish to measure the time variation of the magnetic flux Φ_M passing through a small surface A that moves with the fluid. Then

$$\frac{d\Phi_M}{dt} = \frac{d}{dt}\int_A dS\,\hat{n}\cdot\vec{B}, \tag{B.13}$$

where $\hat{n}$ is the local surface normal and the time derivative is a *convective* derivative $\partial/\partial t + \vec{v}\cdot\nabla$. Applying the convective derivative and using the vector identity

$$\nabla \times (\vec{B}\times\vec{v}) = \vec{v}\cdot\nabla\vec{B} - \vec{v}\left(\nabla\cdot\vec{B}\right), \tag{B.14}$$

yields

$$\frac{d\Phi_M}{dt} = \int_A dS\,\hat{n}\cdot\left[\frac{\partial\vec{B}}{\partial t} + \nabla\times(\vec{B}\times\vec{v})\right]. \tag{B.15}$$

Thus, the infinite conductivity limit of Ohm's law implies that the magnetic flux through a fluid surface does not change.

When R_M is finite, plasma can slip across field lines. Returning to equation B.11, we can use Ampère's law to define the current density in terms of the magnetic field and find that the magnetic field changes according to

$$\frac{\partial \vec{B}}{\partial t} = \nabla \times \left(\vec{v} \times \vec{B} \right) + \frac{c^2}{4\pi\sigma} \nabla^2 \vec{B}. \tag{B.16}$$

Finite conductivity (nonzero resistivity) allows field to decay on a timescale

$$t_d \sim \frac{4\pi\sigma L^2}{c^2} = R_M \frac{L}{V}, \tag{B.17}$$

that is, t_d is approximately the flow-time times R_M. Clearly, magnetic field decay occurs most readily in regions where $\vec{B}$ changes very quickly.

It is also convenient to use Ampère's law to remove explicit mention of the current from the force equation. We then have

$$\rho \left(\frac{\partial \vec{v}}{\partial t} + \vec{v} \cdot \nabla \vec{v} \right) = \frac{1}{4\pi} (\nabla \times \vec{B}) \times \vec{B} - \nabla p + \vec{F}_{\text{ext}}; \tag{B.18}$$

The vector identity

$$\frac{1}{2}\nabla B^2 = \vec{B} \cdot \nabla B + \vec{B} \times \nabla \times \vec{B} \tag{B.19}$$

allows a further simplification to

$$\rho \left(\frac{\partial \vec{v}}{\partial t} + \vec{v} \cdot \nabla \vec{v} \right) = \frac{1}{4\pi} \vec{B} \cdot \nabla \vec{B} - \nabla \left(p + \frac{B^2}{8\pi} \right) + \vec{F}_{\text{ext}}. \tag{B.20}$$

This form also provides the enlightening interpretation that fluids tend to be expelled from regions of high magnetic pressure in exactly the same way that they are pushed away from regions of high gas pressure, although the presence of the $\vec{B} \cdot \nabla B$ term reminds us that the field direction also matters.

The dynamics of any MHD system are then defined by equations B.1 (the mass continuity equation), B.12 (the flux-freezing condition), B.20 (the force equation in the MHD approximation), the condition of zero magnetic flux divergence $\nabla \cdot \vec{B} = 0$, and an equation of state $p = p(\rho)$. For example, the normal modes of oscillation for wavelengths short compared to any gradients in the plasma equilibrium structure are found by introducing small perturbations into these equations and linearizing. Although the dispersion relation for modes traveling in an arbitrary direction is some-what messy, it simplifies significantly in either of two limiting cases: waves traveling parallel to the equilibrium magnetic field direction, and waves traveling perpendicular. In the former case, there are two separate modes, normal sound waves with $\omega = \pm k c_s$ and incompressible Alfvén waves with $\omega = k v_A \equiv k B / \sqrt{4\pi\rho}$. In the latter case, there is only one mode, the compressible "fast" magnetosonic wave, for which $\omega = k\sqrt{c_s^2 + v_A^2}$.

C Shocks and Other Discontinuities

Ordinarily, the physical quantities describing a fluid vary smoothly as functions of time and distance. However, it is also possible for them to take sudden jumps, due to a qualitative transition in the history or character of the fluid across the boundary. Although many different sorts of discontinuities can occur in Nature, they are not totally free in character, for like all other mechanical events, they must satisfy the relevant conseration laws.

In a fluid, any conserved quantity obeys a *continuity equation*. That is, if the volume density of the conserved quantity is σ, and its flux is $\vec{f}$, the two are related by

$$\frac{\partial \sigma}{\partial t} + \frac{\partial f_i}{\partial x_i} = 0, \tag{C.1}$$

where the tensor summation convention for repeated indices is assumed. Suppose there is a discontinuity along the surface $x_1 = 0$. Then we can define what it means for σ to be conserved across that discontinuity by integrating equation C.1 over a short distance near $x_1 = 0$:

$$\frac{\partial}{\partial t} \int_{-\Delta x}^{+\Delta x} dx_1\, \sigma + \int_{-\Delta x}^{+\Delta x} dx_1 \left(\frac{\partial f_1}{\partial x_1} + \frac{\partial f_2}{\partial x_2} + \frac{\partial f_3}{\partial x_3} \right) = 0. \tag{C.2}$$

In the limit that $\Delta x \to 0$, the integrals of σ, $\partial f_2/\partial x_2$, and $\partial f_3/\partial x_3$ all go to zero because their integrands remain finite. However, the discontinuity causes $\partial f_1/\partial x_1$ to be singular at $x_1 = 0$, so that the integrated continuity equation becomes

$$\lim_{\Delta x \to 0} [f_1(+\Delta x) - f_1(-\Delta x)] = 0. \tag{C.3}$$

Thus, conservation of σ across the boundary means that the component of the flux parallel to the surface normal must be continuous there. Written in the conventional notation, this condition is

$$[f_1] = 0. \tag{C.4}$$

Three quantities—mass, momentum, and energy—are conserved in an ordinary fluid. The three associated fluxes are

$$\rho v_i,$$

$$p\delta_{ij} + \rho v_i v_j,$$

$$\rho v_i \left(\frac{1}{2}v^2 + w \right),$$

where ρ is the mass density, v_i is the fluid velocity in the ith direction, p is the pressure, and w is the enthalpy, or $U + p/\rho$ if U is the internal energy per unit mass. The first two forms are obvious; that energy flux should take the form shown is due to the fact that the fluid energy density increases when compressive work is done on it. More formally, if the energy density is $\rho(U + v^2/2)$, we can find its associated flux by rearranging its time derivative into a form that is a divergence of a vector. That vector must then be the opposite of the energy flux. Following this program, we write

$$\frac{\partial}{\partial t}\left(\frac{1}{2}\rho v^2\right) = \frac{1}{2}v^2\frac{\partial \rho}{\partial t} + \rho v_i\frac{\partial v_i}{\partial t}. \tag{C.5}$$

Using the continuity equation for mass and the fluid force equation, the rate of change of the energy density becomes

$$\frac{\partial}{\partial t}\left(\frac{1}{2}\rho v^2\right) = -\frac{1}{2}v^2\frac{\partial}{\partial x_i}(\rho v_i) - v_i\frac{\partial p}{\partial x_i} - \frac{1}{2}\rho v_i\frac{\partial v^2}{\partial x_i}. \tag{C.6}$$

Because $dp = \rho dw - \rho T ds$, where T is the temperature and s is the entropy per unit mass, equation C.6 is also equivalent to

$$\frac{\partial}{\partial t}\left(\frac{1}{2}\rho v^2\right) = -\frac{1}{2}v^2\frac{\partial}{\partial x_i}(\rho v_i) - \rho v_i\frac{\partial}{\partial x_i}\left(\frac{1}{2}v^2 + w\right) + \rho T v_i\frac{\partial s}{\partial x_i}. \tag{C.7}$$

Now the time derivative of the internal energy density is

$$\frac{\partial}{\partial t}(\rho U) = U\frac{\partial \rho}{\partial t} + \rho\frac{\partial U}{\partial t}. \tag{C.8}$$

Using the first law of thermodynamics $dU = T ds + (p/\rho)d\rho$ and the definition of w, this is

$$\frac{\partial}{\partial t}(\rho U) = w\frac{\partial \rho}{\partial t} + \rho T\frac{\partial s}{\partial t}. \tag{C.9}$$

These two pieces combine to form

$$\frac{\partial}{\partial t}\left[\rho\left(\frac{1}{2}v^2 + U\right)\right] = -\frac{\partial}{\partial x_i}\left[\rho v_i\left(\frac{1}{2}v^2 + w\right)\right] + \rho T\frac{ds}{dt}, \tag{C.10}$$

where the total time derivative d/dt is the time derivative following a fluid element, that is, $d/dt = \partial/\partial t + v_i\partial/\partial x_i$. The term $\rho T ds/dt$ is exactly the rate at which heat is added to a fluid element from some external source. Thus, the conserved energy flux is as shown above.

First, consider discontinuities with no cross-boundary flow, that is, those in which the normal component of the velocity $v_n = 0$. Mass and energy are trivially conserved, for they have no flux across the boundary. The only condition that must be satisfied is then that of momentum conservation. Its component parallel to the surface normal reduces to simple pressure continuity. The tangential component is

$$[\rho v_n \vec{v}_t] = 0, \tag{C.11}$$

where v_t is the magnitude of the tangential velocity. Because $v_n = 0$, conservation of this component is satisfied no matter what $\vec{v}_t$ does crossing the boundary. A discontinuity of this sort is called a "contact" or "tangential" discontinuity. It may form if two fluids slide past one another with parallel velocity (and zero viscosity), or if the two fluids have different equations of state, so that the same pressure corresponds to different densities in the different fluids.

Contact discontinuities generally have limited lifetimes. At the very least, transport processes will tend to mix the fluids. In addition, in many circumstances the boundary can be unstable. Two fluids sliding by one another with relative speed Δv are subject to the Kelvin-Helmholtz instability, for example, in which "fluting" of the boundary between them grows exponentially with a rate $\sim k\Delta v$, where k is the wavenumber of the perturbation.

Next consider the more general case, in which there is flow across the boundary. Conservation of mass, momentum, and energy requires satisfying the four "jump conditions"

$$\begin{aligned}
[\rho v_n] &= 0, \\
[p + \rho v_n^2] &= 0, \\
[\rho v_n \vec{v}_t] &= 0, \\
[\rho v_n(v^2/2 + w)] &= 0.
\end{aligned} \tag{C.12}$$

Use of the first condition simplifies the third and fourth:

$$\begin{aligned}
[\vec{v}_t] &= 0, \\
[v^2/2 + w] &= 0.
\end{aligned} \tag{C.13}$$

Thus, unlike in the case of contact discontinuities, in a shock the tangential velocity must be continuous. In fact, this allows us to simplify our analysis of fluid shocks by transforming into a frame of motion in which $\vec{v}_t = 0$, so that $v^2 = v_n^2$.

Inside the shock, some of the fluid's energy of organized motion is dissipated into heat and therefore increases the fluid's entropy. This dissipation is, in fact, essential to the creation of a shock because a sudden compressive change can occur only in association with an increase in entropy (in this regard, shocks and contact discontinuities are exactly complementary: dissipation due to transport processes is necessary to create a shock but in the end destroys a contact discontinuity). However, let us suppose that outside the shock, the fluid is adiabatic so that it obeys the equation of state $p = K\rho^\gamma$. The dissipation inside the shock increases K, but we will make the simplifying assumption that γ stays the same (this is not always the case; consider, e.g., a shock that heats the fluid so strongly that formerly frozen degrees of freedom become available).

In simple situations such as these, it is convenient to nondimensionalize the jump conditions in terms of the sound speed, using the relations

$$p = \frac{1}{\gamma}\rho c_s^2,$$
$$w = \frac{1}{\gamma - 1}c_s^2. \tag{C.14}$$

We also define the Mach number $\mathcal{M} \equiv v/c_s$, and label the upstream side (the fluid that has not yet entered the shock) with a 1, and the downstream side (already-shocked fluid) with a 2. The three conditions then become

$$\mathcal{M}_1 = rC\mathcal{M}_2,$$
$$1/\gamma + \mathcal{M}_1^2 = rC^2\left(1/\gamma + \mathcal{M}_2^2\right),$$
$$\frac{1}{\gamma - 1} + \frac{1}{2}\mathcal{M}_1^2 = C^2\left(\frac{1}{\gamma - 1} + \frac{1}{2}\mathcal{M}_2^2\right). \tag{C.15}$$

Here $r = \rho_2/\rho_1$ is the compression ratio, and $C = c_{s2}/c_{s1}$, which is also the square root of the temperature ratio $(T_2/T_1)^{1/2}$. These equations can be solved for r, C, and $\mathcal{M}_2$ in terms of $\mathcal{M}_1$:

$$r = \frac{(\gamma + 1)\mathcal{M}_1^2}{2 + (\gamma - 1)\mathcal{M}_1^2},$$
$$C^2 = \frac{[2\gamma\mathcal{M}_1^2 - (\gamma - 1)]\left[(\gamma - 1)\mathcal{M}_1^2 + 2\right]}{(\gamma + 1)^2\mathcal{M}_1^2}, \tag{C.16}$$
$$\mathcal{M}_2^2 = \frac{2 + (\gamma - 1)\mathcal{M}_1^2}{2\gamma\mathcal{M}_1^2 - (\gamma - 1)}.$$

In the limit of strong shocks, $\mathcal{M}_1 \gg 1$, these expressions simplify considerably:

$$r = \frac{\gamma + 1}{\gamma - 1},$$

$$C^2 = \frac{2\gamma(\gamma - 1)}{(\gamma + 1)^2} \mathcal{M}_1^2, \qquad \text{(C.17)}$$

$$\mathcal{M}_2^2 = \frac{\gamma - 1}{2\gamma}.$$

That is, no matter how strong the shock, the compression reaches a maximum that depends only on γ, and the postshock flow is always subsonic. The temperature, however, increases $\propto \mathcal{M}_1^2$. Combining the pressure and temperature ratios gives the pressure ratio:

$$\mathcal{P} = \frac{2\gamma}{\gamma + 1} \mathcal{M}_1^2. \qquad \text{(C.18)}$$

Higher pressure behind the shock than in front is, of course, what drives the shock through upstream gas.

In the commonly encountered case of $\gamma = 5/3$ (particles with no internal degrees of freedom), the strong shock relations become

$$r = 4,$$

$$C^2 = \frac{5}{16} \mathcal{M}_1^2,$$

$$\mathcal{M}_2^2 = \frac{1}{5}, \qquad \text{(C.19)}$$

$$\mathcal{P} = \frac{5}{4} \mathcal{M}_1^2.$$

On the other hand, in the limit as $\gamma \to 1$ (this can be interpreted either as a gas with many internal degrees of freedom, or one that quickly radiates away any heat),

$$r \to \infty,$$

$$C^2 \to 0,$$

$$\mathcal{M}_2^2 \to 0, \qquad \text{(C.20)}$$

$$\mathcal{P} \to \mathcal{M}_1^2.$$

So far we have supposed that the incoming gas is a unitary fluid. This is not necessarily a good description for an ionized plasma. When the particles' directed velocities are turned into random motions in the shock, the newly acquired random energy of the ions is greater than that of the electrons by their mass ratio. It is quite possible, therefore, for the immediately postshock ion temperature to be much greater than the electron temperature. Plasma turbulence within the shock proper may help to heat the electrons, but as it is expected to diminish in the quieter conditions outside the shock, it has only a limited time to act. In the absence of plasma turbulence, only Coulomb scattering couples the ion and electron temperatures, and that is relatively slow:

$$t_{equil} \simeq 3 \times 10^{11} n_p^{-1} T_{e7}^{3/2} \text{ s}, \tag{C.21}$$

for n_p in cm^{-3}, and for T_e large enough that the electron thermal speeds are greater than the ion speeds (Spitzer 1978). Consequently, there may be shocks in which $T_e \ll T_i$ for a considerable distance behind the front.

Similar relations can be found for MHD conditions. The momentum flux receives a supplement from the Maxwell stress:

$$p\delta_{ij} + \rho v_i v_j \rightarrow p\delta_{ij} + \rho v_i v_j + \frac{B^2}{8\pi}\delta_{ij} - \frac{B_i B_j}{4\pi}.$$

Likewise, the energy flux must include both the energy density in the time-steady magnetic field that is convected by the flow and the Poynting flux:

$$\rho v_i(v^2/2 + w) \rightarrow \rho v_i(v^2/2 + w) + v_i\frac{B^2}{4\pi} - \frac{B_i B_j v_j}{4\pi}.$$

The associated jump conditions are

$$\left[p + \rho v_n^2 + \frac{B_t^2 - B_n^2}{8\pi} \right] = 0, \tag{C.22}$$

for the normal component of momentum flux,

$$\left[\rho v_n \vec{v}_t - \frac{B_n \vec{B}_t}{4\pi} \right] = 0, \tag{C.23}$$

for the tangential part of the momentum flux, and

$$\left[\rho v_n(v_n^2/2 + v_t^2/2 + w) + v_n\frac{B_n^2 + B_t^2}{4\pi} - \frac{B_n \vec{v} \cdot \vec{B}}{4\pi} \right] = 0$$

for the energy flux.

Two more conditions also appear that are specific to MHD. One is the fact that the magnetic field always has zero divergence:

$$[B_n] = 0. \tag{C.24}$$

In other words, B_n cannot change across the surface of discontinuity. The other is a consequence of the flux-freezing condition (eq. B.12):

$$\left[B_n \vec{v}_t - \vec{B}_t v_n \right] = 0. \qquad (C.25)$$

For this condition to apply, the fluid must have a high enough conductivity that there is no appreciable slippage between the field lines and the fluid. In conditions of weak conductivity, shocklike transitions occur but are spread out over greater thicknesses (Draine and McKee 1993). These additional conditions, combined with the generalizations of the fluid conditions, yield a constraint on the change in the magnetic field. First, the zero divergence of $\vec{B}$ means that the component of $\vec{B}$ normal to the front cannot change; second, unless $[v_n^2 - B_n^2/(4\pi\rho)] = 0$, $\vec{B}_t$ must remain parallel to itself. This last fact allows us to drop the vector $\vec{B}_t$ from the following discussion.

Just as in the case of fluid discontinuities, there can also be MHD discontinuities without flow across the boundary, provided the field also does not cross the boundary (i.e., $B_n = 0$). The tangential velocity and magnetic field can change in either magnitude or direction across a boundary in the limit that viscosity and electrical resistance are negligible. Here, however, we will focus on shocks.

The additional complexity created by the presence of a magnetic field makes solving the jump conditions in the general case rather tedious and also introduces some conceptual subtleties (for a more complete discussion, see Draine and McKee 1993). Two limiting cases, however, afford helpful physical insight.

Consider first what happens when the upstream field is entirely normal to the shock plane (a "quasi-parallel" shock). This condition substantially simplifies the jump conditions:

$$\begin{aligned} \left[p + \rho v_n^2 \right] &= 0, \\ j \left[\frac{1}{2} v^2 + w \right] &= 0, \\ j [v_t] &= 0, \\ B_n [v_t] &= 0, \end{aligned} \qquad (C.26)$$

where $j = \rho v_n$ is the (conserved) mass flux through the shock, and we have restricted ourselves to shocks in which $v_n^2 \neq B_n^2/(4\pi\rho)$, so that $\vec{B}_t = 0$ both upstream and down. Because the compression in the shock is entirely parallel to the field, we see that the field has no effect at all (except in influencing the character of dissipation associated with the shock).

The other interesting limit is one in which there is zero normal field (a "quasi-perpendicular" shock). Here the jump conditions reduce to

$$\left[p + \rho v_n^2 + \frac{B_t^2}{8\pi} \right] = 0,$$

$$j \left[\frac{1}{2} v^2 + w \right] + \left[v_n \frac{B t^2}{4\pi} \right] = 0, \qquad \text{(C.27)}$$

$$j[v_t] = 0,$$

$$[v_n B_t] = 0.$$

Because there is no normal magnetic field, there is no magnetic contribution to the tangential momentum carried across the surface. Conserving tangential momentum flux then reduces to continuity of the tangential velocity. Applying (almost) the same technique to nondimensionalize these equations that we did in the pure fluid case, we transform them to

$$1/\gamma + \mathcal{M}_1^2 + \frac{1}{\gamma\beta_1} = \left(1/\gamma + \mathcal{M}_2^2\right) rC^2 + \frac{b^2}{\gamma\beta_1},$$

$$\mathcal{M}_1 \left(\frac{1}{2}\mathcal{M}_1^2 + \frac{1}{\gamma-1} \right) + 2\frac{\mathcal{M}_1}{\gamma\beta_1} = \mathcal{M}_2 \left(\frac{1}{2}\mathcal{M}_2^2 + \frac{1}{\gamma-1} \right) rC^{3/2} + 2\mathcal{M}_2 \frac{b^2 C}{\gamma\beta_1},$$

$$\mathcal{M}_1 = C^{1/2} b \mathcal{M}_2,$$

$$\mathcal{M}_1 = rC^{1/2} \mathcal{M}_2.$$

$$\text{(C.28)}$$

Two changes were necessary to the earlier scheme to accommodate the magnetic field contributions. Following the usual plasma physics notation, we define $\beta \equiv 8\pi p/B^2$, the ratio of gas to magnetic field pressure. Second, the magnetic field amplification is $b = B_2/B_1$. From the last two equations, we learn (not surprisingly) that

$$b = r. \qquad \text{(C.29)}$$

That is, when the magnetic field is entirely tangential, compression in the fluid is replicated exactly by compression in the magnetic field.

Combining these equations yields a cubic equation for the compression ratio in terms of $\mathcal{M}_1$ and β_1:

$$\left(\frac{2-\gamma}{\gamma\beta_1} \right) r^3 + \left[1 + (\gamma-1)\left(\frac{1}{2}\mathcal{M}_1^2 + \frac{2}{\gamma\beta_1} \right) \right] r^2$$

$$- \left(1 + \gamma\mathcal{M}_1^2 + \frac{1}{\beta_1} \right) r + \frac{1}{2}(\gamma+1)\mathcal{M}_1^2 = 0.$$

$$\text{(C.30)}$$

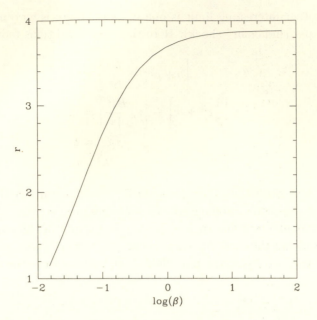

Fig. C.1 The compression ratio r for a shock wave with Mach number 10 in gas with $\gamma = 5/3$ whose magnetic field is parallel to the shock front. As the plasma β (the ratio of gas pressure to magnetic field pressure) declines, the compression becomes weaker. For $\beta < 0.01$, the front is moving slower than the magnetosonic speed, and the shock wave disappears.

Because a null discontinuity, that is, one in which the density is unchanged, is always permitted, we can divide this equation by the factor $r-1$ to obtain

$$\left(\frac{2-\gamma}{\gamma\beta_1}\right) r^2 + \left[1 + \frac{1}{2}(\gamma - 1)\mathcal{M}_1^2 + \frac{1}{\beta_1}\right] r - \frac{1}{2}(\gamma + 1)\mathcal{M}_1^2 = 0. \quad \text{(C.31)}$$

This quadratic equation has the solution

$$r = \frac{\gamma}{2(2-\gamma)} \left(-\beta_1 \left[1 + \frac{1}{2}(\gamma - 1)\mathcal{M}_1^2\right] - 1 \right.$$

$$+ \left\{ \beta_1^2 \left[1 + (\gamma - 1)\mathcal{M}_1^2 + \frac{1}{4}(\gamma - 1)^2 \mathcal{M}_1^4\right] \right. \quad \text{(C.32)}$$

$$\left. + \beta_1 \left(2 + \frac{\gamma + 4 - \gamma^2}{\gamma}\mathcal{M}_1^2\right) + 1 \right\}^{1/2} \right),$$

where adding the square root gives the physical solution.

In the limit that $\beta_1 \to \infty$ (the field-free limit), all magnetic field effects disappear, and the usual solution for the compression ratio emerges. When $\gamma \neq 1$, in order for the field to affect the compression, β_1 must be less than or comparable to $\mathcal{M}_1^2$. This condition is equivalent to requiring that the magnetic field be so strong in the upstream fluid that the shock speed is at most weakly super-Alfvénic. In fact, as can be found from examining equation C.32 in the limit that $r \to 1$, when the shock speed is actually less than the fast magnetosonic speed $\sqrt{v_A^2 + c_s^2}$, no shock can form. Figure C.1 illustrates how the compression is altered as the magnetic field gradually increases in strength.

The postshock temperature behaves in a similar fashion. In terms of the compression ratio,

$$C^2 = 1 + (\gamma - 1) \left(\frac{1}{2} \mathcal{M}_1^2 + \frac{2}{\gamma \beta_1} \right) - \frac{1}{2} (\gamma - 1) \frac{\mathcal{M}_1^2}{r^2} - \frac{2(\gamma - 1)}{\gamma \beta_1} r. \quad \text{(C.33)}$$

Because the magnetic field makes the gas "stiffer" and therefore less compressive, less work is done on the gas, and the heating is reduced when β_1 diminishes.

The presence of a magnetic field has an important effect when $\gamma \to 1$, even if $\beta_1 \gg 1$. In this limit, the compression behind a strong shock is

$$r \to \mathcal{M}_1 \beta_1^{1/2} = \frac{v_{s1}}{v_{A1}}. \quad \text{(C.34)}$$

That is, unlike the purely fluid case in which $\gamma = 1$ leads to divergent compression, when there is any transverse magnetic field, r is limited to the Alfvénic Mach number of the shock.

D   Luminosity Functions

and Other Population Statistics

D.1   More Sophisticated Estimates of the Luminosity Function

In addition to the simple method presented in §3.3, there are other techniques for estimating luminosity functions that, depending on circumstance, may offer advantages (see, e.g., the extensive review by Petrosian 1992). Two are of particular interest, Lynden-Bell's "C^-" method (Lynden-Bell 1971) and the proper use of "censored" data (as reviewed, e.g., by Feigelson 1992).

One objection that might be raised to the simple method of §3.3 is that it places bin boundaries at arbitrary locations in the L–z plane. Discontinuities in the luminosity function (viewed either as a function of L or of z) therefore occur at locations of our choice, rather than where Nature might have put them. Another problem with that method is that choosing the correct effective volume for a bin partially sliced off by the flux limit is often a serious issue, particularly because most of the points in a flux-limited sample will lie near the cutoff. In the statistics literature, such data are deemed *truncated*. Lynden-Bell's C^- method solves both these problems.

All one needs to do to remove the arbitrariness of luminosity bin division is make the cumulative luminosity function the quantity of interest, rather than the differential luminosity function. Given that change in point of view, each object can be regarded as contributing a delta function to the density in the L–z plane at its particular luminosity and redshift. The cumulative luminosity function near any particular redshift can then be found as

$$N(> L; z) = \int dz' \int_L^\infty dL' \sum_i \frac{\delta(L' - L_i)\delta(z' - z_i)}{V_{\text{eff}}(L', z')}, \qquad (D.1)$$

where the integral in z' is over a band of redshifts centered on z, the (L_i, z_i) are the luminosities and redshifts of the objects in the sample, and V_{eff} gives the object-by-object effective volume of the survey, accounting for the flux limit, the cosmology, and any peculiarities of the selection function. In this formulation, any discontinuities are due to the exigencies of where the particular objects of this sample happen to be found.

The name C^- comes from the way this method treats the region near the flux cutoff. Suppose that in this region we (temporarily) abandon interest in the normalization of the luminosity function but wish only to find its shape, that is, let $N(> L; z) = \Psi(> L)N(z)$. A local description of the shape of Ψ comes from its logarithmic differential $d\Psi/\Psi$. Now from sample data we find not $\Psi(> L)$ but merely the number of objects $X(> L)$, a function strongly affected by the flux limit. However, in the vicinity of L, it should be true that

$$\frac{d\Psi}{\Psi} = \frac{dX}{C},$$ (D.2)

where $C(> L)$ is the number of objects in the sample within the effective volume $V_{\rm eff}(L)$. That is, at each value of L, C is the number of objects within a genuine volume-limited subsample; that is the justification for equation D.2.

A formal solution to the problem is then given by

$$\ln \Psi(> L) = \int_L^\infty dL' \frac{dX}{dL'} \frac{1}{C(> L')}.$$ (D.3)

However, because our samples count individual objects, X is an integer-valued function, so dX/dL is not well defined. When the luminosity errors are negligible, the formal solution becomes

$$\ln \Psi(> L) = \int_L^\infty dL' \sum_i \frac{\delta(L' - L_i)}{C(> L')},$$ (D.4)

where the sum is over all the objects in the sample. This wouldn't be so bad, except that the δ functions always occur on the edge of the area in the L–z plane over which $C(> L)$ is defined, so there is an ambiguity over how to evaluate $C(> L_i)$. Lynden-Bell (1971) showed that the correct way is to set

$$\ln \Psi(> L) = \sum_i{}' \ln \left[\frac{C^-(> L_i)}{C^-(> L_i) + 1} \right],$$ (D.5)

where $C^-(L_i)$ is the number of objects in the volume-limited subsample not including the one at L_i, and the prime on the summation indicates that i is limited to those values for which $L_i \leq L$.

Once the shape of the luminosity function has been found by this method, its normalization can be established by continuity with the luminosity function found at higher luminosities.

Thinking of the data points as representing density samples in the L–z plane also lends itself to generalization in several different directions. If, for example, there are substantial uncertainties in either luminosity or redshift, probability densities can be substituted for the δ-functions. Application of any of a variety of smoothing algorithms can give estimates of the differential probability density.

"Censored data" in the parlance of statisticians refers simply to members of a sample for which only upper limits have been obtained. These present a problem for analysis of the sample. On the one hand, ignoring these objects would certainly introduce a bias; on the other, an upper limit, although not locating an object precisely, still conveys some information that can be used. A variety of techniques have been developed for incorporating this information, of which the most noteworthy is the Kaplan-Meier (1958) nonparametric estimator for the cumulative probability distribution $P(L \leq L')$. In the statistics literature, this is known as the *survival function*, and the whole subject of dealing with upper or lower bounds is called *survival analysis*.

In the Kaplan-Meier method, the path to $P(L \leq L')$ when some of the points have only upper bounds (see Isobe, Feigelson, and Nelson 1986 for a generalization to the case of both upper and lower bounds) lies through the conditional probabilities

$$P_i = P\left(L \leq L_{i+1} | L \leq L_i\right), \tag{D.6}$$

that is, the probability that $L \leq L_{i+1}$, given that $L \leq L_i$. We take this route because the cumulative distribution at each sample value L_j follows directly from the P_i:

$$P(L \leq L_j) = \prod_{i=0}^{j-1} P_i. \tag{D.7}$$

So now the question is how to find the P_i. The device suggested by Kaplan and Meier is to first order the data points $L_1 > L_2 > \ldots > L_n$. If we know the ith point has a luminosity smaller than B_i, we include it in the list by setting $L_i = B_i$. Given this ordering, a reasonable estimate of P_i (in fact, Kaplan and Meier derived it from a maximum likelihood analysis) is

$$P_i = \left(1 - \frac{1}{n - i + 1}\right)^{\delta_i}, \tag{D.8}$$

where $\delta_i = 1$ if L_i is an actual measurement, but zero if it is a bound. If L_i is a real measurement, this is a plausible estimate for P_i because, for all those points in the sample known to have $L \leq L_i$, it is the fraction for which $L \leq L_{i+1}$. On the other hand, if the ith point is a bound, all the measured values in the sample at least as small as L_i are also smaller than L_{i+1}, so it is appropriate to set $P_i = 1$ at such a point. Note, however, that for this estimate to be valid, the upper limits must be "random"— that is, the sensitivity of the measurement cannot have been adjusted for individual cases in response to the data obtained. To begin the process, we take the natural step of setting $P(L \leq L_1) = 1$. Starting from that point, the result of computing the recursive product defined in equation D.7 is a function that, as L decreases, takes a step downward at each measured point but remains constant at each bound.

Although the Kaplan-Meier estimator for samples containing bounds, and the Lynden-Bell C^- method for flux-limited samples, are formally quite independent and have somewhat different properties with respect to error analysis, they are in fact at heart closely related. That this is so can be seen by some simple manipulation of the expression for $\Psi(> L)$ given by the C^- method. From equation D.5, we immediately find that

$$\Psi(> L) = \prod_i{}' \frac{C^-(> L_i)}{C^-(L_i) + 1}. \tag{D.9}$$

When $C^- \gg 1$ (i.e., when we have moved away from the region of small number statistics),

$$\Psi(> L) \simeq \prod_i{}' \left[1 - \frac{1}{C^-(L_i)}\right]. \tag{D.10}$$

The connection to the Kaplan-Meier solution is now immediately apparent.

D.2 Correlations and Multivariate Probability Distributions

The search for correlations is a tool widely used to explore possible relationships between different measured quantities. Although easy to use, tests for correlations are also notoriously tricky to interpret. On the purely statistical level, one must be vigilant to avoid being misled by hidden correlations between the variables or sample biases. In addition, even when the existence of a correlation is secure, its interpretation is always ambiguous— does variable A depend on variable B, or the other way around? Or do both depend on variable C?

Correlations between different sorts of light output in flux-limited samples are particularly subject to these difficulties. They also serve as an illustrative testbed for pointing out both how naive analysis can trap the unwary and how proper treatment can rescue one from the trap and (at least partially) cure the problems just enumerated. The key point is to think of the data in terms of the underlying probability distribution (also known as the *luminosity function*) that they sample.

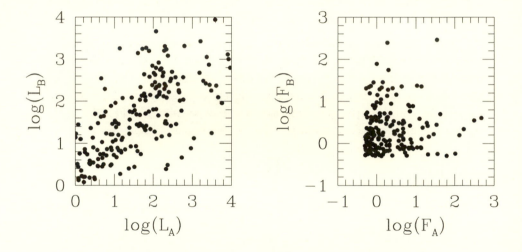

Fig. D.2 A simulation of the spurious luminosity-luminosity correlation created by doubly flux-limited samples. In this case, the luminosity functions and flux limits for both A and B are identical, but completely independent: $dn/dL_{A,B} \propto L_{A,B}^{-1.75}$ from 1 to 10^4 and $F_{\min,A,B} = 0.5$. The flux-flux plot (right) shows the true correlation (nil); the luminosity-luminosity plot (left) shows a strong *artificial* correlation.

Suppose we ask whether two different kinds of light output A and B are related. At first it might seem that the luminosity-luminosity form of the correlation is more physical than the flux-flux correlation. However, when a sample possessing a large dynamic range of luminosity has been selected in a way that is flux-limited for both sorts of luminosity under consideration, a luminosity-luminosity plot will *automatically* exhibit a correlation due solely to the fact that the more luminous objects will also preferentially be the most distant (fig. D.1, left). In other words, in terms of observables, such a plot is really flux × distance versus (the other) flux × distance, and it is the distances that are correlated rather than the fluxes. A correlation of flux versus flux does not contain this hidden connection between the variables and consequently does not create this sort of spurious correlation (fig. D.1, right). For this reason, any correlation based on a luminosity-luminosity plot is automatically suspect, while flux-flux plots are innocent (of this kind of problem). To earn acquittal from this suspicion, there must be good reason to believe that in the luminosity-luminosity diagram few objects are missing from the corners because of the flux limits.

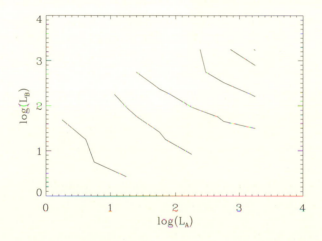

Fig. D.3   Contours illustrating the bivariate luminosity function estimated from the same data shown in fig. D.1. The contours are equally spaced logarithmically. Near the line $L_A = L_B$, where there are many objects in the sample, the luminosity function is well defined; in the corners of the plot, where there are hardly any objects in the sample, it becomes impossible even to define the density.

The mathematical origin of this problem becomes much clearer when phrased in terms of the bivariate probability distribution. In this case, what we would like to find is the bivariate luminosity function $p(L_A, L_B)$. To do so, we count how many objects in the sample have luminosities near L_A and L_B, and divide by the effective volume that we have searched to find

557

them. It is this last step, the normalization of the density to account for the sample definition, that is crucial here. If we rejected any objects with flux in the A band less than $F_{A,\min}$, and any objects with flux in the B band less than $F_{B,\min}$, then the volume searched for luminosities near L_A and L_B is the smaller of $(\Delta\Omega/3)[L_A/(4\pi F_{A,\min})]^{3/2}$ and $(\Delta\Omega/3)[L_B/(4\pi F_{B,\min})]^{3/2}$. Clearly, then, the effective volume varies dramatically over the L_A–L_B plane, and the raw density of points in this plane, which is the product of the effective volume and the bivariate luminosity function, is often a very poor measure of the real density of objects. This is the origin of the spurious correlation described heuristically in the previous paragraph. Flux-flux plots avoid this problem because their "effective volume" is the solid angle of the sample, a quantity that is constant across the entire plane (fig. D.1, right).

Thus, the proper way to study a luminosity-luminosity plot is to cast it as the bivariate luminosity function with a proper normalization for the effective volume (fig. D.2). When this is done, the correct underlying statistic is revealed (but see Walsh et al. 1989 for a particularly insightful discussion of the practical difficulties that may arise).

In practice, samples are often large enough to permit computation of a "global" statistic like a correlation coefficient, but too small to permit evaluation of a "local" statistic like a probability distribution. When this is the case, one has two options: to fall back on the flux-flux plot, or to use more sophisticated measures of the correlation coefficient. In many cases, there is also an additional complication: the sample may have been selected with respect to a flux limit solely for luminosity A. When that is the case, the data for luminosity B may contain a number of upper limits. This situation requires a melding of techniques appropriate to both "truncated" and "censored" data. Methods relevant to both these problems are discussed in Feigelson (1992).

E   The Tensor Virial Theorem

The tensor virial theorem was first developed by Chandrasekhar (for an elegant presentation of its properties, see Chandrasekhar 1981). It provides a ready way to compare the relative global balance of different forces, either in driving changes in the structure of a dynamical system or in determining the character of its equilibrium. This theorem also helps both to distinguish those conditions permitting equilibrium from those in which it is impossible and to estimate the characteristic frequencies of normal modes when the system is in equilibrium. It also has the virtue of applying (albeit in slightly different forms) to both fluid and collisionless systems.

Let us begin with the fluid version. If we wish to describe force balance in a fluid, then the natural place to start is the fluid analog of Newton's Second Law:

$$\rho \frac{dv_i}{dt} = -\frac{\partial p}{x_i} - \rho \frac{\partial(\Phi^{(\text{int})} + \Phi^{(\text{ext})})}{\partial x_i} + \frac{\partial M_{ki}}{\partial x_k}, \qquad (\text{E.1})$$

where ρ is the mass density, p the pressure, $\Phi^{(\text{int,ext})}$ is the gravitational potential due to the intrinsic gravity of the matter under consideration or to some external field, and M is the Maxwell stress tensor. To facilitate manipulation, we write this vector equation using the index notation, so that the position of a fluid element is specified by the three components x_i, its velocity $v_i = dx_i/dt$, and d/dt is a total, that is, Lagrangian time derivative. One way to obtain a global picture of the system's dynamics would be simply to integrate equation E.1 over the entire volume it occupies. However, all that would give is the essentially trivial result that the total momentum of a closed system can change only by virtue of momentum passing across its boundary.

A more interesting global constraint is found by taking the first moment of equation E.1 with respect to position:

$$\int dV \, x_j \rho \frac{dv_i}{dt} = -\int dV \, x_j \frac{\partial p}{x_i}$$
$$-\int dV \, x_j \rho \frac{\partial \Phi^{(\text{int})}}{\partial x_i} - \int dV \, x_j \rho \frac{\partial \Phi^{(\text{ext})}}{\partial x_i} + \int dV \, x_j \frac{\partial M_{ki}}{\partial x_k}.$$
$$(\text{E.2})$$

Multiplication of forces by distances results in quantities with the units of energy, so this procedure is closely related to a study of the amount of energy associated with each of the different forces.

Each of the integrals in equation E.2 can be fruitfully rewritten by the three-dimensional analog of integration by parts: bringing the factor of x_j within the differentiation, subtracting off the necessary correction, and making use of Gauss's theorem. Let us follow this program, operating on each term in equation E.2 separately.

We begin with the first term on the right-hand side, the pressure-gradient term. After the operations just described, it becomes

$$\int dV \, x_j \frac{\partial p}{\partial x_i} = -W \delta_{ij} + \oint dA \, n_i x_j p_{\text{ext}}, \tag{E.3}$$

where

$$W \equiv \int dV \, p(\vec{x}), \tag{E.4}$$

n_i represents the unit normal vector of the surface, and p_{ext} is the pressure at the boundary. In many cases p_{ext} is negligible; when that is so, the surface integral is also negligible.

Next turn to the intrinsic gravitational potential. It can be written in terms of its sources:

$$\int dV \, \rho x_j \frac{\partial \Phi^{(\text{int})}}{\partial x_i} = G \int dV \int dV' \frac{\rho(\vec{x})\rho(\vec{x}')(x_i - x_i')x_j}{|\vec{x} - \vec{x}'|^3}. \tag{E.5}$$

If we merely interchanged the definition of which integration variable was $\vec{x}$ and which $\vec{x}'$, it cannot make any difference to the result, so this term can equally well be written as half the sum of the two versions:

$$\int dV \, \rho x_j \frac{\partial \Phi^{(\text{int})}}{\partial x_i} = \frac{G}{2} \int dV \int dV' \frac{\rho(\vec{x})\rho(\vec{x}')}{|\vec{x} - \vec{x}'|^3} (x_i - x_i')(x_j - x_j') \equiv \Psi_{ij}^{(\text{int})}. \tag{E.6}$$

The tensor $\Psi^{(\text{int})}$ is manifestly symmetric. In addition, as we shall see shortly, it is closely related to the total intrinsic gravitational potential energy.

Third, consider the part due to extrinsic gravity. Similar manipulations transform it as follows:

$$\int dV \, \rho x_j \frac{\partial \Phi^{(\text{ext})}}{\partial x_i} = -U_{\text{ext}} \delta_{ij} - \int dV \, x_j \Phi^{(\text{ext})} \frac{\partial \rho}{\partial x_i} + \oint dA \, n_i x_j \rho \Phi^{(\text{ext})}. \tag{E.7}$$

Here U_{ext} is the integrated potential energy of the system due to the external gravity. To the degree that the density is homogeneous, the term with the gradient (which we will call $D_{ij}^{(\text{ext})}$) will be small. The surface integral is negligible if the density of extrinsic gravitational potential energy falls at least as fast as $|\vec{x}|^{-3}$ at large $|\vec{x}|$.

The last term on the right-hand side of equation E.2 is the moment of the gradient in the Maxwell tensor. Again integrating by parts, we find

$$\int dV\, x_j \frac{\partial M_{ki}}{\partial x_k} = -\int dV\, M_{ij} + \oint dA\, n_k x_j M_{ki}. \qquad \text{(E.8)}$$

We will define $\mathcal{M}_{ki} \equiv \int dV\, M_{ki}$. In this case, the surface integral is negligible when the fields fall at least as rapidly as $|\vec{x}|^{-3/2}$. Because static electric fields from localized charge distributions always weaken at least as fast as $|\vec{x}|^{-2}$, while static magnetic fields diminish $\propto |\vec{x}|^{-3}$ or faster, the surface integral is always negligible for static fields. However, wave fields may be significant on the boundary.

Finally, we now turn to the left-hand side of equation E.2. Writing

$$x_j \frac{dx_i}{dt} = \frac{d}{dt}\left(x_j \frac{dx_i}{dt}\right) - \frac{dx_i}{dt}\frac{dx_j}{dt}, \qquad \text{(E.9)}$$

it becomes

$$\int dV\, \rho x_j \frac{dv_i}{dt} = \int dV\, \rho \frac{d}{dt}\left(x_j \frac{dx_i}{dt}\right) - 2K_{ij}, \qquad \text{(E.10)}$$

where

$$K_{ij} \equiv \frac{1}{2}\int dV\, \rho v_i v_j \qquad \text{(E.11)}$$

is a generalization of the integrated bulk kinetic energy of the system. To further simplify the first term on the right-hand side of equation E.10, we make use of the fact that

$$\frac{d}{dt}\int dV\, \rho f(\vec{x}) = \int dV\, \rho \frac{df}{dt} \qquad \text{(E.12)}$$

for any function $f(\vec{x})$. This identity follows from mass conservation and the assumption that there is no flow across the boundaries of the region. It can be easily proved by making use of the definition of the Lagrangian derivative, that is, $df/dt \equiv \partial f/\partial t + \vec{v}\cdot\nabla f$. Using this identity, we transform the right-hand side of equation E.10 so that

$$\int dV\, \rho x_j \frac{dv_i}{dt} = \frac{d}{dt}\int dV\, \rho x_j \frac{dx_i}{dt} - 2K_{ij}. \qquad \text{(E.13)}$$

561

At this point it is useful to define two new tensor quantities: the *inertia tensor*

$$I_{ij} = \int dV \, \rho x_i x_j \tag{E.14}$$

and what we can call the *angular momentum tensor*

$$\mathcal{J}_{ij} = \int dV \, \rho \left(x_j \frac{dx_i}{dt} - x_i \frac{dx_j}{dt} \right). \tag{E.15}$$

The physical interpretation of I is obvious: it defines the moments of inertia for solid-body rotational motions of the system. $\mathcal{J}$ is related to the actual angular momentum vector $\vec{J}$ by $J_k = \epsilon_{ijk} \mathcal{J}_{ij}$, where ϵ_{ijk} is the completely antisymmetric Levi-Civita symbol, but the usual tensor notation in which repeated indices are summed over is not intended. Using these two new quantities, we find

$$\frac{d}{dt} \int dV \, \rho x_j \frac{dx_i}{dt} = \frac{1}{2} \left(\frac{d^2 I_{ij}}{dt^2} + \frac{dJ_{ij}}{dt} \right). \tag{E.16}$$

We now combine all these pieces and arrive at the tensor virial theorem itself:

$$\frac{1}{2} \frac{d^2 I_{ij}}{dt^2} + \frac{1}{2} \frac{dJ_{ij}}{dt} = 2K_{ij} + W\delta_{ij} - \Psi_{ij}^{(\text{int})} + U_{\text{ext}}\delta_{ij} + D_{ij}^{(\text{ext})} - \mathcal{M}_{ij}$$
$$+ \oint dA \, n_k \left[x_j \mathcal{M}_{ki} - \delta_{ki} x_j \left(\rho \Phi^{(\text{ext})} + p_{\text{ext}} \right) \right]. \tag{E.17}$$

It shows that the rates of change of the moment of inertia and the angular momentum are controlled by the balance between the levels in the different energy reservoirs listed on the right-hand side.

Further insight into the physical content of the tensor virial theorem can be gleaned by observing that its terms divide into subsets that are either symmetric or antisymmetric with respect to interchange of the indices i and j. The inertia tensor is symmetric; the angular momentum tensor is antisymmetric. Thus, the part of the right-hand side of equation E.17 that is symmetric contributes to changes in the inertia tensor, but not to changes in the angular momentum, while the antisymmetric part of the right-hand side contributes only to changes in the angular momentum.

Let us then take a symmetry census of the right-hand side. Four tensors (K_{ij}, δ_{ij}, $\Psi_{ij}^{(\text{int})}$, and $\mathcal{M}_{ij}$) are intrinsically symmetric, while the others have indeterminate symmetry. For example, when the external gravity is directed radially, all its contributions to the right-hand side are necessarily symmetric, but there are certainly cases in which the external gravity is not purely radial.

Because asymmetric tensors can be rewritten as a sum of a symmetric tensor and an antisymmetric one, it is easy to isolate the parts that drive changes in either the angular momentum or the inertia. For any tensor X_{ij},

$$\text{Symm}\,(X_{ij}) = \frac{1}{2}\,(X_{ij} + X_{ji})\,, \tag{E.18}$$

while

$$\text{Anti}\,(X_{ij}) = \frac{1}{2}\,(X_{ij} - X_{ji})\,. \tag{E.19}$$

It is obvious that $\text{Symm}(X) + \text{Anti}(X) = X$. In particular, applying these transformations to the surface integrals, we find

$$\text{Anti}\left[\oint dA\, n_k \delta_{ki} x_j \left(\rho\Phi^{(\text{ext})} + p_{\text{ext}}\right)\right] =$$
$$= \frac{1}{2}\oint dA\,(n_i x_j - n_j x_i)\left(\rho\Phi^{(\text{ext})} + p_{\text{ext}}\right). \tag{E.20}$$

That is, the antisymmetric part of the external pressure and external gravity surface integral can be small for either of two reasons. It may be that the external pressure or gravity are small; but it also approaches zero to the degree that $\hat{n}$ and $\vec{x}$ are parallel, that is, to the degree that the bounding surface is spherical. Similarly, setting the electric field $\vec{E} = 0$ (as is usually appropriate due to the high conductivity generally found in astrophysical plasmas), we obtain

$$\text{Anti}\left(\oint dA\, n_k x_j M_{ki}\right) =$$
$$= \frac{1}{8\pi}\oint dA\left[n_k B_k\,(x_j B_i - x_i B_j) - \frac{1}{2}B^2\,(n_i x_j - x_i n_j)\right]. \tag{E.21}$$

From this form we see that the anti-symmetry arises either from a nonradial magnetic field, or a nonspherical boundary.

Summarizing, we find that in order for a fluid system to experience a net torque, one of three conditions is necessary: that the external gravity is asymmetric, that the magnetic field on the boundary is nonradial, or that the boundary itself is nonspherical. If none of these three conditions applies, angular momentum is conserved, but the moment of inertia of the system may change. If at least one of them is satisfied, the angular momentum will change in time provided the antisymmetric terms don't identically cancel.

A particularly simple summary of the (symmetric parts) of these nine equations can be made by taking their trace:

$$\frac{1}{2}\frac{d^2 I}{dt^2} = 2K + 3W + U_{\text{int}} + 3U_{\text{ext}} + Tr(D^{(\text{ext})})$$

$$+ U_{\text{mag}} + \oint dA\, n_k \left[x_i M_{ki} - \delta_{ki} x_i \left(\rho \Phi^{(\text{ext})} + p_{\text{ext}} \right) \right],$$

$$\text{(E.22)}$$

where I is the sum of the three principal moments of inertia and U_{mag} is the total energy in magnetic fields (again taking $\vec{E} = 0$). $U_{\text{int}} = Tr[\Psi_{ij}^{(\text{int})}]$ is the total intrinsic gravitational potential energy because

$$Tr\left[\Psi_{ij}^{(\text{int})}\right] = \sum_i \frac{G}{2} \int dV \int dV' \frac{\rho(\vec{x})\rho(\vec{x}')}{|\vec{x} - \vec{x}'|^3} (x_i - x_i')^2$$

$$= \frac{G}{2} \int dV \int dV' \frac{\rho(\vec{x})\rho(\vec{x}')}{|\vec{x} - \vec{x}'|}.$$

$$\text{(E.23)}$$

From equation E.22, called the *scalar virial theorem*, we immediately see the relative magnitude of the contributions to an equilibrium. Suppose, for example, that we are examining an equilibrium in which all the surface integrals are negligible. Then, recalling that we are setting the zero point of the gravitational potential at infinity, so that gravitational potential energy is always negative, we can rewrite the scalar virial equation in the form

$$2K + 3W + U_{\text{mag}} = -\left[U_{\text{int}} + 3U_{\text{ext}} + Tr(D^{(\text{ext})}) \right]. \qquad \text{(E.24)}$$

The three terms on the right-hand side summarize the three contributions to gravitational confinement; the three on the left-hand side express how the confining forces are balanced: by kinetic energy (of rotation or turbulent motions), by pressure, and by magnetic pressure.

It is easy to derive an analogous tensor virial theorem for collisionless systems. Instead of starting with the fluid equation of motion, instead we begin with the collisionless Boltzmann equation

$$\frac{\partial f}{\partial t} + v_i \frac{\partial f}{\partial x_i} - \frac{\partial(\Phi^{(\text{int})} + \Phi^{(\text{ext})})}{\partial x_i} \frac{\partial f}{\partial v_i} = 0. \qquad \text{(E.25)}$$

Here $f(\vec{x}, \vec{v})$ is the density of particles in six-dimensional phase space $d^3x\, d^3v$, normalized so that $\int d^3v\, f = \rho$, the spatial mass density (for simplicity, we suppose that all the particles have the same mass). The first moment of this equation with respect to velocity is a close analogue of the fluid equation of motion because it speaks about how the momentum density changes in time. Making use of the fact that $f \to 0$ as $|\vec{v}| \to \infty$, this moment equation can be written in the form

$$\frac{\partial(\rho\langle v_j\rangle)}{\partial t} + \frac{\partial(\rho\langle v_i v_j\rangle)}{\partial x_i} + \rho\frac{\partial(\Phi^{(\text{int})} + \Phi^{(\text{ext})})}{\partial x_i} = 0. \tag{E.26}$$

The analogy to fluid behavior is clear: the only change has been to substitute $\rho\langle v_i v_j\rangle$, the momentum flux tensor, for the fluid pressure and to drop the electromagnetic forces.

With this analogue of the fluid equation of motion in hand, we proceed along a course nearly identical to the one we took in deriving the fluid tensor virial theorem. The only significant conceptual difference is that we must be a bit more careful about the nature of the system's boundary. Although fluid systems, by definition, cannot interpenetrate, there is nothing to stop collisionless systems from doing so. Consequently, the only sensible way to define the edge of a collisionless system for the purpose of applying the virial theorem is to place it far enough from the center that the density of particles from the system of interest is negligible there. This restriction has the side benefit, of course, of eliminating the surface integrals. We then arrive at the collisionless version of the tensor virial theorem

$$\frac{1}{2}\frac{d^2 I_{ij}}{dt^2} + \frac{1}{2}\frac{dJ_{ij}}{dt} = 2K_{ij} + W_{ij} - \Psi_{ij}^{(\text{int})} + U_{\text{ext}}\delta_{ij} + D_{ij}, \tag{E.27}$$

where

$$W_{ij} = \int dV\, \rho\langle v_i v_j\rangle. \tag{E.28}$$

It, too, has a scalar version, found by taking the trace of the tensor equation:

$$\frac{1}{2}\frac{d^2 I}{dt^2} = 2K + W - U_{\text{int}} + 3U_{\text{ext}} + Tr(D^{(\text{ext})}). \tag{E.29}$$

In normal galaxies, the dominant terms in the scalar virial theorem are those due to gravity, pressure, and kinetic energy (in the form of rotation). Typical Alfvén speeds are roughly one-tenth typical orbital speeds, so that $U_{\text{mag}} \sim 0.01(U_{\text{int}} + U_{\text{ext}})$. Whether pressure or rotation is the principal support varies from case to case. With respect to the global equilibrium, rotation dominates in disk galaxies, while pressure (in the form of the three-dimensional orbital motions of the stars) dominates in ellipticals. In AGN hosts a wider range of possibilities may exist.

F Kinematics in an Expanding Universe

When the Universe we live in changes its size as a function of time, we must take special pains to define just what is meant by distance, volume, light flux, and related quantities. Here we present a short summary of the facts of relativistic geometry needed to describe the evolution of AGN density with epoch. Fuller treatments can be found in Weinberg (1972), Carroll, Press, and Turner (1992), and Peebles (1993).

The first step is to establish a coordinate system. Truly inertial frames of reference only exist locally; if that is the case, how do we define a single set of coordinates for the entire Universe? The answer is that we must make the approximation that matter is spread uniformly through the Universe. There is then nothing to distinguish any one location from any other, and it is possible to define a single frame that encompasses the whole Universe. On the grandest scales, that appears to be a reasonable approximation, but there are effects (e.g., deflection of light rays by local perturbations to the gravitational field) for which this approximation may not be valid.

The most convenient coordinates to use are those associated with Friedmann (or Robertson-Walker) cosmologies, the system of *co-moving coordinates*. The meaning of this system is that if objects have no peculiar velocities, that is, no velocity with respect to this mean frame, they retain the same coordinates throughout the history of the Universe. This property makes it a very convenient system to use because distinct objects possess unique labels, and because objects that are neither created nor destroyed retain constant density.

Assuming spherical symmetry, the most general form for the metric of this coordinate system is (see Weinberg 1972 for a proof of this statement)

$$g = \begin{pmatrix} c^2 & 0 & 0 & 0 \\ 0 & -\frac{R^2(t)}{1-kr^2} & 0 & 0 \\ 0 & 0 & -R^2(t)r^2 & 0 \\ 0 & 0 & 0 & -R^2(t)r^2\sin^2\theta \end{pmatrix}, \qquad (\text{F.1})$$

where the components of four-vectors are ordered (t, r, θ, ϕ), $R(t)$ gives the cosmic scale factor as a function of time, and $k = -1, 0, 1$ gives the sign of the spatial curvature ($k = -1$ corresponds to an open Universe, $k = +1$ to a closed Universe, and $k = 0$ indicates a flat Universe). Note that r is defined to be dimensionless. The differential proper time is then

$$ds^2 = c^2 dt^2 - R^2(t)\left(\frac{dr^2}{1 - kr^2} + r^2 d\theta^2 + r^2 \sin^2\theta d\phi^2\right). \qquad (\text{F.2})$$

Note that because the metric is a function of (four-space) location, the definition of proper time must be given locally, that is, in terms of differentials.

The Einstein field equation, which relates curvature in the metric to the sources of gravity, determines how the scale of curvature, that is, $R(t)$, is related to the stress-energy tensor. In the matter-dominated era, where all ordinary astronomy takes place, this equation reduces to

$$\left(\frac{dR}{dt}\right)^2 = \frac{8\pi G\rho}{3}R^2 - k + \frac{\Lambda}{3}R^2, \tag{F.3}$$

where ρ is the mean density of gravitating matter and Λ is the cosmological constant. This equation can be simplified by choosing characteristic dimensions at the present epoch. That is, if we normalize to the current scale R_o, expansion rate $H_o = d(R/R_o)/dt$, and density ρ_o, equation F.3 becomes

$$\left(\frac{da}{d\tau}\right)^2 = \Omega_M/a + \Omega_k + \Omega_\Lambda a^2, \tag{F.4}$$

where

$$a = R/R_o,$$
$$\tau = H_o t,$$
$$\Omega_M = \frac{8\pi G\rho_o}{3H_o^2},$$
$$\Omega_k = -\frac{k}{H_o^2},$$
$$\Omega_\Lambda = \frac{\Lambda R_o^2}{3H_o^2}. \tag{F.5}$$

Because we have not fixed the normalization of R, we are free to set $R_o = c/H_o$. The inverse of a is the factor by which all lengths have expanded since time t, including photon wavelengths. For this reason, we define the redshift $z \equiv 1/a - 1$.

Because the unit of expansion rate is H_o, $da/d\tau = 1$ at the present epoch, and $\Omega_M + \Omega_k + \Omega_\Lambda \equiv 1$. Using this fact, we can remove Ω_k from the expansion equation:

$$\left(\frac{da}{d\tau}\right)^2 = 1 + \Omega_M(1/a - 1) + \Omega_\Lambda(a^2 - 1). \tag{F.6}$$

The three normalized densities can also be related to the deceleration parameter

$$q_o \equiv -R\frac{d^2R/dt^2}{(dR/dt)^2}(t_o)$$
$$\equiv \frac{\Omega_M}{2} - \Omega_\Lambda, \tag{F.7}$$

where t_o is the present time. If $\Omega_\Lambda = 0$, the three statements $q_o < 1/2$, $\Omega_M < 1$, and $k = -1$ are all equivalent and imply that the Universe will expand indefinitely. In the absence of a cosmological constant, the converses of these statements are likewise equivalent. This is not so, however, if $\Omega_\Lambda \neq 0$. $\Omega_M + \Omega_\Lambda < 1$ does imply that $k = -1$, but it does not necessarily imply that $q_o < 1/2$, or that the Universe will expand forever. The criterion for indefinite expansion when there is a nonzero cosmological constant is

$$\Omega_\Lambda \geq \begin{cases} 0 & 0 \leq \Omega_M \leq 1 \\ 4\Omega_M \left\{ \cos \left[\frac{1}{3} \cos^{-1} \left(\frac{1 - \Omega_M}{\Omega_M} \right) + \frac{4\pi}{3} \right] \right\}^3 & \Omega_M > 1 \end{cases} \tag{F.8}$$

(Felten and Isaacman 1986).

Cast in dimensionless form (as in eq. F.6), the expansion equation can be directly solved by quadrature. It is easiest to present this solution in terms of the "lookback time" as a function of z:

$$\tau_o - \tau(z) = \int_{1/(1+z)}^{1} \frac{da}{[1 + \Omega_M(1/a - 1) + \Omega_\Lambda(a^2 - 1)]^{1/2}}. \tag{F.9}$$

In this way, we can compute (parameterized by Ω_M and Ω_Λ) how long it has taken for the Universe to expand from redshift z to the present.

The integral in equation F.9 can be evaluated analytically in several important special cases. When $\Omega_M = \Omega_\Lambda = 0$ (an extreme low-density Universe),

$$\tau_o - \tau(z) = \frac{z}{1 + z}. \tag{F.10}$$

That is, the total age of the Universe in this limit is exactly H_o^{-1}. Alternatively, when $\Omega_M = 1$ but $\Omega_\Lambda = 0$ (also known as an Einstein–de Sitter Universe),

$$\tau_o - \tau(z) = \frac{2}{3} \left[1 - \frac{1}{(1 + z)^{3/2}} \right]. \tag{F.11}$$

When this limit applies, the age of the Universe is $(2/3)H_o^{-1}$. In the general case of a flat Universe (i.e., $\Omega_M + \Omega_\Lambda = 1$, but Ω_Λ not necessarily zero), the lookback time as a function of redshift is

$$\tau_o - \tau(z) = \frac{2/3}{(1 - \Omega_M)^{1/2}} \times$$

$$\left\{ \sinh^{-1} \left[\left(\frac{1 - \Omega_M}{\Omega_M} \right)^{1/2} \right] - \sinh^{-1} \left[\left(\frac{1 - \Omega_M}{\Omega_M} \right)^{1/2} (1 + z)^{-3/2} \right] \right\}. \tag{F.12}$$

Finally, when $\Omega_\Lambda = 0$ but Ω_M is free, the expansion equation has an analytic solution best written in terms of the "development angle" θ:

$$\theta = \begin{cases} \cos^{-1}\left[1 + \frac{2(1-\Omega_M)}{\Omega_M}\frac{1}{1+z}\right] & \Omega_M > 1 \\ \cosh^{-1}\left[1 + \frac{2(1-\Omega_M)}{\Omega_M}\frac{1}{1+z}\right] & \Omega_M < 1 \end{cases}, \tag{F.13}$$

and a function $f(\theta)$ that is

$$f(\theta) = \begin{cases} \sin\theta & \Omega_M > 1 \\ \sinh\theta & \Omega_M < 1 \end{cases}. \tag{F.14}$$

In terms of these quantities, time corresponds to redshift according to

$$\tau = \frac{\Omega_M/2}{|\Omega_M - 1|^{3/2}}|f(\theta) - \theta|. \tag{F.15}$$

When $\Omega_M < 1$, this can also be written as

$$\tau = \frac{\Omega_M/2}{(1 - \Omega_M)^{3/2}}\left[\sqrt{y^2 - 1} - \ln\left(y + \sqrt{y^2 - 1}\right)\right], \tag{F.16}$$

where

$$y = 1 + \frac{1 - \Omega_M}{\Omega_M/2}\frac{1}{1+z}. \tag{F.17}$$

At our present state of knowledge, we do not know how to predict the parameters H_o, Ω_M, and Ω_Λ from fundamental physical theory (although some people claim they can do this for $\Omega_M + \Omega_\Lambda$). They are instead measured quantities, and different groups claim significantly different values for them. For the purpose of AGN studies, it is most appropriate to take an agnostic view and make use of empirical results. As discussed in Chapter 3, we will generally take $H_o = 100h$ km/s Mpc^{-1}, and scale h to 0.75. The mean mass density Ω_M is less well known; current estimates would place it somewhere between 0.05 and 1. We know even less about Ω_Λ. It cannot be too large and positive because that would imply the Universe was never dense enough to have thermalized the cosmological microwave background. Similarly, it cannot be too large and negative because the Universe would have collapsed already. So a plausible range might be $|\Omega_\Lambda| < 1$. Primarily because many formulae simplify greatly when $\Omega_M = 1$ and $\Omega_\Lambda = 0$, and secondarily because these are popular values with many theorists, we will often make numerical estimates for this special case.

Suppose we wish to find the coordinate of origin for light emitted with observed spectral redshift z. Light travels with zero elapsed proper time, so if it was emitted at a time t_1 and received by us at time t_o and co-moving coordinate $r = 0$, it must have been emitted from co-moving coordinate r_1 defined by

$$\int_0^{r_1} \frac{dr}{(1 - kr^2)^{1/2}} = \int_{t_1}^{t_o} c\frac{dt}{R(t)} = \int_{R_1}^{R_o} c\frac{dR}{R\dot{R}}. \qquad \text{(F.18)}$$

The solution to this problem is found directly through use of equation F.6:

$$\int_{a_1}^1 \frac{da/a}{[1 + \Omega_M(1/a - 1) + \Omega_\Lambda(a^2 - 1)]^{1/2}} = \begin{cases} \sin^{-1}(r_1) & k = +1 \\ r_1 & k = 0 \\ \sinh^{-1}(r_1) & k = -1 \end{cases}. \qquad \text{(F.19)}$$

When $\Omega_\Lambda = 0$, the integral in equation F.19 can be computed analytically, and the solution for r_1 as a function of z has the same form for all three values of k:

$$r_1 = \frac{4}{\Omega_M^2(1+z)}\left[z\Omega_M/2 + (\Omega_M/2 - 1)\left(-1 + \sqrt{\Omega_M z + 1}\right)\right]. \qquad \text{(F.20)}$$

Taking the two usual limits (an open Universe with $\Omega_M \to 0$ and an Einstein–de Sitter universe with $\Omega_M = 1$), this expression for the co-moving radial coordinate simplifies further:

$$r_1(z) = \begin{cases} \frac{z + z^2/2 + \cdots}{1 + z} & z\Omega_M \ll 1 \\ 2 - \frac{2}{\sqrt{1+z}} & \Omega_M = 1 \end{cases}. \qquad \text{(F.21)}$$

We are now prepared to ask the question, "Given a luminosity L in the frame of emission, what is the flux received at the present epoch?" In a Euclidean universe, the answer would be $F = L/(4\pi D^2)$, where the distance D is defined unambiguously. However, in an expanding universe, the situation is more complicated.

First, we ask, "What is the solid angle Ω of an object (e.g., a telescope) with cross section πb^2 and located here on Earth, as viewed by a light source at r_1?" Provided $b/(R_1 r_1) \ll 1$ (generally an excellent approximation for any Earthly object), $\Omega \simeq \pi b^2/(R_o^2 r_1^2)$ because light rays emitted from r_1 and separated by an angle ϵ achieve a transverse spatial separation of $\simeq R_o r_1 \epsilon$ at the present time. That is, our telescope intercepts a fraction $(1\pi)[b/(R_o r_1)]^2$ of all the photons emitted. Conversely, if two rays that intersect here and now with an angle ϵ are traced back in time to radial coordinate r_1, their physical separation would then be $\epsilon R_1 r_1$. For this reason, the combination $R_1 r_1$ is called the *angular diameter distance* D_A:

it is the distance such that the length between two objects whose angular separation is ϵ is given by $D_A \epsilon$.

Second, we must apply two Doppler factors to the rate at which energy is received: one affecting the rate at which we receive individual photons, the other affecting their frequency, and hence their energy. Combining these redshifts with the fractional solid angle subtended by our telescope, the relation between flux and luminosity becomes

$$F = \frac{L}{4\pi(1+z)^2 R_o^2 r_1^2}. \tag{F.22}$$

If we are to write $F = L/(4\pi D_L^2)$, where D_L is the *luminosity distance*, then

$$D_L = R_o r_1 (1+z) = (1+z)^2 D_A. \tag{F.23}$$

In the case of zero cosmological constant, use of equation F.20 gives an explicit relation between D_L and the redshift:

$$z(D_L) = \Omega_M D_L/2 + \Omega_M/2 - 1 + (1 - \Omega_M/2)\sqrt{1 + 2D_L}. \tag{F.24}$$

The dependence of surface brightness on redshift follows immediately from the definition of D_L. What we mean by surface brightness S is that

$$S = \frac{F}{\Omega}, \tag{F.25}$$

when a source produces flux F over a solid angle Ω. Now $F = L/(4\pi D_L^2)$ and $\Omega = \pi b^2/D_A^2$. Thus, we obtain

$$S = \frac{L}{4\pi^2 b^2} \frac{D_A^2}{D_L^2} = \frac{L}{4\pi b^2 (1+z)^4}. \tag{F.26}$$

It is the four factors of $1+z$ that make extended objects dim so rapidly once their redshifts become appreciable. One should also note that the scaling is $\propto (1+z)^{-4}$ for *bolometric* surface brightness; if one actually measures the surface brightness only within a particular band, then one must take into account how νF_ν changes from the observed frequency ν to the frequency of original emission $\nu(1+z)$.

Thus, in an expanding Universe there are two different measures of distance: the angular diameter distance and the luminosity distance, that coincide on small lengthscales but differ substantially when applied across cosmological scales. Other cosmological measures of distance also exist, but they are not needed for the arguments presented here.

Finally, we also need to determine the relation between volume and distance (or redshift) in an expanding Universe. Using the metric (eq. F.1), we can immediately write down an integral for the physical volume out to redshift z contained within a solid angle $\Delta\Omega$:

$$V_{\text{phys}}(z) = \Delta\Omega \int_0^z dz' \frac{dr}{dz'} \frac{r^2(z')}{(1+z')^3} \left[1 - kr^2(z')\right]^{-1/2}. \qquad (\text{F.27})$$

The function $r(z)$ can be obtained implicitly from equation F.19 (if there is a nonzero cosmological constant) or explictly from equation F.20 (if $\Lambda = 0$). Typically, however, the *co-moving* volume is more useful, and it is

$$\begin{aligned} V_{\text{co}}(z) &= \Delta\Omega \int_0^z dz' \frac{dr}{dz'} r^2(z') \left[1 - kr^2(z')\right]^{-1/2} \\ &= \Delta\Omega \int_0^{r_1} dr \frac{r^2}{(1 - kr^2)^{1/2}} \end{aligned} \qquad (\text{F.28})$$

The curvature factor remains in the volume element for the co-moving volume because it is necessary for measuring volumes with $dt = 0$. The integral for the co-moving volume can be computed analytically for the case $\Omega_\Lambda = 0$:

$$V_{\text{co}}(r_1) = \frac{\Delta\Omega}{2} \begin{cases} \sin^{-1}(r_1) - r_1\sqrt{1 - r_1^2} & k = +1 \\ 2r_1^3/3 & k = 0 \\ -\sinh^{-1}(r_1) + r_1\sqrt{1 + r_1^2} & k = -1 \end{cases}. \qquad (\text{F.29})$$

In open (closed) Universes, the co-moving volume increases with redshift faster (slower) than in a flat Universe. In the special case of an Einstein–de Sitter Universe, the co-moving volume has an especially simple dependence on redshift:

$$V_{\text{co}}(z) = \frac{32\pi}{3} \left(1 - \frac{1}{\sqrt{1+z}}\right)^3. \qquad (\text{F.30})$$

We rarely have any way to directly measure the coordinate r_1 of a distant source—z or even D_L are the observable quantities. The relations embodied in equations F.19 or F.20 can be used to rewrite equation F.29 in terms of z, and equation F.23 then gives the connection to D_L.

References

Abramowicz, M.A., Bao, G., Lanza, A., and Zhang, X. 1991, *Astron. & Astrop.* **245**, 454

Abramowicz, M.A., Czerny, B., Lasota, J.-P., and Szuszkiewicz, E. 1988, *Ap. J.* **332**, 646

Abramowicz, M.A., Lanza, A., and Percival, M.J. 1997, *Ap. J.* **479**, 179

Adams, T.F. 1977, *Ap. J. Suppl.* 33, 19

Agol, E., and Blaes, O. 1996, *M.N.R.A.S.* **282**, 965

Alexander, T., and Netzer, H. 1994, *M.N.R.A.S.* **270**, 781

Alfvén, H., and Herlofson, N. 1950, *Phys. Rev.* **78**, 616

Allen, C.W. 1973, *Astrophysical Quantities* (London: Athlone Press)

Anderson, S.F., Weymann, R.J., Foltz, C.B., and Chaffee, F.H., Jr. 1987, *Astron. J.* **94**, 278

Angel, J.R.P. 1969, *Ap. J.* **158**, 219

Antonucci, R.R.J. 1983, *Nature* **303**, 158

Antonucci, R.R.J. 1984, *Ap. J.* **278**, 499

Antonucci, R.R.J., Hurt, T., and Miller, J.S. 1994, *Ap. J.* **430**, 210

Antonucci, R.R.J., and Miller, J.S. 1985, *Ap. J.* **297**, 621

Antonucci, R.R.J., and Ulvestad, J. 1985, *Ap. J.* **294**, 158

Arav, N., Korista, K.T., Barlow, T.A., and Begelman, M.C. 1995, *Nature* **376**, 576

Arnett, D. 1996 *Supernovae and Nucleosynthesis* (Princeton: Princeton University Press)

Arons, J., Kulsrud, R.M., and Ostriker, J.P. 1975, *Ap. J.* **198**, 687

Athanassoula, E. 1992, *M.N.R.A.S.* **259**, 345

Baade, W., and Minkowski, R. 1954, *Ap. J.* **199**, 206

Bahcall, J.N., Bergeron, J., Boksenberg, A., Hartig, G.F., Januzzi, B.T., Kirhakos, S., Sargent, W.L.W., Savage, B.D., Schneider, D.P., Turnshek, D.A., Weymann, R.J., and Wolfe, A.M. 1993, *Ap. J. Suppl.* **87**, 1

Bahcall, J.N., Kirhakos, S., Saxe, D.H., and Schneider, D.P. 1997, *Ap. J.* **479**, 642

Bahcall, J.N., Kirhakos, S., and Schneider, D.P. 1994, *Ap. J. Lett.* **435**, L11

Bahcall, J.N., Kirhakos, S., and Schneider, D.P. 1996, *Ap. J.* **457**, 557

Bahcall, J.N., Kirhakos, S., Schneider, D.P., Davis, R.J., Muxlow, T.W.B., Garrington, S.T., Conway, R.G., and Unwin, S.C. 1995, *Ap. J. Lett.* **452**, L91

Bahcall, J.N., Kozlovsky, B.-Z., and Salpeter, E.E. 1972, *Ap. J.* **171**, 467

Balbus, S.A., and Hawley, J.F. 1991, *Ap. J.* **376**, 214

Baldwin, J.A. 1977, *Ap. J.* **214**, 679

Baldwin, J.A., Phillips, M.M., and Terlevich, R. 1981, *P.A.S.P.* **93**, 5

Baldwin, J.A., Wampler, E.J., and Gaskell, C.M. 1989, *Ap. J.* **338**, 630

Band, D.L., Klein, R.I., Castor, J.I., and Nash, J.K. 1990, *Ap. J.* **362**, 90

Bardeen, J.M., and Petterson, J.A. 1975, *Ap. J. Lett.* **195**, 65L

Bardeen, J., Press, W.H., and Teukolsky, S.A. 1972, *Ap. J.* **178**, 347

Barlow, T.A. 1994, *P.A.S.P.* **106**, 548

Barnes, J.E., and Hernquist, L. 1992, *Ann. Rev. Astron. Astrop.* **30**, 705

Barnes, J.E., and Hernquist, L. 1996, *Ap. J.* **471**, 115

Barthel, P. 1989, *Ap. J.* **336**, 606

Bashkin, S., and Stoner, J.O., Jr. 1975, *Atomic Energy Levels and Grotrian Diagrams* (Amsterdam/New York: North-Holland/American Elsevier)

Bassani, I., Butler, R.C., Di Cocco, G., Della Ventura, A., and Perotti, F. 1986, *Ap. J.* **311**, 623

Baum, S.A., and Heckman, T.M. 1989a, *Ap. J.* **336**, 681

Baum, S.A., and Heckman, T.M. 1989b, *Ap. J.* **336**, 702

Baum, S.A., Heckman, T.M., Bridle, A.H., van Bruegel, W., and Miley, G.K. 1988, *Ap. J. Suppl.* **68**, 833

Becker, R.H., White, R.L., and Helfand, D.J. 1995, *Ap. J.* **450**, 559

Begelman, M.C. 1978, *M.N.R.A.S.* **184**, 53

Begelman, M.C. 1991, in *Relativistic Hadrons in Cosmic Compact Objects*, eds. A.A. Zdziarski and M. Sikora (Berlin: Springer-Verlag), p. 1

Begelman, M.C., Blandford, R.D., and Rees, M.J. 1980, *Nature* **287**, 307

Begelman, M.C., Blandford, R.D., and Rees, M.J. 1984, *Rev. Mod. Phys.* **56**, 255

Begelman, M.C., and Chiueh, T. 1988, *Ap. J.* **332**, 872

Begelman, M.C., de Kool, M., and Sikora, M. 1991, *Ap. J.* **382**, 416

Begelman, M.C., McKee, C.F., and Shields, G.A. 1983, *Ap. J.* **271**, 70

Bekenstein, J.D. 1973, *Phys. Rev. D* **7**, 2333

Benn, C.R. 1995, *M.N.R.A.S.* **272**, 699

Bennett, A.S. 1962, *Mem. Roy. Ast. Soc.* **68**, 163

Bergamini, R., Braccesi, A., Colla, G., Fanti, C., Fanti, R., Ficarra, A., Formiggini, L., Gandolfi, E., Gioia, I., Lari, C., Marano, B., Padrielli, L., Tomasi, P., and Vigotti, M. 1973, *Astron. & Astrop.* **23**, 195

Berriman, G., Schmidt, G.D., West, S.C., and Stockman, H.S. 1990, *Ap. J. Suppl.* **74**, 869

Binette, L., Wang, J., Villar-Martin, M., Martin, P.G., and Magris C., Gladis 1993, *Ap. J.* **414**, 535

Binette, L., Wilson, A.S., and Storchi-Bergmann, T. 1996, *Astron. & Astrop.* **312**, 365

Binggeli, B., Sandage, A., and Tammann, G.A. 1988, *Ann. Rev. Astron. Astrop.* **26**, 509

Binney, J., and Mamon, G.A. 1982, *M.N.R.A.S.* **200**, 361

Binney, J., and Tremaine, S.D. 1987, *Galactic Dynamics* (Princeton University Press: Princeton)

Bjornsson, G., and Svensson, R. 1991, *M.N.R.A.S.* **249**, 177

Blaes, O.M. 1985, *M.N.R.A.S.* **216**, 553

Blaes, O.M. 1987, *M.N.R.A.S.* **227**, 975

Blaes, O.M., and Agol, E. 1996, *Ap. J. Lett.* **469**, L41

Blaes, O.M., and Hawley, J.F. 1988, *Ap. J.* **326**, 277

Blandford, R.D. 1976, *M.N.R.A.S.* **176**, 465

Blandford, R.D. 1990, in *Active Galactic Nuclei: 1990 Saas-Fee Lectures*, eds. T.J.-L. Courvoisier and M. Mayor (Berlin: Springer-Verlag)

Blandford, R.D., and Eichler, D. 1987, *Phys. Rep.* **154**, 1

Blandford, R.D., and Königl, A. 1979, *Ap. J.* **232**, 34

Blandford, R.D., and Levinson, A. 1995, *Ap. J.* **441**, 79

Blandford, R.D., and McKee, C.F. 1982, *Ap. J.* **255**, 419

Blandford, R.D., and Payne, D.G. 1982, *M.N.R.A.S.* **199**, 883

Blandford, R.D., and Rees, M.J. 1974, *M.N.R.A.S.* **169**, 395

Blandford, R.D., and Znajek, R.L. 1977, *M.N.R.A.S.* **179**, 433

Blumenthal, G.R., and Gould, R.J. 1970, *Rev. Mod. Phys.* **42**, 237

Bolton, J.G., Gardner, F.F., and Mackey, M.B. 1964, *Aust. J. Phys.* **17**, 340

Bondi, H. 1952, *M.N.R.A.S.* **112**, 195

Bonometto, S., and Rees, M.J. 1971, *M.N.R.A.S.* **152**, 21

References

Born, M., and Wolf, E. 1980, *Principles of Optics*, 6th ed., (Oxford: Pergamon Press)

Boroson, T., and Green, R.F. 1992, *Ap. J. Suppl.* **80**, 109

Bowyer, S., Lampton, M., Mack, J., and de Mendonca, F. 1970, *Ap. J. Lett.* **161**, L1

Boyle, B.J., and Couch, W.J. 1993, *M.N.R.A.S.* **264**, 604

Boyle, B.J., Fong, R., Shanks, R., and Peterson, B.A. 1990, *M.N.R.A.S.* **243**, 1

Boyle, B.J., Griffiths, R.E., Shanks, T., Stewart, G.C., and Georgantopoulos, I. 1993, *M.N.R.A.S.* 260, 49

Braatz, J.A., Wilson, A.S., and Menkel, C. 1994, *Ap. J. Lett.* **437**, L99

Brandenburg, A., Nordlund, A., Stein, R.F., and Torkelsson, U. 1996, *Ap. J. Lett.* **458**, L45

Bridle, A.H., and Perley, R.A. 1984, *Ann. Rev. Astron. Astrop.* **22**, 319

Bridle, A.H., Hough, D.H., Lonsdale, C.J., Burns, J.O., and Laing, R.A. 1994, *Astron. J.* **108**, 766

Brotherton, M.S., Wills, B.J., Steidel, C.C., and Sargent, W.L.W. 1994, *Ap. J.* **423**, 131

Brunner, H., Friedrich, P., Zimmermann, H.-U., and Staubert, R. 1992, in *X-ray Emission from Active Galactic Nuclei and the Cosmic X-ray Background*, eds. W. Brinkmann and J. Trümper, (Garching bei München: Max-Planck-Institut für Extraterrestrische Physik), p. 198

Burbidge, G.R. 1958, *Ap. J.* **129**, 841

Burns, J.O., Norman, M.L., and Clarke, D.A. 1991, *Science* **253**, 522

Cameron, M., Storey, J.W.V., Rotaciuc, V., Genzel, R., Verstraete, L., Drapatz, S., Siebenmorgen, R., and Lee, T.J. 1993, *Ap. J.* **419**, 136

Canizares, C.R., and Kruper, J. 1984, *Ap. J. Lett.* **278**, L99

Capetti, A., Macchetto, F., Axon, D.J., Sparks, W.B., and Boksenberg, A. 1995, *Ap. J. Lett.* **452**, L87

Cardelli, J.A., Clayton, G.C., and Mathis, J.S. 1989, *Ap. J.* **345**, 245

Carilli, C.L., Röttgering, H.J.A., van Ojik, R., Miley, G.K., and van Bruegel, W.J.M. 1997, *Ap. J. Suppl.* **109**, 1

Carroll, S., Press, W.H., and Turner, E.L. 1992, *Ann. Rev. Astron. Astrop.* **30**, 499

Castor, J.I. 1970, *M.N.R.A.S.* **149**, 111

Cavallo, G.,and Rees, M.J. 1978, *M.N.R.A.S.* **183**, 359

Chaffee, F.H., Foltz, C.B., Hewett, P.C., Francis, P.J., and Weymann, R.J. 1991, *Astron. J.* **102**, 461

Chambers, K.C., Miley, G.K., and van Bruegel, W.J.M. 1987, *Nature* **329**, 604

Chandrasekhar, S. 1960, *Radiative Transfer* (New York: Dover)

Chandrasekhar, S. 1981, *Hydrodynamic and Hydromagnetic Stability* (New York: Dover)

Chen, X.-M., and Taam, R.E. 1993, *Ap. J.* **412**, 254

Chokshi, A., and Turner, E.L. 1992, *M.N.R.A.S.* **259**, 421

Clarke, D.A., Norman, M.L., and Burns, J.O. 1986, *Ap. J. Lett.* **311**, L63

Clavel, J., Reichert, G.A., Alloin, D., Crenshaw, D.M., Kriss, G., Krolik, J.H., Malkan, M.A., Netzer, H., Peterson, B.M., Wamsteker, W., Altamore, A., Baribaud, T., Barr, P., Beck, S., Binette, L., Bromage, G.E., Brosch, N., Diaz, A.I., Filippenko, A.V., Fricke, K, Gaskell, C.M., Giommi, P., Glass, I.S., Gondhalekhar, P., Hackney, R.L., Halpern, J.P., Hutter, D.J., Joersaeter, S., Kinney, A.L., Kollatschny, W., Koratkar, A., Korista, K.T., Laor, A., Lasota, J.-P., Leibowitz, E., Maoz, D., Martin, P.G., Mazeh, T., Meurs, E.J.A., Nair, A.D., O'Brien, P.T., Pelat, D., Perez, E., Perola, G.c., Ptak, R.L., Rodriguez-Pascal, P., Rosenblatt, E.I., Sadun, A.C., Santos-Lleo, M., Shaw, R.A., Smith, P.S., Stirpe, G.M., Stoner, R., Sun, W.-H., Ulrich, M.-H., van Groningen, E., and Zheng, W. 1991, *Ap. J.* **366**, 64

Cohen, M.H., Cannon, W., Purcell, G.H., Shaffer, D.B., Broderick, J.J., Kellermannn, K.I., and Jauncey, D.L. 1971, *Ap. J.* **170**, 207

Cohen, M.H., Ogle, P.M., Tran, H.D., Vermeulen, R.C., Miller, J.S., Goodrich, R.W., and Martel, A.R. 1995, *Ap. J. Lett.* **448**, L77

Colla, G., Fanti, C., Fanti, R., Ficarra, A., Formiggini, L., Gandolfi, E., Grueff, G., Lari, C., Padrielli, L., Roffi, G., Tomasi, P., and Vigotti, M. 1970, *Astron. Astrop. Suppl.* **1**, 281

Condon, J.J. 1984, *Ap. J.* **287**, 461

Connors, P.A., Piran, T., and Stark, R.F. 1980 *Ap. J.* **235**, 224

Coppi, P.S., Blandford, R.D., and Rees, M.J. 1993, *M.N.R.A.S.* **262**, 603

Cower, J.F.R., Scott, P.F., and Wills, D. 1967, *Mem. Roy. Ast. Soc.* **71**, 49

Cowie, L.L., Ostriker, J.P., and Stark, A.A. 1978, *Ap. J.* **226**, 1041

Cox, D.P., and Raymond, J.C. 1985, *Ap. J.* **298**, 651

Cox, D.P., and Smith, B.W. 1974, *Ap. J. Lett.* **189**, L105

Cunningham, C.T. 1975, *Ap. J.* **202**, 788

Cunningham, C.T., and Bardeen, J.M. 1973, *Ap. J.* **183**, 237

Dahari, O. 1984, *Astron. J.* **89**, 966

Dallacasa, D., Fanti, C., Fanti, R., Schilizzi, R.T., and Spencer, R.E. 1995, *Astron. & Astrop.* **295**, 27

David, L., Durisen, R.H., and Cohn, H.N. 1987a, *Ap. J.* **313**, 556

David, L., Durisen, R.H., and Cohn, H.N. 1987b, *Ap. J.* **316**, 505

Davidson, K. 1972, *Ap. J.* **171**, 213

Davidson, K. 1977, *Ap. J.* **218**, 20

Davidson, K., and Netzer, H. 1979, *Rev. Mod. Phys.* **51**, 715

de Grijp, M.H.K., Keel, W.C., Miley, G.K., Goudfrooij, P., and Lub, J. 1992, *Astron. & Astrop. Suppl.* **96**, 389

Dermer, C.D., Liang, E.P., and Canfield, E. 1991, *Ap. J.* **369**, 410

de Vaucouleurs, G., de Vaucouleurs, A., and Corwin, H.G., Jr. 1976, *Second Reference Catalog of Bright Galaxies* (Austin: University of Texas Press)

Dey, A., van Bruegel, W.J.M., Vacca, W.D., and Antonucci, R.R.J. 1997, *Ap. J.* **490**, 698

di Serego Alighieri, S., Cimatti, A., and Fosbury, R.A.E. 1994, *Ap. J.* **431**, 123

Djorgovski, S., and Davis, M. 1987, *Ap. J.* **313**, 59

Done, C., and Krolik, J.H. 1996, *Ap. J.* **463**, 144

Done, C., Madejski, G.M., and Smith, D.A. 1996, *Ap. J. Lett.* **463**, L63

Dopita, M.A., and Sutherland, R.S. 1996, *Ap. J. Suppl.* **102**, 161

Draine, B.T., and McKee, C.F. 1993, *Ann. Rev. Astron. Astrop.* **31**, 373

Dressler, A. 1980, *Ap. J.* **236**, 351

Dressler, A., Lynden-Bell, D., Burstein, D., Davies, R.L., Faber, S.M., Terlevich, R.J., and Wegner, G. 1987, *Ap. J.* **313**, 42

Duncan, M.J., and Wheeler, J.C. 1980, *Ap. J. Lett.* **237**, L27

Dunlop, J.S., and Peacock, J.A. 1990, *M.N.R.A.S.* **247**, 19

Eardley, D.M., and Lightman, A.P. 1975, *Ap. J.* **200**, 187

Edelson, R.A., and Malkan, M.A. 1986, *Ap. J.* **308**, 59

Edelson, R.A., Malkan, M.A., and Rieke, G.H. 1987, *Ap. J.* **321**, 233

Edvardsson, B., Andersen, J., Gustafsson, B., Lambert, D.L., Nissen, P.E., and Tomkin, J. 1993, *Astron. & Astrop.* **275**, 101

Efstathiou, A., Hough, J.M., and Young, J. 1995, *M.N.R.A.S.* **277**, 1134

Elvis, M., Maccacaro, T., Wilson, A.S., Ward, M.J., Penston, M.V., Fosbury, R.A.E., and Perola, G.C. 1978, *M.N.R.A.S.* **183**, 129

Elvis, M., Wilkes, B.J., McDowell, J.C., Green, R.F., Bechtold, J., Willner, S.P., Oey, M.S., Polomski, E., and Cutri, R. 1994, *Ap. J. Suppl.* **95**, 1

References

Eracleous, M., and Halpern, J.P. 1994, *Ap. J. Suppl.* **90**, 1

Fabian, A.C., and Crawford, C.S. 1990, *M.N.R.A.S.* **247**, 439

Fabbiano, G., Kim, D.-W., and Trinchieri, G. 1992, *Ap. J. Suppl.* **80**, 531

Falcke, H., Gopal-Krishna, and Biermann, P.L. 1995, *Astron. & Astrop.* **298**, 395

Falomo, R., Scarpa, R., and Bersanelli, M. 1994, *Ap. J. Suppl.* **93**, 125

Fanaroff, B.L., and Riley, J.M. 1974, *M.N.R.A.S.* **167**, 31p

Fanti, R., Fanti, C., Schilizzi, R.T., Spencer, R.E., Nanrendong, Parma, P., van Bruegel, W.J.M., and Venturi, T. 1990, *Astron. & Astrop.* **231**, 333

Feigelson, E.D. 1992, in *Statistical Challenges in Modern Astronomy*, eds. E.D. Feigelson and G.J. Babu (New York: Springer-Verlag), p. 221

Feigelson, E.D., Laurent-Muehleisen, S.A., Kollgaard, R.I., and Fomalont, E.B. 1995, *Ap. J. Lett.* **449**, L149

Felten, J.E., and Isaacman, R. 1986, *Rev. Mod. Phys.* **58**, 689

Ferland, G.J. 1995, in *The Analysis of Emission Lines*, eds. R. Williams and M. Livio (Cambridge: Cambridge University Press), p. 83

Ferland, G.J., and Osterbrock, D.E. 1986, *Ap. J.* **300**, 658

Filippenko, A. 1985, *Ap. J.* **289**, 475

Filippenko, A., and Halpern, J.P. 1984, *Ap. J.* **285**, 458

Field, G.B. 1958, *Proc. I.R.E.* **46**, 240

Field, G.B. 1965, *Ap. J.* **142**, 531

Fireman, E.L. 1974, *Ap. J.* **187**, 57

Foltz, C.B., Chaffee, F.H., Hewett, P.C., MacAlpine, G.M., and Turnshek, D.A. 1987, *Astron. J.* **94**, 1423

Foltz, C.B., Chaffee, F.H., Hewett, P.C., Weymann, R.J., Anderson, S.F., and MacAlpine, G.M. 1989, *Astron. J.* **98**, 1959

Fomalont, E.B., Windhorst, R.A., Kristian, J.A., and Kellermann, K.I. 1991, *Astron. J.* **102**, 1258

Ford, H.C., Harms, R.J., Tsvetanov, Z., Hartig, G.F., Dressel, L.L., Kriss, G.A., Bohlin, R., Davidsen, A.F., Margon, B., and Kochhar, A.K. 1994, *Ap. J. Lett.* **435**, L27

Francis, P.J., Hewett, P.C., Foltz, C.B., Chaffee, F.H., and Weymann, R.J. 1991, *Ap. J.* **373**, 465

Friedman, H., and Byram, E.T. 1967, *Science* **159**, 748

Fukada, Y., Hayakawa, S., Kasahara, I., Makino, F., Tanaka, Y. 1975, *Nature* **254**, 398

Gaskell, C.M. 1982, *Ap. J.* **263**, 79

Gaskell, C.M., and Sparke, L.S. 1986, *Ap. J.* **305**, 175

Gayley, K.G., and Owocki, S.P. 1995, *Ap. J.* **446**, 801

Gendreau, K.C. 1995, unpublished M.I.T. Ph.D. thesis

George, I., and Fabian, A.C. 1991, *M.N.R.A.S.* **249**, 352

Gerhard, O.E., and Binney, J. 1985, *M.N.R.A.S.* **216**, 467

Ghisellini, G., Guilbert, P., and Svensson, R. 1988, *Ap. J. Lett.* **334**, L5

Ghisellini, G., Haardt, F., and Matt, G. 1994, *M.N.R.A.S.* **267**, 743

Ghisellini, G., Padovani, P., Celotti, A., and Maraschi, L. 1993, *Ap. J.* **407**, 65

Gioia, I.M., Maccacaro, T., Schild, R., Wolter, A., and Stocke, J.T. 1990, *Ap. J. Suppl.* **72**, 567

Goldreich, P., Goodman, J., and Narayan, R. 1986, *M.N.R.A.S.* **221**, 39

Goldreich, P., and Lynden-Bell, D. 1965, *M.N.R.A.S.* **130**, 125

Goldreich, P., and Tremaine, S.D. 1978, *Icarus* **34**, 227

Goldstein, H. 1980, *Classical Mechanics* (Reading: Addison-Wesley)

Goodman, J., Narayan, R., and Goldreich, P. 1987 *Ap. J.* **225**, 695

Goodrich, R.W., and Miller, J.S. 1995, *Ap. J. Lett.* **448**, L73

Gopal-Krishna, Kulkarni, V.K., and Wiita, P.J. 1996, *Ap. J. Lett.* **463**, L1

Granato, G.L., and Danese, L. 1994, *M.N.R.A.S.* **268**, 235

Green, P.J., and Mathur, S. 1996, *Ap. J.* **462**, 637

Green, P.J., Schartel, N., Anderson, S.F., Hewett, P.C., Foltz, C.B., Brinkmann, W., Fink, H., Trümper, J., and Margon, B. 1995, *Ap. J.* **450**, 51

Greenhill, L.J., Gwinn, C.R., Antonucci, R.R.J., and Barvainis, R. 1996, *Ap. J.* **472**, L21

Gregory, P.C., and Condon, J.J. 1991, *Ap. J. Suppl.* **75**, 1011

Gruber, D. 1992, in *The X-ray Background*, eds. X. Barcons and A.C. Fabian (Cambridge: Cambridge University Press), p. 44

Guilbert, P.W., and Rees, M.J. 1988, *M.N.R.A.S.* **233**, 475

Haardt, F., and Madau, P. 1996, *Ap. J.* **461**, 20

Haardt, F., and Maraschi, L. 1991, *Ap. J. Lett.* **380**, L51

Harms, R.J., Ford, H.C., Tsvetanov, Z., Hartig, G.F., Dressel, L.L., Kriss, G.A., Bohlin, R., Davidsen, A.F., Margon, B., and Kochhar, A.K. 1994, *Ap. J. Lett.* **435**, L35

Hartwick, F.D.A., and Schade, D. 1990, *Ann. Rev. Astron. Astrop.* **28**, 437

Hawking, S.W. 1974, *Nature* **248**, 30

Hawking, S.W., and Ellis, G.F.R. 1973, *The Large Scale Structure of Space-Time* (Cambridge: Cambridge University Press)

Hawley, J.F. 1987, *Ap. J.* **225**, 677

Heckman, T.M., Blitz, L., Wilson, A.S., Armus, L., and Miley, G.K. 1989, *Ap. J.* **342**, 735

Heckman, T.M., Chambers, K.C., and Postman, M. 1992, *Ap. J.* **391**, 39

Heckman, T.M., O'Dea, C.P., Baum, S.A., Laurikainen, E. 1994, *Ap. J.* **428**, 65

Heckman, T.M., Krolik, J.H., Meurer, G., Calzetti, D., Kinney, A., Koratkar, A., Leitherer, C., Robert, C., and Wilson, A. 1995, *Ap. J.* **452**, 549

Helfer, T., and Blitz, L. 1997, *Ap. J.* **478**, 162

Hewett, P.C., Foltz, C.B., and Chaffee, F.H. 1993, *M.N.R.A.S.* **261**, 185

Hewett, P.C., Foltz, C.B., and Chaffee, F.H. 1995, *Astron. J.* **109**, 1498

Hewett, P.C., Francis, P.J., Foltz, C.B., Chaffee, F.H., and Weymann, R.J. 1991, *Astron. J.* **102**, 1121

Hewitt, A., and Burbidge, G.R. 1993, *Ap. J. Suppl.* **87**, 451

Hill, G.J., and Lilly, S.J. 1991, *Ap. J.* **367**, 1

Ho, L., Filippenko, A., and Sargent, W.L.W. 1997a, *Ap. J.* **487**, 568

Ho, L., Filippenko, A., and Sargent, W.L.W. 1997b, *Ap. J.* **487**, 591

Hoag, A.A., and Smith, M.G. 1977, *Ap. J.* **217**, 362

Hollenbach, D. and McKee, C.F. 1978, *Ap. J. Suppl.* **41**, 555

Hooper, E.J., Impey, C.D., Foltz, C.D., and Hewett, P.C. 1996, *Ap. J.* **473**, 746

Horne, K.D., Welch, W.F., and Peterson, B.M. 1991, *Ap. J. Lett.* **367**, L5

Huchra, J.P., and Burg, R. 1992, *Ap. J.* **393**, 90

Hughes, P.A., Aller, H.D., and Aller, M.F. 1992, *Ap. J.* **396**, 469

Hummer, D.G., and Storey, P.J. 1987, *M.N.R.A.S.* **224**, 801

Hurt, T., Antonucci, R.R.J., Cohen, R.D., Kinney, A.L., and Krolik, J.H. 1998, *Ap. J.* in press

Hutchings, J.B., Crampton, D., and Campbell, B. 1984, *Ap. J.* **280**, 41

Impey, C.D., Malkan, M.A., Webb, W., and Petry, C.E. 1995 *Ap. J.* **440**, 801

Isobe, T., Feigelson, E.D., and Nelson, P.I. 1986, *Ap. J.* **306**, 490

Jackson, J.D. 1975, *Classical Electrodynamics* (New York: John Wiley and Sons)

Jauch, J.M., and Rohrlich, F. 1976, *The Theory of Photons and Electrons*, 2d ed. (New York: Springer-Verlag)

References

Jauncey, D.L. 1975, *Ann. Rev. Astron. Astrop.* **13**, 23

Julian, W.H., and Toomre, A. 1966, *Ap. J.* **146**, 810

Kapahi, V., and Saikia, D. 1982, *J. Astrophys. Astron.* **3**, 465

Kaplan, E.L., and Meier, P. 1958, *J. Am. Statistical Assoc.* **53**, 457

Karzas, W.J., and Latter, R. 1961, *Ap. J. Suppl.* **6**, 167

Kay, L.E. 1994, *Ap. J.* **430**, 196

Kazanas, D. 1989, *Ap. J.* **347**, 74

Keel, W.C. 1980, *Astron. J.* **85**, 198

Keel, W.C., de Grijp, M.H.K., Miley, G.K., and Zheng, W. 1994, *Astron. & Astrop.* **283**, 791

Kellermann, K.I., and Pauliny-Toth, I.I.K. 1969, *Ap. J. Lett.* **155**, L71

Kellermann, K.I., Sramek, R., Schmidt, M., Shaffer, D.B., and Green, R.F. 1989, *Astron. J.* **98**, 1195

Kellermann, K.I., and Wall, J.V. 1987, in *IAU Symposium 124: Observational Cosmology*, eds. A. Hewitt, G.R. Burbidge, and L.-Z. Fang (Dordrecht: Reidel), p. 545

Kennicutt, R.C. 1989, *Ap. J.* **344**, 685

Kennicutt, R.C., Tamblyn, P., and Congdon, C.E. 1994, *Ap. J.* **435**, 22

King, J.R. 1971, *P.A.S.P.* **83**, 199

Kinzer, R.L., Jung, C.V., Gruber, D.E., Matteson, J.L., Peterson, L.E. 1997, *Ap. J.* **475**, 361

Kollgaard, R.I., Wardle, J.F.C., Roberts, D.H., and Gabuzda, D.C. 1992, *Astron. J.* **104**, 1687

Kolykhalov, P.I., and Sunyaev, R.A. 1984, *Adv. Sp. Res.* **3**, 249

Königl, A., and Kartje, J.F. 1994, *Ap. J.* **434**, 446

Koratkar, A.P., Antonucci, R.R.J., Goodrich, R.W., Bushouse, H., and Kinney, A.L. 1995, *Ap. J.* **450**, 501

Koratkar, A.P., Kinney, A.P., and Bohlin, R.C. 1992, *Ap. J.* **400**, 435

Korista, K.T., Alloin, D., Barr, P., Clavel, J., Cohen, R.D., Crenshaw, D.M., Evans, I.N., Horne, K., Koratkar, A.P., Kriss, G.A., Krolik, J.H., Malkan, M.A., Morris, S.L., Netzer, H., O'Brien, P.T., Peterson, B.M., Reichert, G.A., Rodriguez-Pascual, P.M., Wamsteker, W., Anderson, K.S.J., Axon, D.J., Benitez, E., Berlind, P., Bertram, R., Blackwell, J.H., Jr., Bochkarev, N.G., Boisson, C., Carini, M., Carrillo, R., Carone, T.E., Cheng, F.-Z., Christensen, J.A., Chuvaev, K.K., Dietrich, M., Dokter, J.J., Doroshenko, V., Dultzin-Hacyan, D., England, M.N., Espey, B.R., Filippenko, A.V.,, Gaskell, C.M., Goad, M.R., Ho, L.C., Huchra, J.P., Jiang, X.-J., Kaspi, S., Kollatschny, W., Laor, A., Luminet, J.-P., MacAlpine, G.M., Mackenty, J.W., Malkov, Yu.F., Maoz, D., Martin, P.G., Matheson, T., McCollum, B., Merkulova, N., Metik, L., Mignoli, M., Miller, H.R., Pastoriza, M.G., Pelat, D., Penfold, J., Perez, M., Perola, G.-C., Persaud, J.L., Peters, J., Pitts, R., Pogge, R.W., Pronik, I., Pronik, V.I., Ptak, R.L., Rawley, L., Recondo-Gonzalez, M.C., Rodriguez-Espinosa, J.M., Romanishin, W., Sadun, A.C., Salamanca, I., Santos-Lleo, M., Sekiguchi, K., Sergeev, S.G., Shapovalova, A.I., Shields, J.C., Shrader, C., Shull, J.M., Silbermann, N.A., Sitko, M.L., Skillman, D.r., Smith, H.A., Smith, S.M., Snijders, M.A.J., Sparke, L.S., Stirpe, G.M., Stoner, R.E., Sun, W.-H., Thiele, U., Tokarz, S., Tsvetanov, Z.I., Turnshek, D.A., Veilleux, S., Wagner, R.M., Wagner, S.J., Wanders, I., Wang, T., Welsh, W.F., Weymann, R.J., White, R.J., Wilkes, B.J., Wills, B.J., Winge, C., Wu, H., Zou, Z.-L. 1995, *Ap. J. Suppl.* **97**, 285

Korista, K.T., Weymann, R.J., Morris, S.L., Kopko, M., Jr., Turnshek, D.A., Hartig, G.F., Foltz, C.B., Burbidge, E.M., and Junkkarinen, V.T. 1992, *Ap. J.* **401**, 529

Kormendy, J. 1992, in *Testing the AGN Paradigm*, eds. S.S. Holt, S.G. Neff, and C.M. Urry (New York: AIP), p. 23

Kormendy, J., and Richstone, D.O. 1995, *Ann. Rev. Astron. Astrop.* **33**, 581

Kotilainen, J.K., and Ward, M.J. 1994, *M.N.R.A.S.* **266**, 953

Krall, N.A., and Trivelpiece, A.W. 1973, *Principles of Plasma Physics* (New York: McGraw-Hill)

Kriss, G.A., Davidsen, A.F., Blair, W.P., Ferguson, H.C., and Long, K.S. 1992, *Ap. J. Lett.* **394**, L37

Kriss, G.A., Davidsen, A.F., Zheng, W., Kruk, J.W., and Espey, B.R. 1995, *Ap. J. Lett.* **454**, L7

Kriss, G.A., Espey, B.R., Krolik, J.H., Tsvetanov, Z., Zheng, W., and Davidsen, A.F. 1996, *Ap. J.* **467**, 622

Krolik, J.H. 1979, *Ap. J.* **228**, 13

Krolik, J.H. 1988, *Ap. J.* **325**, 148

Krolik, J.H., and Begelman, M.C. 1986, *Ap. J. Lett.* **308**, L55

Krolik, J.H., and Begelman, M.C. 1988, *Ap. J.* **329**, 702

Krolik, J.H., Done, C., and Madejski, G.M. 1993, *Ap. J.* **402**, 432

Krolik, J.H., Horne, K.D., Kallman, T.R., Malkan, M.A., Edelson, R.A., and Kriss, G.A. 1991, *Ap. J.* **371**, 541

Krolik, J.H., and Kallman, T.R. 1984, *Ap. J.* **286**, 366

Krolik. J.H. and Kriss, G.A. 1995, *Ap. J.* **447**, 512

Krolik, J.H., and Lepp, S. 1989, *Ap. J.* **347**, 179

Krolik, J.H., and London, R.A. 1983, *Ap. J.* **267**, 18

Krolik, J.H., Madau, P., and Życki, P. 1994, *Ap. J. Lett.* **420**, L57

Krolik, J.H., and McKee, C.F. 1978, *Ap. J. Suppl.* **37**, 459

Krolik, J.H., McKee, C.F., and Tarter, C.B. 1981, *Ap. J.* **249**, 422

Krolik, J.H., and Voit, G.M. 1998, *Ap. J. Lett.* **497**, L5

Kumar, P., Narayan, R., and Loeb, A. 1995, *Ap. J.* **453**, 480

Kwan, J.Y., and Krolik, J.H. 1979, *Ap. J. Lett.* **233**, L91

Lainela, M., and Valtaoja, E. 1993, *Ap. J.* **416**, 485

Laing, R.A. 1988, *Nature* **331**, 149

Lamers, H.J.G.L.M. 1994, *Astrop. Sp. Sci.* **221**, 41

Landau, L.D., and Lifshitz, E.M. 1960, *Mechanics* (London: Pergamon Press)

Landau, L.D., and Lifshitz, E.M. 1985, *Fluid Mechanics*, 2d ed. (London: Pergamon Press)

Landau, R., Golisch, W., Jones, T.J., Jones, T.W., Pedelty, J., Rudnick, L., Sitko, M.L., Kenney, J., Roellig, T., Salonen, E., Urpo, S., Schmidt, G., Neugebauer, G., Matthews, K., Elias, J.H., Impey, C., Clegg, P., and Harris, S. 1986, *Ap. J.* **308**, 78

Landy, S.D., and Szalay, A.S. 1992, *Ap. J.* **391**, 494

Laor, A. 1990, *M.N.R.A.S.* **246**, 369

Laor, A., and Draine, B.T. 1993, *Ap. J.* **402**, 441

Laor, A., Fiore, F., Elvis, M., Wilkes, B.J., and McDowell, J. 1994, *Ap. J.* **435**, 611

Laor, A., and Netzer, H. 1989, *M.N.R.A.S.* **238**, 897

Laor, A., Netzer, H., and Piran, T. 1990, *M.N.R.A.S.* **242**, 560

Laurikainen, E., and Salo, H. 1995, *Astron. & Astrop.* **293**, 683

Lawrence, A. 1991, *M.N.R.A.S.* **252**, 586

Ledlow, M.J., and Owen, F.N. 1996, *Astron. J.* **112**, 9

Lee, H.-W., Blandford, R.D., and Western, L. 1994, *M.N.R.A.S.* **267**, 303

Liang, E.P.T. 1979, *Ap. J. Lett.* **231**, L111

Lieu, R., and Axford, W.I. 1995, *Ap. J.* **447**, 302

Lightman, A.P., and Eardley, D.M. 1974, *Ap. J. Lett.* **187**, L1

References

Lightman, A.P., and Rybicki, G. 1979, *Ap. J.* **232**, 882

Lightman, A.P., and White, T.R. 1988, *Ap. J.* **335**, 57

Lightman, A.P., and Zdziarski, A.A. 1987, *Ap. J.* **319**, 643

Lin, D.N.C., Papaloizou, J., and Kley, W. 1993, *Ap. J.* **416**, 689

Loeb, A., and Laor, A. 1992, *Ap. J.* **384**, 115

Longair, M.S., Ryle, M., and Scheuer, P.A.G. 1973, *M.N.R.A.S.* **164**, 243

Lovelace, R.V.E. 1976, *Nature* **262**, 649

Lynden-Bell, D. 1969, *Nature* **223**, 690

Lynden-Bell, D. 1971, *M.N.R.A.S.* **155**, 95

Lynden-Bell, D., and Kalnajs, A.J. 1972, *M.N.R.A.S.* **157**, 1

Lynden-Bell, D., and Pringle, J. 1974, *M.N.R.A.S.* **168**, 603

Macchetto, F., Capetti, A., Sparks, W.B., Axon, D.J., and Boksenberg, A. 1994, *Ap. J. Lett.* **435**, L15

Mackenty, J. 1989, *Ap. J.* **343**, 125

Mackenty, J. 1990, *Ap. J. Suppl.* **72**, 231

Madau, P., Ghisellini, G., and Fabian, A.C. 1994, *M.N.R.A.S.* **270**, L17

Maiolino, R., and Rieke, G.H. 1995, *Ap. J.* **454**, 95

Maloney, P.R., Begelman, M.C., and Rees, M.J. 1994, *Ap. J.* **432**, 606

Markarian, B.E. 1967, *Astrofizika* **3**, 24

Markarian, B.E. 1981, Lipovetsky, V.A., and Stepanian, D.A. 1981, *Astrofizika* **16**, p. 609 (English transl. in *Astrophysics* **16**, 353)

Martel, A.R., Baum, S.A., Sparks, W.B., Wyckoff, E., Biretta, J.A., Golombek, D., Macchetto, F.D., de Koff, S., McCarthy, P.J., and Miley, G.K. 1998, *Ap. J. Suppl.* in press

Marzke, R.O., Geller, M.J., Huchra, J.P., and Corwin, H.G. 1994, *Astron. J.* **108**, 437

Maslowski, J., Pauliny-Toth, I.I.K., Witzel, A., and Kühr, H. 1984, *Astron. & Astrop.* **139**, 85

Mathis, J.S. 1990, *Ann. Rev. Astron. Astrop.* **28**, 37

Mathur, S., Elvis, M., and Singh, K.P. 1995, *Ap. J. Lett.* **455**, L9

Mathur, S., Wilkes, B.J., Elvis, M., and Fiore, F. 1994, *Ap. J.* **434**, 493

Matt, G., Brandt, W.N., and Fabian, A.C. 1996, *M.N.R.A.S.* **280**, 823

Maximon, L.C. 1968, *J. Res. NBS* **72B**, 79

McCarthy, P.J., van Bruegel, W.J.M., Spinrad, H.S., and Djorgovski, S. 1987, *Ap. J. Lett.* **321**, L29

McCray, R.M. 1979, in *Active Galactic Nuclei* (Cambridge: Cambridge University Press)

McCray, R.M., and Snow, T.P., Jr. 1979, *Ann. Rev. Astron. Astrop.* **17**, 213

McKee, C.F., and Ostriker, J.P. 1977, *Ap. J.* **218**, 148

McLeod, K.K., and Rieke, G.H. 1994, *Ap. J.* **431**, 137

McLeod, K.K., and Rieke, G.H. 1995, *Ap. J.* **441**, 96

McMahon, R.G. 1991, in *The Space Distribution of Quasars*, ed. D. Crampton (San Francisco: Astronomical Society of the Pacific), p. 129

Merritt, D. 1987, *Ap. J.* **319**, 55

Mihalas, D. 1978, *Stellar Atmospheres* (New York: W.H. Freeman)

Miller, J.S., Goodrich, R.W., and Mathews, W.G. 1991, *Ap. J.* **378**, 47

Mills, B.Y., and Slee, O.B. 1957, *Austr. J. Phys.* **10**, 162

Minkowski, R. 1960, *Ap. J.* **132**, 908

Mirabel, I.F., and Wilson, A.S. 1984, *Ap. J.* **277**, 92

Misner, C., Thorne, K.S., and Wheeler, J.A. 1973, *Gravitation* (San Francisco: W.H. Freeman)

Miyoshi, M., Moran, J., Herrnstein, J., Greenhill, L., Nakai, N., Diamond, P., and Inoue, M. 1995, *Nature* **373**, 127

Morris, S.L., Weymann, R.J., Anderson, S.F., Hewett, P.C., Francis, P.J., Foltz, C.B., Chaffee, F.H., and MacAlpine, G.M. 1991, *Astron. J.* **102**, 1627

Morrison, P. 1969, *Ap. J. Lett.* **157**, L73

Mulchaey, J.S., and Regan, M.W. 1997, *Ap. J. Lett.* **482**, L135

Murdoch, H.S., Crawford, D.F., and Jauncey, D.L. 1973, *Ap. J.* **183**, 1

Murray, N., and Chiang, J. 1997, *Ap. J.* **474**, 91

Murray, N., Chiang, J., Grossman, S.A., and Voit, G.M. 1995, *Ap. J.* **451**, 498

Murray, S.D., and Balbus, S.A. 1992, *Ap. J.* **395**, 99

Mushotzky, R.F., Done, C., and Pounds, K.A. 1993, *Ann. Rev. Astron. Astrop.* **31**, 717

Mushotzky, R.F., Fabian, A.C., Iwasawa, K., Kunieda, H., Matsuoka, M., Nandra, K., and Tanaka, Y. 1995, *M.N.R.A.S.* **272**, L9

Mushotzky, R.F., and Ferland, G.J. 1984, *Ap. J.* **278**, 558

Narayan, R., Piran, T., and Shemi, A. 1991, *Ap. J. Lett.* **379**, L17

Narayan, R., and Yi, I. 1994, *Ap. J. Lett.* **428**, L13

Nelson, C.H., and Whittle, M. 1995, *Ap. J. Suppl.* **99**, 67

Nelson, C.H., and Whittle, M. 1996, *Ap. J.* **465**, 96

Netzer, H. 1978, *Ap. J.* **219**, 822

Netzer, H. 1985, *Ap. J.* **289**, 451

Netzer, H. 1990, in *Active Galactic Nuclei: 1990 Saas-Fee Lectures*, eds. T.J.-L. Courvoisier and M. Mayor (Berlin: Springer-Verlag)

Neufeld, D.A., Maloney, P.R., and Conger, S. 1994, *Ap. J. Lett.* **436**, L127

Neufeld, D.A., and McKee, C.F. 1988, *Ap. J. Lett.* **331**, L87

Norman, C.A., and Ikeuchi, S. 1989, *Ap. J.* **345**, 372

Norman, C.A., May, A., and van Albada, T.S. 1985, *Ap. J.* **296**, 20

Norman, C.A., and Scoville, N.Z. 1988, *Ap. J.* **332**, 124

Norman, C.A., Sellwood, J.A., and Hasan, H. 1996, *Ap. J.* **462**, 114

Novikov, I.D., and Thorne, K.S. 1973, in *Black Holes*, eds. C. De Witt and B. De Witt (New York: Gordon and Breach), p. 343

O'Dea, C.P., Baum, S.A., and Stanghellini, C. 1991, *Ap. J.* **380**, 66

Ogasaka, Y., Inoue, H., Brandt, W.N., Fabian, A.C., Kii, T., Nakagawa, T., Fujimoto, R., and Otani, C. 1997, *P.A.S.J.* **49**, 179

Orr, M.J.W., and Browne, I.W.A. 1982, *M.N.R.A.S.* **200**, 1067

Osterbrock, D.E. 1989, *Astrophysics of Gaseous Nebulae and Active Galactic Nuclei* (Mill Valley CA: University Science Books)

Osterbrock, D.E., and Martel, A. 1993, *Ap. J.* **414**, 552

Osterbrock, D.E., and Shaw, R.A. 1988, *Ap. J.* **327**, 89

Ostriker, J.P., McCray, R.M., Weaver, R., and Yahil, A. 1976, *Ap. J. Lett.* **208**, L61

Ostriker, J.P., and Peebles, P.J.E. 1973, *Ap. J.* **186**, 467

Owocki, S.P., and Rybicki, G.B. 1995, *Ap. J.* **299**, 265

Pacholczyk, A.G. 1970, *Radio Astrophysics* (San Francisco: W.H. Freeman)

Paczyński, B. 1978a, *Acta Astron.* **28**, 111

Paczyński, B. 1978b, *Acta Astron.* **28**, 241

Paczyński, B. and Wiita, P.J. 1980, *Astron. & Astrop.* **88**, 23

Padovani, P., and Urry, C.M. 1992, *Ap. J.* **387**, 449

Padovani, P., and Urry, C.M. 1995, *P.A.S.P.* **107**, 803

Page, D.N., and Thorne, K.S. 1974, *Ap. J.* **191**, 499

References

Papaloizou, J., and Lin, D.N.C. 1995, *Ann. Rev. Astron. Astrop.* **33**, 505

Papaloizou, J., and Pringle, J. 1984, *M.N.R.A.S.* **208**, 721

Pauliny-Toth, I.I.K. and Kellermann, K.I. 1972 *Astron. J.* **77**, 797

Peacock, J.A., Miller, L., and Longair, M.S. 1986, *M.N.R.A.S.* **218**, 265

Pearson, T.J. 1978, *M.N.R.A.S.* **182**, 273

Peebles, P.J.E. 1972, *G.R.G.* **3**, 63

Peebles, P.J.E. 1993, *Principles of Physical Cosmology* (Princeton: Princeton University Press)

Pelat, D., Alloin, D., and Fosbury, R.A.E. 1981, *M.N.R.A.S.* **195**, 787

Penrose, R. 1969, *Riv. Nuovo Cimento* **1**, 252

Perley, R.J., Dreher, J.W., and Cowan, J.J. 1984, *Ap. J. Lett.* **285**, L35

Peterson, B.M., Balonek, T.J., Barker, E.S., Bechtold, J., Bertram, R., Bochkarev, N.G., Bolte, M.J., Bond, D., Boroson, T.A., Carini, M.T., Carone, T.E., Christensin, J.A., Clements, S.D., Cochran, A.L., Cohen, R.D., Crampton, D., Dietrich, M., Elvis, M., Ferguson, A., Filippenko, A.V., Fricke, K.J., Gaskell, C.M., Halpern, J.P., Huchra, J.P., Hutchings, J.B., Kolltaschny, W., Koratkar, A.P., Korista, K.T., Krolik, J.H., Lame, N.J., Laor, A., Leacock, R.J., MacAlpine, G.M., Malkan, M.A., Maoz, D., Miller, H.R., Morris, S.L., Netzer, H., Oliveira, C.L.M., Penfold, J., Penston, M.V., Perez, E., Pogge, R.W., Richmond, M.W., Romanishin, W., Rosenblatt, E.I., Saddlemyer, L., Sadun, A., Sawyer, S.R., Shields, J.C., Shapovalova, A.I., Smith, A.G., Smith, H.A., Smith, P.S., Sun, W.-H., Thiele, U., Turner, T.J., Veilleux, S., Wagner, R.M., Weymann, R.J., Wilkes, B.J., Wills, B.J., Wills, D., and Younger, P.F. 1991, *Ap. J.* **368**, 119

Peterson, B.M., Alloin, D., Axon, D., Balonek, T.J., Bertram, R., Boroson, T.A., Christensen, J.A., Clements, S.D., Dietrich, M., Elvis, M., Filippenko, A.V., Gaskell, C.M., Haswell, C.A., Huchra, J.P., Jackson, N., Kollatschny, W., Korista, K.T., Lame, N.J., Leacock, R.J., Lin, S.-N., Malkan, M.A., Monk, A.S., Penston, M.V., Pogge, R.W., Robinson, A., Rosenblatt, E.I., Shields, J.C., Smith, A.G., Stirpe, G.M., Sun, W.-H., Turner, T.J., Wagner, R.M., Wilkes, B.J., and Wills, B.J. 1992, *Ap. J.* **392**, 470

Peterson, B.M., Berlind, P., Bertram, R., Bochkarev, N.G., Bond, D., Brotherton, M.S., Busler, J.R., Chuvaev, K.K., Cohen, R.D., Dietrich, M., Elvis, M., Filippenko, A.V., Foltz, C.B., Garnavich, P.M., Ho, L.C., Horine, E., Horne, K., Huchra, J.P., Kollatschny, W., Korista, K.T., Malkan, M.A., Matheson, T., Mignoli, M., Morris, S.L., Nazarova, L., Penfold, J., Peters, J., Pogge, R.W., Pronik, V.I., Rush, B., Sergeev, S.G., Shapovalova, A.I., Shields, J.C., Stirpe, G.M., Tokarz, S., Wagner, R.M., Weymann, R.J., White, R.J., Wilkes, B.J., Wills, D., Wills, B.J., Winge, C., and Younger, P.F. 1994, *Ap. J.* **425**, 622

Petrosian, V. 1992, in *Statistical Challenges in Modern Astronomy*, eds. E.D. Feigelson and G.J. Babu (New York: Springer-Verlag), p. 173

Phillips, M.M. 1977, *Ap. J.* **215**, 746

Phinney, E.S. 1983, unpublished Ph.D. thesis, Cambridge University

Phinney, E.S. 1985, in *Astrophysics of Active Galaxies and Quasi-Stellar Objects*, ed. J.S. Miller (Mill Valley CA: University Science Books), p. 453

Phinney, E.S. 1987, in *Superluminal Radio Sources*, eds. J.A. Zensus and T.J. Pearson (Cambridge: Cambridge University Press), p. 301

Phinney, E.S. 1991, *Ap. J. Lett.* **380**, L17

Pian, E., Urry, C.M., Treves, A., Maraschi, L., Penton, S., Shull, J.M., Pesce, J.E., Grandi, P., Kii, T., Kollgaard, R.I., Madejski, G., Marshall, H., Wamsteker, W., Celotti, A., Courvoisier, T.J.-L., Falomo, R., Fink, H., George, I.M., and Ghisellini, G. 1997, *Ap. J.* **486**, 784

Piccinotti, G., Mushotzky, R.F., Boldt, E.A., Holt, S.S., Marshall, F.E., Serlemitsos, P.J., and Shafer, R.A. 1982, *Ap. J.* **253**, 485

Pier, E.A., Antonucci, R.R.J., Hurt, T., Kriss, G.A., and Krolik, J.H. 1994, *Ap. J.* **428**, 124

Pier, E.A., and Krolik, J.H. 1992a, *Ap. J. Lett.* **399**, L23

Pier, E.A., and Krolik, J.H. 1992b, *Ap. J.* **401**, 99

Pietrini, P., and Krolik, J.H. 1995, *Ap. J.* **447**, 526

Pilkington, J.D.H., and Scott, P.F. 1965, *Mem. Roy. Ast. Soc.* **69**, 183

Piran, T. 1978, *Ap. J.* **221**, 652

Portegies Zwaart, S.F., and Spreeuw, H.N. 1996, *Astron. & Astrop.* **312**, 670

Postman, M., and Geller, M. 1984, *Ap. J.* **281**, 95

Pounds, K.A., Nandra, K., Stewart, G.C., George, I.M., and Fabian, A.C. 1990, *Nature* **344**, 132

Poutanen, J., and Svensson, R. 1996, *Ap. J.* **470**, 249

Pozd'nyakov, L.A., Sobol, I.M., and Sunyaev, R.A. 1977, *Sov. Astron.* **21**, 708

Price, R.M., and Milne, D.K. 1965, *Aust. J. Phys.* **18**, 329

Pringle, J.E. 1981, *Ann. Rev. Astron. Astrop.* **19**, 137

Pringle, J.E. 1997, *M.N.R.A.S.* **292**, 136

Pringle, J.E., Rees, M.J., and Pacholczyk, A.G. 1973, *Astron. & Astrop.* **29**, 179

Quirrenbach, A., Witzel, A., Krichbaum, T.P., Hummel, C.A., Wegner, R., Schalinski, C.J., Ott, M., Alberdi, A., and Rioja, M. 1992, *Astron. & Astrop.* **258**, 279

Racah, G. 1934, *Nuovo Cimento* **11**, 477

Readhead, A.C.S. 1994, *Ap. J.* **426**, 51

Rees, M.J. 1966, *Nature* **211**, 468

Rees, M.J. 1977, *Quart. J. Roy. Ast. Soc.* **18**, 429

Rees, M.J. 1987, *M.N.R.A.S.* **228**, 47p

Rees, M.J., Phinney, E.S., Begelman, M.C., and Blandford, R.D. 1982, *Nature* **295**, 17

Reynolds, C.S. 1997, *M.N.R.A.S.* **286**, 513

Reynolds, C.S., and Begelman, M.C. 1997, *Ap. J.* **488**, 109

Rice, W., Lonsdale, C.J., Soifer, B.T., Neugebauer, G., Kopan, E.L., Lloyd, L.A., de Jong, T., and Habing, H.J. 1988, *Ap. J. Suppl.* **68**, 91

Rieke, G.H., and Low, F.J. 1975, *Ap. J. Lett.* **199**, 13

Roberts, W.W., Huntley, J.M., and van Albada, G.D. 1979, *Ap. J.* **233**, 67

Robertson, J.G. 1973, *Aust. J. Phys.* **26**, 403

Robertson, J.G. 1977, *Austr. J. Phys.* **30**, 241

Ross, R.R., Weaver, R., and McCray, R.M. 1978, *Ap. J.* **219**, 292

Ruden, S.P., Papaloizou, J.C., and Lin, D.N.C. 1988, *Ap. J.* **329**, 739

Rybicki, G., and Lightman, A.P. 1979, *Radiative Processes in Astrophysics* (New York: John Wiley and Sons)

Ryu, D., and Goodman, J. 1992, *Ap. J.* **388**, 438

Salpeter, E.E. 1964, *Ap. J.* **140**, 796

Salzer, J.J. 1989, *Ap. J.* **347**, 152

Sambruna, R.M., Barr, P., Giommi, P., Maraschi, L., Tagliaferri, G., and Treves, A. 1994, *Ap. J.* **434**, 468

Sambruna, R.M., George, I.M., Madejski, G.M., Urry, C.M., Turner, T.J., Weaver, K.A., Maraschi, L., and Treves, A. 1997, *Ap. J.* **483**, 774

Sambruna, R.M., Maraschi, L., and Urry, C.M. 1996, *Ap. J.* **463**, 444

Sandage, A. and Tammann, G. 1981, *A Revised Shapley-Ames Catalog of Bright Galaxies* (Carnegie Inst. Washington Pub. 635)

Sanders, D.B., Soifer, B.T., Elias, J.H., Madore, B.F., Matthews, K., Neugebauer, G., and Scoville, N.Z. 1988, *Ap. J.* **325**, 74

Santiago, B.X., and Strauss, M.A. 1992, *Ap. J.* **387**, 9

References

Sarazin, C.L. 1988, *X-ray Emission from Clusters of Galaxies* (Cambridge: Cambridge University Press)

Scheuer, P.A.G. 1957, *Proc. Cam. Phil. Soc.* **53**, 764

Scheuer, P.A.G. 1987, in *Superluminal Radio Sources*, eds. J.A. Zensus and T.J. Pearson (Cambridge: Cambridge University Press), p. 104

Scheuer, P.A.G., and Readhead. A.C.S. 1979, *Nature* **277**, 182

Schmidt, M. 1959, *Ap. J.* **129**, 243

Schmidt, M. 1963, *Nature* **197**, 1040

Schmidt, M. 1968, *Ap. J.* **151**, 393

Schmidt, M., and Green, R.F. 1983, *Ap. J.* **269**, 352

Schmidt, M., Schneider, D.P., and Gunn, J.E. 1986a, *Ap. J.* **306**, 411

Schmidt, M., Schneider, D.P., and Gunn, J.E. 1986b, *Ap. J.* **310**, 516

Schmitt, H.R., Kinney, A.L., Storchi-Bergmann, T., and Antonucci, R.R.J. 1997, *Ap. J.* **477**, 623

Schneider, D.P., Schmidt, M., and Gunn, J.E. 1994, *Astron. J.* **107**, 1245

Schwarz, M.P. 1984, *M.N.R.A.S.* **209**, 93

Schwarzschild, K. 1916, *Sitzungsber. Dtsch. Akad. Wiss. Berlin, Kl. Math. Phys. Tech.*, p. 189

Schwinger, J. 1949, *Phys. Rev.* **75**, 1912

Scoville, N.Z., and Norman, C.A. 1988, *Ap. J.* **332**, 163

Scoville, N.Z., and Sanders, D.B. 1986, in *Interstellar Processes*, eds. D.J. Hollenbach and H.A. Thronson Jr. (Dordrecht: Reidel), p. 21

Seyfert, C. 1943, *Ap. J.* **97**, 28

Shakura, N.I., and Sunyaev, R.A. 1973, *Astron. & Astrop.* **24**, 337

Shakura, N.I., and Sunyaev, R.A. 1976, *M.N.R.A.S.* **175**, 613

Shapiro, S.L., Lightman, A.P., and Eardley, D.M. 1976, *Ap. J.* **204**, 187

Shapiro, S.L., and Teukolsky, S.A. 1983, *Black Holes, White Dwarfs, and Neutron Stars* (New York: John Wiley and Sons)

Shapiro, S.L., and Teukolsky, S.A. 1993, *Ap. J.* **419**, 636

Shields, G.A. 1978, *Nature* **272**, 706

Shklovsky, I.S. 1953, *Dokl. Akad. Nauk SSSR* **90**, 983

Shlosman, I., and Begelman, M.C. 1989, *Ap. J.* **341**, 685

Shlosman, I., Frank, J., and Begelman, M.C. 1989, *Nature* **338**, 45

Shull, J.M., and McKee, C.F. 1979, *Ap. J.* **227**, 131

Sigut, T.A.A., and Pradhan, A.K. 1998, *Ap. J. Lett.* **499**, L139

Simkin, S.M., Su, H.J., and Schwarz, M.P. 1980, *Ap. J.* **237**, 404

Sincell, M.W., and Krolik, J.H. 1994, *Ap. J.* **430**, 550

Sincell, M.W., and Krolik, J.H. 1998, *Ap. J.* **496**, 737

Sobolev, V.V. 1960, *Moving Envelopes of Stars* (Cambridge MA: Harvard University Press)

Soifer, B.T., Sanders, D.B., Neugebauer, G., Danielson, G.E., and Lonsdale, C.J. 1986, *Ap. J. Lett.* **303**, L41

Soifer, B.T., Sanders, D.B., Neugebauer, G., Madore, B.F., Danielson, G.E. 1987, *Ap. J.* **320**, 238

Sołtan, A. 1982, *M.N.R.A.S.* **200**, 115

Spitzer, L., Jr. 1978, *Physical Processes in the Interstellar Medium* (New York: Wiley-Interscience)

Spitzer, L., Jr. 1990, *Ann. Rev. Astron. Astrop.* **28**, 71

Spitzer, L., Jr., and Saslaw, W.C. 1966, *Ap. J.* **143**, 400

Stasińska, G. 1984, *Astron. & Astrop.* **135**, 341

Stecker, F., de Jager, O.C., and Salamon, M.H. 1992, *Ap. J. Lett.* **390**, L49

Stengler-Larrea, E.A., Boksenberg, A., Steidel, C.C., Sargent, W.L.W., Bahcall, J.N., Bergeron, J., Hartig, G.F., Januzzi, B.T., Kirhakos, S., Savage, B.D., Schneider, D.P., Turnshek, D.A., and Weymann, R. J. 1995, *Ap. J.* **444**, 64

Stern, B.E., Begelman, M.C., Sikora, M., and Svensson, R. 1995a, *M.N.R.A.S.* **272**, 291

Stern, B.E., Poutanen, J., Svensson, R., Sikora, M., and Begelman, M.C. 1995b, *Ap. J. Lett.* **449**, L13

Stickel, M., Padovani, P., Urry, C.M., Fried, J.W., and Kühr, H. 1991, *Ap. J.* **374**, 431

Stocke, J., Morris, S.L., Gioia, I.M., Maccacaro, T., and Schild, R. 1991, *Ap. J. Suppl.* **76**, 813

Stocke, J.T., Morris, S.L., Weymann, R.J., and Foltz, C.B. 1992, *Ap. J.* **396**, 487

Stone, J.M., and Balbus, S.A. 1996, *Ap. J.* **464**, 364

Stone, J.M., Hawley, J.F., Gammie, C.F., and Balbus, S.A. 1996, *Ap. J.* **463**, 656

Störzer, H., Hauschildt, P.H., and Allard, F. 1994, *Ap. J. Lett.* **437**, L91

Sun, W.-H., and Malkan, M.A. 1989, *Ap. J.* **346**, 68

Sunyaev, R.A., and Titarchuk, L. 1980, *Astron. & Astrop.* **86**, 121

Svensson, R. 1982a, *Ap. J.* **258**, 321

Svensson, R. 1982b, *Ap. J.* **258**, 335

Svensson, R. 1983, *Ap. J.* **270**, 300

Svensson, R. 1984, *M.N.R.A.S.* **209**, 175

Tanaka, Y. 1989, in *Proceedings of the 23rd ESLAB Symposium*, eds. J. Hunt and B. Battrick (Paris: ESA), vol. 1, p. 3

Tanaka, Y., Nandra, K., Fabian, A.C., Inoue, H., Otani, C., Dotani, T., Hayashida, K., Iwasawa, K., Kii, T., Kunieda, H., Makino, F., and Matsuoka, M. 1995, *Nature* **375**, 659

Thatte, N., Quirrenbach, A., Genzel, R., Maiolino, R., and Tecza, M. 1997, *Ap. J.* **490**, 238

Thompson, D.J., Bertsch, D.L., Dingus, B.L., Esposito, J.A., Etienne, A., Fichtel, C.E., Friedlander, D.P., Hartman, R.C., Hunter, S.D., Kendig, D.J., Mattox, J.R., Mc-Donald, L.M., von Montigny, C., Mukherjee, R., Ramanamurthy, P.V., Sreekumar, P., Fierro, J.M., Lin, Y.C., Michelson, P.F., Nolan, P.L., Shriver, S.K., Willis, T.D., Kanbach, G., Mayer-Hasswelwander, H.A., Merck, M., Radecke, H.-D., Kniffen, D.A., and Schneid, E.J. 1995, *Ap. J. Suppl.* **101**, 259

Thorne, K.S. 1974, *Ap. J.* **191**, 507

Thorne, K.S., Price, R.H., and MacDonald, D.A. 1986, *Black Holes: The Membrane Paradigm* (New Haven: Yale University Press)

Tielens, A.G.G.M., McKee, C.F., Seab, C.G., and Hollenbach, D.J. 1994, *Ap. J.* **431**, 231

Titarchuk, L. 1994, *Ap. J.* **434**, 570

Tomisaka, K., and Ikeuchi, S. 1987, *P.A.S.J.* **38**, 697

Toomre, A. 1964, *Ap. J.* **139**, 1217

Toomre, A., and Toomre, J. 1972, *Ap. J.* **178**, 623

Tornikoski, M., Valtaoja, E., Teräsranta, H., Lainela, M., Bramwell, D., and Botti, L.C.L. 1993, *Astron. J.* **105**, 1680

Tran, H.D. 1995, *Ap. J.* **440**, 597

Tran, H.D., Miller, J.S., and Kay, L.E. 1992, *Ap. J.* **397**, 452

Turner, E.L. 1991, *Astron. J.* **101**, 5

Turner, J., Kirby-Docken, K., and Dalgarno, A. 1977, *Ap. J. Suppl.* **35**, 281

Turnshek, D.A. 1988, in *QSO Absorption Lines: Probing the Universe* (Cambridge and New York: Cambridge University Press), p. 17

References

Ulmschneider, P., Priest, E.R., and Rosner, R. 1991, *Mechanisms of Chromospheric and Coronal Heating*, (Berlin: Springer-Verlag)

Ulrich, M.-H., Boksenberg, A., Penston, M.V., Bromage, G.E., Clavel, J. Elvius, A., Perola, G.-C., and Snijders, M.A.J. 1991, *Ap. J.* **382**, 483

Ulrich, M.-H., and Horne, K. 1996, *M.N.R.A.S.* **283**, 748

Ulrich, M.-H., Maraschi, L., and Urry, C.M. 1997, *Ann. Rev. Astron. Astrop.* **35**, 445

Ulvestad, J., Wilson, A.S., and Sramek, R.A. 1981, *Ap. J.* **247**, 419

Urry, C.M., Padovani, P., and Stickel, M. 1991, *Ap. J.* **382**, 501

Veilleux, S., and Osterbrock, D.E. 1987, *Ap. J. Suppl.* **63**, 295

Velikhov, E.P. 1959, *J.E.T.P.* **36**, 1398

Vermeulen, R.C., and Cohen, M.H. 1994, *Ap. J.* **430**, 467

Veron-Cetty, M. and Veron, P. 1996, *ESO Science Reports* **17**, 1

von Montigny, C., Bertsch, D.L., Chiang, J., Dingus, B.L., Esposito, J.A., Fichtel, C.E., Fierro, J.M., Hartman, R.C., Hunter, S.D., Kanbach, G., Kniffen, D.A., Lin, Y.C., Mattox, J.R., Mayer-Hasswelwander, H.A., Michelson, P.F., Nolan, P.L., Radecke, H.D., Schneid, E., Sreekumar, P., Thompson, D.J., and Willis, T. 1995, *Ap. J.* **440**, 525

Vrtilek, J.M., and Carleton, N.P. 1985, *Ap. J.* **294**, 106

Wall, J.V., and Peacock, J.A. 1985, *M.N.R.A.S.* **216**, 173

Wall, J.V., Wright, A.E. and Bolton, J.G. 1976, *Aust. J. Phys., Astrop. Suppl. No. 39*, p. 1

Walsh, D.E.P., Knapp, G.R., Wrobel, J.M., and Kim, D.-W. 1989, *Ap. J.* **337**, 209

Wanders, I., Goad, M.R., Korista, K.T., Peterson, B.M., Horne, K., Ferland, G.J., Koratkar, A.P., Pogge, R.W., and Shields, J.C. 1995, *Ap. J. Lett.* **453**, L87

Warren, S.J., Hewett, P.C., and Osmer, P.S. 1991, *Ap. J. Suppl.* **76**, **1**

Warren, S.L., Hewett, P.C., and Osmer, P.S. 1994, *Ap. J.* **421**, 412

Weaver, K.A., Nousek, J., Yaqoob, T., Mushotzky, R.F., Makino, F., and Otani, C. 1996, *Ap. J.* **458**, 160

Wehrle, A.E., Pian, E., Urry, C.M., Maraschi, L., McHardy, I.M., Lawson, A.J., Ghisellini, G., Hartman, R.C., Madejski, G.M., Makino, F., Marscher, A.P., Wagner, S.J., Webb, J.R., Aldering, G.S., Aller, M.F., Aller, H.D., Backman, D.E., Balonek, T.J., Boltwood, P., Bonnell, J., Caplinger, J., Celotti, A., Collmar, W., Dalton, J., Drucker, A., Falomo, R., Fichtel, C.E., Freudling, W., Gear, W.K., Gonzalez-Perez, N., Hall, P., Inoue, H., Johnson, W.N., Kazanas, D., Kidger, M.R., Kii, T., Kollgaard, R.I., Kondo, Y., Kurfess, J., Lin, Y.C., McCollum, B., McNaron-Brown, K., Nagase, F., Nair, A.D., Penton, S., Pesce, J.E., Pohl, M., Raiteri, C.M., Renda, M., Robson, E.I., Sambruna, R.M., Schirmer, A.F., Shrader, C., Sikora, M., Sillanpaeae, A., Smith, P.S., Stevens, J.S., Stocke, J., Takalo, L.O., Teraesranta, H., Thompson, D.J., Thompson, R., Tornikoski, M., Tosti, G., Treves, A., Turcotte, P., Unwin, S.C., Valtaoja, E., Villata, M., Xu, W., Yamashita, A., Zook, A. 1998, *Ap. J.* **497**, 178

Weidemann, V. 1987, *Astron. & Astrop.* **188**, 74

Weinberg, S. 1972, *Gravitation and Cosmology: Principles and Applications of the General Theory of Relativity* (New York: John Wiley and Sons)

Weymann, R.J., Morris, S.L., Foltz, C.B., and Hewett, P.C. 1991, *Ap. J.* **373**, 23

White, R.L., and Becker, R.H. 1992, *Ap. J. Suppl.* **79**, 331

Whitney, A.R., Shapiro, I.I., Rogers, A.E.E., Robertson, D.S., Knight, C.A., Clark, T.A., Goldstein, R.M., Marandino, G.E., and Vandenburg, N.R. 1971, *Science* **173**, 225

Whittet, D.C.B. 1992, *Dust in the Galactic Environment* (Bristol: Institute of Physics)

Whittle, M. 1985, *M.N.R.A.S.* **213**, 1

Wilkes, B. 1986, *M.N.R.A.S.* **218**, 331

Wilkes, B.J., Tananbaum, H., Worrall, D.M., Avni, Y., Oey, M.S., and Flanagan, J. 1994, *Ap. J. Suppl.* **92**, 53

Wills, B.J., and Brotherton, M.S. 1995, *Ap. J. Lett.* **448**, L81

Wills, B.J., Netzer, H., Brotherton, M.S., Han, M., Wills, D., Baldwin, J.A., Ferland, G.J., and Browne, I.W.A. 1993, *Ap. J.* **410**, 534

Wills, B.J., Netzer, H., and Wills, D. 1985, *Ap. J.* **288**, 94

Wills, B.J., Thompson, K.L., Han, M.-S., Netzer, H., Wills, D., Baldwin, J.A., Ferland, G.J., Browne, I.W.A., and Brotherton, M.S. 1995, *Ap. J.* **447**, 139

Wilson, A.S. 1994, in *Proceedings of the Oxford Torus Workshop*, ed. M.J. Ward (Oxford: Oxford University), p. 55

Wilson, A.S., and Ulvestad, J. 1982, *Ap. J.* **263**, 576

Wilson, D.B., and Rees, M.J. 1982, *M.N.R.A.S.* **185**, 297

Windhorst, R.A., Fomalont, E.B., Partridge, R.B., and Lowenthal, J.D. 1993, *Ap. J.* **405**, 498

Windhorst, R.A., Miley, G.K., Owen, F.N., Kron, R.G., and Koo, D.C. 1985, *Ap. J.* **289**, 494

Wolter, A., Caccianiga, A., Della Ceca, R., and Maccacaro, T. 1994, *Ap. J.* **433**, 29

Wozniak P. R., Zdziarski A. A., Smith D., Madejski G. M., and Johnson W. N., 1998, *M.N.R.A.S.* in press

Wyse, R.F.G., and Silk, J.I. 1989, *Ap. J.* **379**, 700

Yee, H.K.C., and Green, R.F. 1984, *Ap. J.* **280**, 79

Yee, H.K.C., and Green, R.F. 1987, *Ap. J.* **319**, 28

Yi, I., Field, G.B., and Blackman, E.G. 1994, *Ap. J. Lett.* **432**, L31

Young, P.J. 1980, *Ap. J.* **242**, 1232

Young, S., Hough, J.H., Axon, D.J., Ward, M.J., and Bailey, J.A. 1996, *M.N.R.A.S.* **280**, 291

Zamorani, G., Henry, J.P., Maccacaro, T., Tananbaum, H., Soltan, A., Avni, Y., Liebert, J., Stocke, J., Strittmatter, P.A., Weymann, R.J., Smith, M.G., and Condon, J.J. 1981, *Ap. J.* **245**, 357

Zdziarski, A.A. 1985, *Ap. J.* **289**, 514

Zdziarski, A.A., Johnson, W.N., Done, C., Smith, D., and McNaron-Brown, K. 1995, *Ap. J. Lett.* **438**, L63

Zdziarski, A.A. and Krolik, J.H. 1993, *Ap. J. Lett.* **409**, L33

Zel'dovich, Ya.B. 1964, *Sov. Phys.—Dokl.* **9**, 195

Zensus, J.A., Cohen, M.H., and Unwin, S.C. 1995, *Ap. J.* **443**, 35

Zirbel, E. 1996a, *Ap. J.* **473**, 144

Zirbel, E. 1996b, *Ap. J.* **473**, 713

Zirbel, E. and Baum, S.A. 1995, *Ap. J.* **448**, 548

Zheng, W., Kriss, G.A., Telfer, R.C., Grimes, J.P., and Davidsen, A.F. 1997, *Ap. J.* **475**, 469

Życki, P., Done, C., and Smith, D.A. 1997, *Ap. J. Lett.* **488**, L113

Życki, P., Krolik, J.H., Zdziarski, A.A., and Kallman, T.R. 1994, *Ap. J.* **437**, 597

Index

21 cm line, 9 (fig. 1.1), 480–481, 485

3C sources, individual
   3C 31, plate 4, plate 10
   3C 48, 32
   3C 175, plate 3
   3C 198, plate 10
   3C 234, 460
   3C 273, 17, 32, 35, 74, 259 (fig. 9.2), 263 (fig. 9.5)
   3C 274 (*see* M 87)
   3C 279, 201 (fig. 8.3)
   3C 295, 32
   3C 305, plate 10
   3C 321, plate 8, plate 10
   3C 345, 264 (fig. 9.6)
   3C 403, plate 10
   3C 418,

3C survey. *See under* surveys, radio

α_{ox}, 197–200

absorption lines
   broad
      column densities, 403, 409–410
      list of lines seen, 402–403
      profiles, 401
      variability, 403
   intervening, 4, 23, 25, 68, 71, 73, 399
   narrow
      column densities, 410
      list of lines seen, 399
      profiles, 399–400
      variability, 400

abundances *See under the specific region*

accessible volume, 56

accretion disks
   advection-dominated, 121, 162–163, 169, 498
   angle dependence of emission, 183–185, 428–429
   coronae, 237–238, 250
   dissipation, 148, 149–150, 151, 153, 156, 160, 162, 167–168, 179, 181, 184, 228
   gas density, 159, 161, 169
   gas-pressure-dominated, 157–159, 161–162, 167, 179
   irradiation, 159, 179, 194, 228
   polarization of emitted radiation, 185–188, 429
   radial structure, 147–153
   radiation-pressure-dominated, 159–162, 167–169, 171, 173, 179
   relativistic corrections to dynamics, 151–153
   self-gravity, 141, 520
   Shakura-Sunyaev (α) model, 139, 143, 149, 160, 161, 165, 167–169, 180, 181 (fig. 7.5), 505
   spectra, 154–155, 169–186
   stability, 165–169
   surface brightness, 150
   surface density, 137, 158, 160–161, 164, 165, 168, 169, 180
   temperature, 136–137, 154, 158, 159, 169, 171, 187
   thickness, 135, 136, 158, 160, 167, 214, 429
   torque, 138–147, 150, 151, 160
      viscous, 149
   vertical structure, 157–161
   warps, 432

action-angle variables, 83–85

affine connection, 92

affine parameter, 95

Airy rings, 30, 466

Alfvén speed, 141, 142, 146, 298, 541

Alfvén waves, 241, 516, 541

alignment effect, 462

angular momentum barrier, 135, 494, 502

anisotropic radiation, 256, 424, 492, 531. *See also* accretion disks, angle-dependence of emission; obscuring tori, reradiation by; relativistic beaming
   statistical tests, 425–426
   X-ray, 429

Ariel V, 35

ASCA, 251 (fig. 8.12), 401 (fig. 11.2)

Auger ionization, 243, 245, 325

Balbus-Hawley mechanism. *See* instability, magneto-rotational

Baldwin effect, 312–313

Balmer decrement, 349–350, 371

Balmer edge, 191

BAL quasars, 401–406, 423
   polarization, 406
   X-ray spectra, 404

beaming, relativistic. *See* relativistic beaming

Bernoulli law, 122

"big blue bump", 189–191

black holes, 74, 77, 90–120, 251
   accretion efficiency, 73, 103, 109, 126, 127, 131, 498, 530
   area theorem, 115

black holes (*continued*)
  binary, 500–501
  creation, 77, 497–498
  electric fields near, 111–113, 238, 292
  event horizons, 100, 103, 106
  evidence for existence, 86–89
  growth time (*see under* timescales)
  inevitability, 76–78
  irreducible mass, 115
  membrane paradigm, 111
  particle orbits near, 97, 101–103, 117
    Carter's constant, 107
    Lense-Thirring precession, 109–111
    marginally bound, 108
    marginally stable, 102, 103, 108, 153, 163–164, 250
  Penrose process, 107
  photon orbits near, 97, 98–100, 107, 115–120, 183–184, 187
    photon orbit radius, 107
  spin, 104–105, 152, 154, 184, 287, 425
  static limit, 105, 106 (*see also* frame-dragging)
  surface resistance, 112–113
blazars, 18, 200, 264. *See also* radio-loud AGN, flat spectrum
BL Lac, 18
BL Lac objects, 18, 32, 60, 77, 190, 201, 202, 264, 268, 303 (fig. 9.15), 309, 400, 445, 447 (fig. 12.3), 463
bolometric luminosity, 59 (fig. 3.7), 63, 69, 197, 198, 228, 256, 258
Boltzmann equation, 83, 87
Bondi accretion, 121–124, 162
Born approximation, 171, 330, 350
Bose-Einstein statistics, 213, 215, 283
Bowen mechanism. *See under* emission lines, excitation by fluorescence
bremsstrahlung, 178, 202, 203–206, 224, 334, 385
  electron-electron, 205, 206
  electron-ion, 131–134, 203–205, 206
  electron-positron, 206
  opacity (*see under* opacity)
  thermal instability, 167
broad absorption line region
  covering factor, 406
  dynamics, 533 (*see also under* winds)
  elemental abundances, 410, 411
  filling factor, 410
  ionization state, 407, 410
  mass outflow, 411
  size, 408–409
broad line region, 409
  covering factor, 320–321
  dynamics, 321, 393–396
  elemental abundances, 371–372

filling factor, 370, 421
generic photo-ionization models, 357–362
geometry, 323, 392
ionization transition, 339, 359–360, 362
kinematics, 373, 378
mass, 391
physical conditions, 323, 357, 368–370, 377
size, 370, 372, 377
CIII] 1909, 309, 360
  constraint on density and ionization parameter, 369
  coolant, 359
  profile, 315
  relative strength among broad lines, 310–311
  relative strength among narrow lines, 316, 368
CIV 1549, 12, 309, 310
  absorption, 399–400, 402, 423
  constraint on ionization parameter, 369, 380
  coolant, 359
  equivalent width, 312–313
  profile, 315
  relative strength among broad lines, 310–311, 363
  relative strength among narrow line, 316, 368
catalogs. *See also* surveys
  3C, 17, 29–30, 32
  Hewitt-Burbidge, 41
  Revised Shapley Ames (RSA), 28, 470
  Veron-Cetty and Veron, 41
charge exchange, 328–329, 381
classification schemes, 18, 317–319
collisional ionization, 244, 325, 350, 384
collision strength, 331
co-moving coordinates, 566, 570
co-moving volume, 44, 45 (fig. 3.1), 70, 572
compactness, 225–231, 234–235, 247, 253, 446
Compact Steep Spectrum Sources (CSSs), 265–266
Compton Gamma-Ray Observatory, 37
Compton heating, 131, 133–134, 329, 357, 412, 513
Comptonization, 209–216
  nonthermal,
    single scatter, 210–211
  thermal, 228–229, 250, 252–253
    multiple scatters, 211–212
    single scatter, 209–210
Compton scattering
  broadening, 207, 452

Compton scattering (*continued*)
  equilibrium, 210, 289
  inverse, 178, 190, 206–208, 531
    γ-ray production, 253
    catastrophe, 283
    comparison to synchrotron radiation, 270, 282
    coolant, 224, 234, 513
    on cosmological microwave background, 203, 275–276
    mean amplification, 207–208, 210, 226
    output spectra, 210–216, 234
    polarization, 269
    seeds for, 211, 212, 228, 231, 250, 270, 305
    total energy loss rate, 208–209, 274
  Klein-Nishina cross section, 208–209
  matrix element, 217–218
  opacity (*see under* opacity)
  recoil, 206–209, 242, 246, 248, 284, 285
  reflection (*see* reflection, Compton)
  stimulated (induced), 213, 282–286, 287 (fig. 9.11)
  Thomson-limit cross section, 126, 204, 208
Compton temperature, 210, 356–357, 413
Compton y parameter, 178, 213–215, 227
contact discontinuity, 544
continuity equation, 122, 133, 137, 232, 537
convection, 140, 157. *See also* convective instability
cooling flows, 121, 512–513
cooling function, 129, 334–335
coordinate time, 91, 98, 106
CO rotational lines, 439, 482–484, 485
  luminosity proportionality to H_2 mass, 484
cosmological background
  AGN contribution, 68
  infrared, 200
  microwave, 44, 69, 269, 274, 275–276
  X-ray, 70, 71, 72 (fig. 4.1)
cosmological parameters (H_o, q_o, Ω_M, and Λ), 42, 44–45, 54, 56, 69, 465, 567, 569
Coulomb scattering, 162, 222, 224, 234, 329, 333
covariant form, 90, 91, 534
cyclotron (Larmor) frequency, 187, 539
Cygnus A, plate 2, 31
damping wings, 340, 341, 342
Debye length, 538
de Laval nozzle, 289
detailed balance, 325–326, 328, 346
diffusion equation, 212, 239

dispersion relation
  accretion disk thermal/viscous modes, 166
  Alfvén, 541
  magneto-rotational, 146
  self-gravitational in disk galaxies, 519
  thermal instability, 133
distance
  angular size, 465, 571
  coordinate, 44, 56
  luminosity, 43, 44, 45 (fig. 3.1), 52, 70, 571
dust
  emission, 8, 430
  extinction, 8, 25–26, 44, 68, 335–336, 378, 426, 430, 436
  reddening, 314, 436, 461
  reflection (*see* reflection, dust)
  X-ray opacity, 244
dynamical friction, 499–501
Eddington approximation, 175
Eddington bias. *See under* surveys, biases
Eddington-limited black hole growth, 127
Eddington luminosity, 121, 126, 160, 223, 290
efficiency of energy production
  accretion (*see under* black holes)
  chemical, 71
  nuclear, 72
EGRET, 37
Einstein A coefficient, 332, 480, 483
Einstein Observatory, 35
electric dipole scattering, 433–434
electron degeneracy, 140
electron distribution function, nonthermal, 231–234, 236, 304
  shock acceleration, 238–241
  steepening by radiative losses, 233–234, 271, 302
electron-positron pairs
  annihilation, 202, 220–222, 224, 229, 253
  cascades, 234–236
  creation, 216–220
    electron-electron, 219–220
    electron-photon, 219–220
    photon-photon, 217–220, 229, 253
  dynamics, 222–223
  equilibrium, 229–231
elliptical galaxies
  fundamental plane, 475
  surface brightness profiles, 475
emission lines *For specific lines, see individual line entries.*
  correlation with continuum luminosity, 311–314 (*see also* Baldwin effect)

emission lines (*continued*)
  deexcitation, 332, 345, 347, 369
  excitation by
    electron collisions, 330–333, 361, 364, 385
    fluorescence, 243, 343, 371
    recombination, 344, 349–351, 385, 437–438
  forbidden, 310, 333
  from excited states, 344
  list of those seen, 309, 313
  profiles
    disk, 392
    modeling, 372–373
    observed, 3, 12, 316–317
  ratios observed, 310–311, 314–315
  ratios used as diagnostics, 318–319
  transfer (*see* radiation transfer, line)
energy at infinity, 96
energy equation, 122, 124, 129, 133, 544
energy-momentum tensor, 293
epicyclic (Lindblad) frequency, 519
equation of state, 122, 124, 128, 129, 537
equipartition estimate of the magnetic field energy, 187, 203, 273–276, 283, 291, 298
escape probability, 342, 347
escape temperature, 289, 412, 492, 509–510
Euler-Lagrange equations, 93, 94, 95, 96
excited state population equilibrium. *See* thermalization, excited state population
Exosat, 202
Fanaroff-Riley types. *See under* radio galaxies, morphological types
Faraday rotation and depolarization. *See under* polarization
FeII emission lines, 191, 309, 310, 311, 361, 371
Fe Kα, 12, 197 (fig. 8.1), 198, 250
  fluorescence, 245–246, 248, 438
  observed profiles, 87, 250–251, 531
  relation to Fe abundance, 438–439
Fe K-edge, 246
Fermi-Dirac statistics, 206
fluid force (Euler's) equation, 122, 133, 537, 540, 559
flux-freezing, 111, 292, 301, 539
Fokker-Planck equation, 213, 342
Fornax A, 276
Fourier analysis, 14
four-vectors, 534–535
four-velocity, 96, 105
frame-dragging, 105, 112, 113, 114

freely falling frame, 91–92, 533–534
fueling, 474, 494
galactic bars 493, 503–504, 522. *See also under* host galaxies, morphology; instability, gravitational
  driving mass inflow, 516, 523, 527
  pattern speed, 504
  torque, 504, 523
galactic encounters, *See under* host galaxies, interactions
galaxy groups and clusters, 491
Gaunt factor, 170, 174, 204–205, 225
Gigahertz Peaked Sources (GPSs), 265
Ginga, 197
gravitational collapse, 498, 524
gravitational lenses, 4, 46
gravitational radiation, 498, 501
gravitational radius, 97
gravitational redshift, 98, 109, 112, 116–120, 183–184
gravitational relaxation, 77
H_2 rotational lines, 481–482
H_2 vibrational lines, 439
$H\alpha$, 12, 86, 309, 313, 379
  production, 360, 361, 364
  relative strength among broad lines, 310–311, 363
  relative strength among narrow lines, 316, 368
  use in classification schemes, 318–319
$H\beta$, 12, 309, 313, 452
  constraint on the column density, 380–381
  estimator of the filling factor, 370
  estimator of the ionizing luminosity, 437–438
  fraction of total optical light, 312
  from interstellar medium of host, 485
  production, 360, 361, 364
  relative strength among broad lines, 310–311, 363
  relative strength among narrow lines, 316, 368
  use in classification schemes, 318–319
  width, 317
HeII edge, 183, 358, 380
HEAO-1, 35
heat conduction, 133–134, 157
hoop stress, 292
host galaxies
  elemental abundances, 499
  emission lines, 314
  extinction, 466, 487
  images, plate 9, plate 10, 5

Index

host galaxies (*continued*)
  imaging issues, 464–467
  interactions, plate 9, 475, 489, 491, 502, 506, 524–527
  interstellar medium, 198, 245, 293, 429, 480–486, 494
    heating and cooling, 510–511
    mass exchange with stars, 508–509
  irradiation by nucleus, 492
  luminosity, 476–478
  morphology, plate 9, plate 10, 468, 469–473
    bars, 474
    correlation with radio-loudness, plate 10, 471–473, 532
    correlation with radio morphology, 475–476
  orientation, 486–487
  quasar, plate 9, 470, 474–475
  sample definition, 467–468, 490
  starlight, 437
Hubble Space Telescope, plate 5, plate 9, plate 10, 86
hydrodynamic drag, 396–397, 422
hydrostatic equilibrium equation, 157, 159, 160, 180
imaging
  grazing-incidence, 35
  interferometric, 30, 40, 302
    VLBI, 17, 86, 87, 260, 264, 268
  seeing-limited, 5, 86, 465–466
inertial confinement, 299
inertial frame. *See* freely falling frame
inertia tensor, 562
instability
  convective, 140
  gravitational, 141, 517–522
  Kelvin-Helmholtz, 544
  magneto-rotational, 144–147, 237
  Papaloizou-Pringle, 163
  radiation pressure driven winds, 421
  radiation pressure Rayleigh-Taylor, 396
  relativistic, 77
  thermal, 121, 131–134, 167
  viscous (Lightman-Eardley), 168
inverse Compton
  frequency, 177
  scattering (*see under* Compton scattering)
ionization cones, plate 5, plate 7, 320, 454, 457, 492
ionization equilibrium, 324–329, 354–356
  relevant processes, 324–325
  time to achieve, 352–353
ionization fronts, 352, 388
ionization parameter, 326–328, 369, 380, 381, 407, 410, 433

IRAS, 21, 36–37
Jeans length, 519
jets, 255, 260, 262 (fig. 9.4), 286–308
  acceleration, 289–292, 532
  collimation, 289–292
  content, 288–289
  counter-, 307, 449
  magnetic fields, 288, 301
  morphological classes, plate 3, plate 4, 259–260
  optical, plate 10, 260, 263 (fig. 9.5)
  power, 287
  projected length, 428, 446
  propagation, 297–300
  radiation, 301–308
  radio, plate 3, plate 4, 299
  relativistic motion and its effects, 268, 531
    boosting and beaming, 253–255, 283, 296, 305–308, 427, 442–443, 449
    superluminal motion, 3, 261, 296–297, 428, 446–448
  speed, 298
  X-ray, 260
Kaplan-Meier estimator, 554–555
Kirchhoff's law, 204, 276. *See also* Milne relations
Kompaneets equation, 213–216, 284
Lagrangian dynamics, 92–95, 107
lapse. *See* gravitational redshift
Larmor frequency. *See* cyclotron frequency
Lense-Thirring precession, 110
lifetime of activity, 66, 299, 431, 492
light travel time, 75
limb-darkening, 183–184, 429
LINERs, 21, 28, 87, 318, 321
little blue bump, 191, 310
l-mixing collisions, 350
lookback time, 569–570
LTE, 154–155, 169, 171, 178, 179, 180, 192, 349
luminosity functions, 43, 54–63
  affected by beaming, 442–447
  biases in, 57
  bivariate, 258, 557–558
  BL Lac object, 60–63
  emission line, 458
  estimation, 55–57
    in presence of bounds, 554–555
    Lynden-Bell C^- method, 553, 555
  evolution, 3, 58, 63–67, 70, 71
    density, 63, 65–66
    luminosity, 63, 65–66
  optical, 57–58, 59 (fig. 3.7), 60 (fig. 3.8)
  radio-loud AGN, 58–59, 61 (fig. 3.9)

luminosity functions (*continued*)
  Schechter, 73
  X-ray, 59–60, 62 (fig. 3.10)
luminosity-luminosity correlations, 556–558
Lyα, 12, 309, 313
  absorption, 399, 402, 423
  pressure, 342–343
  production, 351, 360, 362, 384
  relative strength among broad lines, 310–311, 363
  relative strength among narrow line, 368
Lyman continuum, 334, 362
Lyman edge, 182, 185, 189, 191–192, 321, 380, 399

Mach number, 123, 298, 299, 546
magnetic braking, 513–516
magnetic Reynolds number, 538
magneto-hydrodynamics (MHD), 146, 537–541
Malmquist bias. *See under* surveys, biases
Markarian (Mrk) 421, 11, 12 (fig. 1.4)
masers, 87, 439, 454
Maxwell-Boltzmann distribution function, 83, 221, 224, 226, 331, 333, 340
Maxwell's equations, 112, 217, 537
McCray, Richard, 78
MCG -6-30-15, 202 (fig. 8.4), 251 (fig. 8.12)
membrane paradigm, 111
Messier objects, individual
  M 87 (3C 274), 31, 86, plate 10
  M 101, 9 (fig. 1.1)
metric, 91, 96, 534–536
  flat-space, 91, 96, 116, 534
  Friedmann (or Robertson-Walker), 566
  Kerr, 103–109, 116, 186
    Boyer-Lindquist coordinates, 104, 106, 112, 151
    ergosphere, 105–107
    locally non-rotating frame, or ZAMO, 105–106, 111, 112, 117, 118
  Schwarzschild, 96–103, 104
    effective potential in, 97–99, 101
    pseudo-potential in, 103
  spherically symmetric, 95
MgII 2800, 309
  absorption, 399, 402
  constraint on column density, 370
  production, 360
  profile, 315
  relative strength among broad lines, 310–311, 363
  relative strength among narrow lines, 316, 368

MHD turbulence, 237
Milne relations, 325–326
minimum energy. *See* equipartition
"mirrors", plate 6, plate 8, 432–433, 437, 440, 460
  moving, 452
multivariate probability distribution, 468, 476, 555–558

Nv 1240, 309
  absorption, 402, 423, 407
  constraint on ionization parameter, 369, 380
  coolant, 358, 362
  relative strength among broad lines, 310–311, 363
narrow line region
  correlation between velocity and density or ionization state, 382
  covering factor, 321
  dynamics, 383
  filling factor, 382
  generic photo-ionization models, 363–366
  geometry, 323, 392
  ionization transition, 364
  kinematics, 383
  mass, 391
  physical conditions, 363–364, 379–382
  size, 378, 382
neutron stars, 74, 76, 77
N galaxies, 19
NGC objects, individual
  NGC 1068, plate 5, plate 6, 11, 13 (fig. 1.5), 381, 382–383, 450–455
  NGC 1275, 35
  NGC 1566, plate 1
  NGC 4151, 9, 10 (fig. 1.2), 15 (fig. 1.7), 35, 196, 399
  NGC 4168, 9 (fig. 1.1)
  NGC 4258, 87
  NGC 4945, 457 (fig. 12.7)
  NGC 5128, (Centaurus A) 31, 35
  NGC 5252, plate 7
  NGC 5548, 15 (fig. 1.7), 377–378
  NGC 6251, 262 (fig. 9.4)
[OI] 6300, 313
  constraint on column density, 380–382
  coolant, 364
  relative strength, 315
  use in classification schemes, 318–319
[OII] 3727, 313
  relative strength, 315
  sensitivity to density, 379
  use in classification schemes, 318–319
[OIII] 5007, plate 5, plate 7, 12, 313, 452–453

[OIII] 5007 (*continued*)
  constraint on physical conditions, 333, 369, 381
  coolant, 364
  possible isotropic emission, 456
  profile, 317
  relative strength, 315, 316
  sensitivity to density and temperature, 333, 369, 379
  use in classification schemes, 318–319
OVI 1034, 309
  absorption, 399, 402, 407
  constraint on ionization parameter, 369, 380, 381
  coolant, 358, 362
  relative strength among broad lines, 310–311, 363
obscuring tori, 248, 291, 429, 450, 461, 531
  dynamics, 431–432
  opening angles, 456–459, 461
  reradiation by, 430, 439–441, 454, 456, 462, 531
Ohm's law, 537–538
onion-skin model, 495–496
opacity, 125
  bremsstrahlung (free-free), 158, 160, 170–174, 176, 178, 214, 330, 348
  Compton, 170, 173, 245, 430, 436
  CO rotational lines, 483
  dust (*see* dust, extinction)
  electric dipole, 434
  H⁻, 330, 337
  H₂ rotational, 482
  HI hyperfine, 480
  line, 340, 405, 409, 420, 436
  pair-production (γ–γ), 200, 219–220, 253–254
  photo-ionization (bound-free), 158, 170–172, 188, 430
    K-shell, 242–245
  Rosseland mean, 159
  stimulated scattering, 285–286
  synchrotron, 224, 276–277
  Thomson, 125, 149, 158, 160, 162, 176
optical colors, 23, 57, 60
optically violently variable (OVV) quasars, 18, 77, 201, 202, 264
orbits
  box, 493, 504
  marginally bound (*see under* black holes)
  marginally stable (*see under* black holes)
  photon (*see under* black holes)
  tube, 504
OSSE, 197 (fig. 8.1)
particle acceleration, 237–242, 275, 283, 288
pattern speed, 297

PG objects, individual
  PG 0052+251, plate 9
  PG 0743-673, 192 (fig. 7.12)
  PG 1012+008, plate 9, 475
phase space, 84
PHL objects, individual,
  PHL 909, plate 9
  PHL 5200, 404
photo-hadron particle production, 241
photo-ionization, 244, 324–327, 348, 350, 408, 531
  cross sections (*see* opacity, photoionization)
  K-shell, 242, 246, 324–325, 360, 364, 421 (*see also* opacity, K-shell photoionization; soft X-ray absorption)
  modeling, 322, 371, 419
  source of energy for emission lines, 319–321
photo-ionization evaporation, 433, 509
photon density of states, 213, 276
photon occupation number (relativistic invariant), 117, 213, 254, 283–285
photon trapping, 130–131, 163
PKS 2155-304, 16 (fig. 1.8)
plasma frequency, 187
plasma turbulence, 223
point spread function, 466
polarization, 15–17
  accretion disk (*see under* accretion disks, polarization of emitted radiation)
  blazar, 20 (table 1.2), 264
  circular, 16
  created by scattering, 186, plate 6, plate 8, 406, 433–436
  effect of Faraday rotation, 187–188, 266, 288–289, 448–449
  linear, 16, 17, 192, 266
  maps, 440, 461
  optical, 16, 268
  quasar, 16
  radio, 17, 266
  radio galaxy, 460–462
  Seyfert galaxy, 16, 450–454
population bias. *See under* surveys, biases
Poynting-Robertson drag, 507
proper length, 99, 109
proper time, 91, 93, 94, 95, 98, 104, 105, 106, 112, 536
QSO's, 19
quasars, 18, 21, 32, 66, 71, 198, 252
  "type 2", 450, 459–460
radiation drag, 290–291
radiation pressure, 121

radiation pressure (*continued*)
  in accretion disks, 135, 140, 156–157, 158, 160–161
  driving winds (*see* winds, radiation pressure driven)
  in emission line clouds, 342–343, 359, 395–396
  on jets (*see* winds, radiation pressure driven)
  in obscuring tori, 432
  resisting accretion, 125 (*see also* Eddington luminosity)
radiation transfer
  continuum, 338–339
  equation, 125, 156, 174, 180, 415
  line, 339–343 (*see also* escape proability)
radiative diffusion, 156
radiative efficiency. *See* black holes, accretion efficiency
radio galaxies, plate 10, 28, 32, 47
  age, 298
  broad line, 19, 460, 462
  lobes, 258, 275, 278, 298
  morphological types, 18, 259, 261 (fig. 9.3)
    FR1, plate 4, 259, 299, 445–447, 449, 463, 491
    FR2, plate 3, 259, 274, 277, 298, 445–447, 449, 491
  narrow line, 460–462
  unification with radio-loud quasars, plate 8, 460–463
radio-loud AGN, 18, 19, 32, 188 400, 404, 442, 460–463
  absence of broad absorption lines, 404–405
  definition, 257
  evolution, 54
  flat-spectrum core-dominated, 34
    broad-band spectra, 303 (fig. 9.15)
    evolution, 59, 61 (fig. 3.9)
    physical conditions in cores, 281
    radio spectra, 263–265
    unification with steep-spectrum lobe-dominated, 441–449
    variability, 268
    VLBI-scale morphology, 261
    X-ray/γ-ray spectra, 200, 253
  fraction of all AGN, 32, 60, 257
  hosts of, 472
  optical/ultraviolet spectra, 190
  steep-spectrum lobe-dominated, 34
    evolution, 59, 61 (fig. 3.9)
    lobe morphology, 258–260
    radio spectra, 263
    unification with flat-spectrum core-dominated, 441–449
    X-ray spectra, 199
radio-loudness, 256–258

radio-quiet AGN, 18, 19, 193, 257, 400, 404
  definition, 257
  hosts of, 472
  radio morphology, 260
  X-ray spectra, 196–198, 250
recombination
  Cases A, B, and C, 351
  dielectronic, 328
  equilibration time, 353
  radiative, 246, 325, 326, 333, 408
    stimulated, 350
  three-body, 328, 350
Rees, Martin, 74, 78
reflected fraction, 452–453, 461
reflection
  Compton, 228, 248–250, 429
  dust, 436, 451, 461
  emission line, 436
  polarizing, 433–436, 450–453 (*see also* polarization, created by scattering)
relativistic beaming, 7, 75, 183–185, 253, 441. *See also under* jets
relativistic (Doppler) boosting, 116, 117, 183–185, 207–208, 254–255, 270. *See also under* jets
remnants, 73–74, 80, 81
reprocessing, 194, 228, 250, 291
reverberation mapping, 320, 373–378
  effective signal/noise, 375
  response function, 375
  time sampling, 375
Roche criterion. *See* tidal forces
Rosat 35, 36,
rotation curve, 86
Saha equation, 171, 328
scale height, 172, 214
Schmidt, Maarten, 17, 32, 74
Schmidt law (for the rate of star formation), 509
Schwarzschild, Karl, 96
Schwarzschild radius, 97
"screens", 432–433, 437, 440, 454, 462
selection effects, 22, 426. *See also* survey biases
Seyfert, Carl, 18, 26, 32
Seyfert galaxies, 18, 21, 28, 35, 66, 87, 198, 201, 252, 399, 404, 485
  types, 19, 197 (fig. 8.1), 317–318, 449, 454–455
    relative numbers, 455–459
shocks, 294, 542–551
  cooling column density, 386

Index

shocks (*continued*)
    emission of ionizing radiation, 385, 387–389
    jump conditions, 294, 384, 386, 544, 547, 549
    line emission, 384, 389–391
    magnetized, 386, 547–551
    particle acceleration, 239–242, 299
    relativistic, 293–295
    working surfaces, 294, 298
simulations
    hydrodynamic, 299, 300 (fig. 9.13), 525–528
    MHD, 147, 299, 300 (fig. 9.14)
    Monte Carlo, 216, 247
    N-body, 77, 525–528
Sobolev approximation, 415
Sobolev escape probability, 418
Sobolev length, 417, 421
soft X-ray absorption, 198, 248, 252, 454, 459. *See also* opacity, K-shell photo-ionization
sonic point, 123, 124, 130
sound speed, 122–123, 130, 133, 136, 140, 141, 223, 298
source counts, 30, 42–52, 68, 428
    extrapolation by fluctuation statistics ("$P(D)$" method), 46
    optical, 47, 49–50
    radio, 47–49
    X-ray, 51–52, 71
spectrum
    γ-ray, 11, 200–201, 265
    blackbody (Planck), 154–155, 172, 174, 176–177, 178, 180
    broad-band, 3, 8–11, 265, 268, 302, 303 (fig. 9.15)
    composite, 10, 11 (fig. 1.3), 14 (fig. 1.6), 189, 311
    emission line, 11–13, 14 (fig. 1.6)
    EUV, 11 (fig. 1.3), 245
    galaxy, 9 (fig. 1.1)
    infrared, 9, 10, (fig. 1.2), 11 (fig. 1.3), 265, 440, 454
    modified blackbody, 174, 176–177
    optical/ultraviolet, 188–192
    polarized, 436, 450, 453, 460
    radio, 263–265, 279, 281, 301–302, 437
    Wien, 215, 225, 230
    X-ray, 437, 457
        power law, 196–200, 228, 236, 250
        reflection bump, 198–200, 248, 250, 252
        soft excess, 198–200, 251
sphere of influence, 80, 82, 121
starburst galaxies, 34, 319 (fig. 10.4), 454, 487–488, 491, 529
stellar clusters, 77–78, 120, 137, 397–398, 509, 529

stellar distribution function, 81–83, 87
stellar irradiation, 509–510
stellar velocity ellipsoid, 88
stimulated scattering. *See under* Compton scattering
stress tensor, 139, 180
    Maxwell, 143, 147, 294, 506, 547, 559
    Reynold, 147
    viscous, 138
superluminal motion. *See under* jets, relativistic motion and its effects
surface brightness
    dependence on redshift, 5, 465, 571
    nightsky, 40
    self-absorbed synchrotron sources, 278
surface gravity, 179
surveys
    γ-ray (*EGRET*), 37
    all-sky, 33, 38–39
    biases, 22, 27, 28, 32, 459
        Eddington, 57
        Malmquist, 57
        population, 46, 56, 57
        relativistic (Doppler favoritism), 307
        Scott effect, 56
    Center for Astrophysics (CfA), 28, 36, 58, 64, 315, 456, 469, 473
    completeness, 46, 57
    confusion, 29–31, 45–46, 56
    efficiency, 22, 30, 35
    emission line, 26–28, 400, 458, 459
    infrared, 36–37, 459
        IRAS, 36–37, 38
    optical color, 23–26, 52, 60, 302, 400, 459
        BQS or PG, 24, 35, 38, 256, 257 (fig. 9.1), 311, 313 (fig. 10.2), 316 (fig. 10.3)
        LBQS, 27, 188–189, 311
        Markarian, 24, 488
        SDSS, 25, 28
    pencil-beam, 33, 38–39
    radio, 29, 32–34, 52, 302, 459
        2 Jy,
        3C, 17, 29–30, 32, 38, 443, 462, 463, 472
        4C, 32
        FIRST, 33
        Parkes (PKS), 33, 312 (fig. 10.2)
    selection effects, 7, 22, 24, 27, 34, 263, 302
    spurious sources, 30–31
    X-ray, 34–36, 52, 302, 459
        Einstein/Rosat, 35–36, 52, 54, 64, 71 (fig. 3.11)
        Piccinotti, 35, 36

swing amplification, 521

synchro-Compton mechanism, 283, 302–305

synchrotron radiation, 190, 202–203, 234, 269–273
    characteristic frequency for individual electrons, 271, 272
    emission polarization, 269, 272-273, 530
    emission spectrum, 270
    inference of physical conditions in source, 280–281
    optically thick emission, 276–281
        brightness temperature, 278–283, 285, 302, 307, 308
        turnover frequency, 279–281, 301
    total energy loss rate, 270, 274

temperatures, varieties of
    brightness, 260, 285, 345, 483 (*see also under* synchrotron radiation, optically thick emission)
    effective, 7, 176, 179
    excitation, 345, 351, 365, 483

thermal Doppler width, 340, 409

thermal equilibrium, 132, 154, 166, 224–227, 329–334
    multiple co-existing phases, 356, 394, 421–422

thermalization
    electron distribution function, 224, 333
    excited state population, 344–348, 350, 360, 484
        departure coefficients, 346–347
        timescale, 353
    photon, 172–173
    photosphere, 172–174, 176, 178, 187

Thomson cross section. *See under* opacity, Compton

tidal forces
    disrupting stars near black holes, 120
    in galaxy encounters, 475, 526–527
    transferring angular momentum, 137

timescales
    black hole growth, 82–83, 497–498
    Compton scattering, 213, 247
    cooling, 132, 134
        in accretion disks, 162, 164–166
        for nonthermal electrons, 233, 235, 272, 275
        in photo-ionized gas, 353
        post-shock, 385–386
    dynamical, 132, 134, 163, 164–165, 235
    electron distribution function, 233
    ionization, 384–385
    orbital, 82, 164–165
    photon diffusion, 131, 163, 352
    recombination, 353

two-body relaxation, 82–83
    vertical motions in disks, 164
    viscous (disk inflow), 139, 162, 163, 165, 166, 168

Toomre criterion, 520

Toomre Q parameter, 519–522

turbulent mixing, 505

Uhuru, 35

ultra-luminous infrared galaxies, 21, 37, 488–489, 529

unification models, 424, 426
    radio galaxies and blazars, 441–449
    radio galaxies and quasars, 460–463
    Seyfert galaxy types, 449–459

V/V_{max} statistic, 52–54, 62

variability, 13, 46
    γ-ray, 253
    acceleration by relativistic motion, 75, 307
    in blazars, 15, 16 (fig. 1.8), 200, 266, 307
    and causality, 74–75, 253–254, 308
    intraday, 268, 308
    measures, 14, 76, 193, 195, 201, 267–268
    optical/ultraviolet, 193–195
    in quasars, 13–14
    radio, 256, 266–269
    in Seyfert galaxies, 13–14, 193–195
    spectral, 200
    timescales, 76, 201, 253, 268
    wavelength-dependence, 13
    X-ray, 201

velocity dispersion, 88–89

virial temperature, 121, 128, 162, 512

virial theorem, 512, 514, 516, 522, 559–565

viscosity, 138, 140, 160
    photon, 140
    turbulent, 140

VLA, 40, 258, 286

warm absorbers, 252, 399, 407, 414

winds
    magnetized, 141–143, 287, 291–292, 414, 432, 506
    radiation pressure driven
        by Compton scattering, 290–291
        force from multiple lines, 419–420
        force from single line, 414–418
        line-locking, 422–423
    thermally driven, 130, 289, 412–414

WKB approximation, 166

Zero Angular Momentum Observer (ZAMO). *See under* metric, Kerr